APPLIED CALCULUS

FOR BUSINESS, SOCIAL SCIENCES AND LIFE SCIENCES

APPLIED CALCULUS

FOR BUSINESS, SOCIAL SCIENCES, AND LIFE SCIENCES

Produced by the Consortium based at Harvard which was formed under a National Science Foundation Grant. All proceeds from the sale of this work are used to support the work of the Consortium

Deborah Hughes-Hallett
Harvard University

William G. McCallum
University of Arizona

Andrew M. Gleason
Harvard University

Brad G. Osgood
Stanford University

Patti Frazer Lock
St. Lawrence University

Andrew Pasquale
Chelmsford High School

Daniel E. Flath
University of South Alabama

Jeff Tecosky-Feldman
Haverford College

Sheldon P. Gordon
Suffolk County Community College

Joe B. Thrash
University of Southern Mississippi

David O. Lomen
University of Arizona

Karen R. Thrash
University of Southern Mississippi

David Lovelock
University of Arizona

Thomas W. Tucker
Colgate University

with the assistance of

Otto K. Bretscher
Harvard University

John Wiley & Sons, Inc.
New York Chichester Brisbane Toronto Singapore

Recognizing the importance of preserving what has been written, it is a policy of John Wiley & Sons, Inc. to have books of enduring value published in the United States printed on acid-free paper, and we exert our best efforts to that end.

This book was set in Times Roman by the Consortium based at Harvard using TEX, Mathematica, and the package *AsTeX*, which was written by Alex Kasman. Special thanks to S.Alex Mallozzi and Rebeca Rapoport for managing the process. It was printed and bound by R.R. Donnelley & Sons, Company. The cover was printed by The Lehigh Press, Inc.

Photo Credits: Greg Pease

Problems from *Calculus: The Analysis of Functions*, by Peter D. Taylor (Toronto: Wall & Emerson, Inc. 1992). Reprinted with permission of the publisher.

ISBN 0-471-13931-9

Printed in the United States of America

10 9 8 7 6 5 4 3 2

Dedicated to Robin, Kari, Eric, and Dennis,
and
Laura, Hannah, Ben, Natty, Isaac, and Matthias.

PREFACE

Calculus is one of the greatest achievements of the human intellect. Inspired by problems in astronomy, Newton and Leibniz developed the ideas of calculus 300 years ago. Since then, each century has demonstrated the power of calculus to illuminate questions in mathematics, the physical sciences, engineering, and the social and biological sciences.

Calculus has been so successful because of its extraordinary power to reduce complicated problems to simple rules and procedures. Therein lies the danger in teaching calculus: it is possible to teach the subject as nothing but the rules and procedures – thereby losing sight of both the mathematics and of its practical value. With the generous support of the National Science Foundation, our consortium set out to create a new calculus curriculum that would restore that insight. This book brings this new calculus curriculum to the applied calculus course.

Basic Principles

Two principles guided our efforts. The first is our prescription for restoring the mathematical content to calculus:

The Rule of Four: *Every topic should be presented geometrically, numerically, algebraically, and verbally.*

We continually encourage students to think about the geometrical and numerical meaning of what they are doing. It is not our intention to undermine the purely algebraic aspect of calculus, but rather to reinforce it by giving meaning to the symbols. In the homework problems dealing with applications, we continually ask students to explain verbally what their answers mean in practical terms.

The second principle, inspired by Archimedes, is our prescription for restoring practical understanding:

The Way of Archimedes: *Formal definitions and procedures evolve from the investigation of practical problems.*

Archimedes believed that insight into mathematical problems is gained by first considering them from a mechanical or physical point of view.[1] For the same reason, our text is problem driven. Whenever possible, we start with a practical problem and derive the general results from it. By practical problems we usually, but not always, mean real world applications. These two principles have led to a dramatically new curriculum – more so than a cursory glance at the table of contents might indicate.

Technology

We take advantage of computers and graphing calculators to help students learn to think mathematically. For example, using a graphing calculator to zoom in on functions is one of the best ways of seeing local linearity.

[1]. . . I thought fit to write out for you and explain in detail . . . the peculiarity of a certain method, by which it will be possible for you to get a start to enable you to investigate some of the problems in mathematics by means of mechanics. This procedure is, I am persuaded, no less useful even for the proof of the theorems themselves; for certain things first became clear to me by a mechanical method, although they had to be demonstrated by geometry afterwards because their investigation by the said method did not furnish an actual demonstration. But it is of course easier, when we have previously acquired, by the method, some knowledge of the questions, to supply the proof than it is to find it without any previous knowledge. From *The Method*, in *The Works of Archimedes* edited and translated by Sir Thomas L. Heath (Dover, NY).

Furthermore, the ability to use technology effectively as a tool is itself of the greatest importance. Students are expected to use their own judgement to determine where technology is useful.

However, the book does not require any specific software or technology. Test sites have used the materials with graphing calculators, graphing software, and computer algebra systems. Any technology with the ability to graph functions and perform numerical integration will suffice.

What Student Background is Expected?

This book is intended for students in business, life and the social sciences. We have found the material to be thought-provoking for well-prepared students while still accessible to students with weak algebra backgrounds. Providing numerical and graphical approaches as well as the algebraic gives students several ways of mastering the material. This approach encourages students to persist, thereby lowering failure rates.

Content

We began work on this book by talking to faculty in business, economics, biology, and a wide range of other fields, as well as to many mathematicians who teach applied calculus. As a result of these discussions we included some new topics, such as differential equations, and omitted some traditional topics whose inclusion we could not justify. In the process, we also changed the focus of certain topics. In order to meet individual needs or course requirements, topics can easily be added or deleted, or the order changed.

Chapter 1: A Library of Functions

Chapter 1 introduces all the elementary functions to be used in the book. Although the functions are probably familiar, the graphical, numerical, and modeling approach to them is fresh. Our purpose is to acquaint the student with each function's individuality: the shape of its graph, characteristic properties, comparative growth rates, and general uses. We expect to give the student the skill to read graphs and think graphically, to read tables and think numerically, and to apply these skills, along with their algebraic skills, to modeling the real world. We introduce exponential functions at the earliest possible stage, since they are fundamental to the understanding of real-world processes. Further attention is given to constructing new functions from old ones—how to shift, flip, and stretch the graph of any basic function to give the graph of a new, related function.

We encourage you to cover this chapter thoroughly, as the time spent on it will pay off when you get to the calculus.

Chapter 2: The Derivative

Chapter 2 presents the key concept of the derivative according to the Rule of Four. The purpose of this chapter is to give the student a practical understanding of the definition of the derivative and its interpretation as an instantaneous rate of change without complicating the discussion with differentiation rules. After finishing this chapter, a student will be able to find derivatives numerically (by taking arbitrarily fine difference quotients), visualize derivatives graphically as the slope of the graph, and interpret the meaning of first and second derivatives in various applications. The student will also understand the concept of marginality and recognize the derivative as a function in its own right.

Chapter 3: The Definite Integral

Chapter 3 presents the key concept of the definite integral, along the same lines as Chapter 2. It is possible to delay Chapter 3 until after Chapter 5 without any difficulty.

The purpose of this chapter is to give the student a practical understanding of the definite integral as a limit of Riemann sums, and to bring out the connection between the derivative and the definite integral in the Fundamental

Theorem of Calculus. We use the same method as in Chapter 2, introducing the fundamental concept in depth without going into technique. The motivating problem is computing the total distance traveled from the velocity function. The student will finish the chapter with a good grasp of the definite integral as a limit of Riemann sums, with the ability to compute it numerically, and with an understanding of how to interpret the definite integral in various contexts.

Chapter 4: Short-Cuts to Differentiation

Chapter 4 presents the symbolic approach to differentiation. The title is intended to remind the student that the basic methods of differentiation are not to be regarded as the definition of the derivative. The derivatives of all the functions in Chapter 1 are introduced as well as the rules for differentiating the combinations discussed in Chapter 1. The student will finish this chapter with basic proficiency in differentiation and an understanding of why the various rules are true.

Chapter 5: Using the Derivative

Chapter 5 presents applications of the derivative. It includes an investigation of parametrized families of functions according to the Way of Archimedes, using the graphing technology to observe basic properties and calculus to confirm them.

Our aim in this chapter is to enable the student to use the derivative in solving problems, rather than to learn a catalogue of application templates. It is not meant to be comprehensive, and you do not need to cover all the sections. The student should finish this chapter with the experience of having successfully tackled a few problems that required sustained thought over more than one session.

Chapter 6: Reconstructing a Function from Its Derivative

Chapter 6 presents applications of the definite integral. All definite integrals are calculated approximately using numerical methods. We emphasize the graphical interpretation of the definite integral as area, and the use of the definite integral to compute total change.

Our aim in this chapter is to enable the student to use the definite integral in solving problems, rather than to learn a catalogue of application templates. It is not meant to be comprehensive, and you do not need to cover all the sections . The student should finish this chapter understanding when and how to use the definite integral in application.

Chapter 7: Differential Equations

Chapter 7 introduces differential equations without many technicalities. The emphasis is on qualitative solutions, modeling, and interpretation. Slope fields are used to visualize the behavior of solutions of first-order differential equations. We include applications to population models and the growth of a business. The student will finish this chapter knowing what differential equations are, why they are important, and how to obtain qualitative solutions.

Chapter 8: Functions of Many Variables

We introduce functions of many variables from several points of view, using graphs of surfaces, contour diagrams, formulas, and tables. This chapter is as crucial for functions of many variables as Chapter 1 is for single variable functions; it gives students the skills to read graphs and contour diagrams and think graphically, to read tables and think numerically, and to apply these skills, along with their algebraic skills, to modeling the real world. Students should finish this chapter with a thorough understanding of functions of several variables.

Chapter 9: Calculus for Functions of Many Variables

Chapter 9 introduces the idea of a partial derivative from a graphical, numerical, and analytical viewpoint. Partial derivatives are then applied to optimization problems, ending with a discussion of Lagrange multipliers. The student will finish this chapter knowing how to use differential calculus in applications involving functions of many variables.

Appendices

There is an appendix on roots and accuracy.

What is the Relationship Between This Book and the Calculus Books by the Same Consortium?

Much of this book is based on Chapters 1–6, 8, 9 of the text *Calculus* (first edition, 1994) and Chapters 11, 13, and 14 of *Multivariable Calculus* (preliminary edition, 1995), both by the same consortium. However, the content of this book was thought out from scratch and substantial changes made as a result of discussions with faculty in business, life, and social sciences. For example, symbolic antidifferentiation plays a much smaller role in this text; the emphasis is instead on when and how to use a definite integral. Since a firm understanding of graphs and tabular data is especially important for this audience, this book emphasizes the graphical and numerical aspects to an even greater extent than in the original texts.

Supplementary Materials

- **Instructor's Manual containing** containing teaching tips, calculator programs, and some overhead transparency masters.
- **Instructor's Solution Manual** with complete solutions to all problems.
- **Answer Manual** with brief answers to all odd-numbered problems.
- **Student's Solution Manual** with complete solutions to half the odd-numbered problems.

Acknowledgements

First and foremost, we want to express our appreciation to the National Science Foundation for their faith in our ability to produce a revitalized calculus curriculum and, in particular, to Louise Raphael, John Kenelly, John Bradley, Bill Haver, and James Lightbourne. We also want to thank the members of our Advisory Board, Benita Albert, Lida Barrett, Bob Davis, Lovenia DeConge-Watson, John Dossey, Ron Douglas, Don Lewis, Seymour Parter, John Prados, and Steve Rodi for their ongoing guidance and advice.

In addition, we want to thank all the people across the country who encouraged us to write this book and who offered so many helpful comments. We would like to thank the following people, for all that they have done to help our project succeed: Wayne Anderson, Leonid Andreev, Ruth Baruth, Jeffrey Bergen, Yoav Bergner, Shelina Bhojani, J.Curtis Chipman, Dave Chen, David Chua, Dean Chung, Richard Chung, Eric Connally, Bob Condon, Radu Constantinescu, Josh Cowley, Larry Crone, Gene Crossley, Jie Cui, Jane Devoe, Arthur Dobelis, Moon Duchin, Joe Fiedler, Holland Filgo, Hermann Flaschka, David Flath, Ron Frazer, Lynn Garner, Richard Iltis, Adrian Iovita, Jerry Johnson, Georgia Kamvosoulis, Joe Kanapka, Misha Kazhdan, Donna Krawczyk, Theodore Laetsch, Sylvain Laroche, Kurt Lemmert, Suzanne Lenhart, Alex Mallozzi, Alfred Manaster, Megan McCallion, Kurt Mederer, Saadat Moussavi, Eric Olson, Jim Osterburg, Edmund Park, Greg Peters, Rick Porter, Rebecca Rapoport, Harry Row, Virginia Stallings, Brian Stanley, "Suds" Sudholz, Noah Syroid, John.S. Thomas, Tom Timchek, Denise Todd, J.Jerry Uhl, Xianbao Xu, Jennie Yoder, Gang Zhang.

Deborah Hughes-Hallett
Andrew M. Gleason
Patti Frazer Lock
Daniel E. Flath
Sheldon P. Gordon
David O. Lomen
David Lovelock
William G. McCallum
Brad G. Osgood
Andrew Pasquale
Jeff Tecosky-Feldman
Joe B. Thrash
Karen R. Thrash
Thomas W. Tucker

To Students: How to Learn from this Book

- This book may be different from other math textbooks that you have used, so it may be helpful to know about some of the differences in advance. At every stage, this book emphasizes the *meaning* (in practical, graphical or numerical terms) of the symbols you are using. There is much less emphasis on "plug-and-chug" and using formulas, and much more emphasis on the interpretation of these formulas than you may expect. You will often be asked to explain your ideas in words or to explain an answer using graphs.
- The book contains the main ideas of calculus in plain English. Success in using this book will depend on reading, questioning, and thinking hard about the ideas presented. It will be helpful to read the text in detail, not just the worked examples.
- There are few examples in the text that are exactly like the homework problems, so homework problems can't be done by searching for similar–looking "worked out" examples. Success with the homework will come by grappling with the ideas of calculus.
- Many of the problems in the book are open-ended. This means that there is more than one correct approach and more than one correct solution. Sometimes, solving a problem relies on common sense ideas that are not stated in the problem explicitly but which you know from everyday life.
- This book assumes that you have access to a calculator or computer that can graph functions, find (approximate) roots of equations, and compute integrals numerically. There are many situations where you may not be able to find an exact solution to a problem, but can use a calculator or computer to get a reasonable approximation. An answer obtained this way is usually just as useful as an exact one. However, the problem does not always state that a calculator is required, so use your own judgement.

  If you mistrust technology, listen to this student, who started out the same way:

  > Using computers is strange, but surprisingly beneficial, and in my opinion is what leads to success in this class. I have difficulty visualizing graphs in my head, and this has always led to my downfall in calculus. With the assistance of the computers, that stress was no longer a factor, and I was able to concentrate on the concepts behind the shapes of the graphs, and since these became gradually more clear, I got increasingly better at picturing what the graphs should look like. It's the old story of not being able to get a job without previous experience, but not being able to get experience without a job. Relying on the computer to help me avoid graphing, I was tricked into focusing on what the graphs meant instead of how to make them look right, and what graphs symbolize is the fundamental basis of this class. By being able to see what I was trying to describe and learn from, I could understand a lot more about the concepts, because I could change the conditions and see the results. For the first time, I was able to see how everything works together . . . .

  That was a student at the University of Arizona who took calculus in Fall 1990, the first time we used some of the material in this text. She was terrified of calculus, got a C on her first test, but finished with an A for the course.
- This book attempts to give equal weight to three methods for describing functions: graphical (a picture), numerical (a table of values) and algebraic (a formula). Sometimes it's easier to translate a problem given in one form into another. For example, you might replace the graph of a parabola with its equation, or plot a table of values to see its behavior. It is important to be flexible about your approach: if one way of looking at a problem doesn't work, try another.
- Students using this book have found discussing these problems in small groups helpful. There are a great many problems which are not cut-and-dried; it can help to attack them with the other perspectives your colleagues can provide. If group work is not feasible, see if your instructor can organize a discussion session in which additional problems can be worked on.
- You are probably wondering what you'll get from the book. The answer is, if you put in a solid effort, you will get a real understanding of one of the most important accomplishments of the millennium – calculus –

as well as a real sense of how mathematics is used in the age of technology.

Deborah Hughes-Hallett
Andrew M. Gleason
Patti Frazer Lock
Daniel E. Flath
Sheldon P. Gordon
David O. Lomen
David Lovelock
William G. McCallum
Brad G. Osgood
Andrew Pasquale
Jeff Tecosky-Feldman
Joe B. Thrash
Karen R. Thrash
Thomas W. Tucker

CONTENTS

8 FUNCTIONS OF MANY VARIABLES 435

9 CALCULUS FOR FUNCTIONS OF MANY VARIABLES 495

APPENDIX 545

INDEX 555

CHAPTER ONE

A LIBRARY OF FUNCTIONS

Functions are truly fundamental to mathematics. For example, in everyday language we say, "The price of a ticket is a function of where you sit," or "The gas mileage on your car is a function of how fast you drive." In each case, the word *function* expresses the idea that knowledge of one fact tells us another. In mathematics, the most important functions are those in which knowledge of one number tells us another number. If we know the length of the side of a square, its area is determined. If the circumference of a circle is known, its radius is determined.

Calculus starts with the study of functions. This chapter will lay the foundation for calculus by surveying the behavior of the most common functions, including powers, exponentials, and logarithms. Besides the behavior of these functions, we will explore ways of handling the graphs, tables, and formulas that represent them.

1.1 WHAT'S A FUNCTION?

Let's look at an example. In the summer of 1990, the temperatures in Arizona reached an all-time high (so high, in fact, that some airlines decided it might be unsafe to land their planes there). The daily high temperatures in Phoenix for June 19–29 are given in Table 1.1.

TABLE 1.1 *Temperature in Phoenix, Arizona, June 1990*

Date: June (1990)	19	20	21	22	23	24	25	26	27	28	29
Temperature (°F)	109	113	114	113	113	113	120	122	118	118	108

Although you may not have thought of something so unpredictable as temperature as being a function, the temperature *is* a function of date, because each day gives rise to one and only one high temperature. There is no formula for temperature (otherwise we would not need the weather bureau), but nevertheless the temperature does satisfy the definition of a function: Each date, t, has a unique high temperature, H, associated with it.

We define a function as follows:

One quantity, H, is a **function** of another, t, if each value of t has a unique value of H associated with it. We say H is the *value* of the function or the *dependent variable*, and t is the *argument* or *independent variable*. Alternatively, think of t as the *input* and H as the *output*. We write $H = f(t)$, where f is the name of the function.
The **domain** of a function is a set of possible values of the independent variable, and the **range** is the corresponding set of values of the dependent variable.

In the temperature example above, the independent variable is the date, and the dependent variable is the temperature. The domain is all possible dates, and the range is the high temperatures on those dates. The function assigns temperatures to dates. For example, we see in Table 1.1 that $f(25) = 120$.

Functions play an important role in many fields. Frequently, one observes that one quantity is a function of another and then tries to find a reasonable formula to express this function. For example, before about 1590 there was no quantitative idea of temperature. Of course, people understood relative notions like warmer and cooler, and some absolute notions like boiling hot, freezing cold, or body temperature, but there was no numerical measure of temperature. It took the genius of Galileo to realize that the expansion of fluids as they warmed was the key to the measurement of temperature. He was the first to think of temperature as a function of fluid volume.

Finding a function which represents a given situation is called making a *mathematical model.* Such a model can throw light on the relationship between the variables and can thereby help us make predictions.

Representation of Functions: Tables, Graphs, and Formulas

Functions can be represented in at least four different ways: by tables, by graphs, by formulas, and in words. For example, the function giving the temperatures in Phoenix, Arizona, as a function of time can be represented by the graphs in Figure 1.1 as well as by a table.

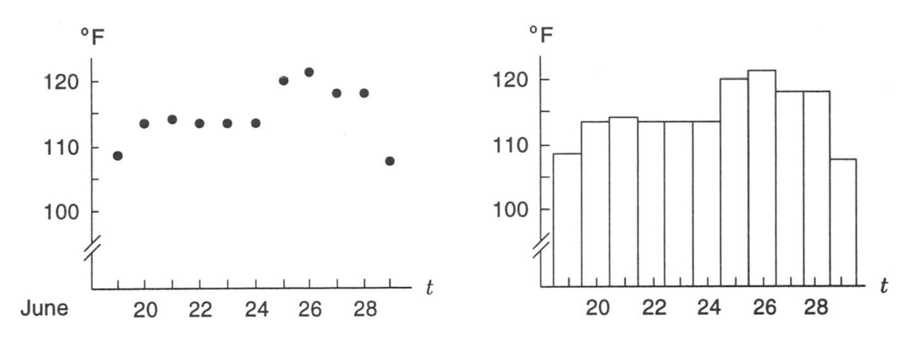

Figure 1.1: Phoenix temperatures, June 1990

Other functions arise naturally as graphs. Figure 1.2 contains electrocardiogram (EKG) pictures showing the heartbeat patterns of two patients, one normal and one not. Although it is possible to construct a formula to approximate an EKG function, this is seldom done. The pattern of repetitions is what a doctor needs to know, and these are much more easily seen from a graph than from a formula. However, each EKG represents a function showing electrical activity as a function of time.

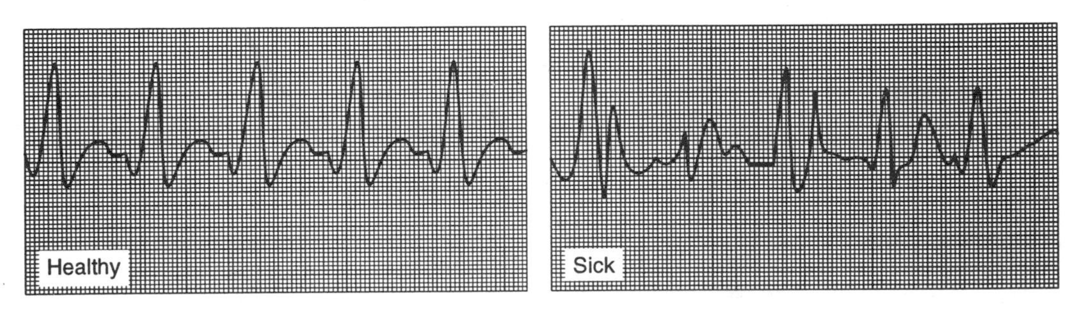

Figure 1.2: EKG readings on two patients

As another example of a function, consider the snow tree cricket. Surprisingly enough, all such crickets chirp at essentially the same rate if they are at the same temperature. That means that the chirp rate is a function of temperature. In other words, if we know the temperature, we can determine the chirp rate. Even more surprisingly, the chirp rate, C, in chirps per minute, increases steadily with the temperature, T, in degrees Fahrenheit, and to a high degree of accuracy can be computed by the formula

$$C = 4T - 160.$$

The formula for C is written $C = f(T)$ to express the fact that we are thinking of C as a function of T. The graph of this function is in Figure 1.3.

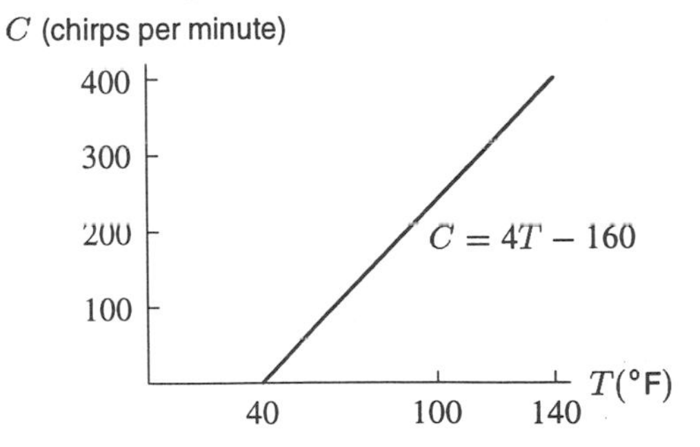

Figure 1.3: Cricket chirp rate versus temperature

Examples of Domain and Range

If the domain of a function is not specified, we will usually take it to be the largest possible set of real numbers. For example, we usually think of the domain of the function $f(x) = x^2$ as all real numbers, whereas the domain of the function $g(x) = 1/x$ is all real numbers except zero since division by zero is undefined. Sometimes, however, we may specify, or restrict, the domain. For example, if the function $f(x) = x^2$ is used to represent the area of a square of side x, we consider only nonnegative values of x and restrict the domain to nonnegative numbers.

Example 1 Consider the function $C = f(T)$ giving chirp rate as a function of temperature. We assume that this equation holds for all temperatures for which the predicted chirp rate is positive, and up to the highest temperature ever recorded at a weather station, namely, 136°F. What is the domain of this function f?

Solution If we consider the equation

$$C = 4T - 160$$

simply as a mathematical relationship between C and T, any T value is possible. However, if we're thinking of it as a relationship between cricket chirps and temperature, then T cannot be less than 40°F, where C falls below the axis and becomes negative. (See Figure 1.3.) In addition, we are told that the formula doesn't hold at temperatures above 136°. Thus, for the function $C = f(T)$ we have

$$\begin{aligned}\text{Domain} &= \text{All } T \text{ values between } 40°\text{F and } 136°\text{F} \\ &= \text{All } T \text{ values with } 40 \leq T \leq 136.\end{aligned}$$

Therefore we say that the function $C = f(T)$ is represented by the formula

$$C = f(T) = 4T - 160 \quad \text{on the domain} \quad 40 \leq T \leq 136.$$

Example 2 Find the range of the function f, given the domain from Example 1. In other words, find all possible values of the chirp rate, C, in the equation $C = f(T)$.

Solution Again, if we consider $C = f(T)$ simply as a mathematical relationship, its range is all real C values. However, thinking of its meaning for crickets, the function will predict cricket chirps per minute between 0 (when $T = 40$°F) and 384 (when $T = 136$°F). Hence,

$$\begin{aligned}\text{Range} &= \text{All } C \text{ values from 0 to 384} \\ &= \text{All } C \text{ values with } 0 \leq C \leq 384.\end{aligned}$$

So far we have used the temperature to predict the chirp rate and thought of the temperature as the *independent variable* and the chirp rate as the *dependent variable.* However, we could do this backwards, and calculate the temperature from the chirp rate. From this point of view, the temperature is dependent on the chirp rate. Thus, which variable is dependent and which is independent may depend on your viewpoint.

Thinking of temperature as a function of chirp rate would enable us (in theory, at least) to use the chirp rate instead of a thermometer to measure temperature. The way we actually do measure temperature is based on another function: the relation between the height of the liquid in a thermometer and temperature. The height of the mercury is certainly a function of temperature; however, we always use this the other way around, and determine the temperature from the height of the mercury, as suggested by Galileo.

Example 3 Figure 1.4[1] shows the birth rate (number of births per 1000 of the population) in developed countries as a function of year, between 1775 and 1977. If B is the birth rate and y is the year, we can write $B = f(y)$.

(a) Describe what this graph tells you about how the birth rate has changed since 1775.
(b) Approximately what is $f(1900)$? What information does this give you?
(c) What is the domain of f? What is the range of f?

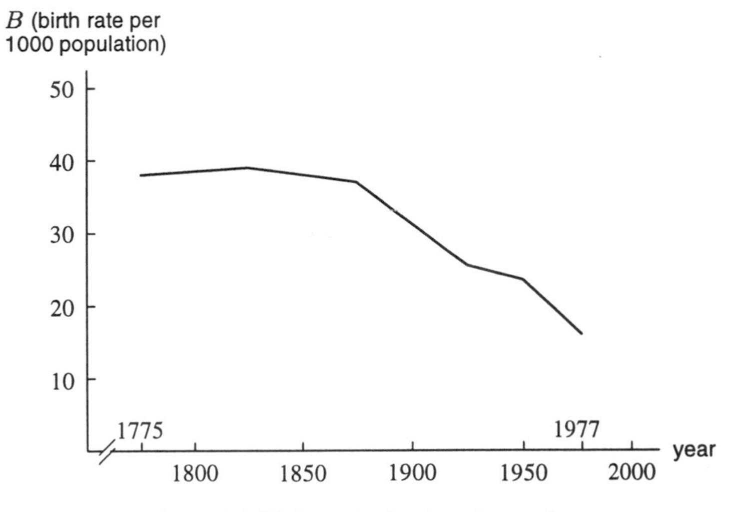

Figure 1.4: Birth rate in developed countries

Solution (a) We can see from the graph in Figure 1.4 that the value of the function stays about the same until about 1875, and then it begins to decrease. Since the value of the function corresponds to the birth rate, we see that the birth rate stayed about the same from 1775 until 1875, and it has been decreasing since 1875.

(b) $f(1900)$ is approximately 32. This means that in 1900, the birth rate was approximately 32 births for every 1000 people.

(c) The domain is the set of all values, y, for which $f(y)$ is defined—in this case all years between 1775 and 1977. The range is the corresponding set of values of the dependent variable B, the birth rate. We see from Figure 1.4 that the birth rate varies between 16 per thousand and about 39 per thousand, so the range of this function is the set of all numbers between 16 and 39.

Supply and Demand Curves

Economists are interested in how the quantity, q, of an item which is manufactured and sold, depends on its price, p. Since manufacturers and consumers react differently to changes in price, there are two functions relating p and q. For a given item, the *supply curve* represents how the quantity of an item that manufacturers are willing to make depends on the price for which the item can be sold. The *demand curve* represents how the quantity of an item demanded by consumers depends on the price of the item. It is usually assumed that as the price increases, the manufacturers will be willing to supply more of the product, so the supply function will go up as price goes up. On the other hand, as price increases, it is assumed that consumer demand will fall, and so we expect the demand function to go down as prices go up.

[1]"Food and Population: A Global Concern", Elaine Murphy, Washington D.C.: Population Reference Bureau Inc., 1984, p.2

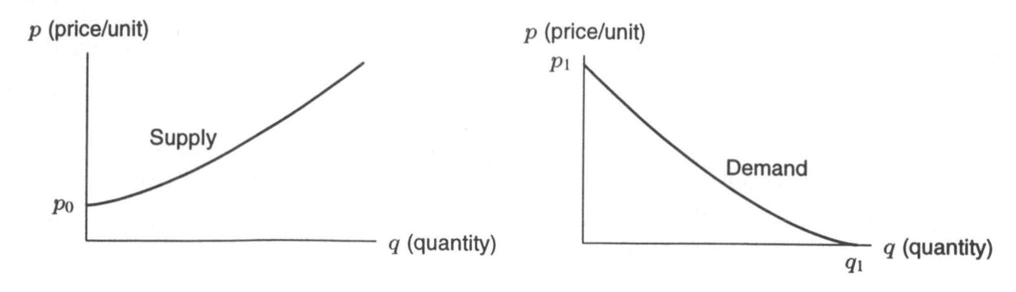

Figure 1.5: Supply and demand curves

Economists think of quantity as a function of price. However, for historical reasons, the economists put price (the independent variable) on the vertical axis and quantity (the dependent variable) on the horizontal axis. (The reason for this state of affairs is that economists originally took price to be the dependent variable and put it on the vertical axis. Unfortunately when the point of view changed, the axes did not.) Thus, graphs of typical supply and demand curves generally look like those shown in Figure 1.5.

Example 4 What is the economic meaning of the prices p_0 and p_1 and the quantity q_1 in Figure 1.5?

Solution The vertical axis corresponds to a quantity of zero; since the price p_0 is the vertical intercept on the supply curve, p_0 is the price at which the quantity supplied is zero. In other words, unless the price is above p_0, the suppliers will not produce anything. The price p_1 is also on the vertical axis, so it corresponds to the price on the demand curve where the quantity demanded is zero. In other words, unless the price is below p_1, consumers won't buy any of the product.

The horizontal axis corresponds to a price of zero, so the quantity q_1 on the demand curve is the quantity that would be demanded if the price were zero—or the quantity which could be given away if the item were free.

If we plot the supply and demand curves on the same axes, as shown in Figure 1.6, there will be exactly one point where the graphs cross. This point (q^*, p^*) is called the ***equilibrium point.***

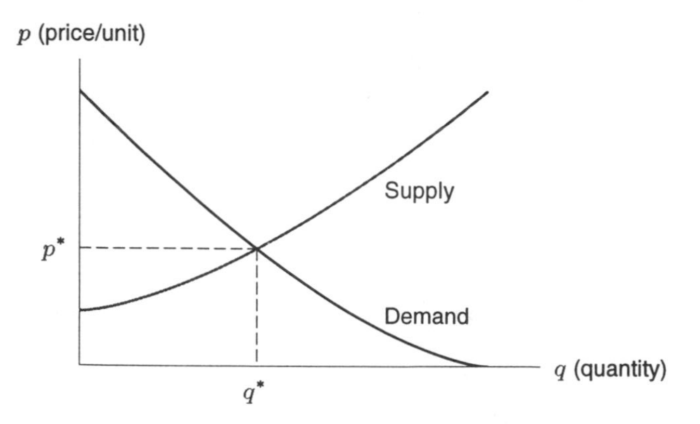

Figure 1.6: The equilibrium point

The values p^* and q^* at this point are called the *equilibrium price* and *equilibrium quantity*, respectively. At this equilibrium point, a quantity q^* of an item is produced and sold for a price of p^*

each. It is assumed that the market will naturally settle to this equilibrium point. (See Problems 13 and 14.)

The equilibrium price and equilibrium quantity are at the point where the supply and demand curves cross or where

$$\text{Demand} = \text{Supply}.$$

Proportionality

A common functional relationship occurs when one quantity is *proportional* to another. For example, if apples are 60 cents a pound, we say the price you pay, p cents, is proportional to the weight you buy, w pounds, because

$$p = f(w) = 60w.$$

As another example, the area, A, of a circle is proportional to the square of the radius, r:

$$A = f(r) = \pi r^2.$$

In general, y is (directly) **proportional** to x if there is a constant k, called *constant of proportionality*, such that

$$y = kx.$$

We also say that one quantity is *inversely proportional* to another if one is proportional to the reciprocal of the other. For example, the speed, v, at which you make a 50-mile trip is inversely proportional to the time, t, taken, because v is proportional to $1/t$:

$$v = \frac{50}{t} = 50\left(\frac{1}{t}\right).$$

Notice that if y is directly proportional to x, then the magnitude of one variable increases (decreases) when the magnitude of the other increases (decreases). If, however, y is inversely proportional to x, then the magnitude of one variable increases when the value of the other decreases.

Problems for Section 1.1

1. Which of the graphs in Figure 1.7 best match the following three stories?[2] Write a story for the remaining graph.
   (a) I had just left home when I realized I had forgotten my books, and so I went back to pick them up.
   (b) Things went fine until I had a flat tire.
   (c) I started out calmly but sped up when I realized I was going to be late.

[2] Adapted from Jan Terwel. "Real Maths in Cooperative Groups in Secondary Education." In *Cooperative Learning in Mathematics*, edited by Neal Davidson, p 234. (Reading: Addison Wesley, 1990).

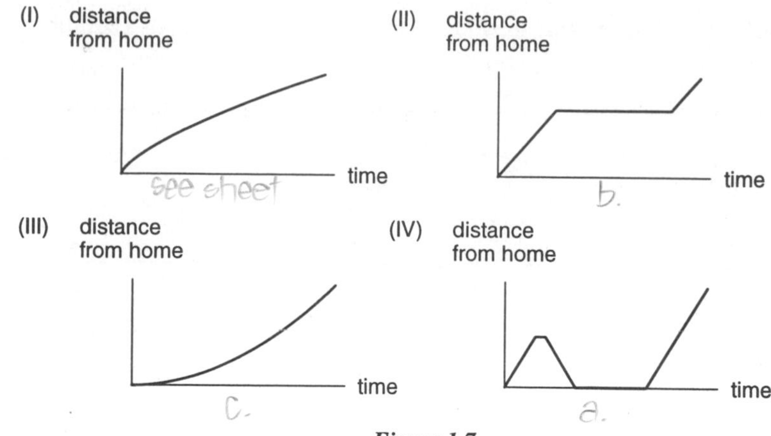

Figure 1.7

2. It warmed up throughout the morning, and then suddenly got much cooler around noon, when a storm came through. After the storm, it warmed up before cooling off at sunset. Sketch a possible graph of this day's temperature as a function of time.
3. Right after a certain drug is administered to a patient with a rapid heart rate, the heart rate plunges dramatically and then slowly rises again as the drug wears off. Sketch a possible graph of the heart rate against time from the moment the drug is administered.
4. Generally, the more fertilizer that is used, the better the yield of the crop. However, if too much fertilizer is applied, the crops become poisoned, and the yield goes down rapidly. Sketch a possible graph showing the yield of the crop as a function of the amount of fertilizer applied.
5. Describe what Figure 1.8 tells you about an assembly line whose productivity is represented as a function of the number of workers on the line.

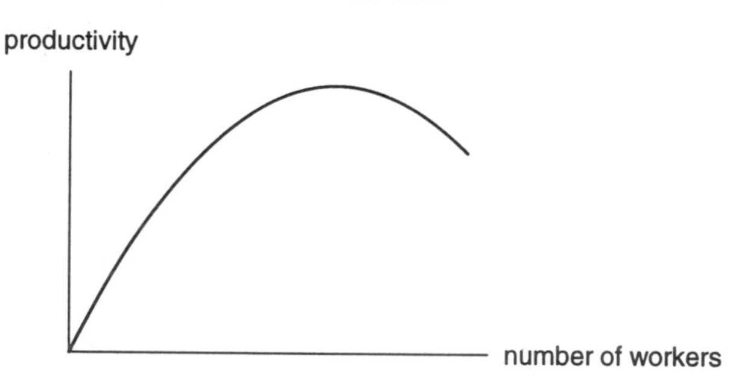

Figure 1.8

6. A flight from Dulles Airport in Washington, D.C., to LaGuardia Airport in New York City has to circle LaGuardia several times before being allowed to land. Plot a graph of distance of the plane from Washington against time, from the moment of takeoff until landing.
7. In her *Guide to Excruciatingly Correct Behavior*, Miss Manners states:

   There are three possible parts to a date of which at least two must be offered: entertainment, food and affection. It is customary to begin a series of dates with a great deal of entertainment, a moderate amount of food and the merest suggestion of affection. As the amount of affection increases, the entertainment can be reduced proportionately. When the affection has replaced the entertainment, we no longer call it dating. Under no circumstances can the food be omitted.

Based on this statement, sketch a graph showing entertainment as a function of affection, assuming the amount of food to be constant. Mark the point on the graph at which the relationship starts, as well as the point at which the relationship ceases to be called dating.

8. Recall that functions can be given by a formula, a table, or a graph. Five different functions are given below. In each case, find $f(5)$.

(a) $f(x) = 2x + 3$

(b)

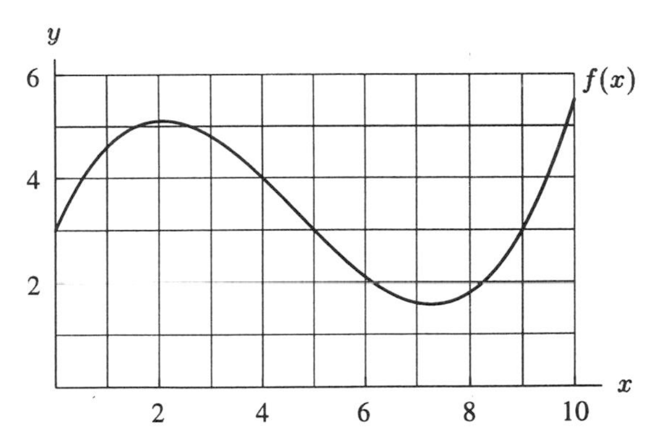

Figure 1.9

(c)

x	1	2	3	4	5	6	7	8
$f(x)$	2.3	2.8	3.2	3.7	4.1	4.9	5.6	6.2

(d)

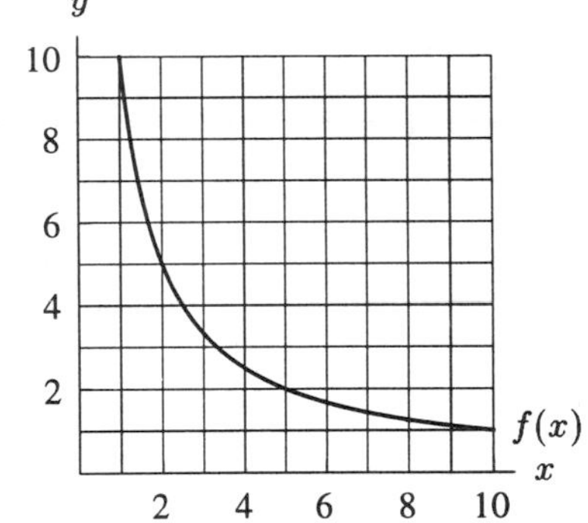

Figure 1.10

(e) $f(x) = 10x - x^2$

9. Specify the domain and range of the function $y = f(x)$ whose graph is given in Figure 1.9 in Problem 8b.

10. The graph of $r = f(p)$ is given in Figure 1.11.

(a) What is the value of r when p is zero?
(b) What is the value of r when p is three?
(c) What is $f(2)$?
(d) What is the domain of f?
(e) What is the range of f?

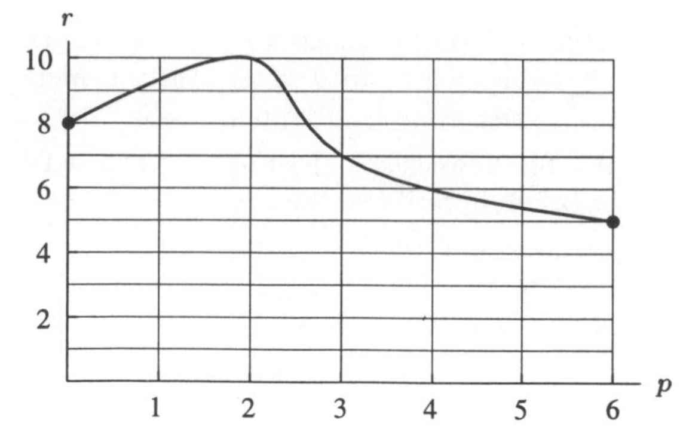

Figure 1.11

11. Let $y = f(x) = x^2 + 2$.
    (a) Find the value of y when x is zero.
    (b) What is $f(3)$?
    (c) What values of x give y a value of 11?
    (d) Are there any values of x that give y a value of 1?

12. One of the graphs in Figure 1.12 is a supply curve, and the other is a demand curve. Which is which? Explain how you made your decision using what you know about the effect of price on supply and demand.

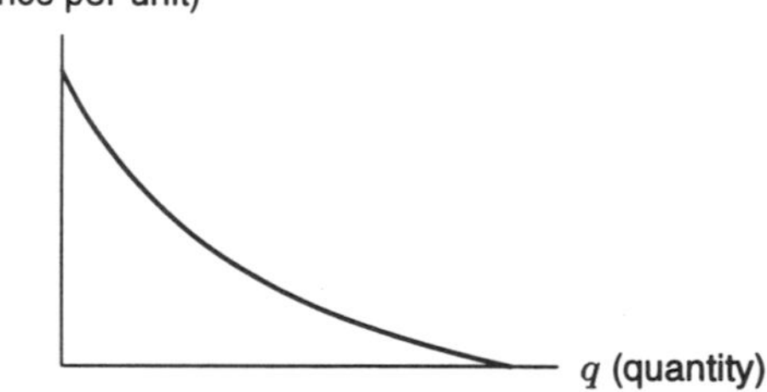

Figure 1.12

13. Figure 1.13 shows the supply and demand curves for a particular product.
    (a) What is the equilibrium price for this product? At this price, what quantity will be produced?
    (b) Choose a price above the equilibrium price—for example, $p = 12$. At this price, how many items will suppliers be willing to produce? How many items will consumers want to buy? Use your answers to these questions to explain why, if prices are above the equilibrium price, the market tends to push prices lower (towards the equilibrium).
    (c) Now choose a price below the equilibrium price—for example, $p = 8$. At this price, how many items will suppliers be willing to produce? How many items will consumers want to buy? Use your answers to these questions to explain why, if prices are below the equilibrium price, the market tends to push prices higher (towards the equilibrium).

P= 8, dem ~ 3,500 · price ↑ cos p'pul have high demand
sup ~ 2,000

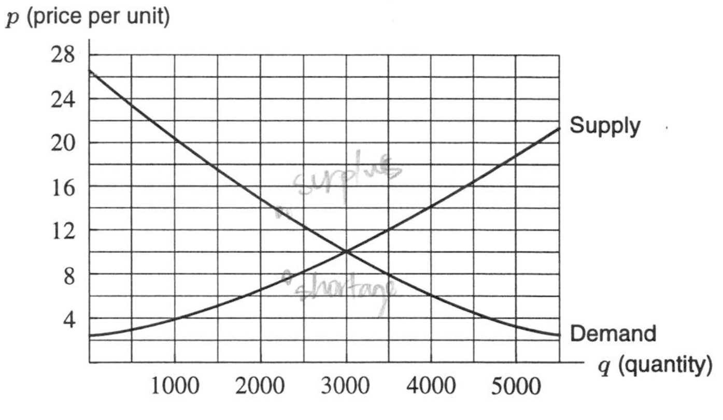

Figure 1.13

14. Figure 1.14 shows the supply and demand curves for a particular product.
   (a) What is the equilibrium price for this product? At this price, what quantity will be produced?
   (b) Choose a price above the equilibrium price—for example, $p = 300$. At this price, how many items will suppliers be willing to produce? How many items will consumers want to buy? Use your answers to these questions to explain why, if prices are below the equilibrium price, the market tends to push prices lower (towards the equilibrium).
   (c) Now choose a price below the equilibrium price—for example, $p = 200$. At this price, how many items will suppliers be willing to produce? How many items will consumers want to buy? Use your answers to these questions to explain why, if prices are below the equilibrium price, the market tends to push prices higher (towards the equilibrium).

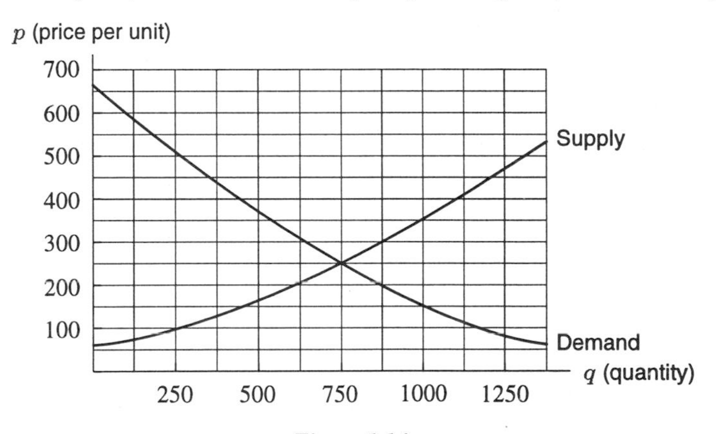

Figure 1.14

15. In a large corporation, one division seemed to be operating smoothly and the members of the division were completely confident in the manager's ability. However, at corporate headquarters the division was judged to be performing inadequately. In a sudden move, the manager was transferred to another division and a new manager was appointed. Confidence in the manager's ability plunged and only gradually recovered. A year later, confidence in the new manager's leadership matched the confidence before the change. Sketch a possible graph of confidence in the manager's ability against time, starting a few days before the change to one year after the change.

16. Based upon its market research, a corporate office provides the demand curve in Figure 1.15 to its ice cream shop franchises. A franchise owner has determined that 240 scoops per day can be sold at a price of \$1.00 per scoop.
    (a) According to the corporate demand curve, estimate how many scoops could be sold per day during a half-price sale.
    (b) The owner is considering an increase in the price per scoop after the half-price sale. Using the corporate demand curve, estimate how many scoops per day could be sold at a price of \$1.50 per scoop.

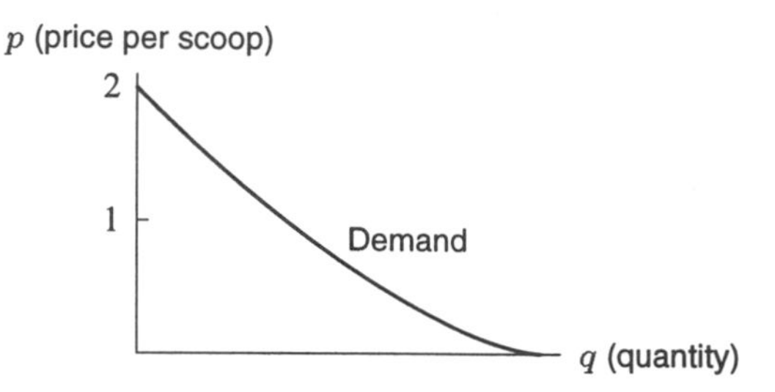

Figure 1.15: Demand curve for ice cream

17. Tables 1.2 and 1.3 are given below. One represents a supply curve and the other represents a demand curve.

TABLE 1.2

p (price/unit)	182	167	153	143	133	125	118
q (quantity)	5	10	15	20	25	30	35

TABLE 1.3

p (price/unit)	6	35	66	110	166	235	316
q (quantity)	5	10	15	20	25	30	35

(a) Which table represents which curve? Why?
(b) At a price of \$155, approximately how many items would consumers purchase?
(c) At a price of \$155, approximately how many items would manufacturers supply?
(d) Will the market push prices higher or lower than \$155?
(e) What would the price have to be if you wanted consumers to buy at least 20 items?
(f) What would the price have to be if you wanted manufacturers to supply at least 20 items?

18. The US production, Q, of copper in metric tons and the value, P, in thousands of dollars per metric ton are given for the years 1984 through 1989 in the following table[3]. Plot these data points with Q, the production, on the horizontal axis and P, the value, on the vertical axis. Sketch a possible supply curve which fits this data reasonably well.

TABLE 1.4

Year	1984	1985	1986	1987	1988	1989
Q	1103	1105	1144	1244	1417	1497
P	1473	1476	1456	1818	2656	2888

[3] *US Copper, Lead, and Zinc Production"*, The World Almanac, 1992, p. 688.

19. Examine the demand curve shown in Figure 1.16.

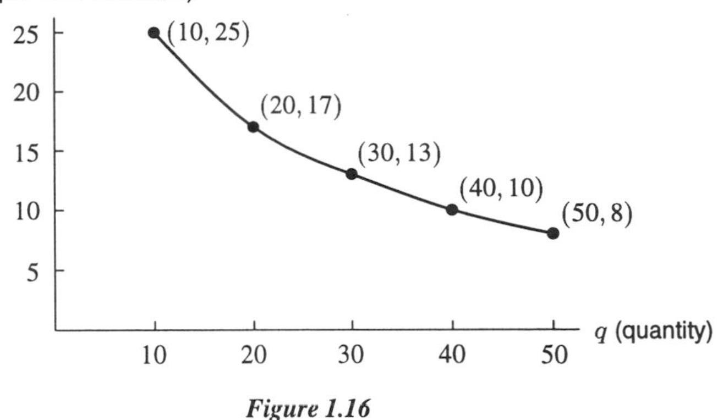

Figure 1.16

(a) If the price is \$17 per item, how many items do consumers purchase?
(b) If the price is \$8 per item, how many items do consumers purchase?
(c) At what price do consumers purchase 30 items?
(d) At what price do consumers purchase 10 items?

20. Examine the supply curve shown in Figure 1.17.

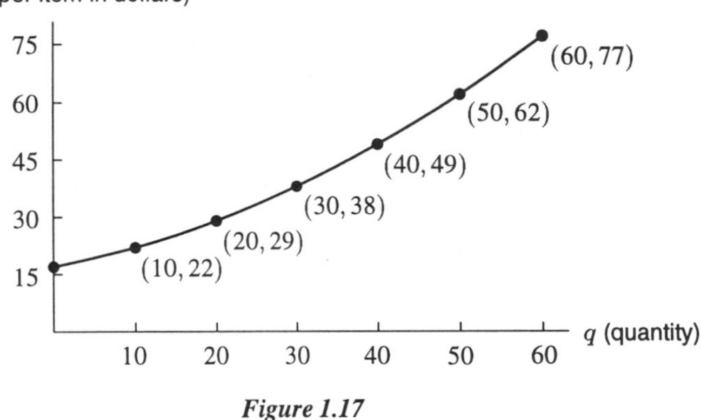

Figure 1.17

(a) If the price is \$49 per item, how many items do manufacturers supply?
(b) If the price is \$77 per item, how many items do manufacturers supply?
(c) At what price are manufacturers willing to supply 20 items?
(d) At what price are manufacturers willing to supply 50 items?

21. When Galileo was formulating the laws of motion, he considered the motion of a body starting from rest and falling under gravity. He originally thought that the velocity of such a falling body was proportional to the distance it had fallen. What light does the data in Table 1.5 shed on Galileo's hypothesis? What alternative hypothesis is suggested by the two sets of data in Table 1.5 and Table 1.6?

TABLE 1.5

Distance (ft)	0	1	2	3	4
Velocity (ft/sec)	0	8	11.3	13.9	16

TABLE 1.6

Time (sec)	0	1	2	3	4
Velocity (ft/sec)	0	32	64	96	128

22. According to the National Association of Realtors[4], the minimum annual gross income, m, in thousands of dollars needed to obtain a 30-year home loan of A-thousand dollars at 9% is given in Table 1.7.

TABLE 1.7

A	50	75	100	150	200
m	17.242	25.863	34.484	51.726	68.968

The minimum annual gross income, m, in thousands needed for a home loan of \$100,000 at various interest rates, r, is given in Table 1.8.

TABLE 1.8

r	8	9	10	11	12
m	31.447	34.484	37.611	40.814	44.084

(a) Is the size of the loan, A, proportional to the minimum annual gross income, m?
(b) Is the percentage rate, r, proportional to the minimum annual gross income, m?

1.2 LINEAR FUNCTIONS

Probably the most commonly used functions are the *linear functions*. These are functions that represent a steady increase or a steady decrease. A function is linear if any change, or increment, in the independent variable causes a proportional change, or increment, in the dependent variable. The graph of a linear function is a line.

The Olympic Pole Vault

During the early years of the Olympics, the height of the winning pole vault increased approximately as shown in Table 1.9. Since the winning height increased regularly by 8 inches every four years, the height is a linear function of time over the period from 1900 to 1912. The height starts at 130 inches and increases by the equivalent of 2 inches every year, so if y is the height in inches and t is the number of years since 1900, we can write

$$y = f(t) = 130 + 2t.$$

The coefficient 2 tells us the rate at which the height increases and is the *slope* of the line $f(t) = 130 + 2t$.

TABLE 1.9 *Olympic pole vault records (approximate)*

Year	1900	1904	1908	1912
Height (inches)	130	138	146	154

[4]"Income needed to get a Mortgage", The World Almanac, 1992, p.720

You can visualize the slope in Figure 1.18 as the ratio

$$\text{Slope} = \frac{\text{Rise}}{\text{Run}} = \frac{8}{4} = 2.$$

Calculating the slope (rise/run) using any other two points on the line gives the same value. It is this fact—that the slope, or rate of change, is the same everywhere—that makes a line straight. For a function that is not linear, the rate of change will vary from point to point. Since $y = f(t)$ increases with t, we say that f is *an increasing function*. What about the constant 130? This represents the initial height in 1900, when $t = 0$. Geometrically, the 130 is the *intercept* on the vertical axis. You may wonder whether the linear trend continues beyond 1912. Not surprisingly, it doesn't exactly. The formula $y = 130 + 2t$ predicts that the height in the 1988 Olympics would be 306 inches or 25 feet 6 inches, which is considerably higher than the actual value of 19 feet 9 inches. In fact, the height does increase at almost every session of the Olympics, but not at a constant rate. Thus, there is clearly a danger in *extrapolating* too far from the given data. You should also observe that the data in Table 1.9 is *discrete*, because it is given only at specific points (every four years). However, we have treated the variable t as though it were *continuous*, because the function $y = 130 + 2t$ makes sense for all values of t. The graph in Figure 1.18 is of the continuous function because it is a solid line, rather than four separate points representing the years in which the Olympics were held.

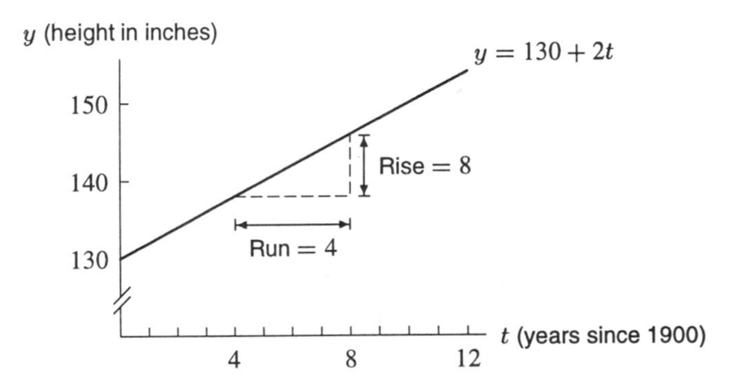

Figure 1.18: Olympic pole vault records

Linear Functions in General

> A **linear function** has the form
>
> $$y = f(x) = b + mx$$
>
> Its graph is a line such that
> - m is the slope, or rate of change of y with respect to x
> - b is the vertical intercept, or value of y when x is zero.

Notice that if the slope is zero, $m = 0$, we have $y = b$, a horizontal line.

> To recognize that a function $y = f(x)$ given by a table of data is linear, look for differences in y values that are constant for equal differences in x.

The slope of a linear function can be calculated from values of the function at two points, given by $x = a$ and $x = c$, using the formula

$$m = \frac{\text{Rise}}{\text{Run}} = \frac{f(c) - f(a)}{c - a}.$$

The quantity $(f(c) - f(a))/(c - a)$ is called a *difference quotient* because it is the quotient of two differences. (See Figure 1.19). In Chapter 2, you will see that difference quotients play an important role in calculus.

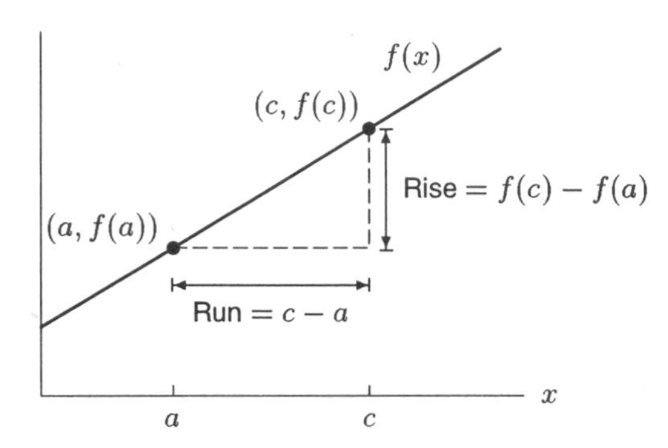

Figure 1.19: Difference quotient $= \dfrac{f(c) - f(a)}{c - a}$

The Success of Search and Rescue Teams

Consider the problem of the "search and rescue" teams working to find lost hikers in remote areas in the West. To search for an individual, members of the search team separate and walk parallel to one another through the area to be searched. Experience has shown that the team's chance of finding a lost individual is related to the distance, d, by which team members are separated. The percentage found[5] for various separations are recorded in Table 1.10.

From the data in the table, you can see that as the separation distance decreases, a larger percentage of the lost hikers is found, which makes sense. Since $P = f(d)$ decreases as d increases, we say that P is a *decreasing function* of d. You can also see that for the data given, each 20-foot increase in distance causes the percentage found to drop by 10. This constant decrease in P as d increases by a fixed amount is a clear indication that the graph of P against d is a line. (See Figure 1.20.) Notice that the slope is $-40/80 = -1/2$. The negative sign shows that P decreases as d increases. The slope is the rate at which P is increasing or decreasing, as d increases.

What about the vertical intercept? If $d = 0$, the searchers are walking shoulder to shoulder and you'd expect everyone to be found, so $P = 100$. This is exactly what you get if the line is continued

[5]From *An Experimental Analysis of Grid Sweep Searching*, by J. Wartes (Explorer Search and Rescue, Western Region, 1974).

TABLE 1.10 *Separation of searchers versus success rate*

Separation distance d (ft)	Percent found, P
20	90
40	80
60	70
80	60
100	50

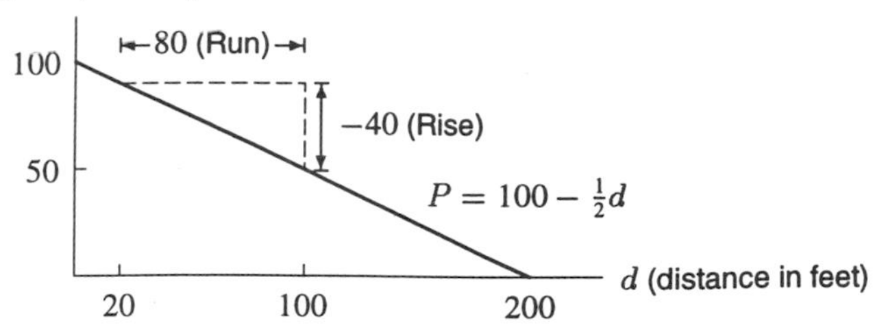

Figure 1.20: Separation of searchers versus success rate

to the vertical axis (a decrease of 20 in d causes an increase of 10 in P). Therefore, the equation of the line is

$$P = f(d) = 100 - \frac{1}{2}d.$$

What about the horizontal intercept? When $P = 0$, or $0 = 100 - \frac{1}{2}d$, then $d = 200$. The value $d = 200$ represents the separation distance at which, according to the model, no one is found. This is unreasonable, because even when the searchers are far apart, the search will sometimes be successful. This suggests that somewhere outside the given data, the linear relationship ceases to hold. As in the pole vault example, extrapolating too far beyond the given data may not give accurate answers.

Increasing versus Decreasing Functions

Let's summarize what we know about increasing and decreasing functions:

> A function f is **increasing** if the values of $y = f(x)$ increase as x increases.
> A function f is **decreasing** if the values of $y = f(x)$ decrease as x increases.
>
> The graph of an *increasing* function *climbs* as you move from left to right.
> The graph of a *decreasing* function *descends* as you move from left to right.

Example 1 In each case below, sketch the graph of a line $y = b + mx$ satisfying the given conditions:

(a) b and m are both positive,
(b) b is negative, m is positive,
(c) b is positive, m is negative,
(d) b and m are both negative,
(e) b is zero, m is positive,
(f) m is zero, b is positive.

Solution Recall that b is the vertical intercept and m is the slope of the line. If b is positive, the graph crosses the y-axis above the origin, and if b is negative, the graph crosses the y-axis below the origin. If $b = 0$, the line goes through the origin. If m is positive, then y increases as x increases and our line is the graph of an increasing function. Similarly, if m is negative, our line is the graph of a decreasing function. If m is zero, the "rise" is zero, and we have a horizontal line. Thus, possible answers are shown in Figure 1.21.

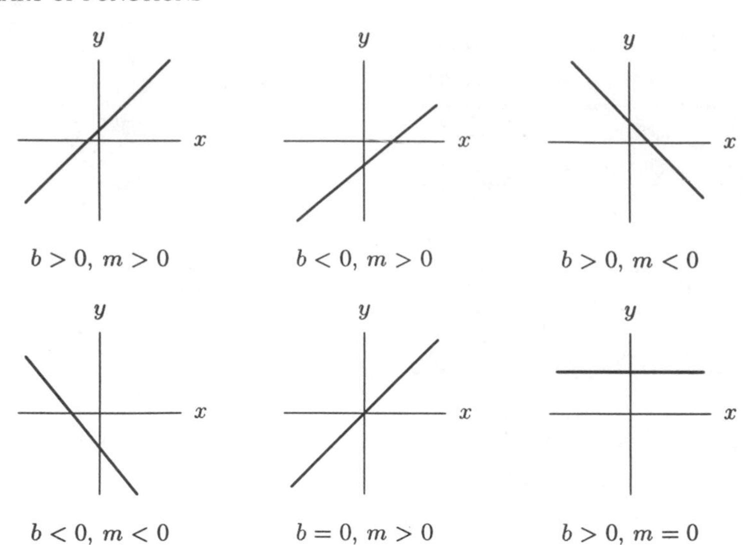

Figure 1.21

Example 2 Which of the following tables of values represent linear functions?

(a)

x	y
0	25
1	30
2	35
3	40
4	45

(b)

x	y
0	10
2	16
4	26
6	40
8	62

(c)

t	s
20	2.4
30	2.2
40	2.0
50	1.8
60	1.6

Solution (a) Since y goes up by 5 for every increase of 1 in x, this is the table of a linear function (with a slope of $5/1 = 5$).

(b) Between $x = 0$ and $x = 2$, y goes up by 6 as x goes up by 2. Between $x = 2$ and $x = 4$, y goes up by 10 as x goes up by 2. Since the slope is not constant, this is not the table of a linear function.

(c) Since s goes down by 0.2 for every increase of 10 in t, this is a table of a linear function (with a slope of $-0.2/10 = -0.02$).

Example 3 We know that two points determine a line. Find the equation of a line through the points $(2, 5)$ and $(4, 1)$.

Solution We first find the slope of this line. Recall that the slope is equal to the difference quotient

$$m = \frac{\text{Rise}}{\text{Run}} = \frac{f(c) - f(a)}{c - a}$$

In this case, the x-coordinates of the points are $a = 2$ and $c = 4$, so

$$m = \frac{1-5}{4-2} = \frac{-4}{2} = -2.$$

The slope of the line is -2. Now we look for the y-intercept. If we substitute the slope $m = -2$, and either one of our points (we use $(x, y) = (2, 5)$) into the equation $y = b + mx$, we can solve for b:

$$\begin{aligned} y &= b + mx \\ 5 &= b + (-2)(2) \\ 5 &= b - 4 \\ 9 &= b. \end{aligned}$$

The y-intercept of the line is 9. The equation of the line is

$$y = 9 - 2x.$$

We can check this answer by sketching the two points, and checking that the slope and y-intercept look right.

Example 4 The solid waste generated each year in the cities of the United States is increasing. The solid waste generated, in millions of tons, was 82.3 in 1960 and 139.1 in 1980.[6]

(a) Construct a model of the amount of municipal solid waste generated in the US by finding the equation of the line through these two points.

(b) Use this model to predict the amount of municipal solid waste generated in the US, in millions of tons, in the year 2000.

Solution (a) We are looking at the amount of municipal solid waste as a function of year, and the two points are $(1960, 82.3)$ and $(1980, 139.1)$. For the model, we assume that the quantity of solid waste is a linear function of year. The slope of the line is

$$m = \frac{139.1 - 82.3}{1980 - 1960} = \frac{56.8}{20} = 2.84.$$

This slope tells us that the amount of solid waste generated in the cities of the US has been going up at a rate of 2.84 million tons per year. To find the equation of the line, we must find the y-intercept. We substitute the point $(1960, 82.3)$ and the slope $m = 2.84$ into the equation $y = b + mx$:

$$\begin{aligned} y &= b + mx \\ 82.3 &= b + (2.84)(1960) \\ 82.3 &= b + 5566.4 \\ -5484.1 &= b. \end{aligned}$$

The equation of the line is $y = -5484.1 + 2.84x$, where y is the amount of municipal solid waste in the US in millions of tons, and x is the year.

[6] *Statistical Abstracts of the US*, 1988, p. 193, Table 333.

(b) How much solid waste does this model predict in the year 2000? We can graph the line and find the y-coordinate when $x = 2000$, or we can substitute $x = 2000$ into the equation of the line, and solve for y:

$$y = -5484.1 + 2.84x$$
$$y = -5484.1 + (2.84)(2000)$$
$$y = -5484.1 + 5680$$
$$y = 195.9.$$

The model predicts that in the year 2000, the solid waste generated by cities in the US will be 195.9 million tons.

Families of Linear Functions

Formulas such as $f(x) = mx$, and $f(x) = b + mx$, containing constants such as m and b which can take on various values, are said to define a *family of functions*. The constants m and b are called *parameters*. Each of the functions in this section belongs to the family $f(x) = b + mx$.

Grouping functions into families which share important features is particularly useful for mathematical modeling. We often choose a family to represent a given situation on theoretical grounds and then use data to determine the particular values of the parameters. The meaning of the parameters m and b in the family $f(x) = b + mx$ is shown in Figures 1.22 and 1.23.

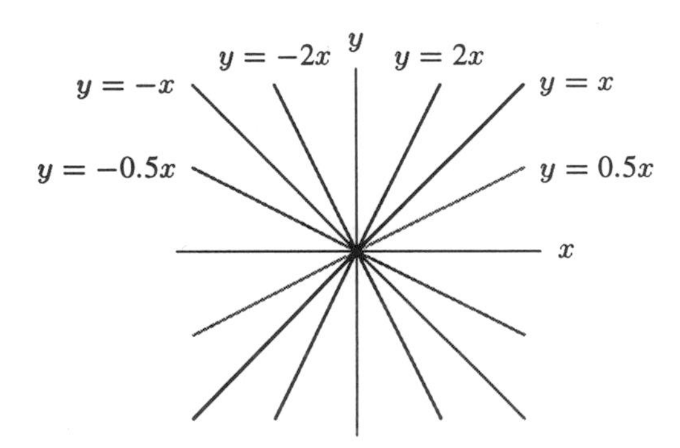

Figure 1.22: The family $y = mx$ (with $b = 0$)

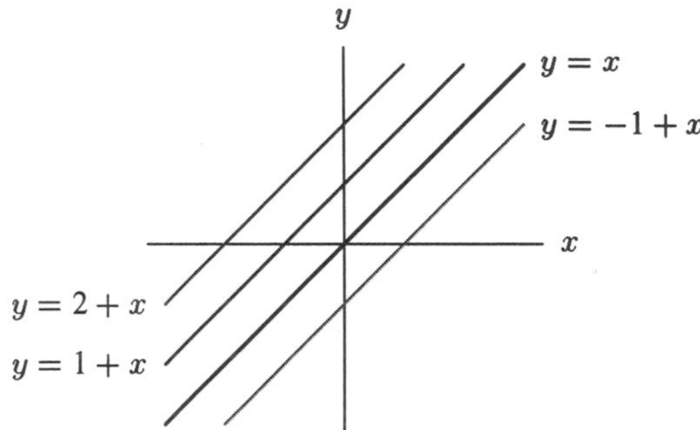

Figure 1.23: The family $y = b + x$ (with $m = 1$)

Problems for Section 1.2

1. Match the graphs in Figures 1.24 with the equations below.
   (a) $y = x - 5$
   (b) $-3x + 4 = y$
   (c) $5 = y$
   (d) $y = -4x - 5$
   (e) $y = x + 6$
   (f) $y = x/2$

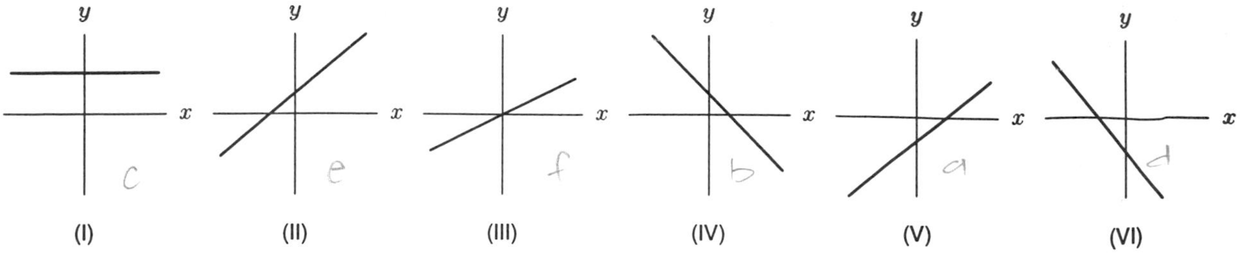

(I) (II) (III) (IV) (V) (VI)

Figure 1.24

2. Match the graphs in Figure 1.25 with the equations below.
   (a) $y = -2.72x$ (c) $y = 27.9 - 0.1x$ (e) $y = -5.7 - 200x$
   (b) $y = 0.01 + 0.001x$ (d) $y = 0.1x - 27.9$ (f) $y = x/3.14$

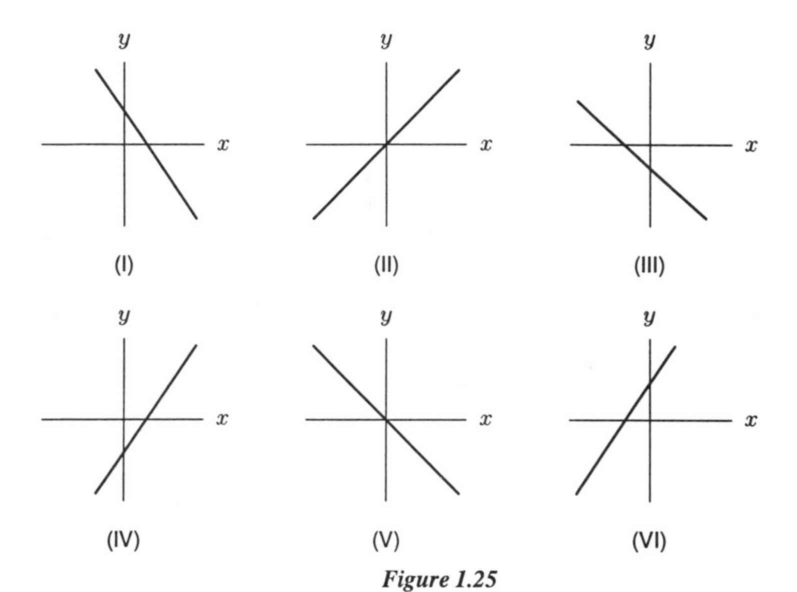

Figure 1.25

3. Find the slope and vertical intercept of the line whose equation is $2y + 5x - 8 = 0$.
4. Find the equation of the line through the points $(-1, 0)$ and $(2, 6)$.
5. Find the equation of the line with slope m through the point (a, c).

For Problems 6–7, use the fact that parallel lines have equal slopes, and that two lines are perpendicular if their slopes are negative reciprocals.

6. Find the equation of the line through the point $(2, 1)$ which is perpendicular to the line $y = 5x - 3$.
7. Find the equations of the lines parallel to and perpendicular to the line $y + 4x = 7$ through the point $(1, 5)$.
8. Estimate the slope of the line shown in Figure 1.26 and use the slope to find an equation for that line. (Note that the x and y scales are unequal.)

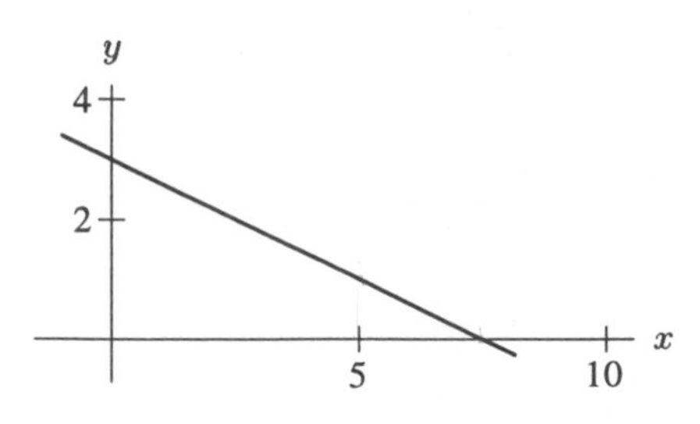

Figure 1.26

9. Which of the following tables of values correspond to linear functions?

(a)

x	y
0	27
1	25
2	23
3	21

(b)

t	s
15	62
20	72
25	82
30	92

(c)

u	w
1	5
2	10
3	18
4	28

10. For each of the tables in Problem 9 which corresponds to a linear function, find a formula for that linear function.

11. Corresponding values of p and q are given in the table below.
    (a) Find q as a linear function of p.
    (b) Find p as a linear function of q.

p	1	2	3	4
q	950	900	850	800

12. A linear equation was used to generate the values in the table below. Find that equation.

x	5.2	5.3	5.4	5.5	5.6
y	27.8	29.2	30.6	32.0	33.4

13. An equation of a line is $3x + 4y = -12$.
    (a) Find the x and y intercepts.
    (b) Find the length of the portion of the line that lies between the x and y intercepts of the equation.

14. Is the demand curve for a product an increasing or decreasing function? Is the supply curve for a product an increasing or decreasing function?

15. A company's pricing schedule is designed to encourage large orders. A table of the price per dozen, p, of a certain item versus the size of the order in gross, q, is given. (A gross is equal to 12 dozen.)
    (a) Find a formula for q as a linear function of p.
    (b) Find a formula for p as a linear function of q.

q	3	4	5	6
p	15	12	9	6

16. The graph of a linear function is given in Figure 1.27. Estimate $f(0)$, $f(1)$, and $f(3)$.

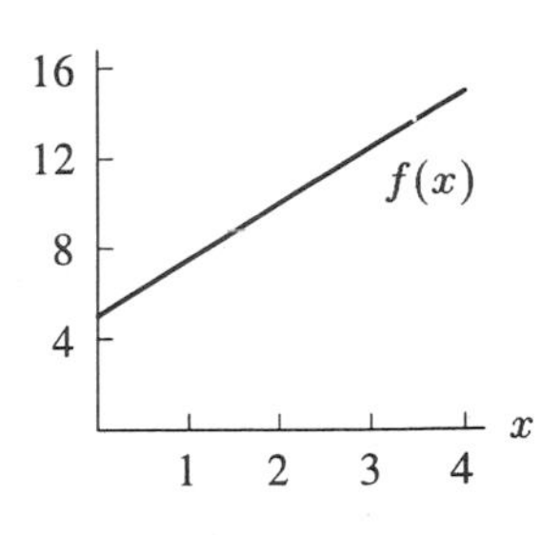

Figure 1.27

17. A linear equation was used to generate the following table.
(a) Find the equation. (b) Sketch the graph of $f(t)$.

t	0	1	2	3	4
$f(t)$	19.72	18.48	17.24	16.00	14.76

18. The Canadian gold reserve, Q, in millions of fine troy ounces is given for the years 1986 through 1990 in the following table.[7] Find a linear formula which approximates this data reasonably well and gives the gold reserve as a function of time measured since 1986.

year	1986	1987	1988	1989	1990
Q	19.72	18.52	17.14	16.10	14.76

19. According to a 1979 *Build and Blood Pressure Study*[8] by the Society of Actuaries, the average weight, w, in pounds, of men in their sixties for various heights, h, in inches, is given in the following table. Find a linear function which gives a reasonable approximation to the average weight as a function of height for men in their sixties.

h	68	69	70	71	72	73	74	75
w	167	172	176	181	186	191	196	200

20. A demand curve is given by $75p + 50q = 300$, where p is the price of the product, in dollars, and q is the quantity demanded at that price. Find the vertical and horizontal intercepts and interpret them in terms of consumer's demand for the product. (Put price p on the vertical axis and quantity q on the horizontal axis.)

21. A tour boat operator found that when the price charged for the scenic boat tour was \$25, the average number of customers per week was 500. When the price was reduced to \$20, the average number of customers per week went up to 650. Find the equation of the demand curve, assuming that it is a line.

[7] "Gold Reserves of Central Banks and Governments", The World Almanac, 1992, p.158
[8] "Average Weight of Americans by Height and Age", The World Almanac, 1992, p.956

22. A car rental company offers cars at \$40 a day and 15 cents a mile. Its competitor's cars are \$50 a day and 10 cents a mile.
    (a) For each company, write a formula giving the cost of renting a car for a day as a function of the distance traveled.
    (b) On the same axes, sketch graphs of both functions.
    (c) How should you decide which company is cheaper?

23. Consider a graph of Fahrenheit temperature, °F, against Celsius temperature, °C, and assume that the graph is a line. You know that 212°F and 100°C both represent the temperature at which water boils. Similarly, 32°F and 0°C both represent water's freezing point.
    (a) What is the slope of the graph?
    (b) What is the equation of the line?
    (c) Use the equation to find what Fahrenheit temperature corresponds to 20°C.
    (d) What temperature is the same number of degrees in both Celsius and Fahrenheit?

24. Suppose you are driving at a constant speed from Chicago to Detroit, about 275 miles away. About 120 miles from Chicago you pass through Kalamazoo, Michigan. Sketch a graph of your distance from Kalamazoo as a function of time.

25. When a cold yam is put into a hot oven to bake, the temperature of the yam rises. The rate, R (in degrees per minute), at which the temperature of the yam rises is governed by Newton's Law of Heating, which says that the rate is proportional to the temperature difference between the yam and the oven. If the oven is at 350°F and the temperature of the yam is H°F:
    (a) Write a formula giving R as a function of H.
    (b) Sketch the graph of R against H.

26. When a cup of hot coffee sits on the kitchen table, its temperature falls. The rate, R, at which its temperature changes is governed by Newton's Law of Cooling, which says that the rate is proportional to the temperature difference between the coffee and the surrounding air. Let's think of the rate, R, as a negative quantity because the temperature of the coffee is falling. If the temperature of the coffee is H°C and the temperature of the room is 20°C:
    (a) Write a formula giving R as a function of H.
    (b) Sketch a graph of R against H.

27. Since the opening up of the West, the US population has moved westward. To observe this, we look at the "population center" of the US, which is the point at which the country would balance if it were a flat plate with no weight, and every person had equal weight. In 1790 the population center was east of Baltimore, Maryland. It has been moving westward ever since, and in 1990 it crossed the Mississippi river to Steelville, Missouri (southwest of St. Louis). During the second half of this century, the population center has moved about 50 miles west every 10 years.
    (a) Express the approximate position of the population center as a function of time, measured in years from 1990. Measure position westward from Steelville, along the line running through Baltimore.
    (b) The distance from Baltimore to St. Louis is a bit over 700 miles. Could the population center have been moving at roughly the same rate for the last two centuries?
    (c) Could the function in part (a) continue to apply for the next three centuries? Why or why not? [Hint: You may want to look at a map. Note that distances are in air miles and are not driving distances.]

1.3 ECONOMIC APPLICATIONS OF FUNCTIONS

Management decisions within a particular firm or industry usually depend on many functions. In this section we will look at some of these functions. One of the most important of these is the cost of production.

The Cost Function

The **cost function**, $C(q)$, gives the total cost of producing a quantity q of some good.

What sort of function do you expect C to be? The more goods that are made, the higher the cost, so C is an increasing function. For most goods, such as cars or cases of soda, q can only be an integer, so the graph of C might look like that in Figure 1.28.

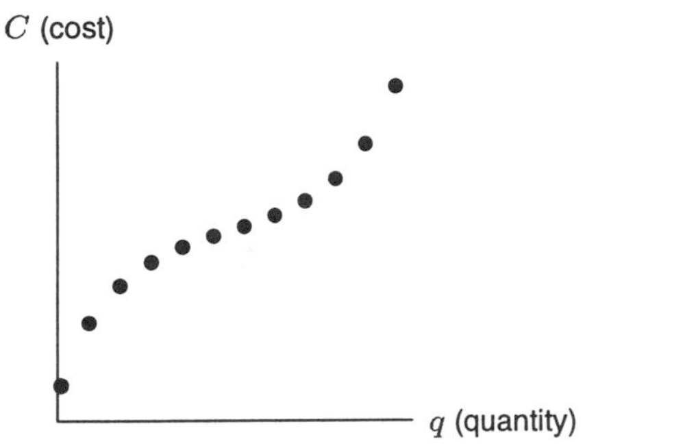

Figure 1.28: Cost: Positive integer values of q

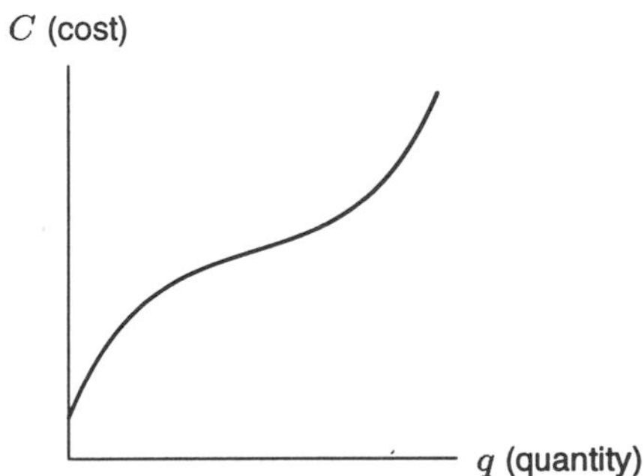

Figure 1.29: Cost: All positive values of q

However, economists usually imagine the graph of C as the smooth curve drawn through these points, as in Figure 1.29. This is equivalent to assuming that $C(q)$ is defined for all nonnegative values of q, not just for integers.

Costs of production can be separated into two parts: the *fixed costs,* which are incurred even if nothing is produced, and the *variable costs* which vary depending on how many units are produced.

A Soccer Ball Example

Let's consider a company that makes soccer balls. The factory and machinery needed to begin production are fixed costs, since these costs are incurred even if no soccer balls are made. The costs of labor and raw materials are variable costs since these quantities depend on how many balls are made. Suppose that the fixed costs for this company are \$24,000 and the variable costs are \$7 per soccer ball. The total costs for the company are seven times the number of soccer balls produced, plus the fixed cost of \$24,000. We have

$$C(q) = 24{,}000 + 7q,$$

where q is the number of soccer balls produced. Notice that this is the equation of a line, with slope equal to the variable costs per ball (7) and vertical intercept equal to the fixed costs (24,000). Our cost function for soccer balls is a linear function of q. (Later, we will see examples of cost functions that are not linear.)

Example 1 Sketch the graph of the cost function $C(q) = 24,000 + 7q$. On the graph, label the fixed costs and variable cost per unit.

Solution The graph is in Figure 1.30. The graph of the cost function is a line with a vertical intercept of 24,000 and a slope of 7. The fixed costs are represented graphically by the vertical intercept. The variable cost per unit is represented graphically by the slope of the line, or equivalently, by the "rise" (the change in cost) corresponding to a horizontal change of one unit.

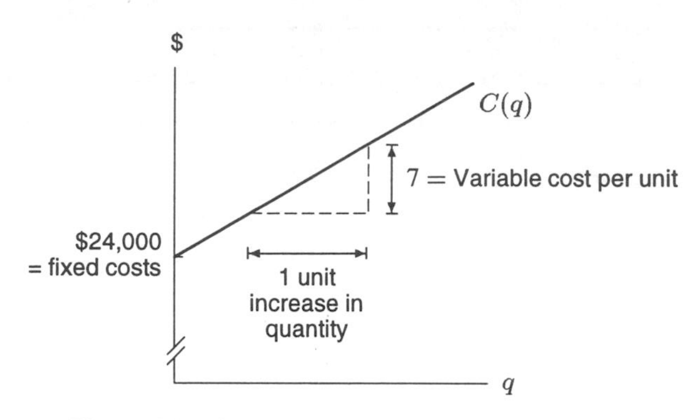

Figure 1.30: Cost function for the soccer ball company

Example 2 In each case, draw a graph of a linear cost function satisfying the given conditions:

(a) Fixed costs are large but variable cost per unit is small.
(b) There are no fixed costs but variable cost per unit is high.

Solution (a) A solution is the graph of any line with a large vertical intercept and a small slope, such as that shown in Figure 1.31.

(b) A solution is the graph of any line with a vertical intercept of zero (so the line goes through the origin) and a large positive slope. See Figure 1.32.

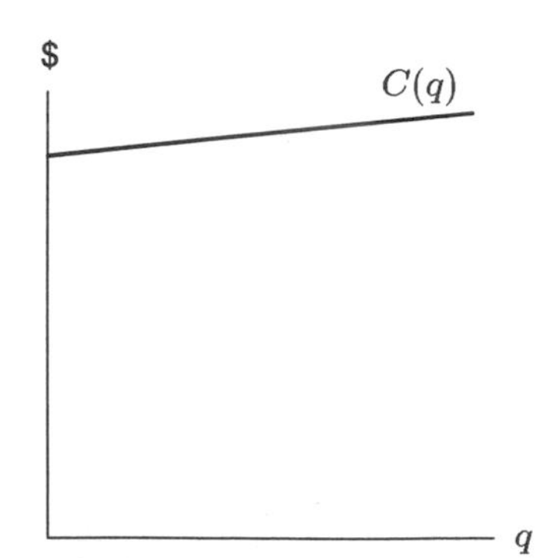

Figure 1.31: Large fixed costs, small variable cost per unit

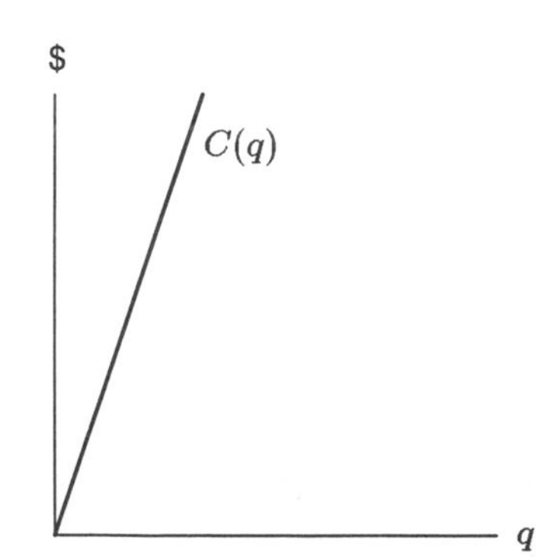

Figure 1.32: No fixed costs, high variable cost per unit

The Revenue Function

The **revenue function**, $R(q)$, represents the total revenue received by the firm from selling a quantity q of some good.

The soccer ball company makes money by selling soccer balls. The revenue function for this company depends on the price of the soccer balls and the quantity sold. If 10 soccer balls are sold at \$15 each, the total revenue is $\$15 \times 10 = \$150$.

If the price, p, is a constant and the quantity sold is q, then

$$\text{Revenue} = \text{Price} \times \text{Quantity}, \quad \text{so} \quad R = pq.$$

The graph of the revenue as a function of q is a line through the origin, with slope equal to the price p. (As with cost functions, we will consider nonlinear revenue functions later.)

Example 3 If the company sells soccer balls for \$15 each, sketch a graph of the company's revenue function and mark on it the price of a soccer ball.

Solution Since $R(q) = pq = 15q$, the revenue graph is a line through the origin with a slope of 15. See Figure 1.33. The price is the slope of the line and represents the increase in revenue corresponding to a unit increase in quantity sold.

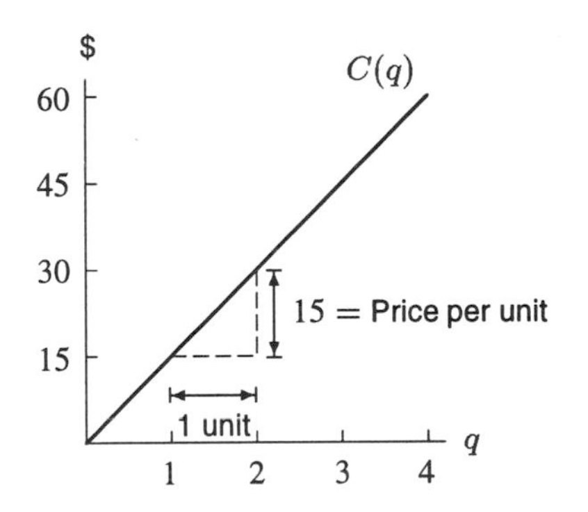

Figure 1.33: Revenue function for the soccer ball company

Example 4 Sketch graphs of $C(q) = 24,000 + 7q$ and $R(q) = 15q$ on the same axes. For what values of q does the firm make money? Explain your answer graphically.

Solution The firm makes money whenever revenues are greater than costs, so we want to find the values of q for which the graph of $R(q)$ lies above the graph of $C(q)$. See Figure 1.34.

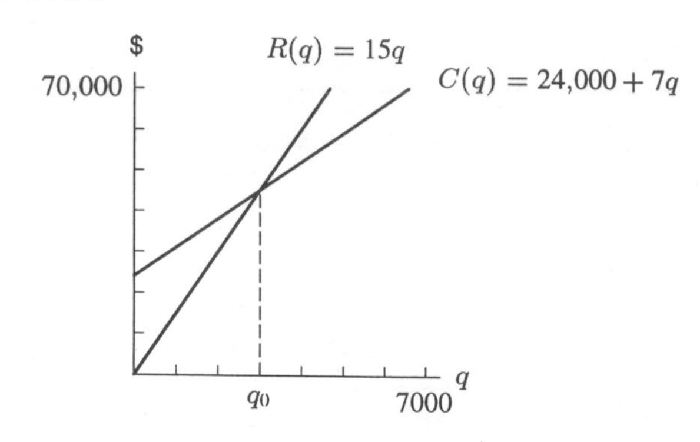

Figure 1.34: Cost and revenue functions for the soccer ball company

In Figure 1.34, the graph of $R(q)$ is above the graph of $C(q)$ for all values of q to the right of the value q_0, where the graphs of $R(q)$ and $C(q)$ cross. What is this point of intersection q_0? It is the point where revenue and cost are equal. We have,

$$\begin{aligned} \text{Revenue} &= \text{Cost} \\ 15q &= 24{,}000 + 7q \\ 8q &= 24{,}000 \\ q &= 3{,}000. \end{aligned}$$

Thus, the company makes a profit if it produces and sells more than 3000 soccer balls.

The Profit Function

Decisions are often made by considering the profit, usually written as π (to distinguish it from the price, p; this π has nothing to do with the area of a circle, and merely stands for the Greek equivalent of the letter "p") and defined by

$$\text{Profit} = \text{Revenue} - \text{Cost} \quad \text{so} \quad \pi = R - C.$$

The *break-even point* for a company is the point where the profit is zero, or, equivalently, the point where revenue equals cost. We saw in Example 4 that the break-even point for the soccer ball company is $q_0 = 3000$. If the company produces less than 3000 balls, it loses money. If it produces exactly 3000 balls, it breaks even. If it produces more than 3000 balls, it makes a profit.

Example 5 A company producing jigsaw puzzles has fixed costs of \$6000 and variable costs of \$2 per puzzle. The company sells the puzzles for \$5 each.

(a) Find a formula for the cost function.
(b) Find a formula for the revenue function.
(c) Find a formula for the profit function.
(d) Sketch a graph of $R(q)$ and $C(q)$ on the same axes.
(e) What is the break-even point for the company?

Solution
(a) $C(q) = 6000 + 2q$
(b) $R(q) = 5q$
(c) $\pi(q) = R(q) - C(q) = 5q - (6000 + 2q) = 3q - 6000$
(d) See Figure 1.35.
(e) The break-even point is labeled q_0 in Figure 1.35. We find q_0 by setting the revenue equal to the cost and solving for q:

$$\begin{aligned} \text{Revenue} &= \text{Cost} \\ 5q &= 6000 + 2q \\ 3q &= 6000 \\ q &= 2000. \end{aligned}$$

Notice that this is the same thing we get if we set the profit function equal to zero. The break-even point is $q_0 = 2000$ puzzles.

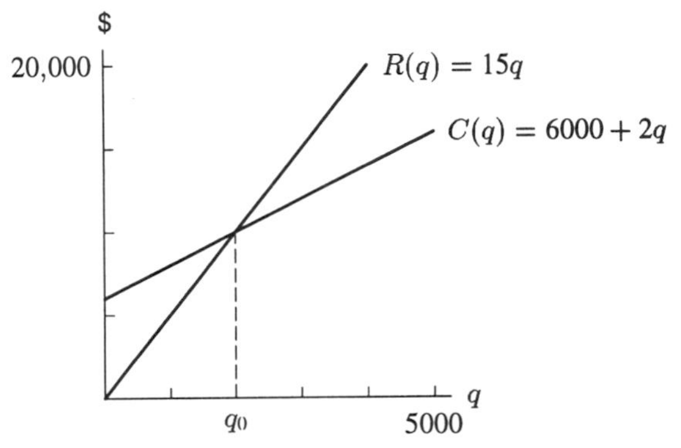

Figure 1.35: Cost and revenue functions for the jigsaw puzzle company

The Depreciation Function

Suppose that the soccer ball company has a big machine that costs $20,000. The managers of the company plan to keep the machine for ten years and then sell it for $3000. They expect that the value of their machine will *depreciate* from $20,000 today to a resale value of $3000 in ten years. For tax purposes, they can depreciate the value of this machine over the ten-year period. The depreciation formula gives the value, $V(t)$, of the machine as a function of the number of years, t, since the machine was purchased. One widely-used depreciation method is known as ***straight-line depreciation***, where we assume that the value of the machine depreciates linearly. The corresponding depreciation formula will be a linear function.

Since the value of the machine when it is new (that is, when $t = 0$) is $20,000, we have $V(0) = 20,000$, and since the resale value at time $t = 10$ is $3000, we have $V(10) = 3000$. The slope of this line is:

$$m = \frac{3000 - 20,000}{10 - 0} = \frac{-17000}{10} = -1700.$$

This slope tells us that the value of the machine is decreasing at a rate of $1700 per year. Since $V(0) = 20,000$, the vertical intercept is 20,000, so the depreciation function for this machine is $V(t) = 20,000 - 1700t$. It makes sense that this function is decreasing since the value of the machine is going down as time passes.

Revisiting the Supply and Demand Curves

In Section 1.1, we learned that a demand function tells us what quantity will be demanded at a given price, and that a supply function tells us what quantity will be produced at a given price. The market tends to settle at the equilibrium price and quantity, which is the point at which demand equals supply.

Example 6 Suppose that the market demand and supply functions are given by

$$D(p) = 100 - 2p \quad \text{and} \quad S(p) = 3p - 50.$$

Find the equilibrium price and quantity.

Solution To find the equilibrium price and quantity, we find the point where supply equals demand. We can do this algebraically or graphically. Algebraically, we have

$$\begin{aligned} \text{Demand} &= \text{Supply} \\ 100 - 2p &= 3p - 50 \\ 150 &= 5p \\ p &= 30 \end{aligned}$$

The equilibrium price is $30. To find the equilibrium quantity, we can use either the demand function or the supply function. At a price of $30, the quantity produced is $100-2(30) = 100-60 = 40$ items. The equilibrium quantity is 40 items.

Alternatively, if we graphed the demand and supply curves on the same axes, we would see that the point of intersection is at a price of $30 and a quantity of 40 items.

The Effect of Taxes on Equilibrium[9]

Suppose that the demand and supply curves for a product are

$$D(p) = 100 - 2p \quad \text{and} \quad S(p) = 3p - 50.$$

We saw in Example 6 that the equilibrium price and quantity in a purely competitive market are

$$\begin{aligned} \text{equilibrium price} &= \$30, \text{ and} \\ \text{equilibrium quantity} &= 40 \text{ items.} \end{aligned}$$

What effect do taxes have on the equilibrium price and quantity? And who (the producers or the consumers) ends up paying for the tax? We will consider two types of taxes. A *specific tax* is a tax that collects a fixed amount per unit of a product sold regardless of the actual selling price. This is

[9]adapted from *A Unified Approach to Mathematical Economics*, by Barry Bressler, Harper & Row.

the case with such items as gasoline, alcohol, and cigarettes. A *sales tax* is a tax that collects a fixed percentage of the product price. Many cities and states collect sales tax on a wide variety of items. We consider the case of a specific tax here. The effect of a sales tax is considered in Problems 20 and 21.

Suppose a specific tax of \$5 per unit is imposed upon suppliers. This means that a selling price of p dollars will not bring forth the same quantity supplied, since suppliers will only receive $p - 5$ dollars. The amount supplied will depend on $p - 5$, while the amount demanded will still depend on p (the price the consumers have to pay). We have:

$$D(p) = 100 - 2p \quad \text{and} \quad \begin{aligned} S(p) &= 3(p-5) - 50 \\ &= 3p - 15 - 50 \\ &= 3p - 65 \end{aligned}$$

What are the equilibrium price and quantity in this situation? We have

$$\begin{aligned} \text{Demand} &= \text{Supply} \\ 100 - 2p &= 3p - 65 \\ 165 &= 5p \\ p &= 33 \end{aligned}$$

The equilibrium price when there is a specific tax is \$33. We saw earlier that the equilibrium price in a purely competitive market is \$30, so the equilibrium price increases by \$3 as a result of the tax. Notice that this is less than the amount of the tax. The consumer ends up paying \$3 more than if the tax did not exist. However the government must receive \$5 per item. Thus the producer pays the other \$2 of the tax and is able to retain \$28 of the amount paid per item.

The equilibrium quantity at a price of \$33 is 34 units. Not surprisingly, the tax has reduced the number of items sold. The government, when imposing a \$5 per unit tax on an industry selling 40 units, should not expect to receive $(\$5)(40) = \$200$ in tax revenue. In this example, tax revenue would be only $(\$5)(34) = \$170$.

A Budget Constraint

An ongoing debate in the federal government concerns the allocation of money between defense and social programs. In general, the more that is spent on defense, the less that is available for social programs, and vice versa. Let's simplify the example to guns and butter. Assuming a constant budget, we will show that the relationship between the number of guns and the quantity of butter is linear. Suppose there is \$12,000 to be spent and that it is to be divided between guns, costing \$400 each, and butter, costing \$2000 a ton. Suppose the number of guns bought is g, and the number of tons of butter is b. Then the amount of money spent on guns is $\$400g$ (because each one is \$400), and the amount spent on butter is $\$2000b$. Assuming all the money is spent,

$$\begin{matrix}\text{Amount spent} \\ \text{on guns}\end{matrix} + \begin{matrix}\text{Amount spent} \\ \text{on butter}\end{matrix} = \$12{,}000$$

or

$$400g + 2000b = 12{,}000$$

or

$$g + 5b = 30.$$

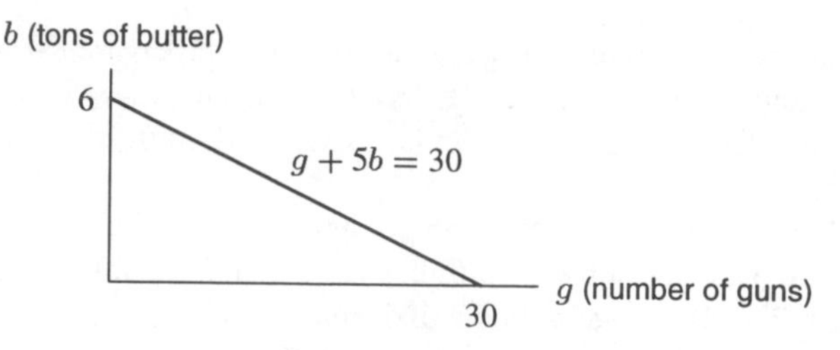

Figure 1.36: Budget constraint

This equation is the budget constraint. Its graph is the line shown in Figure 1.36, which can be found by plotting points. We will calculate the points at which the graph crosses the axes. If $b = 0$, then

$$g + 5(0) = 30 \quad \text{so} \quad g = 30.$$

If $g = 0$, then

$$0 + 5b = 30 \quad \text{so} \quad b = 6.$$

Since the number of guns bought determines the amount of butter bought (because all the money which doesn't go to guns goes to butter), b is a function of g. Similarly, the amount of butter bought determines the number of guns, so g is a function of b. The budget constraint represents an *implicitly defined function*, because neither quantity is given explicitly in terms of the other. If we solve for g, giving

$$g = 30 - 5b$$

we have an *explicit* formula for g in terms of b. Similarly,

$$b = \frac{30 - g}{5} \quad \text{or} \quad b = 6 - 0.2g$$

gives b as an explicit function of g. Since the explicit functions

$$g = 30 - 5b \quad \text{and} \quad b = 6 - 0.2g$$

are linear, the graph of the budget constraint must be a line.

Problems for Section 1.3

1. An amusement park charges an admission fee of \$7 as well as an additional \$1.50 for each ride.
   (a) Find the total cost $C(n)$ as a function of the number of rides, n.
   (b) Find $C(2)$ and $C(8)$.
   (c) Plot the points $(2, C(2))$ and $(8, C(8))$ and use them to graph the function $C(n)$.

2. The cost function for a certain company is $C(q) = 4000 + 2q$ and the revenue function is $R(q) = 10q$.
   (a) What are the fixed costs for the company?
   (b) What is the variable cost per unit?
   (c) What price is the company charging for its product?
   (d) Graph $C(q)$ and $R(q)$ on the same axes and label the break-even point, q_0. Explain graphically how you know the company will make a profit if the quantity produced is greater than q_0.
   (e) Find the break-even point q_0.

3. A table of values for a linear cost function is given below. What are the fixed costs and the variable cost per unit? Find a formula for the cost function.

q	0	5	10	15	20
$C(q)$	5000	5020	5040	5060	5080

4. A company has a cost function of $C(q) = 6000 + 10q$ and a revenue function of $R(q) = 12q$.
   (a) Find the revenue and cost if the company is producing 500 units. Does the company make a profit?
   (b) Find the revenue and cost if the company is producing 5000 units. Does the company make a profit?
   (c) Find the break-even point and illustrate it graphically.

5. A company that makes Adirondack chairs has fixed costs of \$5000 and variable costs of \$30 per chair. The company plans to sell the chairs for \$50 each.
   (a) Find formulas for the cost function and the revenue function.
   (b) Graph the cost and the revenue functions on the same axes.
   (c) Find the break-even point.

6. The Quick-Food company wants to provide students with an alternative to the college food service plan. Quick-Food has fixed costs of \$350,000 per term that are incurred regardless of how many students are served. In addition, the cost of providing and serving the food amounts to \$400 per student served. Quick-Food plans to charge \$800 per student. How many students must sign up with the Quick-Food plan in order for the company to make a profit?

7. A company manufactures and sells a new type of running shoe. Total production costs to the company consist of fixed overhead of \$650,000 plus production costs of \$20 per pair of shoes. Each pair of shoes produced is sold for \$70.
   (a) Find the total cost, $C(q)$, as a function of the number of pairs of shoes produced, q.
   (b) Find the total revenue, $R(q)$, as a function of the number of pairs of shoes produced, q.
   (c) Find the total profit, $\pi(q)$, as a function of the number of pairs of shoes produced, q.
   (d) How many pairs of shoes must be produced and sold in order for the company to make a profit?

8. A graph of the cost function for a certain company is shown in Figure 1.37.
   (a) Estimate the fixed costs for this company.
   (b) Estimate the variable cost per unit.
   (c) Estimate $C(10)$.

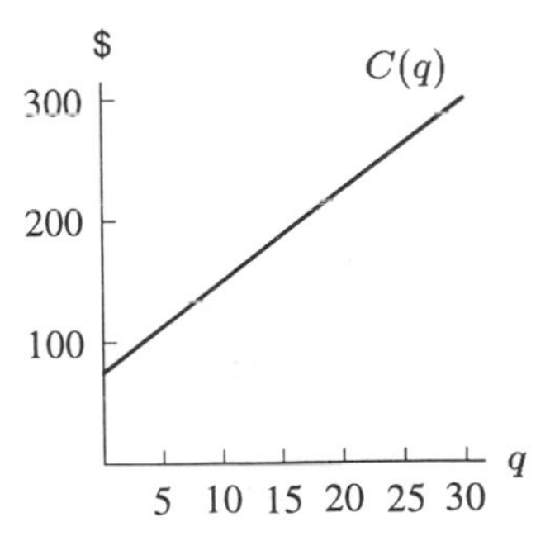

Figure 1.37

9. A graph of the cost function for a certain company is shown in Figure 1.38.
   (a) What are the fixed costs for this company?
   (b) What is the variable cost per unit?
   (c) We see from the graph that $C(100) = 2500$. Explain in words what this is telling you about costs.

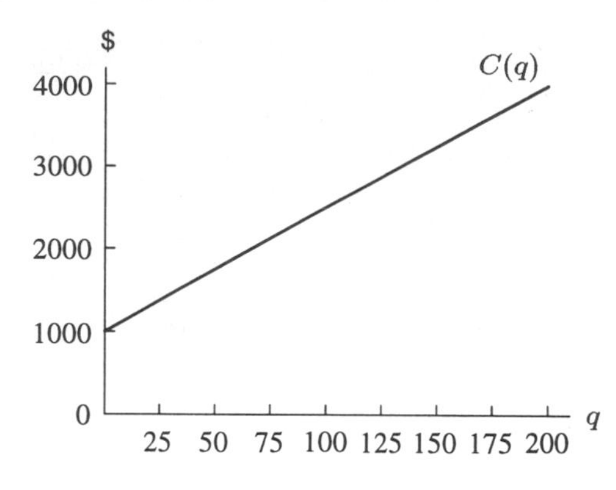

Figure 1.38

10. (a) Give an example of a possible company where the fixed costs would be zero (or very small).
    (b) Give an example of a possible company where the variable cost per unit would be zero (or very small).
11. Graphs of the costs and revenue functions for a certain company are given in Figure 1.39. Approximately what quantity of goods does this company have to produce in order to make a profit?

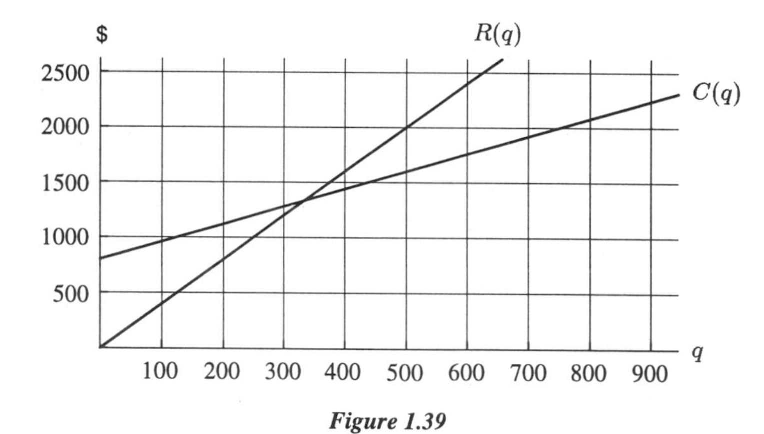

Figure 1.39

12. For large orders, a photocopying company allows customers to choose from two different price lists. The first price list is \$100 plus 3 cents per copy, whereas the second price list is \$200 plus 2 cents per copy.
    (a) For each price list, find the total cost as a function of the number of copies needed.
    (b) Determine which price list is cheaper if you need 5000 copies.
    (c) For what number of copies do both price lists charge you the same amount?

13. A $15,000 robot depreciates completely (down to a value of zero) in 10 years. Assume that this depreciation is linear.
    (a) Find a formula for its value as a function of time.
    (b) How much is the robot worth three years after it is purchased?

14. A $50,000 tractor has a resale value of $10,000 twenty years after it was purchased. Assume that the value of the tractor depreciates linearly and that time is measured from the time of purchase.
    (a) Find a formula for the value of the tractor as a function of the time since it was purchased.
    (b) Sketch a graph of the value of the tractor against time.
    (c) Find the horizontal and vertical intercepts, give units for them, and interpret them.

15. Suppose that you have a budget of $1000 for the school year to cover your books and social outings, and suppose that books cost (on average) $40 each and social outings cost (on average) $10 each. Let b denote the number of books purchased per year and s denote the number of social outings in a year.
    (a) What is the equation of your budget constraint?
    (b) Graph the budget constraint. (It doesn't matter which variable you put on which axis.)
    (c) Find the vertical and horizontal intercepts and give a financial interpretation for each one.

16. A company has a total budget of $500,000 and spends this budget on raw materials and personnel. The company uses m units of raw materials, at a cost of $100 per unit, and hires r employees, at a cost of $25,000 each.
    (a) What is the equation of the company's budget constraint?
    (b) Solve for m, the quantity of raw materials the company can buy, as a function of r, the number of employees.
    (c) Solve for r, the number of employees the company can hire, as a function of m, the quantity of raw materials used.

17. Hot peppers have been rated according to Scoville units, with a maximum human tolerance level of 14,000 Scovilles per dish. The West Coast Restaurant, known for spicy dishes, promises a daily special to satisfy the most avid spicy-dish fans. The restaurant imports Indian peppers rated at 1200 Scovilles each and Mexican peppers with a Scoville rating of 900 each.
    (a) Determine the Scoville constraint equation relating the maximum number of Indian and Mexican peppers the restaurant should use for their speciality dish.
    (b) Solve the equation from part (a) to show explicitly the number of Indian peppers needed in the hottest dishes as a function of the number of Mexican peppers.

18. You have a fixed budget of $\$k$ to spend on soda and suntan oil, which cost $\$p_1$ per liter and $\$p_2$ per liter respectively.
    (a) Write an equation expressing the relationship between the number of liters of soda and the number of liters of suntan oil that you can buy if you exhaust your budget. This is your *budget constraint*.
    (b) Graph the budget constraint, assuming that you can buy fractions of a liter. Label the intercepts.
    (c) Suppose your budget is suddenly doubled. Graph the new budget constraint on the same axes.
    (d) With a budget of $\$k$, the price of suntan oil suddenly doubles. Sketch the new budget constraint on the same axes.

19. The demand and supply functions for a certain product are given by

$$D(p) = 2500 - 20p \quad \text{and} \quad S(p) = 10p - 500.$$

 (a) Find the equilibrium price and quantity.
 (b) If a specific tax of \$6 is imposed on this product, find the new equilibrium price and quantity.
 (c) How much of the \$6 tax is paid by consumers, and how much by producers?
 (d) What is the total tax revenue received by the government?

20. In Example 6, we saw that when the demand and supply curves are given by $D(p) = 100 - 2p$ and $S(p) = 3p - 50$, the equilibrium price is \$30 and the equilibrium quantity is 40 units. Suppose that a sales tax of 5% is imposed on the supplier, so that the supplier only keeps $p - 0.05p$ of the price.
 (a) Find the new equilibrium price and quantity.
 (b) How much is paid in taxes on each unit? How much of this is paid by the consumer, and how much by the producer?

21. Suppose that the sales tax in Problem 20 is imposed on the consumer instead of the producer, so that the consumer's price is $p+0.05p$, while the producer's price is p. Answer the questions in Problem 20 under these circumstances, and compare your answers with those for Problem 20.

22. Linear supply and demand curves are shown in Figure 1.40, with price on the vertical axis.
 (a) Label the equilibrium price p_0 and the equilibrium quantity q_0 on the axes.
 (b) Explain the effect on equilibrium price and quantity if the slope of the supply curve increases. Illustrate your answer graphically.
 (c) Explain the effect on equilibrium price and quantity if the slope of the demand curve becomes more negative. Illustrate your answer graphically.

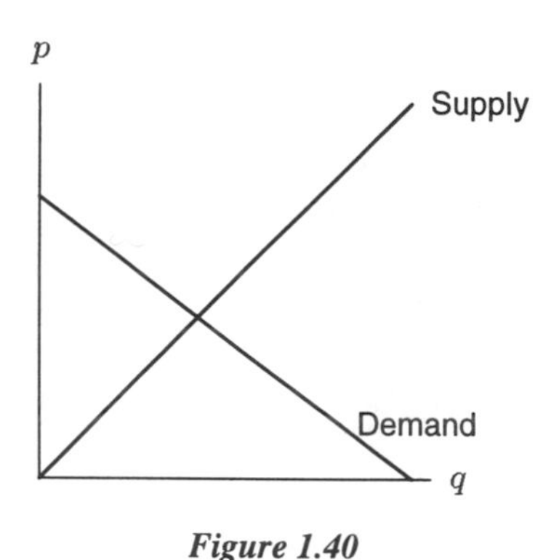

Figure 1.40

1.4 EXPONENTIAL FUNCTIONS

Population Growth

Consider the data for the population of Mexico in the early 1980s in Table 1.11. To see how the population is growing, you might look at the increase in population from one year to the next, as shown in the third column. If the population had been growing linearly, all the numbers in the third column would be the same. But populations usually grow faster as they get bigger, because there are more people to have babies. So you shouldn't be surprised to see the numbers in the third column increasing.

TABLE 1.11 *Population of Mexico (estimated), 1980–1986*

Year	Population (millions)	Change in population (millions)
1980	67.38	
		1.75
1981	69.13	
		1.80
1982	70.93	
		1.84
1983	72.77	
		1.89
1984	74.66	
		1.94
1985	76.60	
		1.99
1986	78.59	

Suppose we divide each year's population by the previous year's population. We get, approximately,

$$\frac{\text{Population in 1981}}{\text{Population in 1980}} = \frac{69.13 \text{ million}}{67.38 \text{ million}} = 1.026$$

$$\frac{\text{Population in 1982}}{\text{Population in 1981}} = \frac{70.93 \text{ million}}{69.13 \text{ million}} = 1.026.$$

The fact that both calculations give 1.026 shows the population grew by about 2.6% between 1980 and 1981 *and* between 1981 and 1982. If you do similar calculations for other years, you will find that the population grew by a factor of about 1.026, or 2.6%, every year. Whenever you have a constant *growth factor* (here 1.026), you have *exponential growth.* If t is the number of years since 1980,

$$\begin{aligned}
&\text{When } t = 0, \quad \text{population} = 67.38 = 67.38(1.026)^0 \\
&\text{When } t = 1, \quad \text{population} = 69.13 = 67.38(1.026)^1 \\
&\text{When } t = 2, \quad \text{population} = 70.93 = 69.13(1.026) = 67.38(1.026)^2 \\
&\text{When } t = 3, \quad \text{population} = 72.77 = 70.93(1.026) = 67.38(1.026)^3
\end{aligned}$$

and so t years after 1980, the population is given by

$$P = 67.38(1.026)^t.$$

This is an *exponential function* with *base* 1.026. It is called exponential because the variable, t, is in the exponent. The base represents the factor by which the population grows each year.

If we assume that the same formula will hold for the next 50 years or so, the population will have the shape shown in Figure 1.41. Since the population is growing, the function is increasing. Notice also that the population grows faster and faster as time goes on. This behavior is typical of an exponential function. You should compare this with the behavior of a linear function, which climbs at the same rate everywhere and so has a straight-line graph. Because this graph is bending upward, we say it is *concave up*. Even exponential functions which climb slowly at first, such as this one, climb extremely quickly eventually. That is why exponential population growth is such a threat to the world.

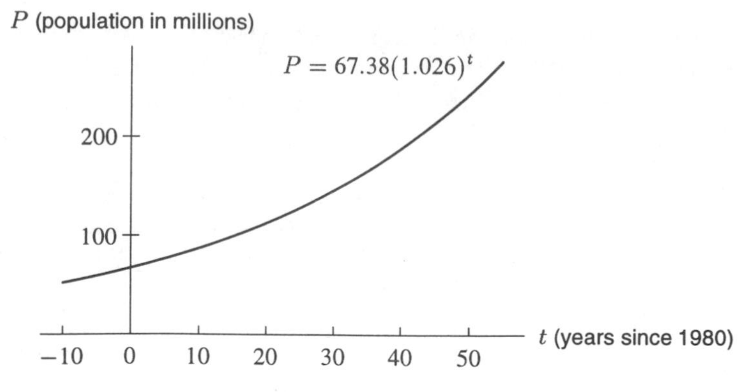

Figure 1.41: Population of Mexico (estimated): Exponential growth

Even if it represents reliable data, the smooth graph in Figure 1.41 is actually only an approximation to the true graph of the population of Mexico. Since we can't have fractions of people, the graph should really be jagged, jumping up or down by one each time someone is born or dies. However, with a population in the millions, the jumps are so small as to be invisible at the scale we are using. Therefore, the smooth graph is an extremely good approximation.

Example 1 Predict the population of Mexico in the year
(a) 2007 (when $t = 27$). (b) 2034 (when $t = 54$). (c) 2061 (when $t = 81$).

Solution Extrapolating so far into the future can be risky because it assumes that the population continues to grow exponentially with the same constant growth factor. (There could, for example, be a medical breakthrough that would increase the growth factor, or an epidemic that would decrease it.) Writing "$\approx$" to represent approximately equal, the model we are using predicts

(a) $P = 67.38(1.026)^{27} \approx 67.38(2) = 134.76$ million.
(b) $P = 67.38(1.026)^{54} \approx 67.38(4) = 269.52$ million.
(c) $P = 67.38(1.026)^{81} \approx 67.38(8) = 539.04$ million.

If you look at the answers to Example 1, you will see something that may surprise you. After 27 years the population has doubled; after another 27 years (at $t = 54$), it has doubled again. Another 27 years later (when $t = 81$), the population has doubled again. As a result, we say that the ***doubling time*** of the population of Mexico is 27 years.

Every exponentially growing population has a fixed doubling time. The world's population currently has a doubling time of about 38 years. Notice what this means: If you live to be 76, the world's population is expected to quadruple in your lifetime.

Musical Pitch

The pitch of a musical note is determined by the frequency of the vibration which causes it. Middle C on the piano, for example, corresponds to a vibration of 263 hertz (cycles per second). A note one octave above middle C vibrates at 526 hertz, and a note two octaves above middle C vibrates at 1052 hertz. (See Table 1.12.)

TABLE 1.12 *Pitch of notes above middle C*

Number, n, of octaves above middle C	Number of hertz $V = f(n)$
0	263
1	526
2	1052
3	2104
4	4208

TABLE 1.13 *Pitch of notes below middle C*

n	$V = 263 \cdot 2^n$
-3	$263 \cdot 2^{-3} = 263(1/2^3) = 32.875$
-2	$263 \cdot 2^{-2} = 263(1/2^2) = 65.75$
-1	$263 \cdot 2^{-1} = 263(1/2) = 131.5$
0	$263 \cdot 2^0 = 263$

Notice that

$$\frac{526}{263} = 2 \quad \text{and} \quad \frac{1052}{526} = 2 \quad \text{and} \quad \frac{2104}{1052} = 2$$

and so on. In other words, each value of V is twice the value before, so

$$\begin{aligned} f(1) &= 526 = 263 \cdot 2 = 263 \cdot 2^1 \\ f(2) &= 1052 = 526 \cdot 2 = 263 \cdot 2^2 \\ f(3) &= 2104 = 1052 \cdot 2 = 263 \cdot 2^3. \end{aligned}$$

In general

$$V = f(n) = 263 \cdot 2^n,$$

where n is the number of octaves above middle C. The base 2 represents the fact that as we go up an octave, the frequency of vibrations doubles. Indeed, our ears hear a note as one octave higher than another precisely because it vibrates twice as fast. For the negative values of n in Table 1.13, this function represents the octaves below middle C. The notes on a piano are represented by values of n between -3 and 4, and the human ear finds values of n between -4 and 7 audible.

Although $V = f(n) = 263 \cdot 2^n$ makes sense in musical terms only for certain values of n, values of the function $f(x) = 263 \cdot 2^x$ can be calculated for all real x, and its graph has the typical exponential shape, as can be seen in Figure 1.42. It is concave up, climbing faster and faster as x increases.

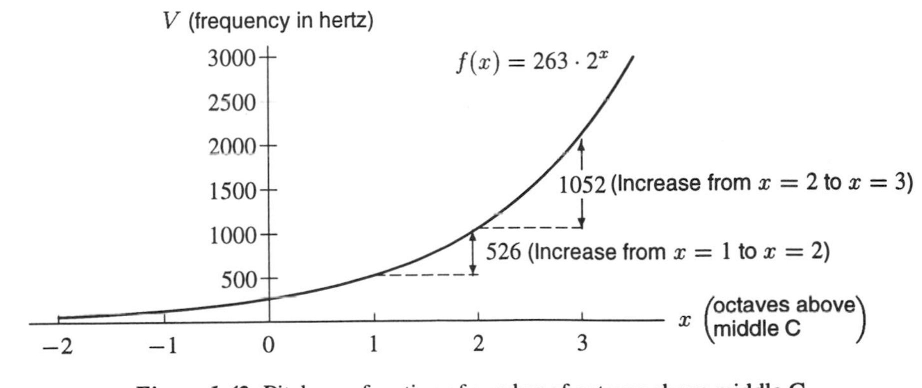

Figure 1.42: Pitch as a function of number of octaves above middle C

Removal of Pollutants from Jet Fuel

Now we will look at an example in which a quantity is decreasing instead of increasing. Before kerosene can be used as jet fuel, federal regulations require that the pollutants in it be removed by passing the kerosene through clay. We will suppose the clay is in a pipe and that each foot of the pipe removes 20% of the pollutants that enter it. Therefore each foot leaves 80% of the pollution. If P_0 is the initial quantity of pollutant and $P = f(n)$ is the quantity left after n feet of pipe:

$$\begin{aligned} f(0) &= P_0 \\ f(1) &= (0.8)P_0 \\ f(2) &= (0.8)(0.8)P_0 = (0.8)^2 P_0 \\ f(3) &= (0.8)(0.8)^2 P_0 = (0.8)^3 P_0 \end{aligned}$$

and so, after n feet,

$$P = f(n) = P_0(0.8)^n.$$

In this example, n must be non-negative. However, the *exponential decay function*

$$P = f(x) = P_0(0.8)^x$$

makes sense for any real x. We'll plot it with $P_0 = 1$ in Figure 1.43; some values of the function are in Table 1.14.

Notice the way the function in Figure 1.43 is decreasing: each downward step is smaller than the one before. This is because as the kerosene gets cleaner, there's less dirt to remove, and so each foot of clay takes out less pollutant than the previous one. Compare this to the exponential growth in Figures 1.41 and 1.42 on pages 38 and 39, where each step upward is larger than the one before. Notice, however, that all three graphs mentioned are concave up.

TABLE 1.14 *Values of decay function*

x	$P = (0.8)^x$
-2	1.56
-1	1.25
0	1
1	0.8
2	0.64
3	0.51
4	0.41

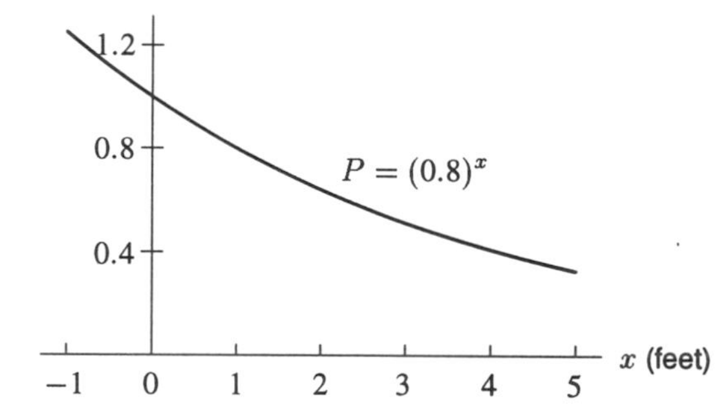

Figure 1.43: Pollutant removal: Exponential decay

The General Exponential Function

> P is an **exponential function** of t with base a if
>
> $$P = P_0 a^t$$
>
> where P_0 is the initial quantity (when $t = 0$) and a is the factor by which P changes when t increases by 1.
> If $a > 1$, we have exponential growth; if $0 < a < 1$, we have exponential decay.

The largest possible domain for the exponential function is all real numbers, provided $a > 0$. (Why do we not want $a \leq 0$?)

> To recognize that a function $P = f(t)$ given by a table of data is exponential, look for ratios of P values that are constant for equally spaced t values.

Example 2 Determine whether each of the function values in Table 1.15 could correspond to an exponential function, a linear function, or neither. For those which could correspond to an exponential or linear function, find a formula for the function.

TABLE 1.15

(a)

x	$f(x)$
0	16
1	24
2	36
3	54
4	81

(b)

x	$g(x)$
0	14
1	20
2	24
3	29
4	35

(c)

x	$h(x)$
0	5.3
1	6.5
2	7.7
3	8.9
4	10.1

Solution (a) We see that $f(x)$ cannot be a linear function, since $f(x)$ increases by different amounts ($24 - 16 = 8$ and $36 - 24 = 12$) as x goes up by one. Could it be an exponential function? We look at the ratios of $f(x)$ values:

$$\frac{24}{16} = 1.5 \qquad \frac{36}{24} = 1.5 \qquad \frac{54}{36} = 1.5 \qquad \frac{81}{54} = 1.5.$$

The ratios of $f(x)$ values are all equal to 1.5, and so this table of values could correspond to an exponential function with a base of 1.5. Since $f(0) = 16$, a formula for $f(x)$ is

$$f(x) = 16(1.5)^x.$$

A good way to check this answer is by substituting $x = 0, 1, 2, 3, 4$ into this function to check that it produces the function values given for $f(x)$ in Table 1.15a.

(b) Now we look at the table of values for $g(x)$. As x goes up by one, g goes up by 6 (from 14 to 20), then 4 (from 20 to 24), and so $g(x)$ is not linear. Now we check to see if $g(x)$ could be an exponential function: $\frac{20}{14} = 1.43$ and $\frac{24}{20} = 1.2$. Since these ratios (1.43 and 1.2) are different, $g(x)$ is not an exponential function either.

(c) How about $h(x)$? As x increases by one, the value of $h(x)$ increases by 1.2 each time, so $h(x)$ could be a linear function with a slope of 1.2. Since $h(0) = 5.3$, a formula for $h(x)$ is

$$h(x) = 5.3 + 1.2x$$

Radioactive Decay

Radioactive substances, such as uranium, decay by a certain percentage of their mass in a given unit of time. The most common way to express this rate of decay is to give the time period it takes for half the mass to decay. This period of time is called the *half-life* of the substance. The important thing to remember about radioactive decay is that two half-lives do not make a whole life! Rather, in the time period of two half-lives, the substance will decay to $(1/2) \cdot (1/2) = 1/4$ of its original mass.

One of the most well-known radioactive substances is carbon-14, which is used to date organic objects. When the object, such as a piece of wood or bone, was part of a living organism, it accumulated small amounts of radioactive carbon-14, so that a certain proportion of the carbon in the object was carbon-14. Once the organism dies, it no longer picks up carbon-14 through interaction with its environment (for example, through respiration).

By measuring the proportion of carbon-14 in the object and comparing that to the proportion in living material, we can estimate how much of the original carbon-14 has decayed. The half-life of carbon-14 is about 5730 years. Thus, after roughly 5000 years, we would find the object had about $1/2$ as much carbon-14 as when it was alive. After 10,000 years, we would find about $1/4$ as much, and after 15,000 years, about 1/8 as much. We can write an exponential function for the amount of carbon-14 left after a period of time t. First, suppose we measure time in units of 5730 years. Then if C_0 was the original amount of carbon-14, the amount, C, of carbon left after T "units" of time (namely, T half-lives), would be $C = C_0(1/2)^T$. However, we usually do not measure time in units of 5730 years, so if we let t be time measured in years (units of one year), then $T = t/5730$, and

$$C = C_0 \left(\frac{1}{2}\right)^{(t/5730)}.$$

In general, if a substance has a half-life of h years (or minutes or seconds), then the quantity, Q, of the substance left after t units of time, if there was Q_0 of the substance originally, is

$$Q = Q_0 \left(\frac{1}{2}\right)^{(t/h)}.$$

In summary, we use the following definitions:

> The **doubling time** of an exponentially increasing quantity is the time for it to double.
> The **half-life** of an exponentially decaying quantity is the time for it to be reduced to half.

Drug Buildup

Suppose that we want to model the amount of a certain drug in the body. Imagine that initially there is none, but that the quantity slowly starts to increase via a continuous intravenous injection. As the quantity in the body increases, so does the rate at which the body excretes the drug, so that eventually the quantity levels off at a saturation value, S. The graph of quantity against time will look something like that in Figure 1.44.

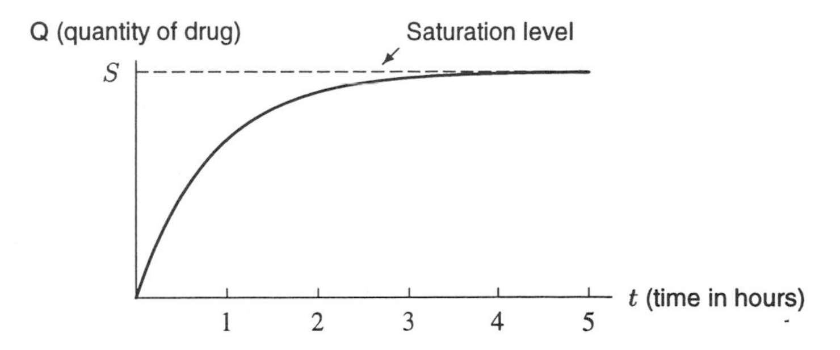

Figure 1.44: Buildup of drug in body

Notice that the quantity, Q, starts at zero and increases toward S. We say that the line representing the saturation level is a *horizontal asymptote*, because the graph gets closer and closer to it as time increases. Since the rate at which the quantity of the drug increases slows as it approaches S, this graph is bending downward; we say that it is *concave down*.

Suppose we want to make a mathematical model of this situation; that is, suppose we want to find a formula giving the quantity, Q, in terms of time, t. Making a mathematical model often involves looking at a graph and deciding what kind of function has that shape. The graph in Figure 1.44 looks like an exponential decay function, upside down. What actually decays is the difference between the saturation level, S, and the quantity, Q, in the blood. Suppose the difference between the saturation level and the quantity in the body is given by the formula

$$\text{Difference} = (\text{Initial difference}) \cdot (0.3)^t$$

with t in hours. Since the difference is $S - Q$, and the initial value of this difference is $S - 0 = S$:

$$S - Q = S(0.3)^t.$$

Solving for Q as a function of t gives

$$Q = S - S(0.3)^t.$$

Factoring out S, we get

$$Q = f(t) = S\left(1 - (0.3)^t\right).$$

Notice that the graph of this function has the shape of an exponential reversed. As t gets larger, $(0.3)^t$ gets smaller, so Q gets closer to S. Using "$\rightarrow$" to mean "tends to," we can say $(0.3)^t \rightarrow 0$ as $t \rightarrow \infty$. This shows that

$$Q = S(1 - (0.3)^t) \rightarrow S(1 - 0) = S \quad \text{as} \quad t \rightarrow \infty$$

confirming that the graph of $Q = S(1 - (0.3)^t)$ has a horizontal asymptote at $Q = S$.

Concavity

We have used the term *concave up* to describe the graphs in Figures 1.41, 1.42, and 1.43, and have used the term *concave down* to describe the graph in Figure 1.44. In general:

> The graph of any function is **concave up** if it bends upward, and it is **concave down** if it bends downward. A line is neither concave up nor concave down.

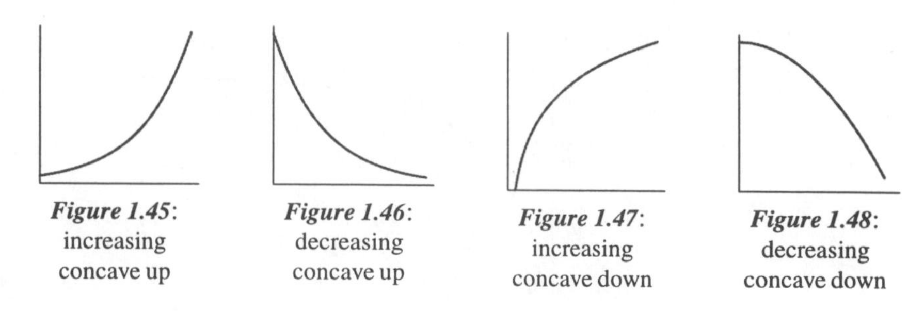

Figure 1.45: increasing concave up

Figure 1.46: decreasing concave up

Figure 1.47: increasing concave down

Figure 1.48: decreasing concave down

Figures 1.45–1.48 shows more examples of curves which are concave up or concave down. Of course, a function may be concave up for part of its graph and concave down for another part, as we will see in the next example.

Example 3 The graph of a function $y = f(x)$ is shown in Figure 1.49.

a) Over what intervals is the function increasing? decreasing?
b) Over what intervals is the function concave up? concave down?

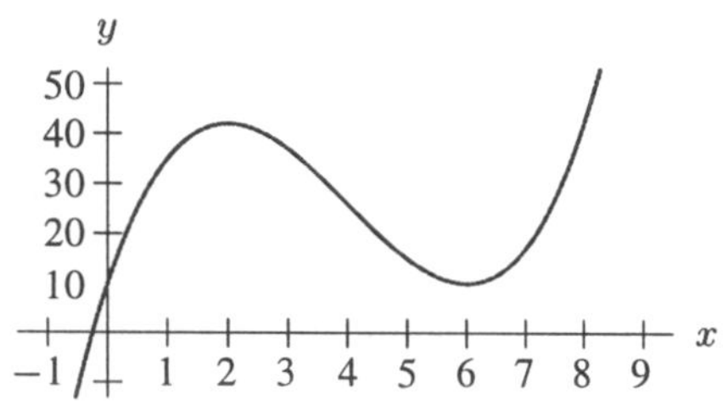

Figure 1.49: Discuss the concavity of this graph

Solution a) By looking at the graph, we can see that the function is increasing for $x < 2$ and for $x > 6$. It is decreasing for $2 < x < 6$.

b) We see that the graph is concave down on the left and concave up on the right. It is difficult to tell exactly where the graph changes concavity, although it appears to be close to $x = 4$. (In Chapter 5, we will learn how to determine the exact value at which a function changes concavity.) Roughly, the graph is concave down for $x < 4$ and concave up for $x > 4$.

The Family of Exponential Functions

The formula $P = P_0 a^t$ gives a family of exponential functions with parameters P_0 (the initial quantity) and a (the base, or growth factor). The base is as important for an exponential function as the slope is for a linear function. We assume $a > 0$ and $a \neq 1$. The base tells you whether the exponential function is increasing ($a > 1$) or decreasing ($0 < a < 1$). Since a is the factor by which P changes when t is increased by 1, large values of a mean fast growth; values of a near 0 mean fast decay. (See Figures 1.50 and 1.51.)

Alternative Formula for the Exponential Function

Exponential growth is often described in terms of percentages. For example, the population of Mexico is growing at a rate of 2.6% per year; in other words, the growth factor is $a = 1.026$. Similarly, each foot of clay removes 20% of the pollution from jet fuel, so the decay factor is $a = 1 - 0.20 = 0.8$. In general, the following formulas apply.

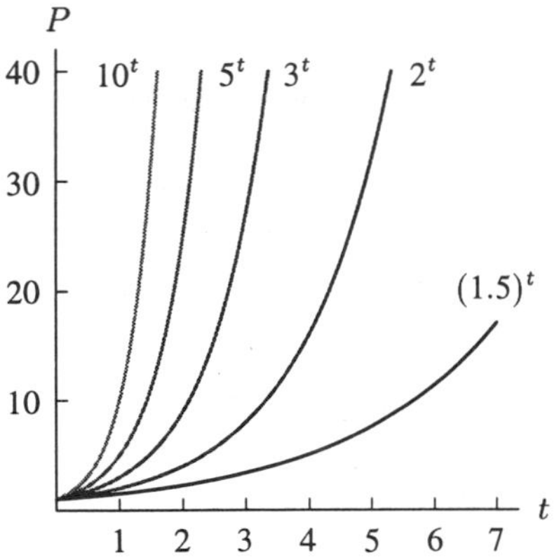

Figure 1.50: Exponential growth: $P = a^t, a > 1$

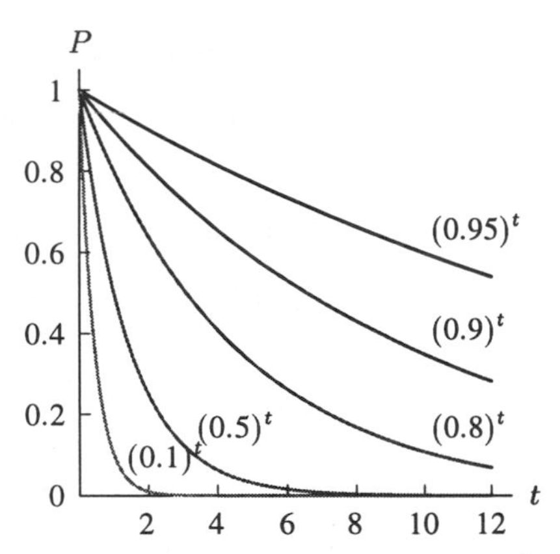

Figure 1.51: Exponential decay: $P = a^t$, $0 < a < 1$

If r is the *growth* rate, then $a = 1 + r$, and

$$P = P_0 a^t = P_0(1 + r)^t.$$

If r is the *decay* rate, then $a = 1 - r$, and

$$P = P_0 a^t = P_0(1 - r)^t.$$

Note, for example, that $r = 0.20$ when the percentage decay rate is 20%.

Example 4 Suppose that $Q = f(t)$ is an exponential function of t. If $f(20.1) = 88.2$ and $f(20.3) = 91.4$:
(a) Find the base. (b) Find the percentage growth rate. (c) Evaluate $f(21.4)$.

Solution (a) Let

$$Q = Q_0 a^t.$$

Then

$$88.2 = Q_0 a^{20.1} \quad \text{and} \quad 91.4 = Q_0 a^{20.3}.$$

Dividing gives

$$\frac{91.4}{88.2} = \frac{Q_0 a^{20.3}}{Q_0 a^{20.1}} = a^{0.2}.$$

Solve for the base, a:

$$a = \left(\frac{91.4}{88.2}\right)^{1/0.2} = 1.195.$$

(b) Since $a = 1.195$, the growth rate is $r = 0.195 = 19.5\%$.

(c) We want to find $f(21.4) = Q_0 a^{21.4} = Q_0(1.195)^{21.4}$. First let's find Q_0 from the equation $91.4 = Q_0(1.195)^{20.3}$. Solving gives $Q_0 = 2.457$. Thus,

$$f(21.4) = 2.457(1.195)^{21.4} = 111.19.$$

Definition and Properties of Exponents

Below we list the definitions and rules that are used to manipulate exponents.

Definition of Zero, Negative, and Fractional Exponents

$$a^0 = 1, \quad a^{-1} = \frac{1}{a}, \quad \text{and, in general, } a^{-x} = \frac{1}{a^x}$$

$$a^{1/2} = \sqrt{a}, \quad a^{1/3} = \sqrt[3]{a}, \quad \text{and, in general, } a^{1/n} = \sqrt[n]{a}.$$

Rules for Computing Using Exponents

1. $a^x \cdot a^t = a^{x+t}$ For example, $2^4 \cdot 2^3 = (2\cdot 2\cdot 2\cdot 2)\cdot(2\cdot 2\cdot 2) = 2^7$.
2. $\dfrac{a^x}{a^t} = a^{x-t}$ For example, $\dfrac{2^4}{2^3} = \dfrac{2\cdot 2\cdot 2\cdot 2}{2\cdot 2\cdot 2} = 2^1$.
3. $(a^x)^t = a^{xt}$ For example, $(2^3)^2 = 2^3 \cdot 2^3 = 2^6$.

Problems for Section 1.4

1. The number of cancer cells grows slowly at first but then grows with increasing rapidity. Draw a possible graph of the number of cancer cells against time.
2. Each year the world's annual consumption of electricity rises. In addition, each year the increase in annual consumption also rises. Sketch a possible graph of the annual world consumption of electricity as a function of time.
3. A drug is injected into a patient's bloodstream over a five-minute interval. During this time, the quantity in the blood increases linearly. After five minutes the injection is discontinued, and the quantity then decays exponentially. Sketch a graph of the quantity versus time.
4. When there are no other steroid hormones (for example, estrogen) in a cell, the rate at which steroid hormones diffuse into the cell is fast. The rate slows down as the amount in the cell builds up. Sketch a possible graph of the quantity of steroid hormone in the cell against time, assuming that initially there are no steroid hormones in the cell.
5. Each of the functions in Table 1.16 is increasing, but each increases in a different way. Which of the graphs in Figure 1.52 below best fits each function?

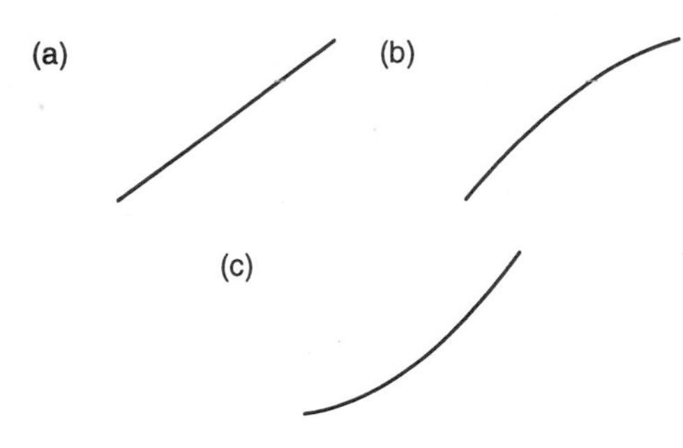

Figure 1.52

TABLE 1.16

t	$g(t)$	$h(t)$	$k(t)$
1	23	10	2.2
2	24	20	2.5
3	26	29	2.8
4	29	37	3.1
5	33	44	3.4
6	38	50	3.7

6. Each of the functions in Table 1.17 decreases, but each decreases in a different way. Which of the graphs in Figure 1.53 below best fits each function?

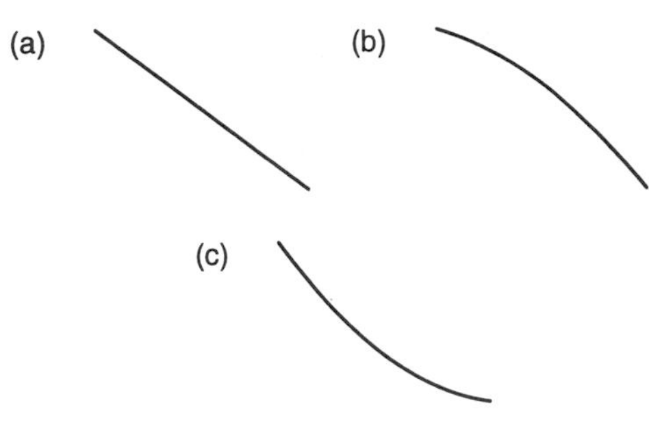

Figure 1.53

TABLE 1.17

x	$f(x)$	$g(x)$	$h(x)$
1	100	22.0	9.3
2	90	21.4	9.1
3	81	20.8	8.8
4	73	20.2	8.4
5	66	19.6	7.9
6	60	19.0	7.3

7. Match up the function values in Table 1.18 with the formulas

$$y = a(1.1)^s, \quad y = b(1.05)^s, \quad y = c(1.03)^s,$$

assuming a, b, and c are constants. Note that the function values have been rounded to two decimal places.

TABLE 1.18

s	$h(s)$	s	$f(s)$	s	$g(s)$
2	1.06	1	2.20	3	3.47
3	1.09	2	2.42	4	3.65
4	1.13	3	2.66	5	3.83
5	1.16	4	2.93	6	4.02
6	1.19	5	3.22	7	4.22

For Problems 8–9, find a possible formula for the functions represented by the data.

8.

x	0	1	2	3
$f(x)$	4.30	6.02	8.43	11.80

9.

t	0	1	2	3
$g(t)$	5.50	4.40	3.52	2.82

10. Simplify each of the following: (a) $8^{2/3}$ (b) $9^{-3/2}$

11. Determine whether each of the following tables of values could correspond to a linear function, an exponential function, or neither. For each table of values that could correspond to a linear or an exponential function, find a formula for the function.

(a)

x	$f(x)$
0	10.5
1	12.7
2	18.9
3	36.7

(b)

t	$s(t)$
−1	50.2
0	30.12
1	18.072
2	10.8432

(c)

u	$g(u)$
0	27
2	24
4	21
6	18

12. Total per capita health expenditures in the United States for the years 1970 to 1982 are given in Table 1.19.[10]
    (a) Check to see that the increase in health expenditures is approximately exponential. Show your work.
    (b) Estimate the doubling time of US per capita health expenditures.

TABLE 1.19

Year	Yearly Health Expenditures ($ per capita)
1970	349
1972	428
1974	521
1976	665
1978	822
1980	1054
1982	1348

Find possible equations for the graphs in Problems 13–16.

13.

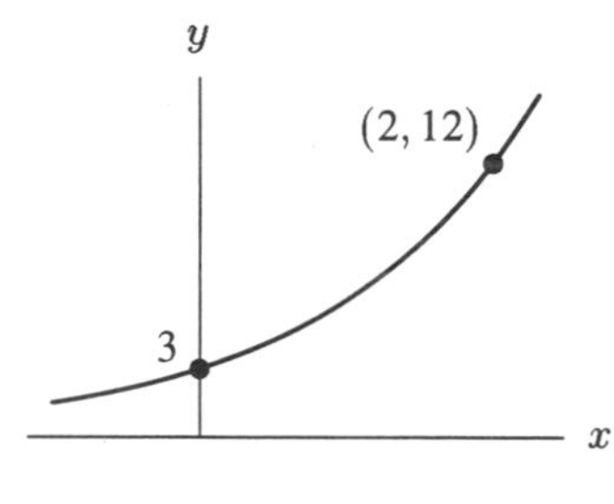

14.

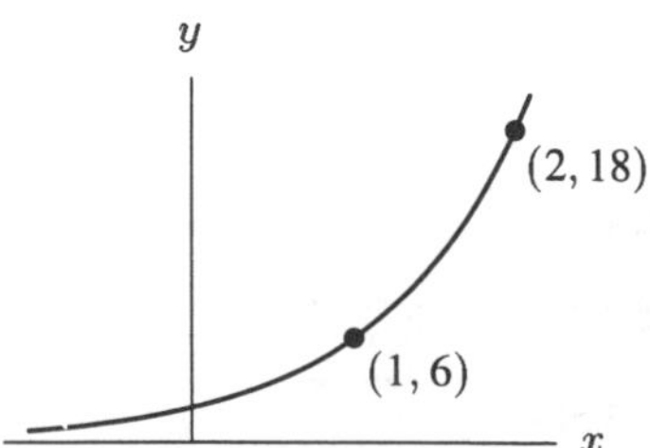

[10] *Statistical Abstracts of the US,* 1988, p.86, Table 129

15.

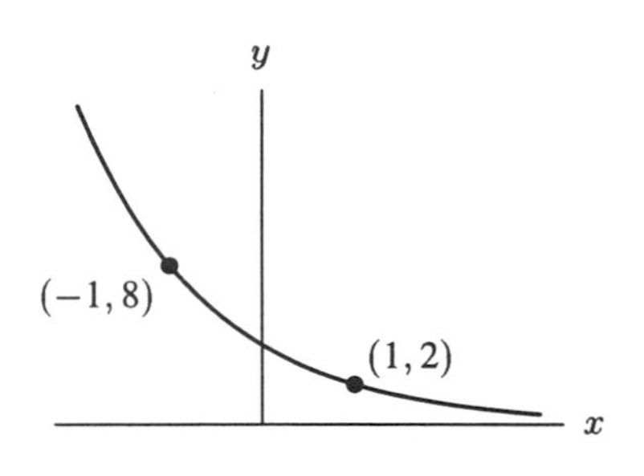

16.

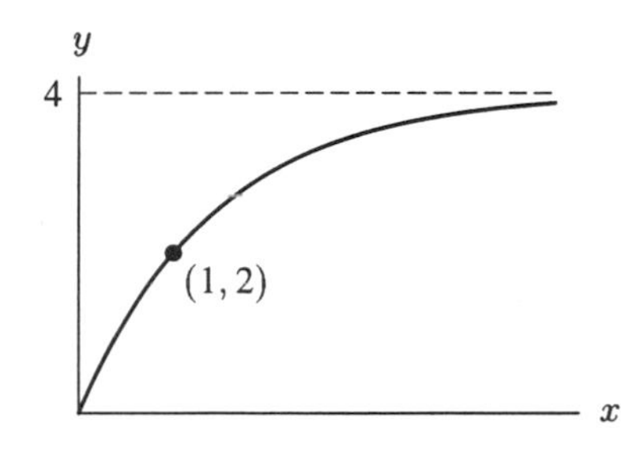

17. (a) The half-life of radium-226 is 1620 years. Write a formula for the quantity, Q, of radium left after t years, if the initial quantity is Q_0.
    (b) What percentage of an original amount of radium is left after 500 years?

18. In the early 1960s, radioactive strontium-90 was released during atmospheric testing of nuclear weapons and got into the bones of people alive at the time. If the half-life of strontium-90 is 29 years, what fraction of the strontium-90 absorbed in 1960 remained in people's bones in 1990?

19. When the Olympic Games were held outside Mexico City in 1968, there was much discussion about the effect the high altitude (7340 feet) would have on the athletes. Assuming air pressure decays exponentially at 0.4% every 100 feet, by what percentage is air pressure reduced by moving from sea level to Mexico City?

20. A population is known to be growing exponentially. Estimate the doubling time of the population shown by the graph in Figure 1.54, and verify graphically that the doubling time is independent of where you start on the graph.

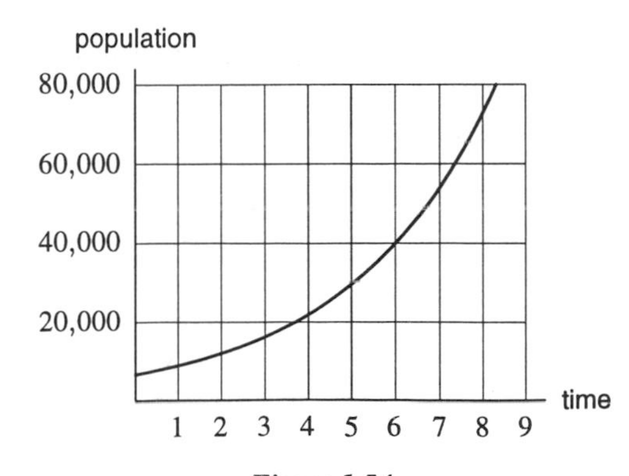

Figure 1.54

21. A certain region has a population of 10,000,000 and an annual growth rate of 2%. Estimate the doubling time by trial and error.

22. A certain radioactive substance decays exponentially in such a way that after 10 years, 70% of the initial amount remains. Find an expression for the quantity remaining after any number of years t. How much will be present after 50 years? What is the half-life? How long will it be before only 20% of the original amount is left? Before only 10% is left? (Use trial and error where necessary.)

23. Assume that the median price, P, of a home rose from $50,000 in 1970 to $100,000 in 1990. Let t be the number of years since 1970.
    (a) Assume the increase in housing prices has been linear. Find an equation for the line representing price, P, in terms of t. Use this equation to complete column (a) of Table 1.20. Work with the price in units of $1000.
    (b) If instead the housing prices have been rising exponentially, determine an equation of the form $P = P_0a^t$ which would represent the change in housing prices from 1970–1990, and complete column (b) of Table 1.20.
    (c) On the same set of axes, sketch the functions represented in column (a) and column (b) of Table 1.20.

TABLE 1.20

t	(a) Linear growth price in $1000 units	(b) Exponential growth price in $1000 units
0	50	50
10		
20	100	100
30		
40		

Countries with very high inflation rates often publish monthly rather than yearly inflation figures, because monthly figures are less alarming. Problems 24–25 involve such high rates, which are called *hyperinflation*.

24. In 1989, US inflation was 4.6% a year. In 1989 Argentina had an inflation rate of about 33% a month.
    (a) What is the yearly equivalent of Argentina's 33% monthly rate?
    (b) What is the monthly equivalent of the US 4.6% yearly rate?

25. Between December 1988 and December 1989, Brazil's inflation rate was 1290% a year. (This means that between 1988 and 1989, prices increased by a factor of $1 + 12.90 = 13.90$.)
    (a) What would an article which cost 1000 cruzados (the Brazilian currency unit) in 1988 cost in 1989?
    (b) What was Brazil's monthly inflation rate during this period?

1.5 POWER FUNCTIONS

Power functions are an important family of functions. A *power function* is one in which the dependent variable is proportional to a power of the independent variable. For example, the area, A, of a square of side s is given by

$$A = f(s) = s^2.$$

The volume, V, of a sphere of radius r is

$$V = f(r) = \frac{4}{3}\pi r^3.$$

Both of these are *power functions.* So is the function which describes how the gravitational attraction of the earth varies with distance. If g is the force of gravitational attraction on a unit mass at a distance r from the earth, Newton's Inverse Square Law of Gravitation says that

$$g = \frac{k}{r^2} \quad \text{or} \quad g = kr^{-2}$$

where k is a positive constant.

> In general, a **power function** has the form
>
> $$y = f(x) = kx^p$$
>
> where k and p are any constants.

In this section we will compare various power functions with one another and with the exponential functions. Since $y = f(x) = mx$ (with m a constant) is also a power function (because $x = x^1$, the first power), linear functions are included in the comparison too.

Positive Integral Powers: $y = x, y = x^2, y = x^3, \ldots$

First, we'll look at functions of the form $f(x) = x^n$, with n a positive integer. Figures 1.55 and 1.56 show that the graphs of these functions fall into two groups: the odd powers and the even powers. All the odd powers (x, x^3, x^5, and so on) are increasing everywhere and their graphs are symmetric about the origin. All odd powers above $n = 1$ have a bend, or "seat," at the origin. The even powers, on the other hand, are first decreasing and then increasing, making them $\cup$-shaped with symmetry about the y-axis. The even powers are concave up everywhere, whereas the odd ones (greater than 1) are concave down for negative x and concave up for positive x. All odd and even powers, however, go through the points $(0, 0)$ and $(1, 1)$.

Figure 1.57 shows that the higher the power of x, the faster the function climbs. For large values of x (in fact, for all $x > 1$), $y = x^5$ is above $y = x^4$, which is above $y = x^3$, and so on. Not only are the higher powers larger, but they are *much* larger. This is because if $x = 100$, for example, 100^5 is one hundred times as big as 100^4 which is one hundred times as big as 100^3. As x gets larger (written as $x \to \infty$), any positive power of x completely swamps all lower powers of x. We say that, as $x \to \infty$, higher powers of x *dominate* lower powers.

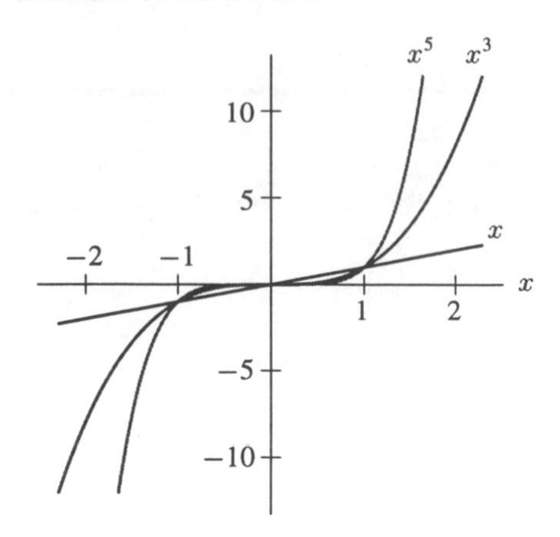

Figure 1.55: Odd powers of x

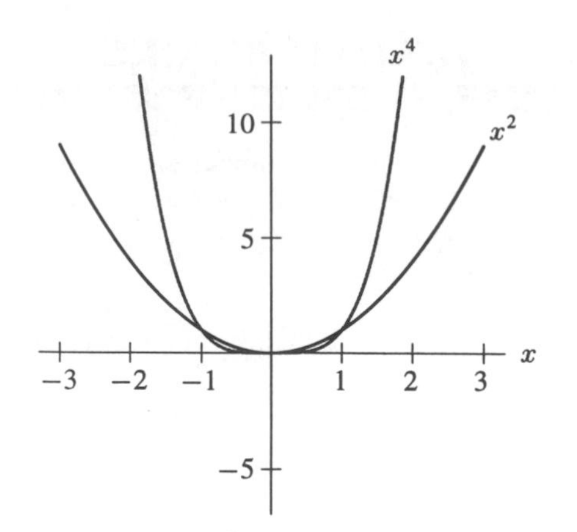

Figure 1.56: Even powers of x

The close-up view near the origin in Figure 1.58 shows an entirely different story. For x between 0 and 1, the order is reversed: x^3 is bigger than x^4, which is bigger than x^5. (Try $x = 0.1$ to confirm this.) The fact that higher powers of x climb faster is true for large values of x but not for small. For big values of x, the highest powers are largest; for values of x near zero, smaller powers dominate.

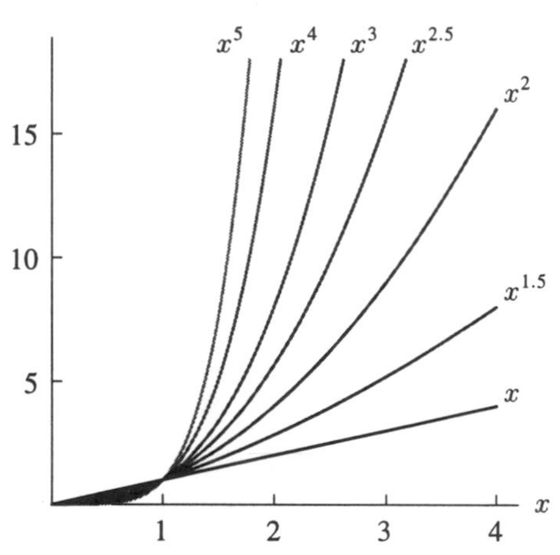

Figure 1.57: Powers of x: Which is largest for large values of x?

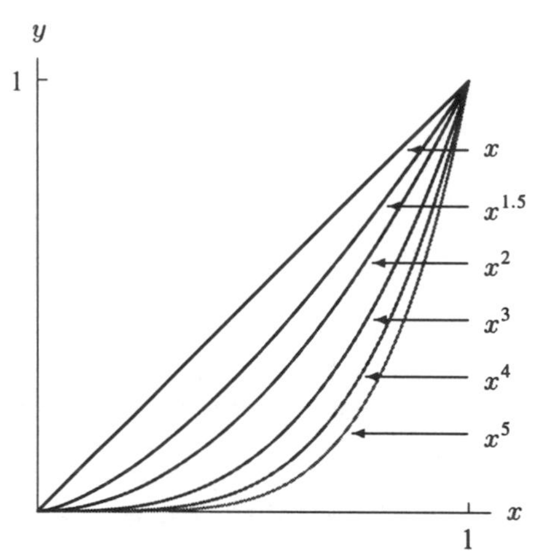

Figure 1.58: Between 0 and 1: Small powers of x dominate

Zero and Negative Integral Powers: $y = x^0, y = x^{-1}, y = x^{-2}, \ldots$

The function $y = x^0 = 1$ has a graph that is a horizontal line. Rewriting

$$y = x^{-1} = \frac{1}{x} \quad \text{and} \quad y = x^{-2} = \frac{1}{x^2}$$

makes it easier to see that as x increases, the denominators increase and the functions decrease. The graphs of $y = x^{-1}$ and $y = x^{-2}$ have both the x and y axes as asymptotes. (See Figure 1.59.) For $x > 1$, the graph of $y = x^{-2}$ is below that of $y = x^{-1}$, and both must stay below $y = x^0 = 1$.

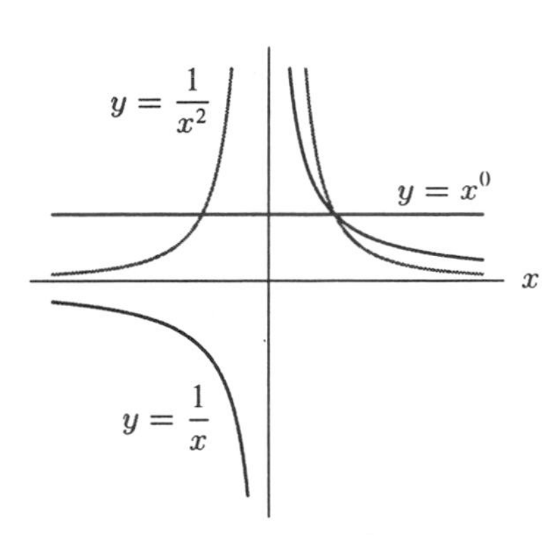

Figure 1.59: Comparison of zero and negative powers of x

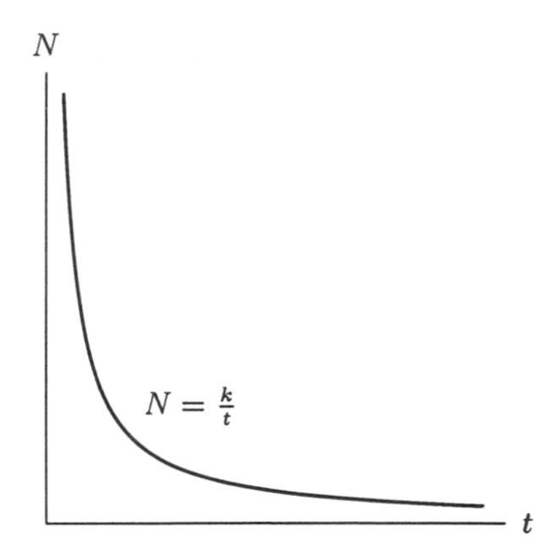

Figure 1.60: Number of nonsense syllables recalled, N, against time, t

Example 1 Psychologists test memory by presenting a person with a list of nonsense syllables (such as kab, nuv, mip) to memorize[11]. After the list has been removed, the psychologist will wait a certain amount of time and then ask the subject how many of the syllables he or she remembers. These experiments have shown that the number of nonsense syllables, N, that the subject can remember is a decreasing function of the time, t, since the list was removed. We can model this by saying that N is inversely proportional to t. Write a formula for N as a function of t and plot a graph of N against t.

Solution As more time elapses between the memorization and the recall, we expect subjects to remember fewer and fewer of the nonsense syllables, so it makes sense that N is a decreasing function of t. When we say N is inversely proportional to t, we are saying that N is proportional to the reciprocal of t. In other words,

$$N = k\left(\frac{1}{t}\right)$$

with k positive. Notice that, as we expect, when t is small, N is large, and when t is large, N is small. For any (positive) value of k, the graph has the shape shown in Figure 1.60. Both axes are asymptotes, showing that as the time tends to infinity, the number remembered tends to zero. The power function $N = k(1/t) = kt^{-1}$ differs from exponential decay in that it is undefined for $t = 0$, so this graph does not cross the vertical axis. In addition, it approaches the horizontal axis more slowly than the exponential decay function.

Positive Fractional Powers: $y = x^{1/2}, y = x^{1/3}, y = x^{3/2}, \ldots$

The function giving the side of a square, s, in terms of its area, A, involves a root, or fractional power:

$$s = \sqrt{A} = A^{1/2}.$$

Similarly, the equation relating the average number of species found on an island and the size of the island involves a fractional power. If N is the number of species and A is the area of the island,

[11] *Memory from a Broader Perspective*, A. Searleman and D. Herrmann, McGraw-Hill, 1994

observations have shown[12] that approximately

$$N = k\sqrt[3]{A} = kA^{1/3}$$

where k is a constant depending on the region of the world in which the island is found.

We will now look at functions of the form $y = x^{m/n} = \sqrt[n]{x^m}$. Since some fractional powers such as $x^{1/2}$ involve roots and are defined only for positive x and 0, we frequently restrict the domain of positive fractional powers of x to $x \geq 0$. Many calculators will not allow you to raise a negative number to a fractional power.

Figure 1.61 shows that for large x (in fact, all $x > 1$), the graph of $y = x^{1/2}$ is below the graph of $y = x$, and $y = x^{1/3}$ is below $y = x^{1/2}$. This is reasonable since, for example, $10^{1/2} = \sqrt{10} \approx 3.16$ and $10^{1/3} = \sqrt[3]{10} \approx 2.15$, so $10^{1/3} < 10^{1/2} < 10$. Between $x = 0$ and $x = 1$, the situation is reversed, and $y = x^{1/3}$ is on top. (Why?) Not surprisingly, $y = x^{3/2}$ is between $y = x$ and $y = x^2$ for all x.

The other important feature to notice about the graphs of $y = x^{1/2}$ and $y = x^{1/3}$ is that they bend in a direction opposite to that of the graphs of $y = x^2$ and x^3. For example, the graph of $y = x^2$ is climbing faster and faster as x increases; it is concave up. On the other hand, the graphs of $y = x^{1/2}$ and $y = x^{1/3}$ are climbing slower and slower; they are concave down. Despite this, all these functions do become infinitely large as x increases.

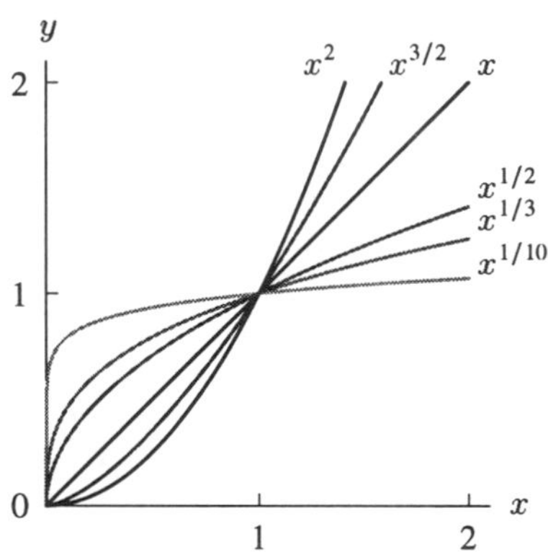

Figure 1.61: Comparison of some fractional powers of x

What Effect Do Coefficients Have?

We know that $x^2 < x^3$ for all $x > 1$. But which is larger, $50x^2$ or x^3? Eventually, $50x^2 < x^3$ too. In fact, $50x^2 < x^3$ for all $x > 50$. (See Figure 1.62.) The graphs of $y = x^2$ and $y = x^3$ cross at $x = 1$, whereas graphs of $y = 50x^2$ and $y = x^3$ cross at $x = 50$. Thus, the effect of the factor of 50 is to change the point at which the graphs cross. However, x^3 ends up on top in both cases: provided the coefficients are positive, as $x \to \infty$, the higher power is always larger eventually.

[12] *Scientific American*, September 1989, p. 112.

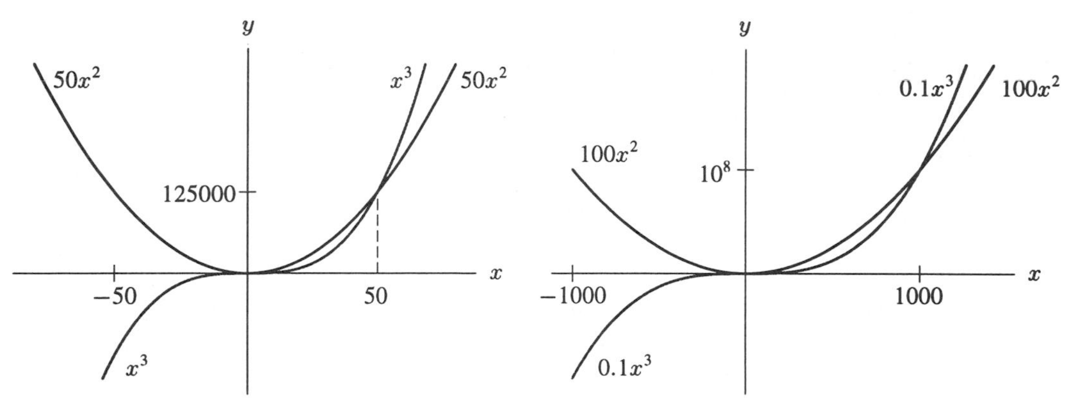

Figure 1.62: Graph of $y = x^3$ lies above graph of $y = 50x^2$ for large positive x

Figure 1.63: For large positive x, $y = 0.1x^3$ dominates $y = 100x^2$

Example 2 Which of $y = 100x^2$ and $y = 0.1x^3$ is larger as $x \to \infty$?

Solution Since $x \to \infty$, we are looking at large positive values of x, where the higher power will eventually be larger, or dominate. Thus, $y = 0.1x^3$ will be larger. (See Figure 1.63.)

Example 3 Sketch a global view of $f(x) = -x^3$, $g(x) = 40x^4$, and $h(x) = -0.1x^5$. Which function has the largest positive values as $x \to \infty$? Which function has the largest positive values as $x \to -\infty$ (i.e., as x gets more and more negative)?

Solution As $x \to \infty$, $g(x) = 40x^4$ is the only function which is positive. As $x \to -\infty$, the graph of $h(x) = -0.1x^5$ is eventually (for $x < -400$) above the graph of the other functions, so $h(x)$ has the largest positive values as $x \to -\infty$ (See Figure 1.64.) Notice that, for large x, the values of $f(x) = -x^3$ are so much smaller in magnitude than the values of the other functions that the graph of $f(x)$ cannot be seen in the far-away view.

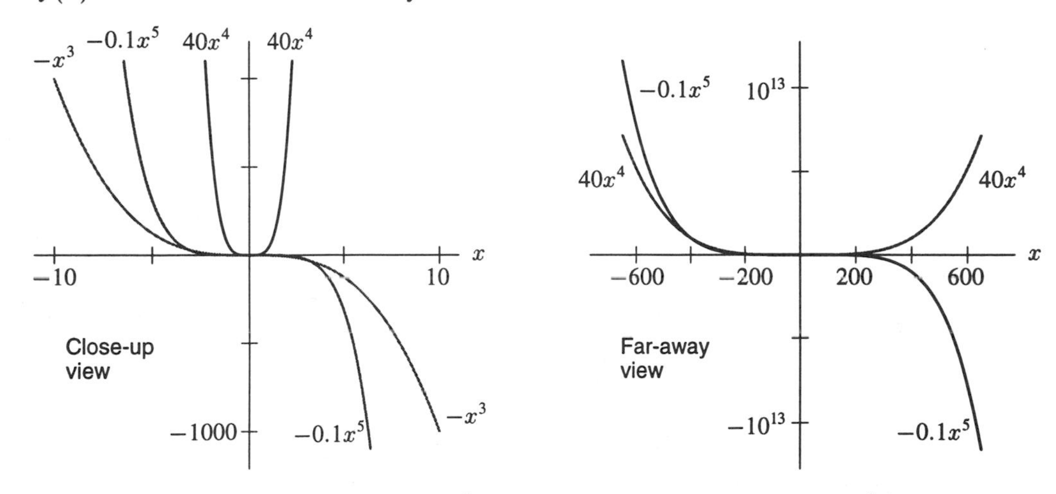

Figure 1.64: As $x \to \infty$, $g(x) = 40x^4$ dominates; as $x \to -\infty$, $h(x) = -0.1x^5$ dominates

---

Exponentials and Power Functions: Which Dominate?

In everyday language, *exponential* is often used to imply very fast growth. But do exponential functions always grow faster than power functions?

Let's consider $y = 2^x$ and $y = x^3$. The close-up, or local, view in Figure 1.65(a) shows that between $x = 2$ and $x = 5$, the graph of $y = 2^x$ lies below the graph of $y = x^3$. But the more global, or far–away, view in Figure 1.65(b) shows that the exponential function $y = 2^x$ eventually overtakes $y = x^3$. And Figure 1.65(c), which gives a very far–away, or global view, shows that, for large x, x^3 is insignificant compared to 2^x. Indeed, 2^x is growing so much faster than x^3 that its graph appears almost vertical—in comparison to the more leisurely climb of x^3.

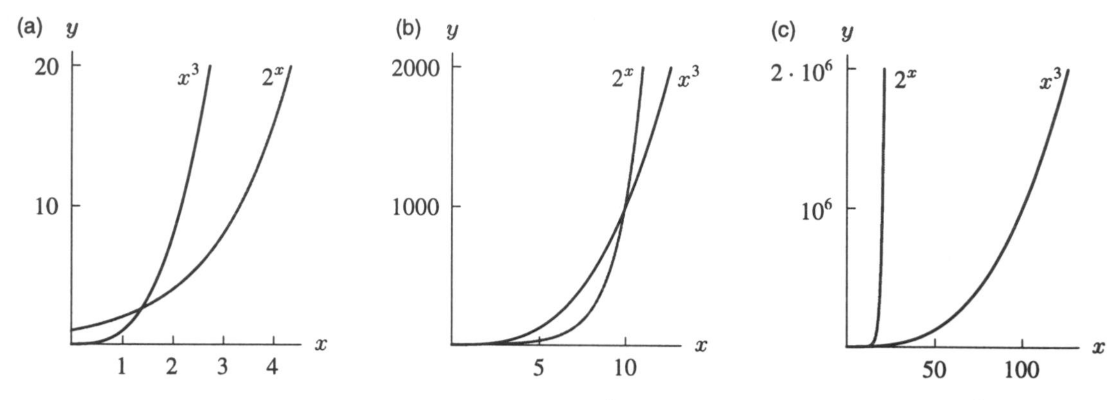

Figure 1.65: Comparison of $y = 2^x$ and $y = x^3$: $y = 2^x$ eventually dominates $y = x^3$

In fact, *every* exponential growth function eventually dominates *every* power function. Although an exponential function may be below a power function for some values of x, if you look at large enough x values, a^x (with $a > 1$) will eventually dominate x^n, no matter what n is. Two more examples are presented in Figure 1.66 and Table 1.21.

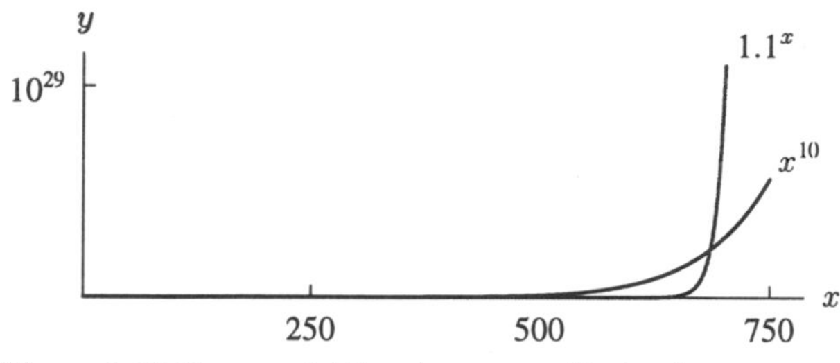

Figure 1.66: Exponential function eventually dominates power function

TABLE 1.21 *Comparison of x^{100} and 1.01^x*

x	x^{100}	1.01^x
10^4	10^{400}	$1.6 \cdot 10^{43}$
10^5	10^{500}	$1.4 \cdot 10^{432}$
10^6	10^{600}	$2.4 \cdot 10^{4321}$

Can you guess what happens in the case of negative powers and negative exponents? For example, consider $y = 2^{-x}$ and $y = x^{-2}$. Since $y = 2^{-x} = 1/2^x$ and $y = x^{-2} = 1/x^2$, knowing that 2^x is eventually larger than x^2 tells you that 2^{-x} is eventually smaller than x^{-2}. Hence $y = 2^{-x}$ is eventually below $y = x^{-2}$. (See Figure 1.67.) This behavior is also typical: Every exponential decay function will eventually approach 0 faster than every power function with a negative exponent.

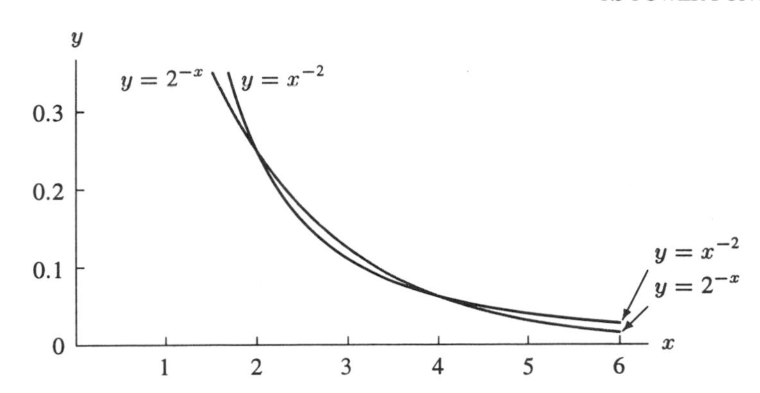

Figure 1.67: Comparison of $y = 2^{-x}$ and $y = x^{-2}$: Exponential dies away faster

Problems for Section 1.5

1. Sketch graphs of $y = x^{1/2}$ and $y = x^{2/3}$ on the same axes. Which function has larger values as $x \to \infty$?
2. What happens to the value of $y = x^4$ as $x \to \infty$? As $x \to -\infty$?
3. What happens to the value of $y = -x^7$ as $x \to \infty$? As $x \to -\infty$?
4. Sketch a graph of $y = x^{-4}$.
5. On a graphing calculator or computer, plot graphs of the following functions, first for $-5 \le x \le 5, -100 \le y \le 100$, and then for $-1.2 \le x \le 1.2, -2 \le y \le 2$.
   (a) $y = x, y = x^3, y = x^6, y = x^9$
   (b) $y = x, y = x^4, y = x^7, y = x^{10}$
   Observe the general shape of these functions: Do the odd powers have the same general shape? What about the even powers? Which function is largest in magnitude for big x? For x near 0? Is this what you expected?
6. Do some calculations using specific values of x to verify that $y = x^{1/3}$ is above $y = x^{1/2}$ and that $y = x^{1/2}$ is above $y = x$ for $0 < x < 1$.
7. (a) Use a graphing calculator (or a computer) to plot the graphs of x^3, x^4, and x^5 on the interval $-0.1 \le x \le 0.1$. Determine an appropriate range for y so that all powers will be distinguishable in the viewing rectangle.
   (b) Plot the same graphs for $-100 \le x \le 100$, and determine an appropriate range for y.
8. By hand, sketch global pictures of $f(x) = x^3$ and $g(x) = 20x^2$ on the same axes. Which function has larger values as $x \to \infty$?
9. By hand, sketch pictures of $f(x) = x^5$, $g(x) = -x^3$, and $h(x) = 5x^2$ on the same axes. Which has larger positive values as $x \to \infty$? As $x \to -\infty$?
10. By trial and error, use a calculator to find to two decimal places the point near $x = 10$ at which $y = 2^x$ and $y = x^3$ cross.
11. Use a graphing calculator to find the point(s) of intersection of the graphs of $y = (1.06)^x$ and $y = 1 + x$.
12. For what values of x is $4^x > x^4$?
13. For what values of x is $3^x > x^3$? (Note: You will need to think about how to deal with the fact that the graphs of 3^x and x^3 are relatively close together for values of x near 3.)

14. According to the April 1991 issue of *Car and Driver*, an Alfa Romeo going at 70 mph requires 177 feet to stop. Assuming that the stopping distance is proportional to the square of velocity, find the stopping distances required by an Alfa Romeo going at 35 mph and at 140 mph (its top speed).

15. Suppose that the demand equation for a product shows that the quantity demanded is inversely proportional to the price of the product.
    (a) Sketch a graph of this demand function.
    (b) There should be no vertical intercept on your graph. What does this tell you about the relationship between price and quantity?
    (c) There should also be no horizontal intercept on your graph. What does this tell you about the relationship between price and quantity?

16. The values of three functions are given in Table 1.22. One function is of the form $y = ab^t$, one is of the form $y = at^2$, and one is of the form $y = bt^3$. Which function is which?

TABLE 1.22

t	$f(t)$	t	$g(t)$	t	$h(t)$
2.0	4.40	1.0	3.00	0.0	2.04
2.2	5.32	1.2	5.18	1.0	3.06
2.4	6.34	1.4	8.23	2.0	4.59
2.6	7.44	1.6	12.29	3.0	6.89
2.8	8.62	1.8	17.50	4.0	10.33
3.0	9.90	2.0	24.00	5.0	15.49

17. Values of three functions are contained in Table 1.23. (The numbers have been rounded to two decimal places.) Two are power functions and one is an exponential. One of the power functions is a quadratic and one a cubic. Which one is exponential? Which one is quadratic? Which one is cubic?

TABLE 1.23

x	$f(x)$	x	$g(x)$	x	$k(x)$
8.4	5.93	5.0	3.12	0.6	3.24
9.0	7.29	5.5	3.74	1.0	9.01
9.6	8.85	6.0	4.49	1.4	17.66
10.2	10.61	6.5	5.39	1.8	29.19
10.8	12.60	7.0	6.47	2.2	43.61
11.4	14.82	7.5	7.76	2.6	60.91

18. Owing to improved seed types and new agricultural techniques, the grain production of a region has been increasing. Over a 20-year period, annual production (in millions of tons) was as follows:

1970	1975	1980	1985	1990
5.35	5.90	6.49	7.05	7.64

At the same time the population (in millions) was:

1970	1975	1980	1985	1990
53.2	56.9	60.9	65.2	69.7

(a) Find a linear or exponential function which approximately fits each set of data. (Pick whichever type of function fits better.)
(b) If this region was self-supporting in this grain in 1970, was it self-supporting between 1970 and 1990? (Being self-supporting means that each person has enough of the grain. How does the amount of grain each person has in later years compare?)
(c) What are your predictions for the future if the trends continue?

19. Use a graphing calculator (or a computer) to graph $y = x^4$ and $y = 3^x$. Determine the appropriate domains and ranges that will give each of the graphs in Figure 1.68.

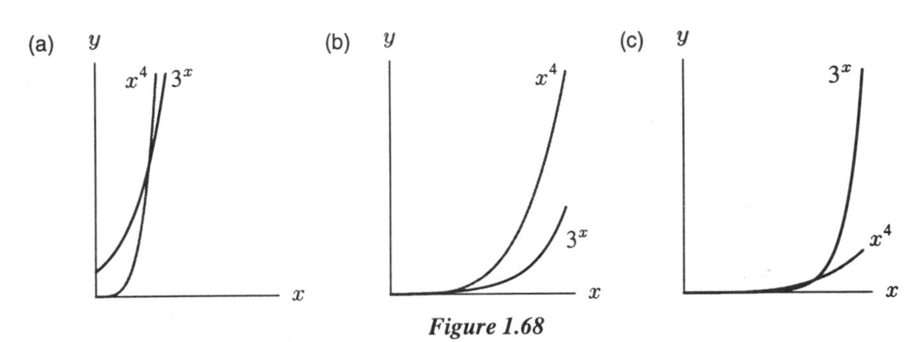

Figure 1.68

1.6 COMPOUND INTEREST AND THE NUMBER e

If you have some money, you may decide to invest it to earn interest. The interest can be paid in many different ways — for example, once a year or many times a year. If the interest is paid more frequently than once per year and the interest is not withdrawn, there is a benefit to the investor since the interest earns interest. This effect is called *compounding*. You may have noticed banks offering accounts that differ both in interest rates and in compounding methods. Some offer interest compounded annually, some quarterly, and others daily. Some even offer continuous compounding.

What is the difference between a bank account advertising 8% compounded annually (once per year) and one offering 8% compounded quarterly (four times per year)? In both cases 8% is an annual rate of interest. The expression 8% *compounded annually* means that at the end of each year, 8% of the current balance is added. This is equivalent to multiplying the current balance by 1.08. Thus, if \$100 is deposited, the balance, B, in dollars, will be

$$B = 100(1.08) \quad \text{after one year,}$$
$$B = 100(1.08)^2 \quad \text{after two years,}$$
$$B = 100(1.08)^t \quad \text{after } t \text{ years.}$$

The expression 8% *compounded quarterly* means that interest is added four times per year (every three months) and that $\frac{8}{4} = 2\%$ of the current balance is added each time. Thus, if \$100 is deposited, at the end of one year, four compoundings will have taken place and the account will contain $\$100(1.02)^4$. Thus, the balance, B, in dollars, will be

$$
\begin{aligned}
B &= 100(1.02)^4 && \text{after one year,}\\
B &= 100(1.02)^8 && \text{after two years,}\\
B &= 100(1.02)^{4t} && \text{after } t \text{ years.}
\end{aligned}
$$

Note that 8% is *not* the rate used for each three month period; the annual rate is divided into four 2% payments. Calculating the total balance after one year under each method shows that

$$
\begin{aligned}
\text{Annual compounding:} \quad & B = 100(1.08) = 108.00\\
\text{Quarterly compounding:} \quad & B = 100(1.02)^4 = 108.24
\end{aligned}
$$

Thus, more money is earned from quarterly compounding, because the interest earns interest as the year goes by. In general, the more often interest is compounded, the more money will be earned (although the increase may not be very large).

We can measure the effect of compounding by introducing the notion of *effective annual yield.* Since $100 invested at 8% compounded quarterly grows to $108.24 by the end of one year, we say that the *effective annual yield* in this case is 8.24%. We now have two interest rates which describe the same investment: the 8% compounded quarterly and the 8.24% effective annual yield. Banks call the 8% the *annual percentage rate*, or *APR*. We may also call the 8% the *nominal rate* (nominal means "in name only"). However, it is the effective yield which tells you exactly how much interest the investment really pays. Thus, to compare two bank accounts, simply compare the effective annual yields. The next time that you walk by a bank, look at the advertisements, which should (by law) include both the APR, or nominal rate, and the effective annual yield. We will often abbreviate *annual percentage rate* to *annual rate.*

Using the Effective Annual Yield

Example 1 Which is better: Bank X paying a 7% annual rate compounded monthly or Bank Y offering a 6.9% annual rate compounded daily?

Solution We will find the effective annual yield for each bank.

Bank X: There are 12 interest payments in a year, each payment being $0.07/12 = 0.005833$ times the current balance. If the initial deposit were $100, then the balance B will be

$$
\begin{aligned}
B &= 100(1.005833) && \text{after one month,}\\
B &= 100(1.005833)^2 && \text{after two months,}\\
B &= 100(1.005833)^t && \text{after } t \text{ months.}
\end{aligned}
$$

To find the effective annual yield, we need to look at one year, or 12 months, which gives $B = 100(1.005833)^{12} = 100(1.072286)$, so the effective annual yield $\approx 7.23\%$.

Bank Y: There are 365 interest payments in a year (assuming it is not a leap year), each being $0.069/365 = 0.000189$ times the current balance. Then the balance B in the account is

$$
\begin{aligned}
B &= 100(1.000189) && \text{after one day,}\\
B &= 100(1.000189)^2 && \text{after two days,}\\
B &= 100(1.000189)^t && \text{after } t \text{ days.}
\end{aligned}
$$

so at the end of one year we have multiplied the initial deposit by

$$(1.000189)^{365} = 1.0714$$

so the effective annual yield for Bank Y $\approx 7.14\%$.

Comparing effective annual yields for the banks, we see that Bank X is offering a better investment, by a small margin.

Example 2 If $1000 is invested in each bank in Example 1, write an expression for the balance in each bank after t years.

Solution For Bank X, the effective annual yield $\approx 7.23\%$, so after t years the balance, in dollars, will be

$$B = 1000(1.0723)^t.$$

For Bank Y, the effective annual yield $\approx 7.14\%$, so after t years the balance, in dollars, will be

$$B = 1000(1.0714)^t.$$

(Again, we are ignoring leap years.)

If interest at an annual rate of r is compounded n times a year, then r/n times the current balance is added n times a year. Thus, with an initial deposit of $\$P$, the balance t years later is

$$B = P\left(1 + \frac{r}{n}\right)^{nt}$$

Note that r is the nominal rate and that, for example, $r = 0.05$ when the annual rate is 5%.

Increasing the Frequency of Compounding: Continuous Compounding

Example 3 Find the effective annual yield for a 7% annual rate compounded
(a) 1000 times a year. (b) 10,000 times a year.

Solution (a)

$$\left(1 + \frac{0.07}{1000}\right)^{1000} \approx 1.0725056$$

giving an effective annual yield of about 7.25056%.

(b)

$$\left(1 + \frac{0.07}{10{,}000}\right)^{10{,}000} \approx 1.0725079$$

giving an effective annual yield of about 7.25079%.

You can see that there's not a great deal of difference between compounding 1000 times each year (about three times per day) and 10,000 times each year (about 30 times per day). What happens if we compound more often still? Every minute? Every second? You may be surprised that the effective annual yield does not increase indefinitely, but tends to a finite value. The benefit of increasing the frequency of compounding becomes negligible beyond a certain point.

For example, if you were to compute the effective annual yield on a 7% investment compounded n times per year for values of n larger than 100,000, you would find that

$$\left(1+\frac{0.07}{n}\right)^n \approx 1.0725082.$$

So the effective annual yield is about 7.25082%. Even if you take $n = 1{,}000{,}000$ or $n = 10^{10}$, the effective annual yield will not change appreciably. The value 7.25082% is an upper bound which is approached as the frequency of compounding increases.

When the effective annual yield is at this upper bound, we say that the interest is being *compounded continuously*. (The word *continuously* is used because the upper bound is approached by compounding more and more frequently.) Thus, when a 7% nominal annual rate is compounded so frequently that the effective annual yield is 7.25082%, we say that the 7% is compounded *continuously*. This represents the most one can get from a 7% nominal rate.

The Number e

We saw exponential functions with many different bases in Section 1.4. However, the most frequently used base is the famous number $e = 2.71828\ldots$. This base is used so often that you will find an $\boxed{e^x}$ button on most calculators. One of the reasons that the number e is so important is that many calculus formulas come out neater when e is used as the base rather than any other base, as we will see in Chapter 4. In the next paragraph, we see another reason why the number e is so important.

It turns out that e is intimately connected to continuous compounding. To see this, use your calculator to check that $e^{0.07} \approx 1.0725082$, which is the same number we obtained when we compounded 7% a large number of times. So you have discovered that for very large n

$$\left(1+\frac{0.07}{n}\right)^n \approx e^{0.07}.$$

As n gets larger, the approximation gets better and better, which we write as

$$\left(1+\frac{0.07}{n}\right)^n \longrightarrow e^{0.07}$$

meaning that as n increases, the value of $\left(1+0.07/n\right)^n$ gets closer and closer to $e^{0.07}$.

Thus, if $\$P$ is deposited at an annual rate of 7% compounded continuously, the balance, $\$B$, will be given by

$$\begin{aligned} B &= P(1.0725082) = Pe^{0.07} && \text{after one year,}\\ B &= P(1.0725082)^2 = P\left(e^{0.07}\right)^2 = Pe^{(0.07)2} && \text{after two years,}\\ B &= P(1.0725082)^t = P\left(e^{0.07}\right)^t = Pe^{0.07t} && \text{after } t \text{ years.} \end{aligned}$$

If interest on an initial deposit of $\$P$ is *compounded continuously* at an annual rate r, the balance t years later can be calculated using the formula

$$B = Pe^{rt}.$$

Again, r is the nominal rate, and, for example, $r = 0.05$ when the annual rate is 5%.

In solving a problem involving compound interest, it is important to be clear whether interest rates are nominal rates or effective yields, as well as whether compounding is continuous or not.

Example 4 Find the effective annual yield of a 6% annual rate, compounded continuously.

Solution In one year, an investment of P becomes $Pe^{0.06}$. Using a calculator, we see that

$$Pe^{0.06} = P(1.0618365)$$

So the effective annual yield is about 6.18%.

Example 5 If \$1000 is deposited in a bank account paying 5% annual interest compounded continuously, how much will be in the account 10 years later?

Solution We use the formula $B = Pe^{rt}$. The annual rate is 5% so $r = 0.05$, the length of time is $t = 10$, and the initial deposit is $P = 1000$. We have

$$B = Pe^{rt} = 1000e^{(0.05)(10)} = 1000e^{0.5} = 1648.72.$$

The amount in the account after 10 years is \$1648.72.

Example 6 Suppose that a bank advertises a nominal annual rate of 8%. If you deposit \$5000, how much will be in the account 3 years later if the interest is compounded (a) Annually? (b) Continuously?

Solution (a) Since $n = 1$, we have $B = P(1 + r)^t = 5000(1.08)^3 = 6298.56$. If interest is compounded annually, the amount in the account after 3 years is \$6298.56.

(b) For interest compounded continuously, we use $B = Pe^{rt} = 5000e^{(0.08)(3)} = 6356.25$. As expected, the amount in the account 3 years later is larger if the interest is compounded continuously (\$6356.25) than if the interest is compounded annually (\$6298.56).

Example 7 Suppose you want to invest money in a certificate of deposit (CD) for your child's education. You need it to be worth \$12,000 in 10 years. How much should you invest if the CD pays interest at a 9% annual rate compounded quarterly? Continuously?

Solution Suppose you invest $\$P$ initially. A 9% annual rate compounded quarterly has an effective annual yield given by $(1+0.09/4)^4 = 1.0930833$, or 9.30833%. So after 10 years you will have

$$P(1.0930833)^{10} = 12000.$$

Thus, you should invest

$$P = \frac{12000}{(1.0930833)^{10}} = \frac{12000}{2.4351885} = 4927.75$$

On the other hand, if the CD pays 9% compounded continuously, after 10 years you will have

$$Pe^{(0.09)10} = 12000.$$

So you would need to invest

$$P = \frac{12000}{e^{(0.09)10}} = \frac{12000}{2.4596031} = 4878.84$$

Notice that to achieve the same result, continuous compounding requires a smaller initial investment than quarterly compounding. This is to be expected since the effective annual yield is higher for continuous than for quarterly compounding.

Problems for Section 1.6

1. Use a graph of $y = (1+0.07/x)^x$ to find the value of $(1+0.07/x)^x$ as $x \to \infty$. Confirm that the value you get is $e^{0.07}$.
2. If you deposit $10,000 in an account earning interest at an 8% annual rate compounded continuously, how much money is in the account after five years?
3. (a) Find the effective annual yield for a 5% annual interest rate compounded (i) 1000 times/year, (ii) 10,000 times/year, (iii) 100,000 times/year.
   (b) Look at the sequence of answers in part (a), and deduce the effective annual yield for a 5% annual rate compounded continuously.
   (c) Compute $e^{0.05}$. How does this confirm your answer to part (b)?
4. (a) Find $(1+0.04/n)^n$ for $n = 10{,}000$, and 100,000, and 1,000,000. Use the results to deduce the effective annual yield of a 4% annual rate compounded continuously.
   (b) Confirm your answer by computing $e^{0.04}$.
5. Use the number e to find the effective annual yield of a 6% annual rate, compounded continuously.
6. Suppose $1000 is invested in an account paying a nominal annual rate of 5.5%. How much is in the account after 8 years if the interest is compounded
   (a) Annually? (b) Quarterly? (c) Continuously?
7. Suppose $1000 is invested at 6% annual interest compounded continuously. Use trial and error, or a graph, to determine how long it will take for the balance to double.
8. What is the effective annual yield of an investment paying at a 12% annual rate, compounded continuously?

9. If you need $10,000 in your account 3 years from now and the annual interest rate on your account is 8% compounded continuously, how much should you deposit now?

10. (a) The Banque Nationale du Zaïre pays 100% nominal interest on deposits, compounded monthly. You invest 1 million zaïre. (The "zaïre" is the unit of currency of the Republic of Zaïre.) How much money do you have after one year?
    (b) How much money do you have after one year if you invest 1 million zaïre with interest compounded daily? Hourly? Each minute?
    (c) Does this amount increase without bound as interest is compounded more and more often, or does it level off? If it levels off, provide a close "upper" estimate for the total after one year.

11. Explain how you can match the interest rates (a)–(e) with the effective annual yields (I)–(V) without doing any calculations.

(a) 5.5% annual rate, compounded continuously.	(I) 5%
(b) 5.5% annual rate, compounded quarterly.	(II) 5.06%
(c) 5.5% annual rate, compounded weekly.	(III) 5.61%
(d) 5% annual rate, compounded yearly.	(IV) 5.651%
(e) 5% annual rate, compounded twice a year.	(V) 5.654%

12. When you rent an apartment, you are often required to give the landlord a security deposit which is returned if you leave the apartment undamaged. In Massachusetts the landlord is required to pay the tenant interest on the deposit once a year, at a 5% annual rate, compounded annually. The landlord, however, may invest the money at a higher (or lower) interest rate. Suppose the landlord invests a $1000 deposit at an annual rate of
    (a) 6%, compounded continuously (b) 4%, compounded continuously.

    In each case, determine the net gain or loss by the landlord at the end of the first year. (Give your answer to the nearest cent.)

13. The newspaper article below is from *The New York Times*, May 27, 1990. Fill in the three blanks. (For the last blank, assume the interest has been compounded yearly, and give your answer in dollars. Exclude the occurrence of leap years.)

213 Years After Loan, Uncle Sam Is Dunned

By LISA BELKIN

Special to The New York Times

SAN ANTONIO, May 26 — More than 200 years ago, a wealthy Pennsylvania merchant named Jacob DeHaven lent $450,000 to the Continental Congress to rescue the troops at Valley Forge. That loan was apparently never repaid.

So Mr. DeHaven's descendants are taking the United States Government to court to collect what they believe they are owed. The total: ____ in today's dollars if the interest is compounded daily at 6 percent, the going rate at the time. If compounded yearly, the bill is only ____.

Family Is Flexible

The descendants say that they are willing to be flexible about the amount of a settlement and that they might even accept a heartfelt thank you or perhaps a DeHaven statue. But they also note that interest is accumulating at ____ a second.

1.7 LOGARITHMS

In Section 1.4, we set up a function approximating the population of Mexico (in millions) as

$$P = f(t) = 67.38(1.026)^t$$

where t is the number of years since 1980. Writing the function this way shows that we are thinking of the population as a function of time, and that we believe the population to be 67.38 million in 1980 and to grow by 2.6% every year.

Now suppose that instead of calculating the population, we want to find when the population is expected to reach 100 million. This means we want to find the value of t for which

$$100 = f(t) = 67.38(1.026)^t.$$

Since the exponential function is always increasing and is eventually more than 100, there's exactly one value of t making $P = 100$. How should we find it? A reasonable way to start is trial and error. Taking $t = 10$ and $t = 20$ we get

$$P = f(10) = 67.38(1.026)^{10} = 87.1\ldots \quad (\text{so } t = 10 \text{ is too small})$$
$$P = f(20) = 67.38(1.026)^{20} = 112.58\ldots \quad (\text{so } t = 20 \text{ is too large})$$

Some more experimenting leads to

$$P = f(15) = 67.38(1.026)^{15} \approx 99.0$$
$$P = f(16) = 67.38(1.026)^{16} \approx 101.6$$

so t is between 15 and 16. In other words, the population is projected to reach 100 million sometime during 1995.

Although it is always possible to approximate t by trial and error like this, it would clearly be better to have a formula that gives t in terms of P. The *logarithm* function will enable us to do this.

Definition of Logs to Base 10:

We define the *logarithm* function, $\log_{10} x$, as follows:

$$\log_{10} x = c \quad \text{means} \quad 10^c = x.$$

We call 10 the *base*. (We will use $\log x$ to mean $\log_{10} x$, which is the notation your calculator uses.) In other words,

The **logarithm** to base 10 of x is the power of 10 you need to get x.

Thus, for example, $\log 1000 = \log 10^3 = 3$ since 3 is the power of 10 needed to get 1000. Similarly, $\log(0.1) = -1$ because $0.1 = 1/10 = 10^{-1}$. However, $\log(-3)$ is undefined since no power of 10 is negative or 0. Therefore, in general

$\log_{10} x$ is not defined if x is negative or 0.

A logarithm can be defined for any positive base. Logarithms to base 10 are used for such things as the Richter scale and determining pH. However, the most frequently used base is the famous number $e = 2.71828\ldots$. In fact, this base is used so often that the logarithm to base e is called the *natural logarithm* and is denoted by "ln." You will find a ln button on most scientific calculators as well as an e^x button; that should be some indication of how important the base e is. At first glance, this is all somewhat mysterious. What can possibly be natural about using logarithms to the base 2.71828? The full answer to that question must wait until Chapter 4, where we show that many calculus formulas come out neater when e is used as the base rather than any other base.

An exponential function can be expressed either using base e or using any other base a (provided $a > 0$, $a \neq 1$). However, the base e is generally used whenever calculus is involved. Logarithms can be defined for any base a (with $a > 1$), although most calculators contain logs only to base 10 and e. We will only use logs to these two bases.

Definition and Properties of the Natural Logarithm

The natural logarithm of x, written $\ln x$, is defined as follows:

$$\ln x = \log_e x = c \quad \text{means} \quad e^c = x$$

so

$\ln x$ is the power of e needed to get x.

Thus, for example, $\ln e^3 = 3$ since 3 is the power of e needed to get e^3. Similarly, $\ln \frac{1}{e^2} = -2$ since $\frac{1}{e^2} = e^{-2}$. Use your calculator to find $\ln 5$. You will see that $\ln 5 \approx 1.6094$. This is because $5 \approx e^{1.6094}$. (Use the $\boxed{e^x}$ button on your calculator to check!) Since no power of e is negative or zero, it follows that

$\ln x$ is not defined if x is negative or zero.

In working with logarithms, you will need to use the following properties:

Rules For Computing Using Natural Logarithms

1. $\ln (AB) = \ln A + \ln B$
2. $\ln \left(\dfrac{A}{B}\right) = \ln A - \ln B$
3. $\ln (A^p) = p \ln A$
4. $\ln e^x = x$
5. $e^{\ln x} = x$

In addition, $\ln 1 = 0$ because $e^0 = 1$.

Solving Equations Using Natural Logarithms

Logarithms are useful when we have to solve for unknown exponents, as in the following examples.

Example 1 Find t such that $3^t = 10$.

Solution First, notice that we expect t to be between 2 and 3, because $3^2 = 9$ and $3^3 = 27$. To find t exactly, we take the natural logarithm of both sides and then use the rules of logarithms to solve for t:

$$\ln(3^t) = \ln 10.$$

Then using the third log rule, we get

$$t \ln 3 = \ln 10$$
$$t = \frac{\ln 10}{\ln 3}.$$

Using a calculator to find the natural logs gives

$$t \approx \frac{2.3026}{1.0986} \approx 2.096.$$

Example 2 Find t such that $12 = 5e^{3t}$.

Solution Since t is in the exponent, in order to solve for t we will use logarithms. It is easiest to begin by isolating the exponential, so we divide both sides of the equation by 5:

$$2.4 = e^{3t}.$$

Now take the natural logarithm of both sides:

$$\ln 2.4 = \ln(e^{3t}).$$

Using the fourth log rule gives

$$\ln 2.4 = 3t,$$

so

$$t = \frac{\ln 2.4}{3}.$$

Using a calculator, we get $t \approx \frac{0.8755}{3} \approx 0.2918$.

Example 3 Suppose that $5000 is deposited in an account paying 8% annual interest. How long will it take for the money to double if the interest is compounded (a) Annually? (b) Continuously?

Solution (a) We use the formula $B = P(1+r)^t$. We have $B = 5000(1.08)^t$, and we wish to find t when $B = 10{,}000$. We use natural logarithms:

$$10000 = 5000(1.08)^t$$

Dividing both sides by 5000 gives

$$2 = (1.08)^t.$$

Taking the natural log of both sides, we get

$$\ln 2 = \ln(1.08^t).$$

Using the third log rule gives

$$\ln 2 = t \ln 1.08.$$

so

$$t = \frac{\ln 2}{\ln 1.08}.$$

Using a calculator gives

$$t \approx \frac{0.6931}{0.0770} \approx 9.001.$$

Thus, if the interest is compounded annually, it takes about 9 years for the balance to double.

(b) Since the interest is compounded continuously, we use the formula

$$B = Pe^{rt},$$

with

$$B = 5000e^{0.08t}.$$

We wish to find t when $B = 10{,}000$. We have

$$10{,}000 = 5000e^{0.08t}.$$

Dividing both sides by 5000 gives

$$2 = e^{0.08t}.$$

Taking the natural log of both sides, we get

$$\ln 2 = \ln(e^{0.08t}).$$

Using the fourth log rule, we have

$$\ln 2 = 0.08t,$$

so

$$t = \frac{\ln 2}{0.08}.$$

Using a calculator gives

$$t \approx \frac{0.6931}{0.08} \approx 8.664.$$

Hence, if the interest is compounded continuously, it takes about eight years and eight months for the balance to double. As we would expect, the balance is doubled faster when the interest is compounded continuously.

Example 4 Find the half-life of the decaying exponential $P = P_0(0.8)^x$ that we used to model the removal of pollutants in jet fuel on page 40. What does your answer mean in practical terms?

Solution We seek the value of x such that

$$P = \frac{1}{2}P_0.$$

Thus, we must solve the equation

$$\frac{1}{2}P_0 = P_0(0.8)^x$$

Dividing both sides by P_0 leaves

$$(0.8)^x = \frac{1}{2}.$$

Taking logs of both sides gives

$$\begin{aligned} x(\ln 0.8) &= \ln\left(\frac{1}{2}\right) \\ -0.223x &= -0.693 \\ x &= 3.1 \end{aligned}$$

The half-life is 3.1. In practical terms, this tells us that forcing kerosene through 3.1 feet of clay pipe removes half of its impurities.

The Graph of $\ln x$

Using the LN button on a calculator to plot a graph of $f(x) = \ln x$ for $0 < x \leq 10$, we get Figure 1.69. Because no power of e gives 0, $\ln 0$ is undefined. Also, no power of e is negative, so $\ln x$ is only defined for $x > 0$, as we see in the graph. The log function has a vertical asymptote at $x = 0$. As x gets closer and closer to 0 from the right, the natural logarithm graph drops toward $-\infty$. Check this on a calculator by finding $\ln(0.1)$, $\ln(0.01)$, $\ln(0.001)$, and so on. In addition, you should notice that the natural logarithm function crosses the x-axis at $x = 1$. This is because $e^0 = 1$ and so $\ln 1 = 0$.

We know that the exponential function grows extremely quickly. The logarithm function, on the other hand, grows extremely slowly. This is because $\ln x$ is the power of e you need to get x and you only need a relatively small power of e to get a pretty large x. In fact, to make $\ln x$ large, you will need a gigantic value of x, as we see in Table 1.24. However, though it grows slowly, the ln function does go to infinity as x increases.

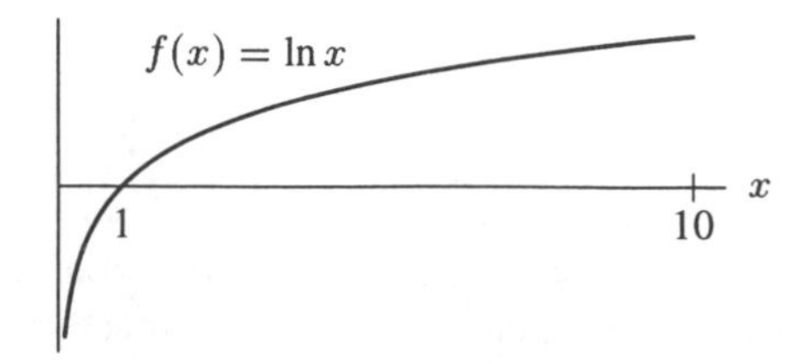

Figure 1.69: Graph of the natural logarithm

TABLE 1.24 *Values for* $\ln x$

x	1	10	100	1000	10000	100000	1000000
$\ln x$	0	2.3	4.6	6.9	9.2	11.5	13.8

Comparison Between Logarithms and Power Functions

The graph of $\ln x$ in Figure 1.69 might remind you of the graph of $y = \sqrt{x}$. Both are concave down and increasing, and both go to infinity. We see in Table 1.24 that $\ln x$ increases very slowly. But how does it compare to $y = \sqrt{x}$? Figure 1.70 compares the graphs of $\sqrt{x}$ and $\ln x$, and we see that indeed $\ln x$ grows substantially slower than $\sqrt{x}$. Recall that the exponential function climbs faster than any positive power of x. The logarithm function is at the other extreme: it climbs slower than any positive power of x.

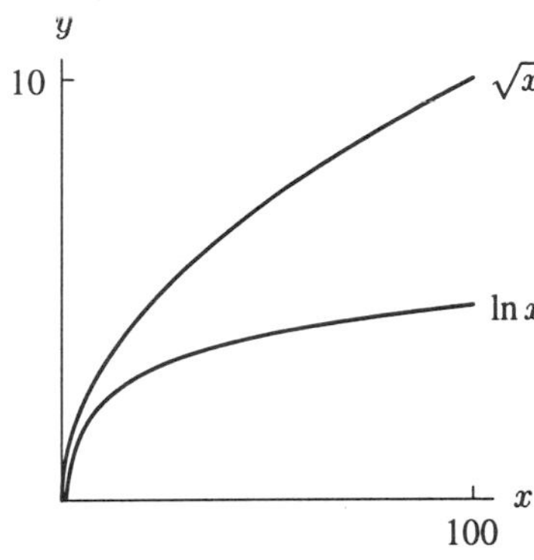

Figure 1.70: Comparison of $y = \sqrt{x}$ and $y = \ln x$.

Problems for Section 1.7

For Problems 1–2, plot a graph of the given function on a calculator or computer. Describe and explain what you see.

1. $y = \ln e^x$
2. $y = e^{\ln x}$
3. Construct a table of values to compare the values of $f(x) = \ln x$ and $g(x) = \sqrt{x}$ for $x = 1, 2, \ldots, 10$. Round to two decimal places. Use these values to graph both functions.

For Problems 4–18, solve for t using natural logarithms.

4. $5^t = 7$
5. $130 = 10^t$
6. $2 = (1.02)^t$
7. $10 = 2^t$
8. $100 = 25(1.5)^t$
9. $50 = 10(3^t)$
10. $a = b^t$
11. $10 = e^t$
12. $5 = 2e^t$
13. $c^{3t} = 100$
14. $10 = 6e^{0.5t}$
15. $B = Pe^{rt}$
16. $2P = Pe^{0.3t}$
17. $7(3^t) = 5(2^t)$
18. $5e^{3t} = 8e^{2t}$

For Problems 19–24, simplify the expression as much as possible.

19. $\ln e^2$
20. $\ln(\frac{1}{e^3})$
21. $e^{\ln(x+1)}$
22. $e^{2\ln x}$
23. $3\ln e + \ln(\frac{1}{e})$
24. $\ln(Ae^{2B})$

25. A fishery stocks a pond with 1000 young trout. The number of the original trout still alive after t years is given by $P(t) = 1000e^{-0.5t}$.
    (a) How many trout are left after six months? After 1 year?
    (b) Find $P(3)$ and interpret it in terms of trout.
    (c) At what time will there be 100 of the original trout left?
    (d) Sketch a graph of the number of trout against time, and describe how the population is changing. What might be causing this?

26. A rumor spreads among a group of 400 people in an enclosed region. Sociologists who study the propagation of rumors have found that $N(t)$, the number of people who have heard the rumor by time t (where t is the time in hours since the rumor started to spread) can be approximated by a function of the form

$$N(t) = \frac{400}{1 + 399e^{-0.4t}}.$$

    (a) Find $N(0)$ and interpret it.
    (b) How many people will have heard the rumor after 2 hours? After 10 hours?
    (c) Sketch a graph of $N(t)$.
    (d) Approximately how long will it take until half the people in the region have heard the rumor?
    (e) Approximately how long will it take until virtually everyone in the region has heard the rumor?

27. It is believed by some that the earth's population cannot exceed 40 billion people. If this is true, then the population, P, in billions, t years after 1990, can be modeled by the function

$$P = \frac{40}{1 + 11e^{-0.08t}}.$$

    (a) Sketch the graph of P against t.
    (b) According to this model, approximately when will the earth's population reach 20 billion people?
    (c) According to this model, approximately when will the earth's population reach 39.9 billion people?
    (d) According to this model, by how many people will the earth's population increase between 1990 and 2000?

28. What is the doubling time of prices which are increasing by 5% a year?

29. In 1980, there were about 170 million vehicles (cars and trucks) and about 227 million people in the United States. If the number of vehicles was growing at 4% a year, while the population was growing at 1% a year, in what year was there, on average, one vehicle per person?

30. The population of a region is growing exponentially. If there were 40,000,000 people in 1980 ($t = 0$) and 56,000,000 in 1990, find an expression for the population at any time t. What would you predict for the year 2000? What is the doubling time?

1.8 EXPONENTIAL GROWTH AND DECAY

In Section 1.4, we saw the family of exponential functions

$$P = P_0 a^t,$$

where P_0 is the initial value of P and a is the growth factor. The case $a > 1$ represents exponential growth; $0 < a < 1$ represents exponential decay. For any positive number a, we can write $a = e^k$ for $k = \ln a$. If $a > 1$, k is positive, and if $0 < a < 1$, k is negative. [13] Thus, the function representing an exponentially growing population can be rewritten as

$$P = P_0 a^t = P_0 (e^k)^t = P_0 e^{kt}$$

with k positive. In the case when $0 < a < 1$, we can use another positive constant, k, and write

$$a = e^{-k}.$$

Thus if Q is a quantity that is decaying exponentially and Q_0 is the initial quantity, at time t we will have

$$Q = Q_0 a^t = Q_0 (e^{-k})^t = Q_0 e^{-kt} = \frac{Q_0}{e^{kt}}.$$

Since e^{kt} is now in the denominator, Q will decrease as time goes on—as you'd expect if Q is decaying.

> Any **exponential growth** function can be written in the form
>
> $$P = P_0 e^{kt}$$
>
> and any **exponential decay** function can be written as
>
> $$Q = Q_0 e^{-kt}$$
>
> where P_0 and Q_0 are the initial quantities and k is positive.
>
> We say that P and Q are growing or decaying at a *continuous rate* of k. (Note that, for example, $k = 0.02$ corresponds to a continuous growth rate of 2%.)

Example 1 Sketch the graphs of $P = e^{0.5t}$ and $Q = e^{-0.2t}$.

Solution

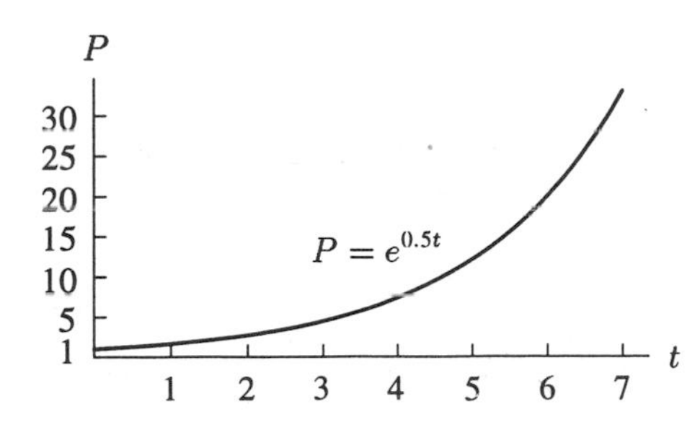

Figure 1.71: An exponential growth function

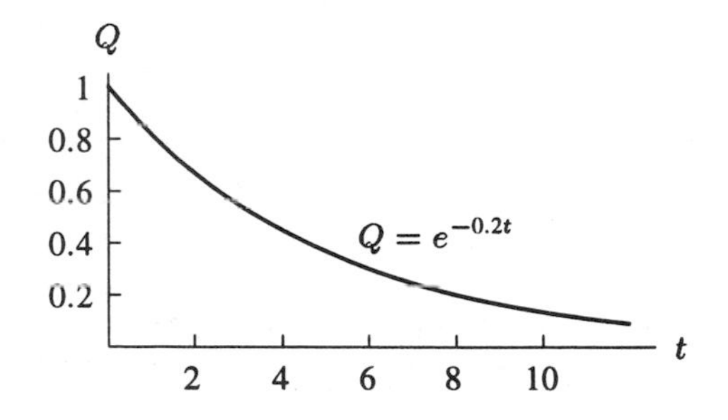

Figure 1.72: An exponential decay function

[13] An alternative notation for e^k is $\exp(k)$.

The graph of $P = e^{0.5t}$ is in Figure 1.71. Notice that the graph has the same shape as the previous exponential growth curves: increasing and concave up. Indeed, the families a^t with $a > 1$ and e^{kt} with $k > 0$ are one and the same. The graph of $Q = e^{-0.2t}$ is in Figure 1.72; it too has the same shape as other exponential decay functions. The families a^t with $0 < a < 1$ and e^{kt} with $k < 0$ are also one and the same.

Example 2 The release of chlorofluorocarbons used in air conditioners and, to a lesser extent, in household sprays (hair spray, shaving cream, etc.) destroys the ozone in the upper atmosphere. At the present time, the amount of ozone, Q, is decaying exponentially at a continuous yearly rate of 0.25%. What is the half-life of ozone? In other words, at this rate, how long will it take for half the ozone to disappear?

Solution If Q_0 is the initial quantity of ozone, then

$$Q = Q_0 e^{-0.0025t}.$$

We want to find T, the value of t making $Q = Q_0/2$, so

$$\frac{Q_0}{2} = Q_0 e^{-0.0025T}.$$

Divide by Q_0 to obtain

$$\frac{1}{2} = e^{-0.0025T}.$$

Taking natural logs yields

$$-0.0025T = \ln\left(\frac{1}{2}\right) = -0.6931,$$

so

$$T \approx 277 \text{ years.}$$

In Example 2 the decay rate was given. However, in many situations where we expect to find exponential growth or decay, the rate is not given. To find it, we must know the quantity at two different times and then solve for the growth rate, as in the next example.

Example 3 The population of Kenya was 19.5 million in 1984 and 21.2 million in 1986. Assuming it increases exponentially, find a formula for the population of Kenya as a function of time.

Solution If we measure the population, P, in millions and time, t, in years since 1984, we can say

$$P = P_0 e^{kt} = 19.5e^{kt}$$

where $P_0 = 19.5$ is the initial value of P. We find k by using the fact that $P = 21.2$ when $t = 2$, so

$$21.2 = 19.5e^{k \cdot 2}.$$

To find k, we divide both sides by 19.5, giving

$$\frac{21.2}{19.5} = 1.0872 = e^{2k}.$$

Now take natural logs of both sides:

$$\ln(1.0872) = \ln(e^{2k}).$$

Using a calculator and the fact that $\ln(e^{2k}) = 2k$, this becomes

$$0.0836 = 2k.$$

So

$$k \approx 0.042,$$

and therefore,

$$P = 19.5e^{0.042t}.$$

Since $k = 0.042 = 4.2\%$, we say the population of Kenya was growing continuously at 4.2% a year.

Relationship between a^t and e^{kt}

In Example 3 we chose to use e for the base of the exponential function representing Kenya's population, making clear that the continuous growth rate was 4.2%. If, however, we had wanted to emphasize the annual growth rate, we could have expressed the exponential function in the form

$$P = P_0a^t.$$

Since the population grew from 19.5 to 21.2 million in 2 years, we know that

$$21.2 = 19.5a^2$$

so

$$a = \left(\frac{21.2}{19.5}\right)^{(1/2)} \approx 1.043.$$

Thus,

$$P = P_0(1.043)^t.$$

Notice that $a \approx 1.043$ corresponds to an annual growth rate of about 4.3%, which is just slightly more than the continuous growth rate of 4.2% found in Example 3.

In general, an exponential function of the form P_0e^{kt} can always be written in the form P_0a^t, and vice versa, simply by letting $e^k = a$ or $k = \ln a$. The two different formulas, $P = P_0e^{kt}$ and $P = P_0a^t$, have the same graph and represent the same function.

To convert between a^t and e^{kt} use:

$$a^t = (e^k)^t = e^{kt} \quad \text{where} \quad k = \ln a$$

If a comes from a percentage growth rate r, that is, $a = 1 + r$, then the continuous growth rate $k = \ln(1 + r)$ will be slightly less than, but very close to, r, provided r is small.

Example 4 In Section 1.4, we saw that radioactive decay of carbon-14 could be modeled by the function

$$C = C_0 \left(\frac{1}{2}\right)^{t/5730}$$

where C is the quantity of carbon-14 at time t, C_0 is the initial quantity, and the half-life of carbon-14 is 5730 years. Express this function in terms of e.

Solution We want to rewrite the function as

$$C = C_0 \left(\frac{1}{2}\right)^{t/5730} = C_0 e^{kt}.$$

Canceling C_0, we have

$$\left(\frac{1}{2}\right)^{t/5730} = e^{kt}.$$

Taking natural logs of both sides gives

$$\frac{t}{5730} \ln\left(\frac{1}{2}\right) = \ln(e^{kt}) = kt.$$

Canceling the t's gives

$$k = \frac{1}{5730} \ln\left(\frac{1}{2}\right) = -0.000121.$$

Thus,

$$C = C_0 e^{-0.000121t}.$$

Using Logarithms to Solve Compound Interest Problems

The natural logarithm may be used to help us answer questions about compound interest.

Example 5 If \$10,000 is deposited in an account paying an annual interest rate of 5%, compounded continuously, how long will it take for the balance in the account to reach \$15,000?

Solution Since the interest is being compounded continuously, we use the formula $B = Pe^{rt}$, where the interest rate is $r = 0.05$, and the initial amount is $P = 10{,}000$. We wish to find t if $B = 15{,}000$. Our equation is

$$15{,}000 = 10{,}000e^{0.05t}.$$

Since we want to solve for t, and t is in the exponent, we need to use logarithms. First, we divide both sides of the equation by 10,000 to isolate the exponential, and then we take the natural logarithm of both sides and solve for t.

$$\begin{aligned} 15{,}000 &= 10{,}000e^{0.05t} \\ 1.5 &= e^{0.05t} \\ \ln(1.5) &= \ln(e^{0.05t}) \\ \ln(1.5) &= 0.05t \\ t &= \frac{\ln(1.5)}{0.05} \approx 8.1093. \end{aligned}$$

It will take about 8.1 years for the balance in the account to reach \$15,000.

Example 6 (a) Find the doubling time, D, for annual growth rates, $i\%$, of 2%, 3%, 4%, and 5%.

(b) Since D decreases as i increases, we might guess that D is inversely proportional to i, that is, $D = \frac{k}{i}$ for some constant k. Use your answers to part (a) to confirm that, approximately,

$$D = \frac{70}{i}.$$

This is the "rule of 70" used by bankers. To compute the approximate doubling time of an investment, the banker divides 70 by the yearly interest rate.

Solution (a) We begin by finding the doubling time for an annual growth rate of 2%. We use the formula $B = P(1.02)^t$. We wish to find t when $B = 2P$, so we are solving

$$\begin{aligned} 2P &= P(1.02)^t \\ 2 &= (1.02)^t \\ \ln 2 &= \ln(1.02)^t \\ \ln 2 &= t\ln(1.02) \qquad \text{(using properties of logarithms)} \\ t &= \frac{\ln 2}{\ln 1.02} \approx 35.002. \end{aligned}$$

We see that if the annual interest rate is 2%, it will take about 35 years for an investment to double in value. Similarly, we can find the doubling times for 3%, 4%, and 5%. The results are given in Table 1.25.

TABLE 1.25

$i\%$	2	3	4	5
D (in years)	35.002	23.450	17.673	14.207

(b) We compute $(70/i)$ for $i = 2, 3, 4, 5$. The results are shown in Table 1.26.

TABLE 1.26

i	2	3	4	5
$(70/i)$	35.000	23.333	17.500	14.000

Comparing the values in Table 1.25 and Table 1.26, we see that the approximation $(70/i)$ given by the "rule of 70" gives a reasonably accurate approximation to the exact doubling time D.

Problems for Section 1.8

1. If $12,000 is deposited in an account paying 8% interest per year, compounded continuously, how long will it take for the balance to reach $20,000?

2. If an investment of $5000 grows to $8080 in four years, what was the annual rate of return on the investment? (Assume continuous compounding.)
3. In 1994, the world's population was 5.6 billion and the population is projected to reach 8.5 billion by the year 2030. What annual rate of growth is assumed in this prediction?
4. Find the doubling time of a quantity that is increasing by 7% per year.
5. If the quantity of a certain substance decreases by 4% in 10 hours, find the half-life of the substance.
6. You invest $5000 in an account which pays interest compounded continuously.
   (a) How much money is in the account after 8 years, if the annual interest rate is 4%?
   (b) If you want the account to contain $8000 after 8 years, what yearly interest rate is needed?
7. The half-life of a certain radioactive substance is 12 days. If there are 10.32 grams initially:
   (a) Write an equation to determine the amount, A, of the substance as a function of time.
   (b) When will the substance be reduced to 1 gram?
8. Owing to an innovative rural public health program, infant mortality in Senegal, West Africa, is being reduced at a rate of 10% per year. How long will it take for infant mortality to be reduced by 50%?
9. Assume that the rate of inflation continues at the 1991 rate of 4.6% a year. If the price of postage stamps goes up, on average, at the rate of inflation, when will it cost $1 to mail a letter? [Note: That cost rose to 29 cents in 1990.]
10. (a) Use the "rule of 70" to predict the doubling time of an investment which is earning 8% interest per year.
    (b) Find the doubling time exactly, and compare your answer to part (a).
11. Figure 1.73 shows several possible graphs of a city's population against time. Match each description below to a graph and write a description to match each of the remaining graphs.
    (a) The city's population increased at 5% per year.
    (b) The city's population increased at 8% per year.
    (c) The city's population increased by 5000 people per year.
    (d) The city's population was stable.

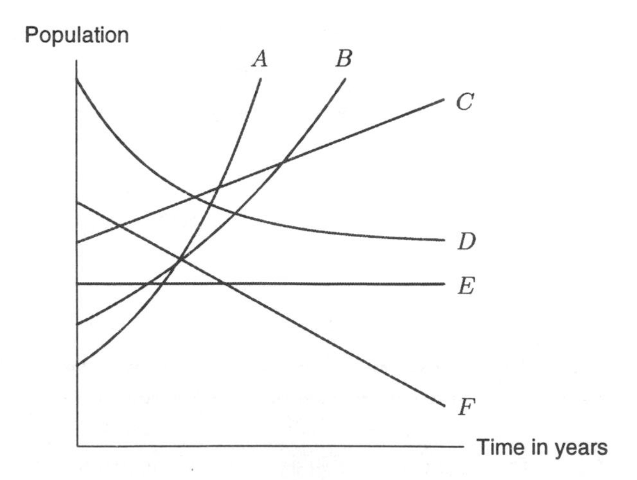

Figure 1.73

Convert the functions in Problems 12–16 into the form $P = P_0 a^t$. Which represent exponential growth and which represent exponential decay?

12. $P = P_0 e^{0.2t}$ 13. $P = 10e^{0.917t}$ 14. $P = P_0 e^{-0.73t}$

15. $P = 79e^{-2.5t}$ 16. $P = 7e^{-\pi t}$

Convert the functions in Problems 17–20 into the form $P = P_0 e^{kt}$.

17. $P = P_0 2^t$ 18. $P = 10(1.7)^t$ 19. $P = 5.23(0.2)^t$ 20. $P = 174(0.9)^t$

21. (a) A population grows according to the equation $P = P_0 e^{kt}$ (with P_0, k constants). Find the population as a function of time, t, if it grows at a continuous rate of 2% a year and starts at 1 million.
    (b) Plot a graph of the population you found in part (a) against time.

22. The island of Manhattan was sold for $24 in 1626. Suppose the money had been invested in an account which compounded interest continuously.
    (a) How much money would be in the account in the year 2000 if the yearly interest rate was 5%?
    (b) How much money would be in the account in the year 2000 if the yearly interest rate was 7%?
    (c) If the yearly interest rate was 6%, in what year would the account be worth one million dollars?

23. A savings account at a local bank is offering a yearly interest rate of 6.85%, compounded quarterly. What should be the yearly interest rate of an account whose interest is compounded continuously, but which has the same effective annual yield?

24. The solid waste generated in cities in the US is increasing. The solid waste generated, in millions of tons, was 82.3 in 1960 and 139.1 in 1980. In Example 4 in Section 1.2, we found the equation of a line through these two points and used this line to predict the amount of municipal solid waste in the year 2000. Since we only have the two points, it is not clear that the line gives the correct relationship between solid waste and year. In this problem, we assume that solid waste is increasing exponentially rather than linearly, and we find an exponential curve to fit the data.
    (a) Using the information about municipal solid waste in the years 1960 and 1980, find the equation of an exponential growth curve through the two points.
    (b) Use your answer to part (a) to predict US municipal solid waste in the year 2000. Compare your answer to the earlier prediction of 195.9 million tons, which we found using the assumption that the growth was linear.

25. Air pressure, P, decreases exponentially with the height above the surface of the earth, h:

$$P = P_0 e^{-0.00012h}$$

where P_0 is the air pressure at sea level and h is in meters.
    (a) If you go to the top of Mount McKinley, height 6,198 meters (about 20,320 feet), what is the air pressure, as a percent of the pressure at sea level?
    (b) The maximum cruising altitude of an ordinary commercial jet is around 12,000 meters (about 40,000 feet). At that height, what is the air pressure, as a percent of the sea level value?

26. Under certain circumstances, the velocity, V, of a falling raindrop is given by $V = V_0(1-e^{-t})$, where t is time and V_0 is a positive constant.
    (a) Sketch a rough graph of V against t, for $t \geq 0$.
    (b) What does V_0 represent?

27. The air in a factory is being filtered so that the quantity of a pollutant, P (measured in mg/liter), is decreasing according to the equation $P = P_0e^{-kt}$, where t represents time in hours. If 10% of the pollution is removed in the first five hours:
    (a) What percentage of the pollution is left after 10 hours?
    (b) How long will it take before the pollution is reduced by 50%?
    (c) Plot a graph of pollution against time. Show the results of your calculations on the graph.
    (d) Explain why the quantity of pollutant might decrease in this way.

28. The population, P, in millions, of Nicaragua was 3.6 million in 1990 and growing at 3.4% per year. Let t be time in years since 1990.
    (a) Express P as a function in the form $P = P_0a^t$.
    (b) Express P as an exponential function using the base e.
    (c) Compare the annual and continuous growth rates.

29. One of the main contaminants of a nuclear accident, such as that at Chernobyl, is strontium-90, which decays exponentially at a continuous rate of approximately 2.47% per year. Preliminary estimates after the Chernobyl disaster suggested that it would be about 100 years before the region would again be safe for human habitation. What percent of the original strontium-90 would still remain by this time?

30. The quantity, Q, of radioactive carbon-14 remaining t years after an organism dies is given by the formula

$$Q = Q_0e^{-0.000121t},$$

where Q_0 is the initial quantity.
    (a) A skull uncovered at an archeological dig has 15% of the original amount of carbon-14 present. Estimate its age.
    (b) Show how you can calculate the half-life of carbon-14 from this equation.

31. A picture supposedly painted by Vermeer (1632–1675) contains 99.5% of its carbon-14 (half-life 5730 years). From this information can you determine whether or not the picture is a fake? Explain your reasoning.

32. Geological dating of rocks is done using potassium-40 rather than carbon-14 because potassium has a longer half-life. The potassium decays to argon, which remains trapped in the rocks and can be measured; thus the original quantity of potassium can be calculated. The half-life of potassium-40 is $1.28 \cdot 10^9$ years. Find a formula giving the quantity, P, of potassium-40 remaining as a function of time in years, assuming the initial quantity is P_0:
    (a) Using base 1/2
    (b) Using base e

33. A bank account is earning interest at 6% per year compounded continuously.
    (a) By what percentage has the bank balance in the account increased over one year? (This is the effective annual yield.)
    (b) How long does it take the balance to double?
    (c) Assuming now that the interest rate is r, find a formula giving the doubling time in terms of the interest rate.

1.9 NEW FUNCTIONS FROM OLD

Shifts

Consider the function $y = x^2 + 4$. The y-coordinates for this function will all be exactly 4 units larger than the y-coordinates of the function $y = x^2$. It follows that the graph of $y = x^2 + 4$ should be the graph of $y = x^2$ with 4 added to each y value. We see in Figure 1.74 that the graph of $y = x^2 + 4$ is the graph of $y = x^2$ moved up four units.

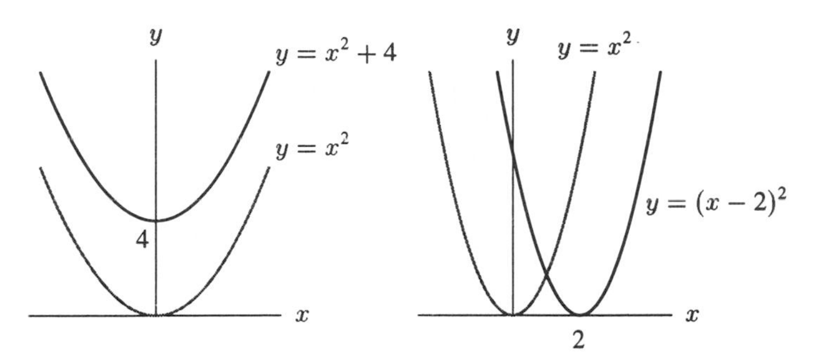

Figure 1.74: Graphs of $y = x^2$ with $y = x^2 + 4$ and $y = (x - 2)^2$

A graph can also be shifted to the left or to the right. In Figure 1.74, we see that the graph of $y = (x - 2)^2$ is the graph of $y = x^2$ shifted to the right 2 units. In general,

- The graph of $y = f(x) + k$ is the graph of $y = f(x)$ moved up k units (down if k is negative).
- The graph of $y = f(x - k)$ is the graph of $y = f(x)$ moved to the right k units (to the left if k is negative).

Example 1 Use your knowledge of graphs of functions to sketch a rough graph of each of the following:
$y = x^3 + 1$
$y = e^x - 3$
$y = (x + 2)^6$
$y = (x - 2)^2 - 1$

Solution The four graphs are shown in Figure 1.75.

(a) The graph of $y = x^3 + 1$ is the graph of $y = x^3$ moved up 1 unit.
(b) The graph of $y = e^x - 3$ is the graph of $y = e^x$ moved down 3 units.
(c) The graph of $y = (x + 2)^6$ is the graph of $y = x^6$ moved to the left 2 units.
(d) The graph of $y = (x - 2)^2 - 1$ is the graph of $y = x^2$ moved to the right 2 units and down 1 unit.

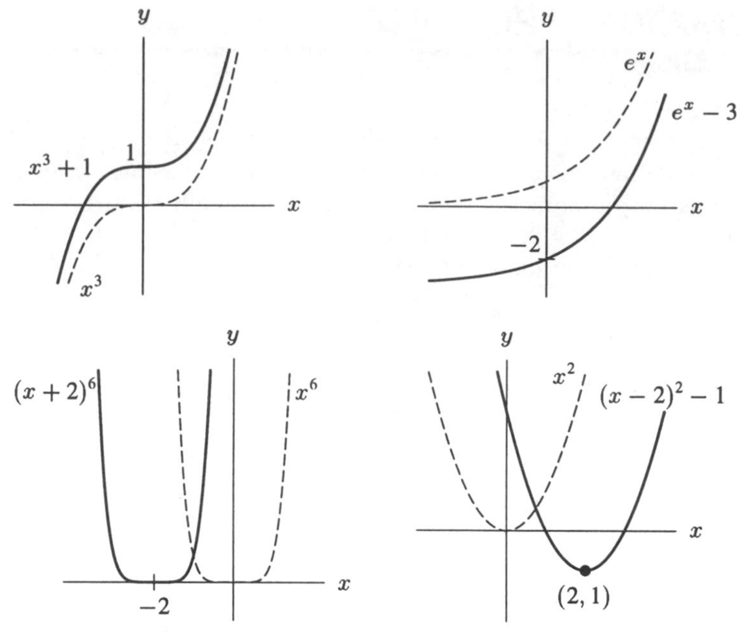

Figure 1.75: Examples of graphs shifted by a constant

Stretches

The graph of a constant multiple of a given function is easy to visualize: each y value is stretched or shrunk by that multiple. For example, consider the function $f(x)$ and its multiples $y = 3f(x)$ and $y = -2f(x)$, whose graphs are in Figure 1.76. The factor 3 in the function $y = 3f(x)$ stretches each $f(x)$ value by multiplying it by 3; the factor -2 in the function $y = -2f(x)$ stretches $f(x)$ by multiplying by 2 and reflecting it about the x-axis. You can think of the multiples of a given function as a family of functions.

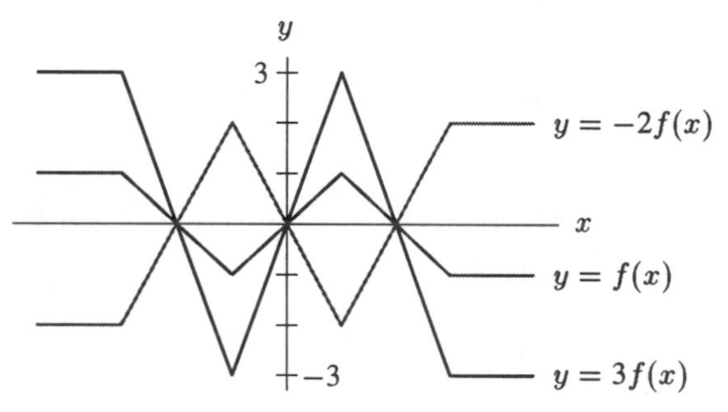

Figure 1.76: The family of multiples of the function $f(x)$

When we looked at the effect of coefficients on power functions in Section 1.5, we were studying the effect of multiplying a function by a constant. In general,

> Multiplying a function by a constant stretches or shrinks its graph vertically; a negative sign reflects the graph about the x-axis.

Sums of Functions

The right-hand graph in Figure 1.77 shows the number of US students majoring in science and engineering, broken down into men and women.[14] The top line represents the total; the upper curve represents the women, and the bottom curve represents the men. It is really a picture of three functions: the number of men majoring in science and engineering, $m(t)$, the number of women majoring in science and engineering, $w(t)$, and the total number of students, $n(t)$, where t is the year. The men's and women's graphs are shown separately to the left.

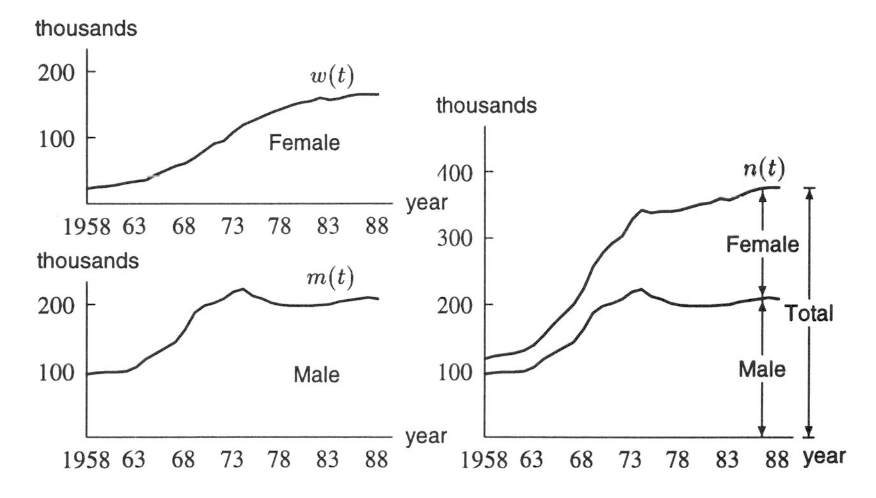

Figure 1.77: Numbers of US students majoring in science and engineering (1958–1988)

Now Total number = Number of men + Number of women, so, in functional notation,

$$n(t) = m(t) + w(t).$$

Figure 1.77 makes it clear that the graph of the sum function is obtained by stacking the other graphs one on top of the other. Let's look at the same idea in a more mathematical setting.

Example 2 Graph the function $y = 2x^2 + 1/x$ for $x > 0$.

Solution We graph both of the functions $y = 2x^2$ and $y = 1/x$ separately first. Now imagine the corresponding y values stacked on top of one another, and you will get the graph of $y = 2x^2 + 1/x$. (See Figure 1.78.)

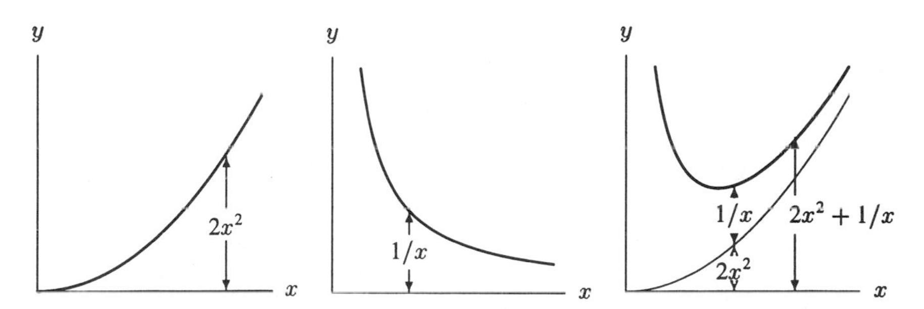

Figure 1.78: Summing two functions from their graphs

[14]Data from the National Science Foundation, Washington D.C.

Notice that, for x near 0, the graph of $y = 2x^2 + 1/x$ looks much like the graph of $y = 1/x$ (because $2x^2$ is near zero). For large x, the graph of $y = 2x^2 + 1/x$ looks like the graph of $y = 2x^2$.

Composite Functions

If oil is spilled from a tanker, the area of the oil slick will grow with time. Suppose that the oil slick is always a perfect circle. (In practice, this does not happen because of winds and tides and the location of the coastline.) The area of the slick is a function of its radius

$$A = f(r) = \pi r^2.$$

The radius is also a function of time, because the radius increases as more oil spills. Thus, the area, being a function of the radius, is also a function of time. If, for example, the radius is given by

$$r = g(t) = 1 + t$$

then the area is given as a function of time by substitution

$$A = \pi r^2 = \pi(1 + t)^2.$$

We say that A is a *composite function* or a "function of a function," which is written

$$A = \underbrace{f(g(t))}_{\substack{\text{Composite function;} \\ f \text{ is outside function} \\ g \text{ is inside function}}} = \pi(g(t))^2 = \pi(1 + t)^2.$$

Here the *inside function*, g, represents the calculation which is done first, and the *outside function*, f, represents the calculation done second. Look at the example, and think about how you would calculate A using the formula $\pi(1+t)^2$. For any given t, the first step is to find $1+t$, and the second step is to square and multiply by π. So the first step corresponds to the inside function $g(t) = 1 + t$, and the second step corresponds to the outside function $f(r) = \pi r^2$.

Example 3 Express each of the following functions as a composition:

(a) $h(t) = (1 + t^3)^{27}$ (b) $k(x) = \ln x^2$ (c) $l(x) = \ln^2 x$ (d) $n(y) = e^{-y^2}$

Solution In each case think about how you would calculate a value of the function. The first stage of the calculation will give you the inside function, and the second stage will give you the outside one.

(a) The first stage is cubing and adding 1, so the inside function is $g(t) = 1 + t^3$. The second stage is taking the 27th power, so the outside function is $f(y) = y^{27}$. Then

$$f(g(t)) = f(1 + t^3) = (1 + t^3)^{27}.$$

In fact, there are lots of different answers: $g(t) = t^3$ and $f(y) = (1+y)^{27}$ is another possibility.

(b) By convention, $\ln x^2$ means $\ln(x^2)$, so evaluating this function involves squaring x first and then taking the log. So if $g(x) = x^2$ is the inside and $f(y) = \ln y$ is the outside, then $f(g(x)) = \ln x^2$.

(c) By convention, $\ln^2 x$ means $(\ln x)^2$, so this function involves taking the log first and then squaring. Using the same definitions of f and g as in part (b), namely $f(x) = \ln x$ and $g(t) = t^2$, the composition is $g(f(x)) = (\ln x)^2$. By evaluating the functions in parts (b) and (c) for $x = 2$, say, giving $\ln(2^2) = 1.386$, and $\ln^2(2) = 0.480$, you can see that the order in which you compose two functions certainly can make a difference.

(d) To calculate e^{-y^2} you square y, take its negative, and then take e to that power. So $g(y) = -y^2$ and $f(z) = e^z$. Then $f(g(y)) = e^{-y^2}$. Alternatively, you could take $g(y) = y^2$ and $f(z) = e^{-z}$.

Example 4 If $f(x) = x^2$ and $g(x) = x + 1$, find each of the following:
(a) $f(g(2))$ (b) $g(f(2))$ (c) $f(g(x))$ (d) $g(f(x))$

Solution
(a) Since $g(2) = 3$, we have $f(g(2)) = f(3) = 9$.
(b) Since $f(2) = 4$, we have $g(f(2)) = g(4) = 5$. Notice that $f(g(2)) \neq g(f(2))$.
(c) $f(g(x)) = f(x + 1) = (x + 1)^2$.
(d) $g(f(x)) = g(x^2) = x^2 + 1$. Again, notice that $f(g(x)) \neq g(f(x))$.

Odd and Even Functions

There is a certain symmetry apparent in the graphs of $f(x) = x^2$ and $g(x) = x^3$ in Figure 1.79. Namely, for each point (x, x^2) on the graph of f, the point $(-x, x^2)$ is also on the graph; and for each point (x, x^3) on the graph of g, the point $(-x, -x^3)$ is also on the graph (see Figure 1.79). The graph of $f(x) = x^2$ is symmetric in the y-axis, whereas the graph of $g(x) = x^3$ is symmetric about the origin. We say that the squaring function is *even* and the cubing function is *odd*. The names come from the fact that the even powers are even functions and odd powers are odd functions.

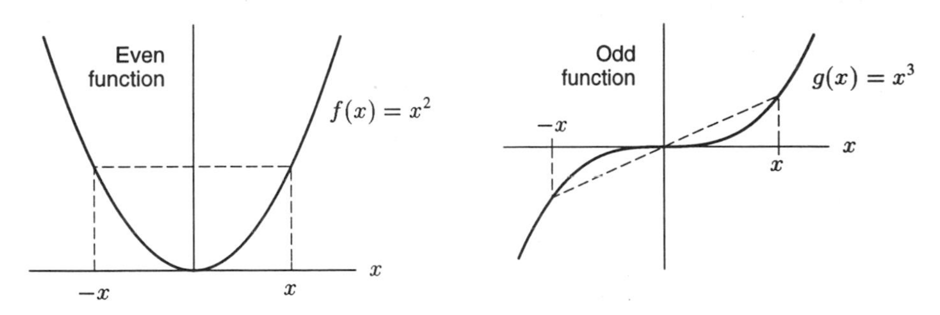

Figure 1.79: Symmetry of odd and even functions

In general, for any function f,

f is an **even** function if $f(-x) = f(x)$ for all x.

f is an **odd** function if $f(-x) = -f(x)$ for all x.

You should be aware that many functions do not have any symmetry and so are neither even nor odd.

Problems for Section 1.9

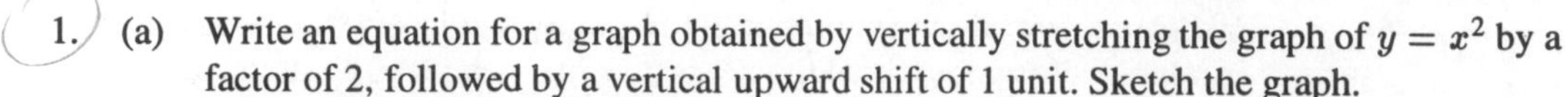

1. (a) Write an equation for a graph obtained by vertically stretching the graph of $y = x^2$ by a factor of 2, followed by a vertical upward shift of 1 unit. Sketch the graph.
   (b) What is the equation if the order of the transformations (stretching and shifting) in part (a) is interchanged?
   (c) Are the two graphs the same? Explain the effect of reversing the order of transformations.
2. Let $f(x) = 2x + 3$ and $g(x) = \ln x$. Find formulas for each of the following functions.
   (a) $g(f(x))$ (b) $f(g(x))$ (c) $f(f(x))$
3. What is the difference (if any) between $\ln[\ln(x)]$ and $\ln^2(x)$ [$= (\ln x)^2$]?
4. If $h(x) = x^3 + 1$ and $g(x) = \sqrt{x}$, find
   (a) $g(h(x))$ (b) $h(g(x))$ (c) $h(h(x))$ (d) $g(x) + 1$ (e) $g(x + 1)$
5. If $f(t) = (t + 7)^2$ and $g(t) = 1/(t + 1)$, find
   (a) $f(g(t))$ (b) $g(f(t))$ (c) $f(t^2)$ (d) $g(t - 1)$
6. If $f(z) = 10^z$ and $g(z) = \log(z)$, find
   (a) $f(g(100))$ (b) $g(f(3))$ (c) $f(g(x))$ (d) $g(f(x))$

For Problems 7–10, let $m(z) = z^2$. Find and simplify the following quantities:

7. $m(z + 1) - m(z)$
8. $m(z + h) - m(z)$
9. $m(z) - m(z - h)$
10. $m(z + h) - m(z - h)$

For Problems 11–14 determine functions f and g such that $h(x) = f(g(x))$. [There is more than one correct answer. Do not choose $f(x) = x$ or $g(x) = x$.]

11. $h(x) = x^3 + 1$
12. $h(x) = (x + 1)^3$
13. $h(x) = \ln^3 x$
14. $h(x) = \ln(x^3)$
15. The Heaviside step function, H, is graphed in Figure 1.80. Sketch graphs of the following functions.
    (a) $2H(x)$
    (b) $H(x) + 1$
    (c) $H(x + 1)$
    (d) $-H(x)$
    (e) $H(-x)$

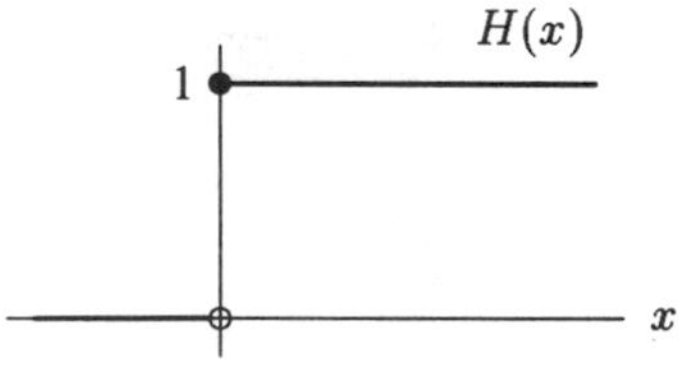

Figure 1.80

For each of the functions $y = f(x)$ whose graphs are in Problems 16 and 17, sketch graphs of
(a) $y = 2f(x)$ (b) $y = f(x + 1)$ (c) $y = f(x) + 1$
Where possible, identify x and y intercepts and horizontal and vertical asymptotes for each graph.

16.

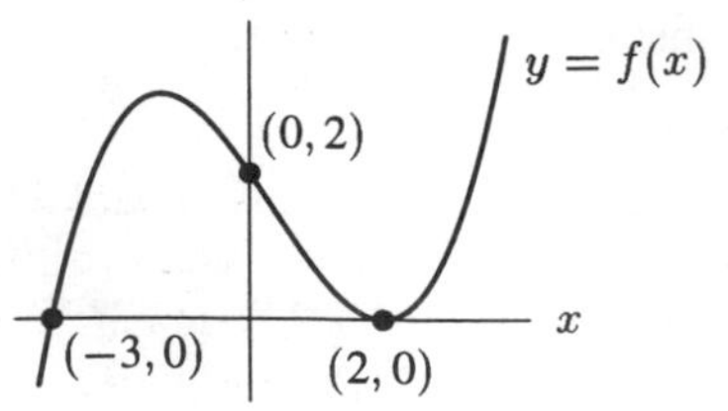

17.

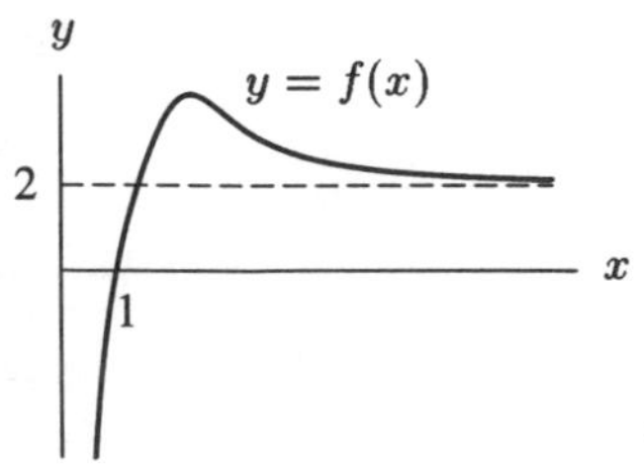

18. What symmetries do the graphs of even and odd functions have?
19. Are the functions $f(x) = 1/x$, $g(x) = \ln(x^2)$, $h(x) = e^x$ even, odd, or neither?

For Problems 20–25, use the graph of $y = f(x)$ in Figure 1.81 to sketch the graph indicated:

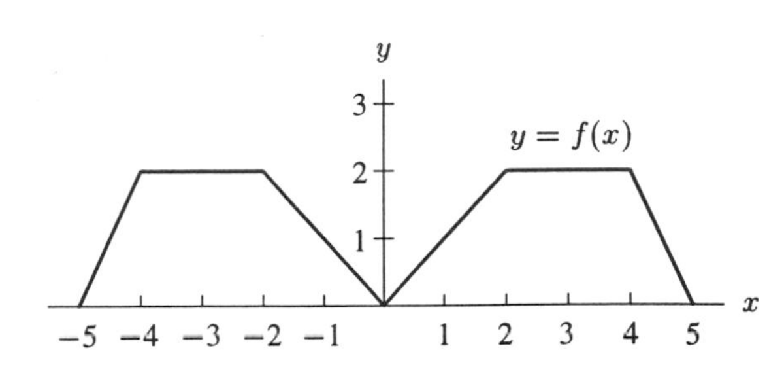

Figure 1.81

20. $y = 2f(x)$
21. $y = f(x) + 2$
22. $y = -3f(x)$
23. $y = f(x - 1)$
24. $y = 2 - f(x)$
25. $y = \dfrac{1}{f(x)}$

For Problems 26–31, suppose that f and g are given by the graphs in Figure 1.82:

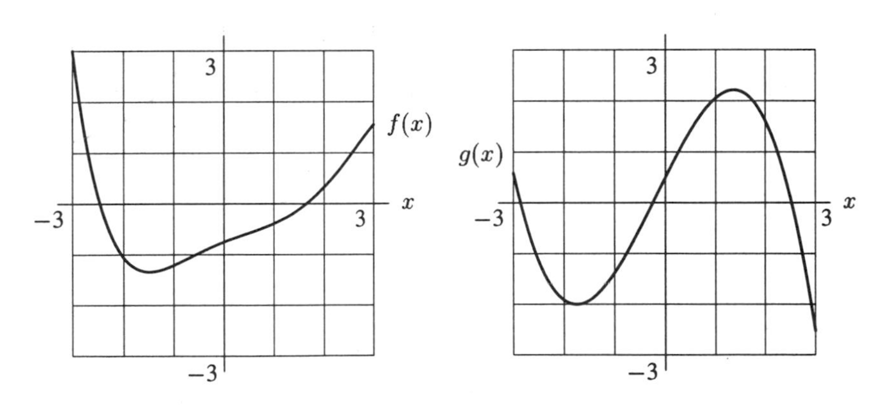

Figure 1.82

26. Find $f(g(1))$.
27. Find $g(f(2))$.
28. Find $f(f(1))$.
29. Sketch a graph of $f(g(x))$.
30. Sketch a graph of $g(f(x))$.
31. Sketch a graph of $f(f(x))$.
32. Complete Table 1.27 to show values for functions f, g, and h, given the following conditions:
    (a) f is symmetric about the y-axis.
    (b) g is symmetric about the origin.
    (c) h is the composition of f with g [that is, $h(x) = g(f(x))$].

TABLE 1.27

x	$f(x)$	$g(x)$	$h(x)$
-3	0	0	
-2	2	2	
-1	2	2	
0	0	0	
1			
2			
3			

33. Complete the graphs of $f(x)$ and $g(x)$ for $-10 \leq x \leq 10$ in Figure 1.83 given that $f(x)$ is even and $g(x)$ is odd.

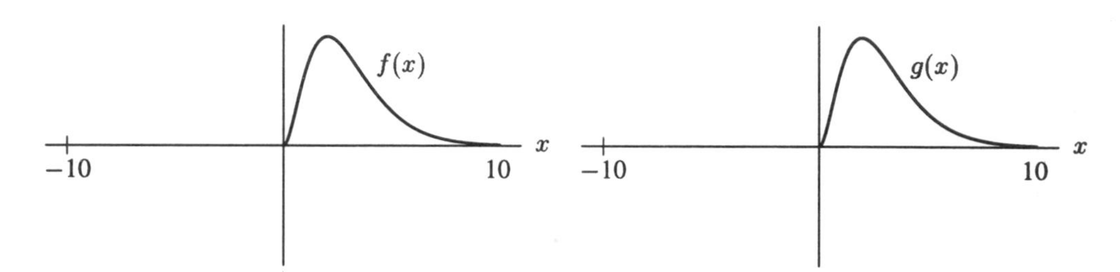

Figure 1.83

34. Graph each of the following functions on a graphing calculator or computer and use the graphs to determine which ones are odd, even, or neither.

(a) $y = x^2$ (b) $y = x^3$ (c) $y = 5x^4$
(d) $y = -10x^5$ (e) $y = x^3 + 3x^2$ (f) $y = x^4 - x^2$
(g) $y = x^5 - 2x^3$ (h) $y = 2x^4 + 5$ (i) $y = 7x + 5$

1.10 POLYNOMIALS

Some of the best-known functions for which there are formulas are the polynomials:

$$y = p(x) = a_n x^n + a_{n-1}x^{n-1} + \cdots + a_1 x + a_0,$$

where n is a positive integer, called the *degree* of the polynomial (provided $a_n \neq 0$). The shape of the graph of a polynomial depends on its degree, as shown in Figure 1.84. These graphs correspond to a positive coefficient for x^n; a negative coefficient flips the graph over. Notice that the graph of the quadratic "turns around" once, the cubic "turns around" twice, and the quartic (fourth degree) "turns around" three times. An nth degree polynomial "turns around" at most $n - 1$ times (where n is a positive integer), but there may be fewer turns.

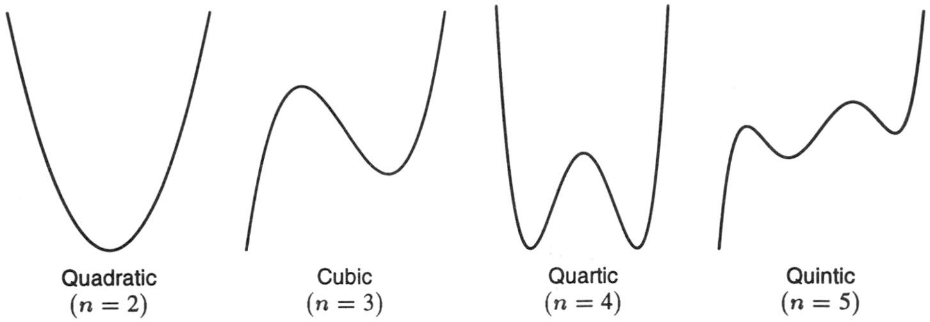

Quadratic
$(n = 2)$
Cubic
$(n = 3)$
Quartic
$(n = 4)$
Quintic
$(n = 5)$

Figure 1.84: Graphs of typical polynomials of degree n

Example 1 Find possible formulas for the polynomials whose graphs are in Figure 1.85.

(a)
4
$f(x)$
x
-2
2
(b)
$g(x)$
x
-3
1
2
-12
(c)
$h(x)$
x
-3
2

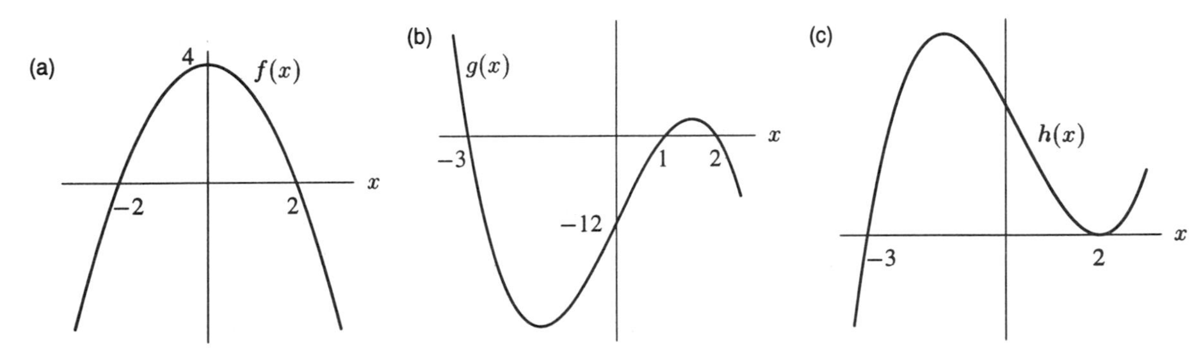

Figure 1.85: Graphs of polynomials

Solution (a) This graph appears to be a parabola, turned upside down and moved up by 4, so

$$f(x) = -x^2 + 4.$$

The minus sign turns the parabola upside down and the $+4$ moves it up by 4. You should notice that this formula does give the correct x-intercepts since $0 = -x^2 + 4$ has solutions $x = \pm 2$. You can also solve this problem by looking at the x-intercepts first: the function has zeros at $x = -2$ and $x = +2$, so $f(x)$ must have factors of $(x + 2)$ and $(x - 2)$, so

$$f(x) = k(x + 2)(x - 2).$$

To find k, use the fact that the graph has a y intercept of 4, so $f(0) = 4$, giving

$$4 = k(0 + 2)(0 - 2)$$

so $k = -1$. Therefore, $f(x) = -(x + 2)(x - 2)$, which multiplies out to $-x^2 + 4$. Note that

$$f(x) = 4 - \frac{x^4}{4}$$

also fits the requirements, but its "shoulders" are sharper. There are many possible answers to these questions.

(b) This looks like a cubic with factors $(x+3)$, $(x-1)$, and $(x-2)$, one for each x-intercept:

$$g(x) = k(x+3)(x-1)(x-2).$$

Since the y intercept is -12,

$$-12 = k(0+3)(0-1)(0-2)$$

so $k = -2$, and

$$g(x) = -2(x+3)(x-1)(x-2).$$

(c) This also looks like a cubic with zeros at $x = 2$ and $x = -3$. Notice that at $x = 2$ the graph of $h(x)$ touches the x-axis but doesn't cross it, whereas at $x = -3$ the graph crosses the x-axis. We say that $x = 2$ is a *double zero*, but that $x = -3$ is a single zero.
To find a formula for $h(x)$, first imagine the graph of $h(x)$ to be slightly lower down, so that the graph has one x intercept near $x = -3$ and two near $x = 2$, say at $x = 1.9$ and $x = 2.1$. Then

$$h(x) \approx k(x+3)(x-1.9)(x-2.1).$$

Now move the graph back to its original position. The zeros at $x = 1.9$ and $x = 2.1$ move toward $x = 2$, giving

$$h(x) = k(x+3)(x-2)^2.$$

Thus the double zero leads to a repeated factor, $(x-2)^2$. Notice that when $x > 2$, the factor $(x-2)^2$ is positive, and when $x < 2$, $(x-2)^2$ is still positive. This reflects the fact that $h(x)$ doesn't change sign near $x = 2$. Compare this with the behavior near the single zero, where h does change sign.
You cannot find k, as no coordinates are given for points off of the x-axis. Inserting any positive value of k will stretch the graph but not change the zeros and therefore will still work.

Example 2 Using a calculator or computer, sketch graphs of $y = x^4$ and $y = x^4 - 15x^2 - 15x$ for $-4 \le x \le 4$ and for $-20 \le x \le 20$. Set the y range to $-100 \le y \le 100$ for the first domain, and to $-100 \le y \le 200{,}000$ for the second. What do you observe?

Solution From the graphs in Figure 1.86 you can see that close up (for $-4 \le x \le 4$) the graphs look different; from far away, however, they are almost indistinguishable. The reason is that the leading terms (those with the highest power of x) are the same, namely x^4, and for large values of x, the leading term dominates the other terms.

TABLE 1.28 *Numerical values of* $y = x^4$ *and* $y = x^4 - 15x^2 - 15x$

x	$y = x^4$	$y = x^4 - 15x^2 - 15x$	Difference
-20	160,000	154,300	5700
-15	50,625	47,475	3150
15	50,625	47,025	3600
20	160,000	153,700	6300

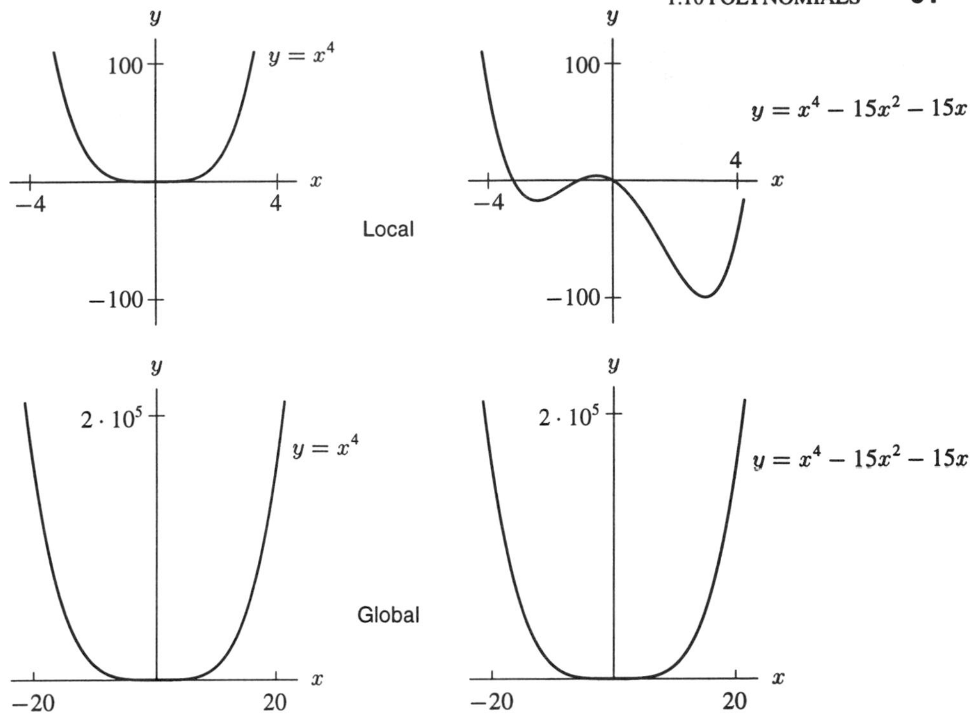

Figure 1.86: Local and global views of $y = x^4$ and $y = x^4 - 15x^2 - 15x$

Looked at numerically in Table 1.28, the differences in the values of the two functions when $x = \pm 20$, although large, are tiny compared with the vertical scale (0 to 200,000) and so cannot be seen on the graph.

Example 3 A company producing compact disk players has determined that the profit function for the company is given by $\pi(p) = -p^2 + 170p - 125$, where p is the price of the product in dollars. Use a graph to determine the price that should be charged to maximize profits.

Solution

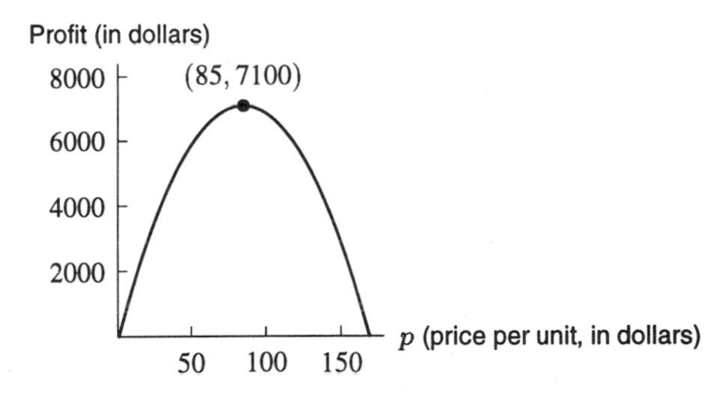

Figure 1.87: Profit function for compact disk players

The graph of this profit function is shown in Figure 1.87. Since the highest point on this graph is the point (85, 7100), we see that the profit function is maximized when the price is \$85. At this price, the profit is \$7100.

Example 4 A company offers dinner cruises on the St. Lawrence River. The company has found that the average number of passengers per night is 75 if the price is \$50 per person. At a price of \$35 per person, the average number of passengers per night is 120.

(a) If we assume that the demand function is linear, write the demand, q, as a function of price, p.

(b) Since revenue is given by $R = pq$, use your answer to part (a) to write the revenue function R as a function of price, p.

(c) Use a graph of the revenue function to determine what price should be charged to obtain the greatest revenue.

Solution (a) The two points given are $(p, q) = (50, 70)$ and $(p, q) = (35, 120)$. The slope of the line is

$$m = \frac{120 - 75}{35 - 50} = \frac{45}{-15} = -3\frac{\text{passengers}}{\text{dollar}}.$$

To find the vertical intercept of the line, we use the slope and one of the points:

$$75 = b + (-3)(50)$$
$$225 = b$$

The demand function is $q = 225 - 3p$.

(b) We see that $R = pq = p(225 - 3p) = 225p - 3p^2$.

(c) The graph of the revenue function is given in Figure 1.88. We see that the maximum revenue is obtained when $p = 37.5$. To maximize revenue, the company should charge \$37.50.

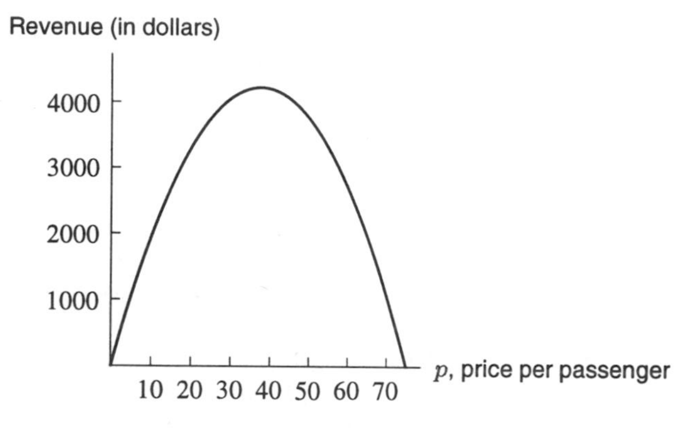

Figure 1.88: Revenue function for dinner cruises

Example 5 If a tomato is thrown vertically into the air at time $t = 0$ with velocity 48 feet per second, its distance, y (in feet), above the surface of the earth at time t (in seconds) is given by the equation

$$y = -16t^2 + 48t.$$

Sketch a graph of position against time, and mark on the graph the coordinates of the points corresponding to instants when the tomato (a) hits the ground and (b) reaches its highest point.

Solution This graph is a parabola opening downward. Its t-intercepts are given by

$$0 = -16t^2 + 48t.$$

Dividing by -16 and factoring out t gives,

$$0 = t(t-3) \quad \text{so } t = 0, 3 \text{ seconds.}$$

The tomato's journey occurs between $t = 0$ and $t = 3$. The point representing the instant when the tomato hits the ground is where $y = 0$ and $t = 3$. Its highest point occurs halfway through the journey, at $t = 1.5$, when $y = -16(1.5)^2 + 48(1.5) = 36$ feet. See Figure 1.89. You should notice that although the graph is arch-shaped, the tomato in fact moves vertically up and down and will land at exactly the spot from which you threw it. The horizontal axis in Figure 1.89 represents time, not horizontal displacement in space.

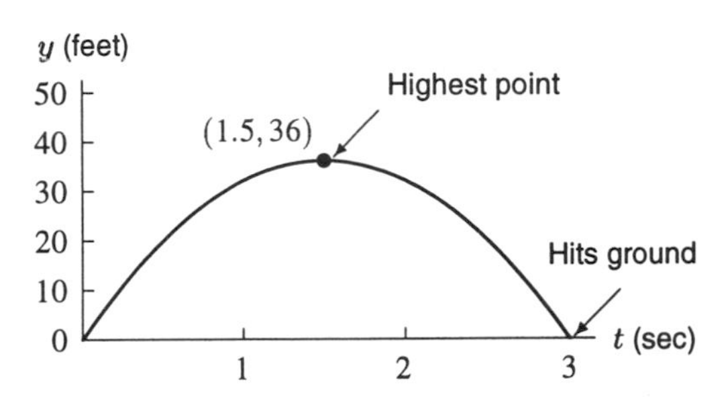

Figure 1.89: The height of the tomato

Problems for Section 1.10

1. Assume that each of the graphs in Figure 1.90 is of a polynomial. For each graph:
   (a) What is the minimum possible degree of the polynomial?
   (b) Is the *leading coefficient* of the polynomial positive or negative? (The leading coefficient is the coefficient of the highest power of x.)

(I) (II) (III) (IV) (V)

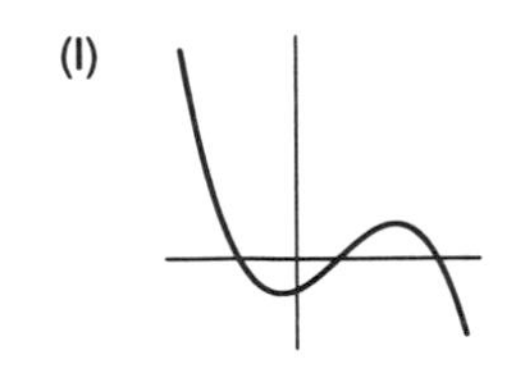

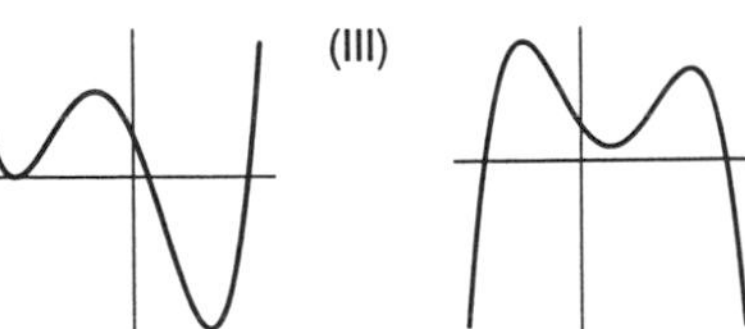

Figure 1.90

For Problems 2–9, sketch graphs of the polynomials:

2. $f(x) = (x-2)(x+3)$
3. $f(x) = -3(x+1)(x-4)$
4. $f(x) = (x+2)(x-1)(x-3)$
5. $f(x) = (x-2)(3-x)$
6. $f(x) = -5(x+2)(x-1)(x-5)$
7. $f(x) = 5(x^2-4)(x^2-25)$
8. $f(x) = -5(x^2-4)(25-x^2)$
9. $f(x) = 5(x-4)^2(x^2-25)$

10. A dog food company finds that its profit function (in dollars) is given by

$$\pi(p) = -p^2 + 130p - 225$$

, where p is the price per pound of the dog food (in cents).

(a) Sketch a graph of the profit function.
(b) What price should be charged to maximize the profit? What is the profit at this price?
(c) For what prices is the profit function positive?

11. A sporting goods wholesaler finds that when the price of one of the products is \$25, the company sells, on the average, 500 units per week. When the price is \$30, the number sold per week decreases to 460 units.

(a) If we assume that the demand function for this product is linear, find the demand, q, as a function of price, p.
(b) Use your answer to part (a) to write revenue as a function of price.
(c) Sketch a graph of the revenue function found in part (b), and find the price that should be charged to maximize revenues. What is the revenue at this price?

12. A private health club has determined that its cost and revenue functions are given by $C = 10{,}000 + 35q$ and $R = pq$ respectively, where q represents the number of annual club members and p represents the price of a one-year membership. The demand function for the health club is known to be $q = 3000 - 20p$.

(a) Use the demand function to write cost and revenue as functions of the price of membership in the club.
(b) Sketch graphs of cost and revenue as a function of price, on the same axes. (In order to get a good view of the graphs, it may help to know that price will not go above \$170 and that the annual costs of running the club can reach \$120,000.)
(c) Explain why the graph of the revenue function has the shape it does.
(d) For what prices does the club make a profit?
(e) Estimate the annual membership fee that will maximize profit for the club. Illustrate this point on your graph.

13. For which positive integers n is $f(x) = x^n$ even? Odd?

14. Which polynomials are even? Odd? Are there polynomials which are neither?

15. If $f(x) = ax^2 + bx + c$, what do you know about the values of a, b, and c if:

(a) $(1, 1)$ is on the graph of $f(x)$?
(b) $(1, 1)$ is the vertex of the graph of $f(x)$? (You may want to use the fact that the equation for the axis of symmetry of the parabola of $y = ax^2 + bx + c$ is $x = -b/2a$.)
(c) The y intercept of the graph is $(0, 6)$?
(d) Find a quadratic function that satisfies all three conditions.

16. A pomegranate is thrown from ground level straight up into the air at time $t = 0$ with velocity 64 feet per second. Its height at time t will be $f(t) = -16t^2 + 64t$. Find the time it hits the ground and the instant that it reaches its highest point. What is the maximum height?

17. The rate, R, at which a population in a confined space increases is proportional to the product of the current population, P, and the difference between the *carrying capacity*, L, and the current population. (The carrying capacity is the maximum population the environment can sustain.)

(a) Write R as a function of P.
(b) Sketch R as a function of P.

For Problems 18–21:

(a) Find a possible formula for the graph.

(b) For each graph, read off approximate intervals over which the function is increasing and over which it is decreasing.

18.

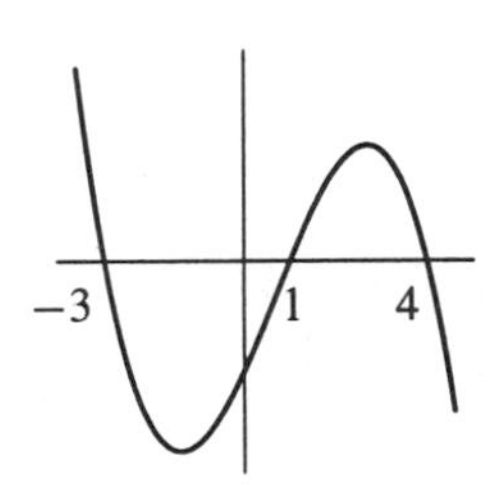

19.

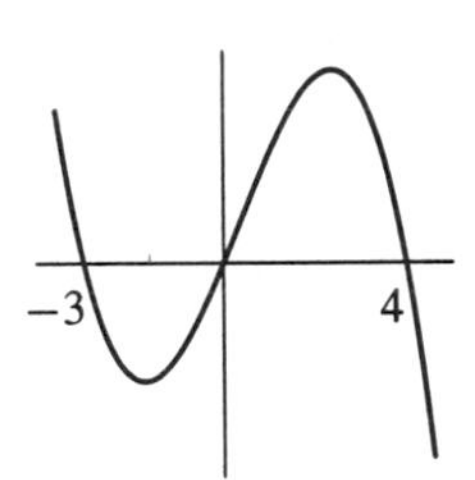

20.

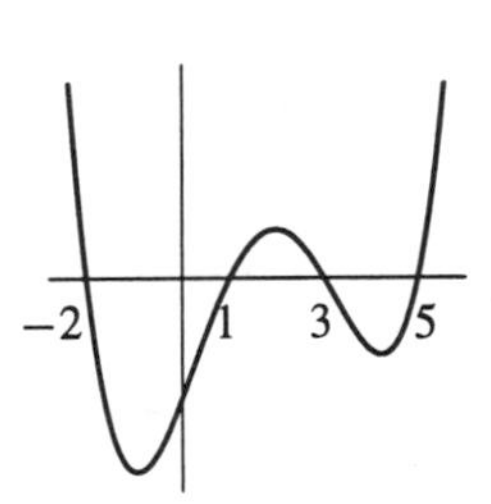

21.

Determine cubic polynomials that represent each of the graphs in Problems 22–23.

22.

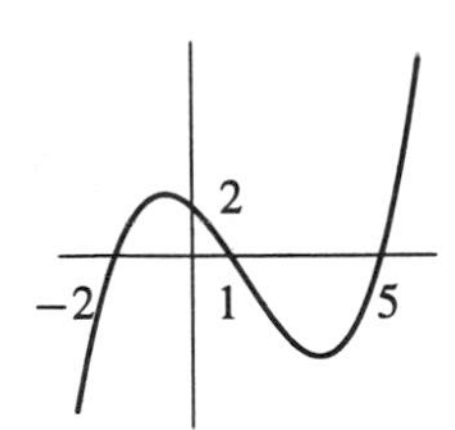

23.

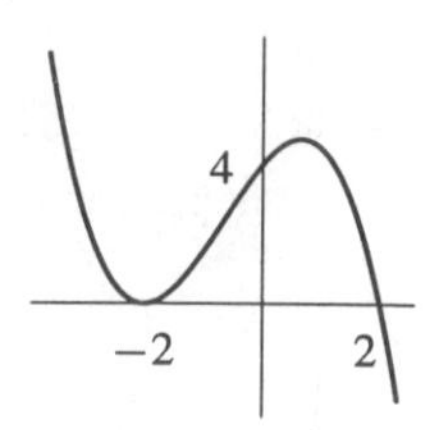

1.11 THE PERIODIC FUNCTIONS

What Are Periodic Functions?

Figure 1.91 shows a graph of the number of new housing construction starts in the United States, where t is time measured in quarter-years. The data is from 1977, 1978, and 1979. Notice that very few new homes begin construction during the first quarter of a year (January, February, and March), whereas a great many new homes are begun in the second quarter (April, May, and June). Presumably, if we continued through the current year this pattern of oscillating up and down would continue.

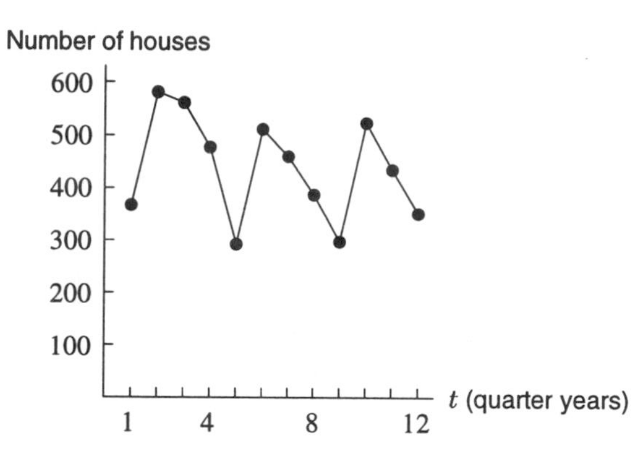

Figure 1.91: Housing construction starts, 1977–1979

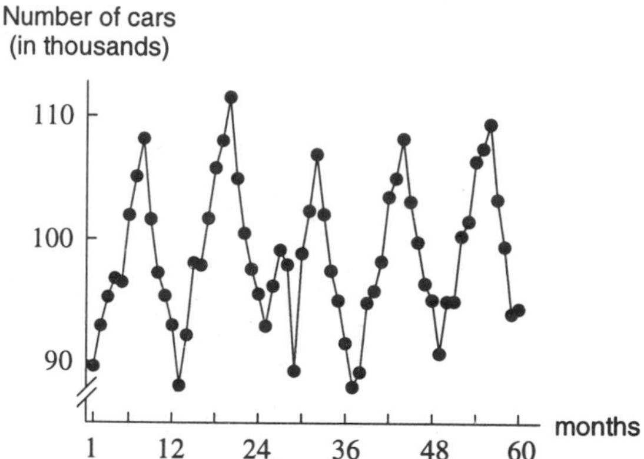

Figure 1.92: Traffic on the Golden Gate Bridge

Many functions have this general shape of oscillating up and down, which looks like a wave. Let's look at another example. Figure 1.92 is a graph of the number of cars (in thousands) traveling across the Golden Gate Bridge per month, from 1976-1980. Notice that traffic is at its minimum in January of each year (except 1978) and reaches its maximum in August of each year. Again, the graph looks like a wave. Functions such as these are called *periodic*, or repeating. Many naturally occurring processes are periodic. The water level in a tidal basin, the blood pressure in a heart, retail sales in the United States, and the position of air molecules transmitting a musical note are all periodic functions of time.

The Sine and Cosine

Many periodic functions can be represented using the functions called sine and cosine. You will find keys for the sine and cosine on your calculator. These two functions can be defined using a unit circle, which is a circle of radius 1 centered at the origin. In Figure 1.93, a length t is measured counterclockwise around the circle from the point $(1, 0)$ to the point P. If the point P has coordinates (x, y), we define

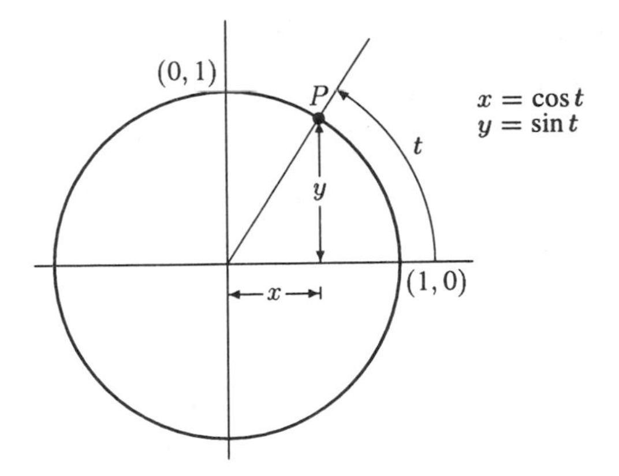

Figure 1.93: The definitions of $\sin t$ and $\cos t$

$$x = \cos t \text{ ("cosine of } t\text{")},$$
$$\text{and} \quad y = \sin t \text{ ("sine of } t\text{")},$$

The units for t are called "radians" and are usually omitted. Make sure your calculator is in "radian" mode when working the sine and cosine problems in this book.

Since the equation of the unit circle is $x^2 + y^2 = 1$, we have the following fundamental identity

$$\cos^2 t + \sin^2 t = 1.$$

As t increases and P moves around the circle, the values of $\sin t$ and $\cos t$ oscillate between 1 and -1, and eventually repeat as P moves through points where it has been before. If t is negative, the length is measured clockwise around the circle.

Graphs of the Sine and Cosine

The graphs of the sine and the cosine are shown in Figure 1.94. Both functions oscillate, and both functions start to repeat at $t = 2\pi$. The maximum and minimum values of the sine and cosine are $+1$ and -1. Notice that the sine is an odd function and the cosine is even. Notice also that the graph of the cosine is identical to the graph of the sine, except that the graph of the cosine function is shifted $\frac{\pi}{2}$ to the left.

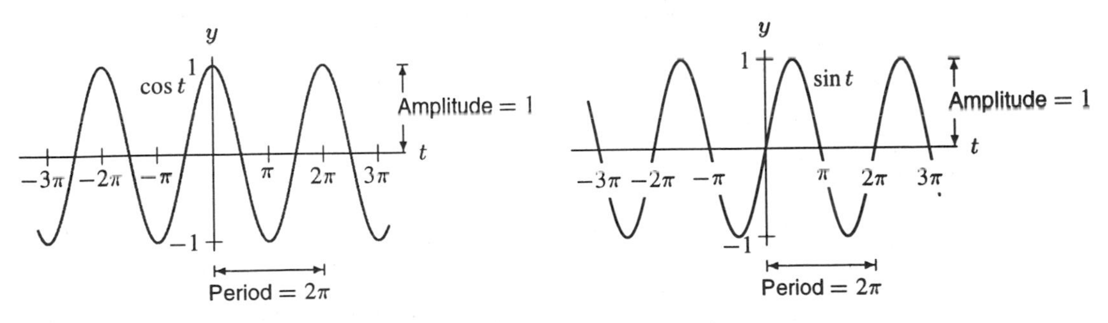

Figure 1.94: Graphs of $\cos t$ and $\sin t$

Amplitude and Period

The graph of $y = \sin t$ moves through a complete cycle between $t = 0$ and $t = 2\pi$; all the rest of the graph is just a repeat of this portion. Such functions that repeat forever are called *periodic*.

> The **amplitude** of an oscillation is half the difference between the maximum and minimum values.
> The **period** of an oscillation is the time needed for the oscillation to execute one complete cycle.

The amplitude of $\cos t$ and $\sin t$ is 1, and the period is 2π.

Example 1 Sketch a graph of $y = 3\sin 2t$ and use the graph to determine the amplitude and period.

Solution

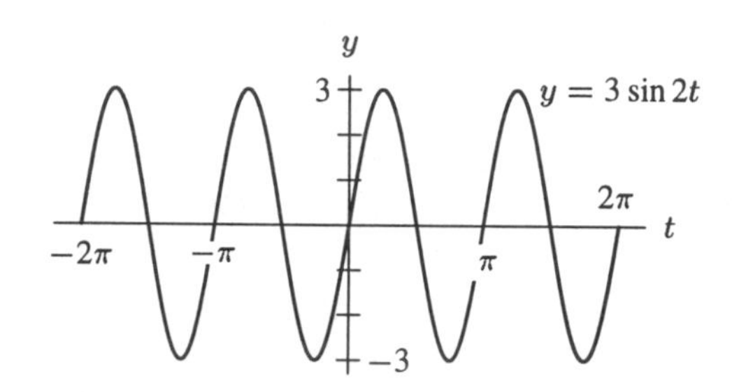

Figure 1.95: What are the amplitude and period?

The waves have a maximum of $+3$ and a minimum of -3, and so the amplitude is 3. The graph completes one complete cycle between $t = 0$ and $t = \pi$, so the period is π.

Example 2 Estimate the amplitude and period of the function for new housing starts shown in Figure 1.91.

Solution Figure 1.91 is not a perfect periodic function since the maximum and minimum are not the same for each wave. Nonetheless, we approximate the minimum to be 300 and the maximum to be about 550. The distance between them is 250, so the amplitude is $\frac{1}{2}(250) = 125$.

The wave completes a cycle between $t = 1$ and $t = 5$, so the period is $t = 4$ quarter-years. The business cycle for new housing construction is 4 quarter-years, or one year.

Example 3 What is the period of the hour hand on a clock?

Solution The hour hand completes one cycle in 12 hours, so the period is 12 hours.

Example 4 Figure 1.96 shows the temperature in an unopened freezer. Estimate the temperature in the freezer at 12:30 and at 2:45.

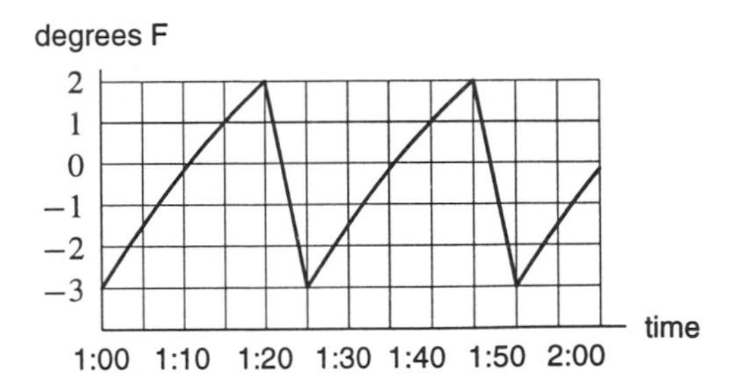

Figure 1.96: What is the temperature at 2:45?

Solution The maximum and minimum values occur every 25 minutes, so the period is 25 minutes. The temperature at 12:30 should be the same as at 12:55 and at 1:20. We see that the temperature at these times is 2°F. Similarly, the temperature at 2:45 should be the same as at 2:20 and 1:55, about −1.5°F.

Understanding the Graphs of A sin Bt and A cos Bt

Example 5

a) Let $y = A \sin t$. Substitute different values for A and sketch the graphs. Explain the effect of A on the graph.

b) Let $y = \sin Bt$. Substitute different values for B and sketch the graphs. Explain the effect of B on the graph.

Solution

a) Graphs of $y = A \sin t$ for different values of A are shown in Figure 1.97. We see that A is the amplitude of the function.

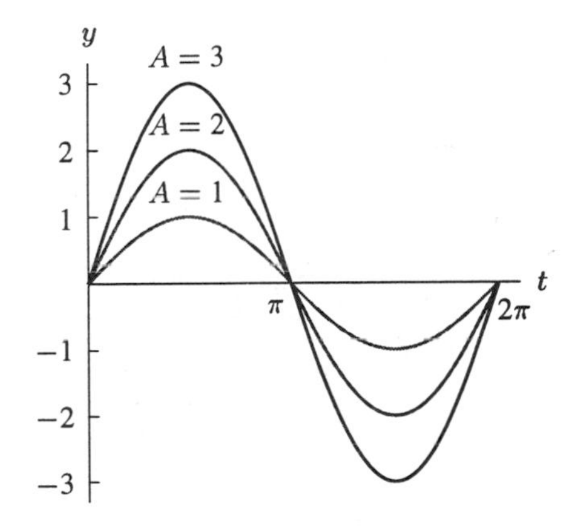

Figure 1.97: Graphs of $y = A \sin t$

b) The graphs of $y = \sin Bt$ for $B = \frac{1}{2}$, $B = 1$, and $B = 2$ are shown in Figure 1.98. We see that B influences the period of the graph. When $B = 1$, the period is 2π; when $B = 2$, the period is π; and when $B = \frac{1}{2}$, the period is 4π.

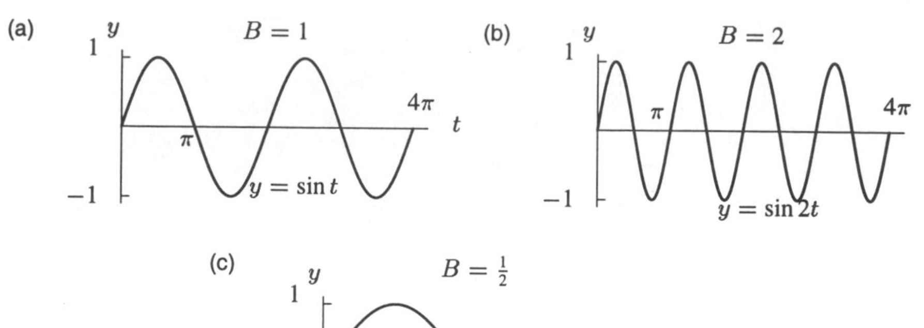

Figure 1.98: Graphs of $y = \sin Bt$

> To describe arbitrary amplitudes and periods, we use functions of the form
> $$f(t) = A \sin Bt \qquad \text{and} \qquad g(t) = A \cos Bt,$$
> where A is the amplitude and $2\pi/B$ is the period.

Example 6 Find and show on a graph the amplitude and period of the functions

(a) $y = 5 \sin 2t$ (b) $y = -5 \sin\left(\dfrac{t}{2}\right)$ (c) $y = 1 + 2 \sin t$

Solution

(a) From Figure 1.99, you can see that the amplitude of $y = 5 \sin 2t$ is 5 because the factor of 5 stretches the oscillations up to 5 and down to -5. The period of $y = \sin 2t$ is π, because when t changes from 0 to π, the quantity $2t$ changes from 0 to 2π, so the sine function will have gone through one complete oscillation.

(b) Figure 1.100 shows that the amplitude of $y = -5 \sin(t/2)$ is again 5, because the negative sign reflects the oscillations in the t-axis but does not change how far up or down they go. The period of $y = -5 \sin(t/2)$ is 4π because when t changes from 0 to 4π, the quantity $t/2$ changes from 0 to 2π, so the sine function goes through one complete oscillation.

(c) The 1 shifts the graph $y = 2 \sin t$ up by 1. Since $y = 2 \sin t$ has an amplitude of 2 and a period of 2π, the graph of $y = 1 + 2 \sin t$ goes up to 3 and down to -1, and has a period of 2π. (See Figure 1.101.) Thus, $y = 1 + 2 \sin t$ has amplitude 2 and period 2π.

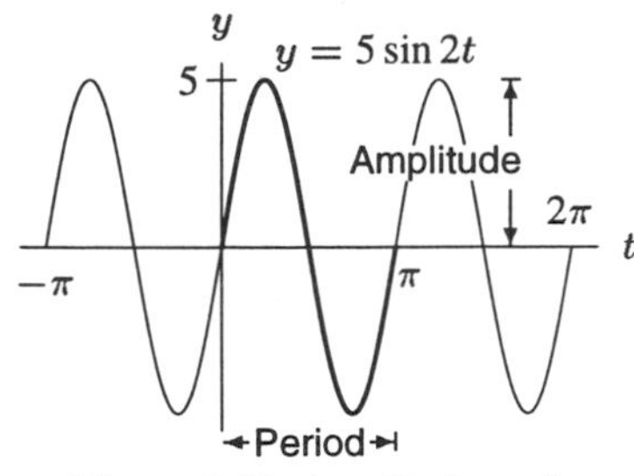

Figure 1.99: Amplitude = 5, period = π

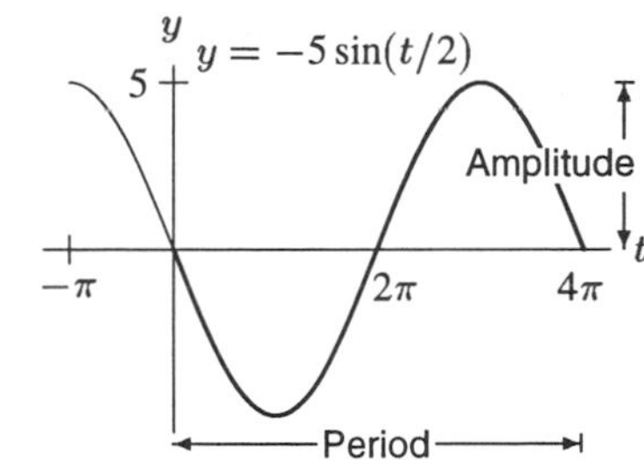

Figure 1.100: Amplitude = 5, period = 4π

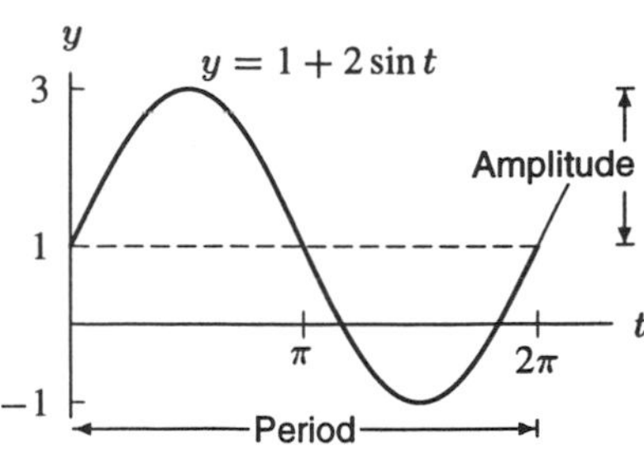

Figure 1.101: Amplitude = 2, period = 2π

Example 7 Find formulas for the functions describing the oscillations in Figures 1.102–1.103.

(a)

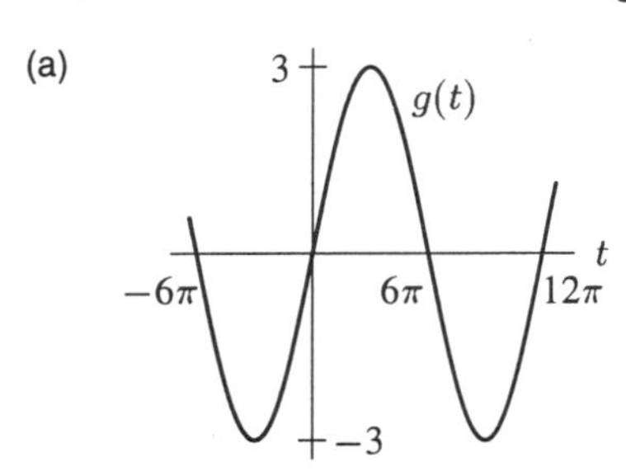

Figure 1.102

(b)

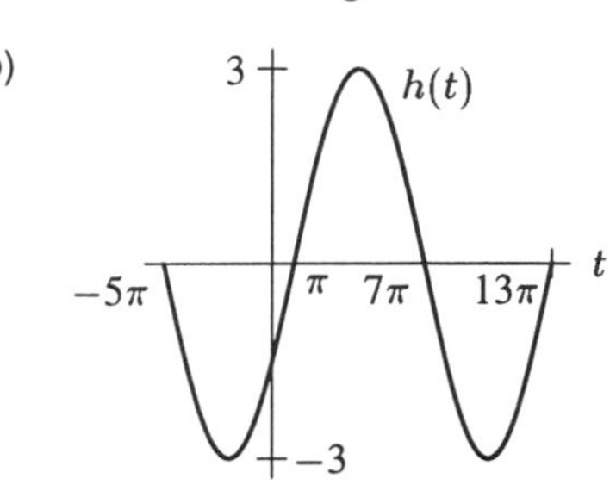

Figure 1.103

Solution (a) The function in Figure 1.102 looks like a sine function of amplitude 3, so $g(t) = A\sin Bt$ with $A = 3$. Since the function executes one full oscillation between $t = 0$ and $t = 12\pi$, when t changes by 12π, the quantity Bt changes by 2π. This means $B \cdot 12\pi = 2\pi$, so $B = 1/6$. Therefore, $g(t) = 3\sin(t/6)$ has the graph shown.

(b) The function in Figure 1.103 looks like the function $g(t)$ in Figure 1.102, but shifted a distance of π to the right. Since $g(t) = 3\sin(t/6)$, we replace t by $(t - \pi)$ to obtain $h(t) = 3\sin[(t - \pi)/6]$.

There are many possible equations for these graphs; you may be able to find others.

Example 8 On February 10, 1990, high tide in Boston was at midnight. The water level at high tide was 9.9 feet; later, at low tide, it was 0.1 feet. Assuming the next high tide is exactly 12 hours later and that the height of the water is given by a sine or cosine curve, find a formula for water level in Boston as a function of time.

Solution Suppose the height of the water level is y feet, and let t be the time measured in hours from midnight. Then the oscillations are to have amplitude 4.9 feet ($= (9.9 - 0.1)/2$) and period 12, so $12B = 2\pi$ and $B = \pi/6$. Since the water is highest at midnight, when $t = 0$, the oscillations are best represented by a cosine, because the cosine is at its maximum at the beginning of the cycle. (See Figure 1.104.) Thus we can say

$$\text{Height above average} = 4.9\cos\left(\frac{\pi}{6}t\right).$$

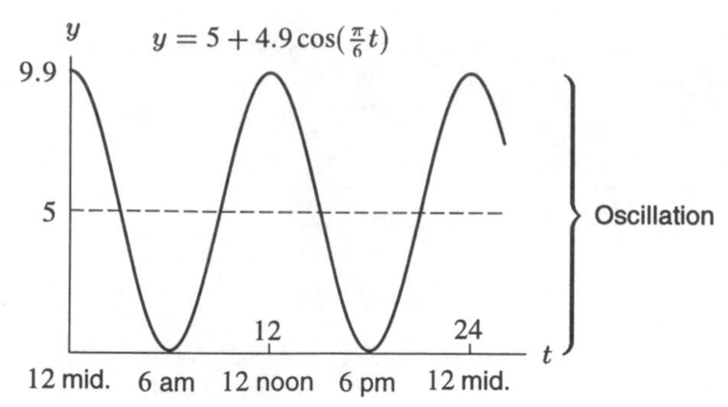

Figure 1.104: Function approximating the tide in Boston on February 10, 1990

Since the average depth of the water was 5 feet ($= (9.9 + 0.1)/2$), we want a cosine curve shifted up by 5. We get this by adding 5:

$$y = 5 + 4.9\cos\left(\frac{\pi}{6}t\right).$$

Example 9 Of course, there's something wrong with the assumption in Example 8 that the next high tide will be at noon. If so, the high tide would always be at noon or midnight, instead of progressing slowly through the day, as in fact it does. The interval between successive high tides actually averages about 12 hours 24 minutes. Using this, give a more accurate formula for the height of the water as a function of time.

Solution The period is 12 hours 24 minutes = 12.4 hours, so $B = 2\pi/12.4$, giving

$$y = 5 + 4.9\cos\left(\frac{2\pi}{12.4}t\right) = 5 + 4.9\cos(0.507t).$$

Example 10 Use the information from Example 9 to write a formula for the water level in Boston on a day when the high tide is at 2 pm.

Solution When the high tide is at midnight

$$y = 5 + 4.9\cos(0.507t).$$

Since 2 pm is 14 hours after midnight, we replace t by $(t - 14)$. Thus, on a day when the high tide is at 2 pm,

$$y = 5 + 4.9\cos[0.507(t - 14)].$$

Problems for Section 1.11

1. Use the solution to Example 8 on page 101 to estimate the water level in Boston Harbor at 3:00 am, 4:00 am, and 5:00 pm on February 10, 1990.
2. What is the period of the earth's revolution around the sun?
3. What is the approximate period of the moon's revolution around the earth?
4. What is the period of the motion of the minute hand of a clock?
5. An LP record rotates $33\frac{1}{3}$ times in a minute. What is the period of its motion?
6. (a) Match the functions f, g, h, k, whose values are given in the table, with the functions whose formulas are:
   (i) $w = 1.5+\sin t$ (ii) $w = 0.5+\sin t$ (iii) $w = -0.5+\sin t$ (iv) $w = -1.5+\sin t$.

t	$w = f(t)$	t	$w = g(t)$	t	$w = h(t)$	t	$w = k(t)$
6.0	−0.78	3.0	1.64	5.0	−2.46	3.0	0.64
6.5	−0.28	3.5	1.15	5.1	−2.43	3.5	0.15
7.0	0.16	4.0	0.74	5.2	−2.38	4.0	−0.26
7.5	0.44	4.5	0.52	5.3	−2.33	4.5	−0.48
8.0	0.49	5.0	0.54	5.4	−2.27	5.0	−0.46

   (b) Based on the table, what is the relationship between the values of $g(t)$ and $k(t)$? Explain this relationship using the formulas you chose for g and k.
   (c) Using the formulas you chose for g and h, explain why all the values of g are positive, whereas all the values of h are negative.

For Problems 7–12, sketch graphs of the functions. What are their amplitudes and periods?

7. $y = 3\sin x$
8. $y = 3\sin 2x$
9. $y = -3\sin 2\theta$
10. $y = 4\cos 2x$
11. $y = 4\cos(\frac{1}{2}t)$
12. $y = 5 - \sin 2t$

For Problems 13–16, find a possible formula for each graph.

13.
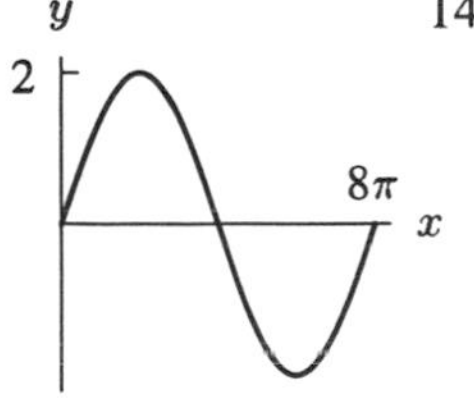

14.
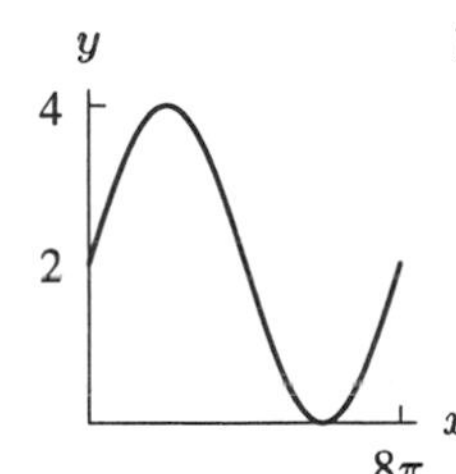

15.
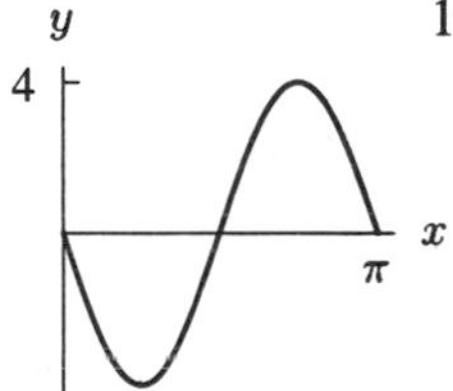

16.
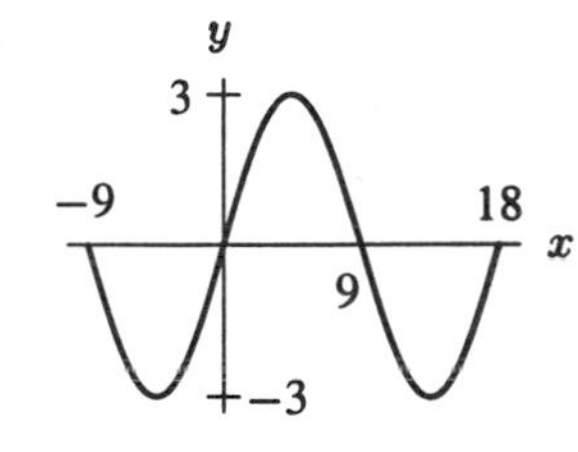

17. A population of animals varies periodically between a low of 700 on January 1 and a high of 900 on July 1.
   (a) Graph the population against time.
   (b) Using the sine function, find a formula for the population as a function of time, t, measured in months since the start of the year.

18. The Bay of Fundy in Canada is reputed to have the largest tides in the world, with the difference between low and high water level being as much as 15 meters (nearly 50 feet). Suppose at a particular point in the Bay of Fundy, the depth of the water, y meters, as a function of time, t, in hours since midnight on January 1, 1994, is given by

$$y = y_0 + A\cos[B(t - t_0)].$$

(a) What is the physical meaning of y_0?
(b) What is the value of A?
(c) What is the value of B? Assume the time between successive high tides is $12\frac{1}{2}$ hours.
(d) What is the physical meaning of t_0?

19. The visitors' guide to St. Petersburg, Florida, contains the chart shown in Figure 1.105 to advertise their good weather. Fit a trigonometric function approximately to the data. The independent variable should be time in months. In order to do this, you will need to find the amplitude and period of the data, and when the maximum occurs. (There are many possible answers to this problem, depending on how you read the graph.)

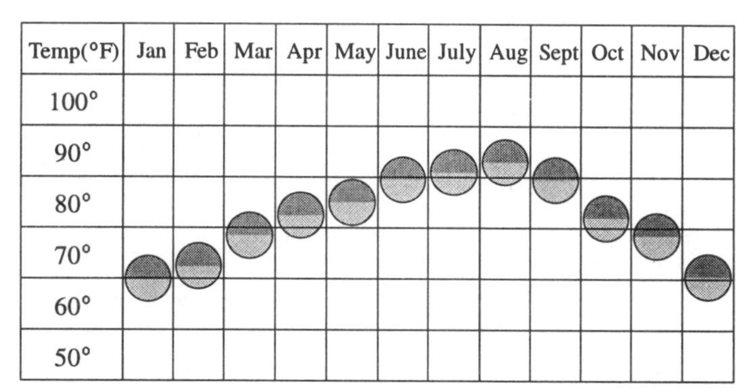

Figure 1.105: "St. Petersburg...where we're famous for our wonderful weather and year-round sunshine." (Reprinted with permission)

20. You are told that two trigonometric functions each have period π and that their graphs intersect at $x = 3.64$, but you are told nothing else about the functions.

(a) Can you say whether these two graphs intersect at any smaller positive x value? If so, what is it?
(b) Find an x value greater than 3.64 at which these graphs intersect.
(c) Find a negative x value at which these graphs intersect.

21. (a) Use a graphing calculator or computer to find the period of $2\sin 3t + 3\cos t$.
(b) What is the period of $\sin 3t$? Of $\cos t$?
(c) Use your answers to part (b) to explain your answer to part (a).

1.12 FITTING FORMULAS TO DATA

We have seen many formulas in this chapter, and you may wonder where these formulas come from. For example, how does a company determine the demand curve for a product? Some of the formulas we have seen are exact. If you deposit \$1000 in the bank and it earns 5% interest compounded continuously, then the amount in the bank at time t is exactly given by $P(t) = 1000e^{0.05t}$. This is an example where we know the exact answer. Most of the time, however, the formulas we use are approximations, often constructed from tables of data.

A company wants to understand the relationship between the amount spent on advertising, a, in thousands of dollars, and the total sales, S, in thousands of dollars. The company has been experimenting, and has collected data that shows, for example, that when they spend \$3000 on advertising, total sales are \$100,000. All of the data is shown in Table 1.29.

TABLE 1.29 *Find a formula to fit this data*

a(\$1000s)	3	4	5	6
S(\$1000s)	100	120	140	160

You can see that the data in Table 1.29 is linear and so we got lucky this time: it is easy to find a formula to fit the data. The slope of the line is 20 and we can determine that the y-intercept is 40, so the line is

$$S = 40 + 20a.$$

Now suppose that the data collected by the company is shown in Table 1.30. We are not so lucky in this case: the data is not linear. In general, it is a very difficult problem to find a formula to fit the data exactly. We must be satisfied with a formula that is close to the data.

TABLE 1.30 *Find a formula to approximate this data*

a(\$1000s)	3	4	5	6
S(\$1000s)	105	117	141	152

Fitting a Linear Function To Data

The data in Table 1.30 has been plotted in Figure 1.106. The relationship is not linear, since not all the data points fall on one line, but we see that it is nearly linear. It seems reasonable to approximate this relationship with a line. One line that fits the data reasonably well is

$$S = 40 + 20a.$$

Figure 1.107 shows this line and the data.

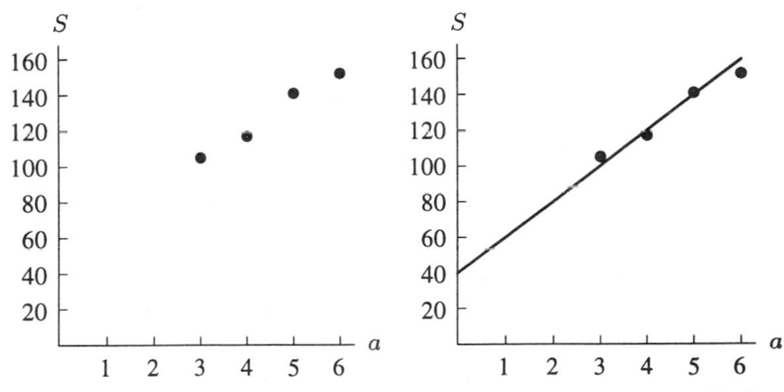

Figure 1.106: The sales data from Table 1.30

Figure 1.107: How well does this line fit the data?

The Regression Line

Is there a line that fits the data better than the one in Figure 1.107? If so, how do we find it? The process of fitting a line to a set of data is called *linear regression*, and the line of best fit is called the *regression line.* There is a formula to find this line, and it will be discussed in Chapter 9. Fortunately, many calculators and computer programs will give you the regression line directly if you enter the data points.

For the data in Table 1.30, the regression line is

$$S = 54.5 + 16.5a.$$

This line fits the data better than any other line. See Figure 1.108. If you have technology available that will find a regression line, you should learn how to use it.

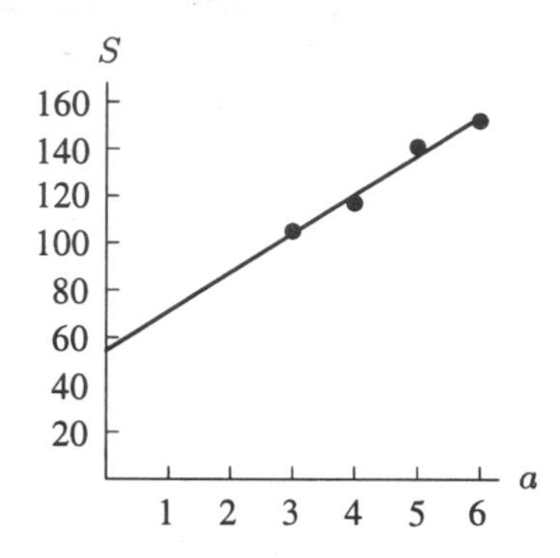

Figure 1.108: The regression line

Using the Regression Line to Make Predictions

Now that we have a formula for sales, we can use it to make predictions. Suppose we want to predict total sales if we spend \$3500 on advertising. We substitute $a = 3.5$ into the regression line:

$$S = 54.5 + 16.5(3.5) = 112.25.$$

The regression line predicts sales of \$112,250. To see that this is a reasonable estimate, we can compare it to the entries in Table 1.30. When $a = 3$, we have $S = 105$, and when $a = 4$, we have $S = 117$. Our predicted sales of $S = 112.25$ when $a = 3.5$ makes sense because it falls between 105 and 117. Figure 1.109 shows where our predicted point is in relation to the data points. Of course, if we spent \$3500 on advertising, sales would probably not be exactly \$112,250. The regression equation allows us to make predictions, but does not provide exact results.

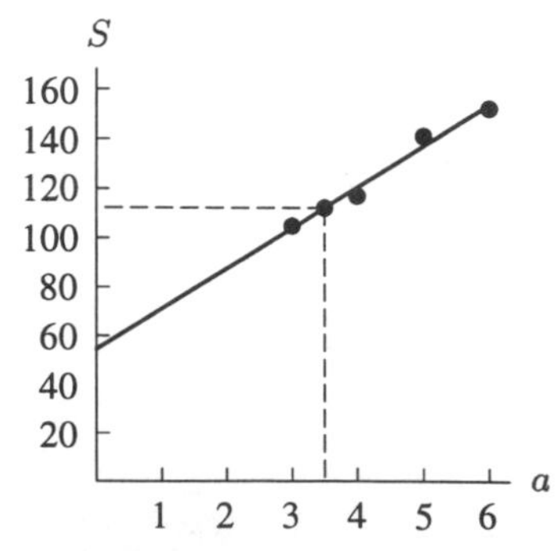

Figure 1.109: Predicting sales when $a = 3.5$

Example 1 Predict total sales for $a = 4.8$ and $a = 10$.

Solution When $a = 4.8$, we have

$$S = 54.5 + 16.5(4.8) = 133.7.$$

Sales are predicted to be $133,700. When $a = 10$, we have

$$S = 54.5 + 16.5(10) = 219.5.$$

Sales are predicted to be $219,500.

Consider the two predictions made in Example 1, at $a = 4.8$ and $a = 10$. We should have much more confidence in the accuracy of the prediction when $a = 4.8$, because we are making a prediction on an interval we already know something about. The prediction for $a = 10$ is very questionable, since we are using the regression line beyond the limits of our knowledge from the data values in Table 1.30.

When we make a prediction between two data points, our estimate is called an *interpolation*. If instead we estimate the value of S for a value of a beyond those given in the data, our estimate is called an *extrapolation*. In general interpolation is safer than extrapolation.

Interpreting the Slope

Slope is the change in the dependent variable divided by the change in the independent variable. In the regression line above, the slope is 16.5, and this tells us the change in sales divided by the change in advertising expenditures. Since 16.5 is the same as $\frac{16.5}{1}$, the slope tells us that S will go up by about 16.5 whenever a goes up by 1. If we increase advertising expenses by $1000, we expect sales to increase by about $16,500. In general, the slope tells you the expected change in the dependent variable given a unit change in the independent variable.

Example 2 A company has collected some data on the cost of producing different quantities. The data is given in Table 1.31. Use technology to find the regression line. Plot the points and the line to check your answer. Interpret the slope of the line. (If you do not have technology to find a regression line, plot the points and visually estimate the regression line.)

TABLE 1.31 *Cost to produce different quantities*

q(quantity)	25	50	75	100	125
C(cost)	500	625	689	742	893

Solution Using technology, we see that the regression line is

$$C = 418.9 + 3.612q.$$

The regression line is shown with the data in Figure 1.110, and it appears to fit the data well. The slope of the line is 3.612. We expect cost to increase about 3.612 dollars for every additional unit produced.

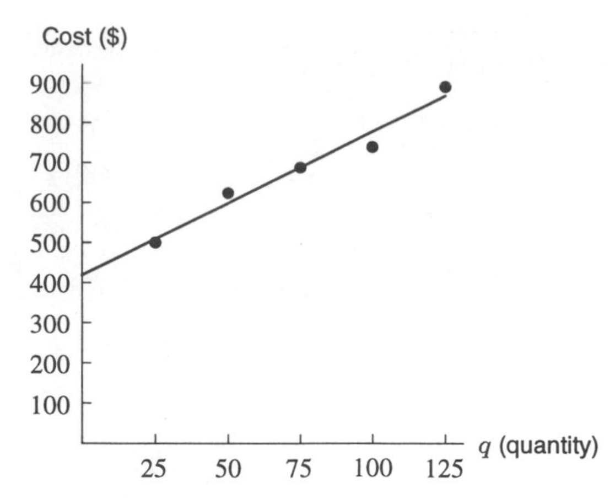

Figure 1.110: The regression line fits the data well

Regression When the Relationship Is Not Linear

Table 1.32[15] shows the number of cars imported into the United States from Japan between 1964 and 1971. These points are plotted in Figure 1.111.

TABLE 1.32 *New passenger cars imported into the U.S., from Japan.*

Year since 1964	0	1	2	3	4	5	6	7
Cars	16,023	23,538	56,050	70,304	169,849	260,005	381,338	703,672

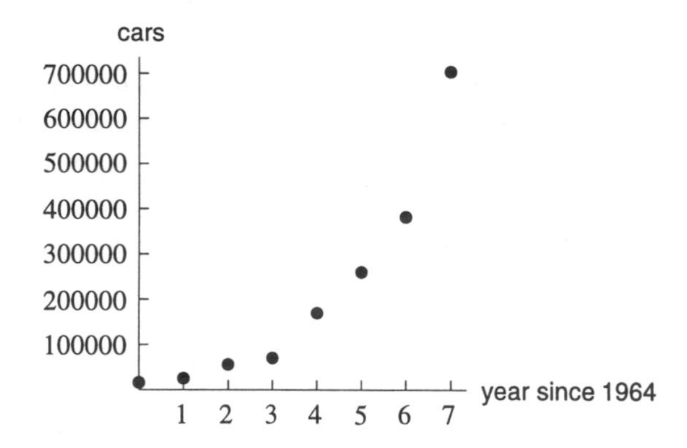

Figure 1.111: Is this data linear?

Does the data in Figure 1.111 look linear? Not really. In fact, it may remind you of an exponential function. With this data, it makes more sense to fit an exponential function than a linear function. Finding the exponential function of best fit is called *exponential regression*. Many calculators and

[15]*The World Almanac, 1995*

computer software packages will give you the exponential regression function when you enter the data.

Using a graphing instrument, we learn that the best exponential function to fit the data in Table 1.32 is

$$C = 16299.733 \cdot (1.7184)^t,$$

where C is the number of imported Japanese cars and t is years since 1964. The graph of the data and this exponential function is given in Figure 1.112. It appears to fit the data very well.

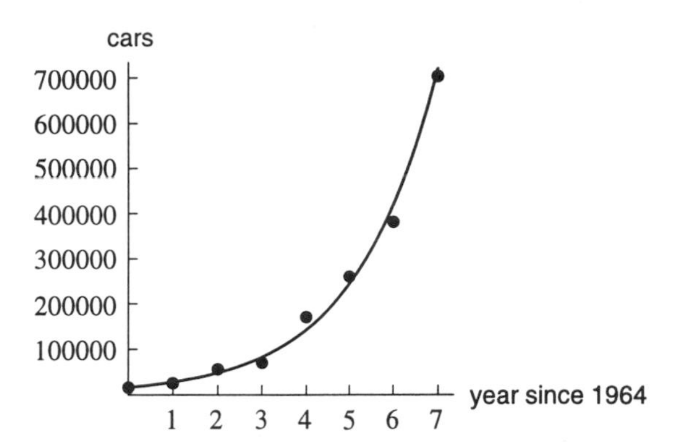

Figure 1.112: The exponential regression function

Since the base of this exponential function is 1.7184, we see that during this time period, sales of Japanese cars in the U.S. were increasing at the rate of about 72% per year.

Calculators and computers will do linear regression, exponential regression, logarithmic regression, quadratic regression, and more. When you are trying to find the best formula for a set of data, the first step is to look at a graph of the data and try to identify the appropriate type of formula to use.

Problems for Section 1.12

1. The Gross National Product of the United States for several different years is given in Table 1.33, in constant 1982 dollars.

TABLE 1.33

Year	1960	1970	1980	1989
GNP (in billions)	1665	2416	3187	4118

(a) Plot this data, with GNP on the vertical axis. Does a line seem to fit the data reasonably well?
(b) Find the regression line for this data, and graph it with the data.
(c) Use the regression line to estimate the GNP in 1985. In 2000. Which estimate should you have more confidence in, and why?

2. The acidity of a solution is measured by its pH, with lower pH values indicating that the solution is more acid. A study of acid rain was undertaken in Colorado between 1975 and 1978, in which the acidity of precipitation was measured for 150 consecutive weeks. The data followed a generally linear pattern, and the regression line was determined to be

$$P = 5.43 - 0.0053t,$$

where P is the pH of the rain, and t is the number of weeks into the study.[16]

   (a) Is the pH level increasing or decreasing over the period of the study? What does this tell you about the level of acidity in the precipitation?
   (b) According to the line, what was the pH at the beginning of the study? At the end of the study ($t = 150$)?
   (c) What is the slope of the regression line? Clearly explain what this slope is telling you about the change in pH over time.

3. In a study of 21 of the best American female runners, researchers measured the average stride rate, S, (number of steps per second) at different speeds, v, measured in feet per second. The data is given in Table 1.34.[17]

TABLE 1.34

Speed (ft/sec)	15.86	16.88	17.50	18.62	19.97	21.06	22.11
Stride rate	3.05	3.12	3.17	3.25	3.36	3.46	3.55

   (a) Use a calculator or computer to find the regression line for this data, using stride rate as the dependent variable.
   (b) Plot the regression line and the data on the same axes. Does the line appear to fit the data well?
   (c) Use the regression line to predict the stride rate when the speed is 18 ft/sec. Use the regression line to predict the stride rate when the speed is 10 ft/sec. Which prediction should you have more confidence in? Why?

4. The level of carbon dioxide in the atmosphere has been increasing, due largely to increased energy consumption and to deforestation of the earth. The CO_2 concentration in parts per million (ppm) is given for four different years in Table 1.35.[18] The measurements were made at the Mauna Loa Observatory in Hawaii.

TABLE 1.35

Year	1965	1970	1980	1988
CO_2 concentration (in ppm)	319.9	325.3	338.5	351.3

   (a) Plot this data. Does it look approximately linear?

[16]Lewis, William M., and Grant, Michael C., "Acid precipitation in the western United States," *Science*, 207 (1980), pp.176-177.

[17]R.C. Nelson, C.M. Brooks, and N.L. Pike, "Biomechanical comparison of male and female distance runners," *The Marathon: Physiological, Medical, Epidemiological, and Psychological Studies*, P. Milvy (ed.), New York Academy of Sciences, 1977, pp.793-807.

[18]Christopher Schaufele and Nancy Zumoff, *Earth Algebra*, Harper-Collins, 1993

(b) Use a calculator or computer to find the regression line for this data. (If you do not have technology available, use your graph of the data to visually estimate the best line.)
(c) What CO_2 concentration does your line predict for the four years 1965, 1970, 1980, and 1988? Compare these predictions to the actual values shown in Table 1.35. Is the line relatively accurate for these four years?
(d) What carbon dioxide concentration does the line predict for the year 1995?

5. In Problem 4, we modeled the increase in carbon dioxide concentration as a linear function of time. However, if we include data for carbon dioxide concentration from as far back as 1900, it appears that the data is more exponential than linear. (It looked linear in Problem 4 because we were only looking at a small piece of the graph.) If we use an exponential regression function to model CO_2 concentration since 1900, we obtain the function

$$C = 272.27(1.0026)^t,$$

where C is the CO_2 concentration in ppm, and t is in years since 1900.
(a) What is the annual growth rate during this period? Interpret this rate in terms of the increase in the CO_2 concentration.
(b) What CO_2 concentration is given by the model for 1900? For 1980? Compare the estimate for 1980 to the actual value given in Table 1.35

6. Climbing health care costs continue to be a concern. Table 1.36 shows the average yearly per-capita (i.e. per person) health care expenditures for various years. Does a linear or an exponential model appear to fit this model best? Find the linear or exponential regression function (whichever you decide is best) for this data. Graph the function with the data and assess how well it fits the data.

TABLE 1.36 *Health care costs*

Year	Per capita expenditure ($)
1970	349
1975	591
1980	1055
1985	1596
1987	1987[21]

7. The number, N, of passenger cars in the United States (in millions) is shown in Table 1.37[22], where t is in years since 1940.

TABLE 1.37

t (years since 1940)	0	10	20	30	40	46
# of cars (in millions)	27.5	40.3	61.7	89.3	121.6	135.4

(a) Plot the data, with number of passenger cars as the dependent variable.

[21] Yes, the last number is correct. The expenditures in 1987 were $1987 per person.
[22] *Statistical Abstracts of the United States*

(b) Does a linear or exponential model appear to fit the data better?
(c) We use a linear model first: Find the regression line for this data. Graph it with the data. Use the regression line to predict the number of passenger cars in the year 2000 ($t = 60$).
(d) Interpret the slope of the regression line found in part (c) in terms of number of passenger cars.
(e) Now we try an exponential model: Find the exponential regression function for this data. Graph it with the data. Use the exponential function to predict the number of passenger cars in the year 2000 ($t = 60$). Compare your prediction with the prediction obtained from the linear model.
(f) What annual rate of growth in number of US passenger cars does your exponential model show?

8. Table 1.38 gives the population of the world in billions, in years since 1950.

TABLE 1.38

Year (since 1950)	0	10	20	30	40	44
World Population (in billions)	2.6	3.1	3.7	4.5	5.4	5.6

(a) Plot this data. Does a linear or exponential model seem to fit it best?
(b) Use a calculator or computer to find the exponential regression function.
(c) What annual growth rate does the exponential function show?
(d) Use the exponential regression function to predict the population of the world in the year 2000. In the year 2050. Comment on the relative confidence you should have in these two estimates.

9. All field goal attempts and successes were analyzed in the National Football League and American Football League in 1969, and the percentage of successes is shown in Table 1.39.

TABLE 1.39

X = yards from the goal line:	14.5	24.5	34.5	44.5	52.0
Y = percentage of tries that were successful:	0.90	0.75	0.54	0.29	0.15

(a) Graph the data, treating the field goal success rate as the dependent variable, and discuss whether a linear or an exponential model provides the best fit.
(b) Find the best linear regression function, and graph it with the data. Interpret the slope of the regression line in terms of success in kicking field goals and distance from the goal line.
(c) Find the best exponential regression function, and graph it with the data. What success rate does this function predict from a distance of 50 yards?
(d) Now that you have looked at the graphs, which model seems to fit the data best?

10. In this chapter, we have developed "a library of functions". Any of these functions can be used to model data. We have discussed linear and exponential regression in this chapter, but it is also possible to fit a logarithmic function or a power function such as x^2 to data. Most graphing calculators will do regression using any of these families of functions. The job of the person analyzing the data, therefore, is to determine which type of function is best for a given data set. Graphs of several different data sets are shown in Figure 1.113. In each case, indicate whether the best function for the data appears to be a *linear function*, an *exponential function*, a *logarithmic function*, or a *quadratic function*.

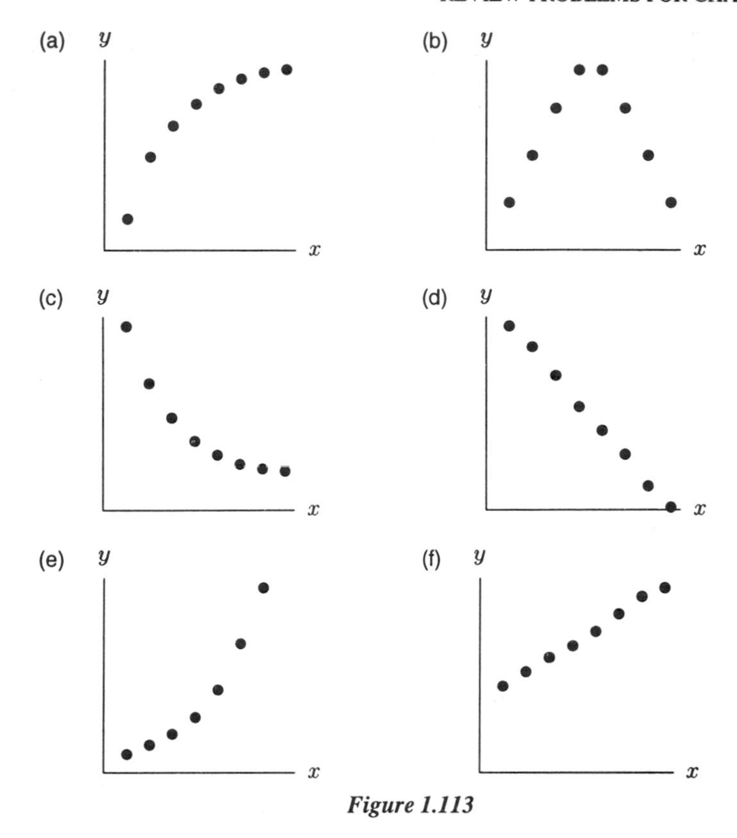

Figure 1.113

REVIEW PROBLEMS FOR CHAPTER ONE

1. A car starts out slowly and then goes faster and faster until a tire blows out. Sketch a possible graph of the distance the car has traveled as a function of time.
2. Having left home in a hurry, I'd only gone a short distance when I realized I hadn't turned off the washing machine, and so I went back to do so. I then set out again immediately. Sketch my distance from home as a function of time.
3. The graph in Figure 1.114 shows how the usage of household gas, e.g., for cooking, varies with the time of day in Ankara, the capital city of Turkey. Give a possible explanation for the shape of the graph.

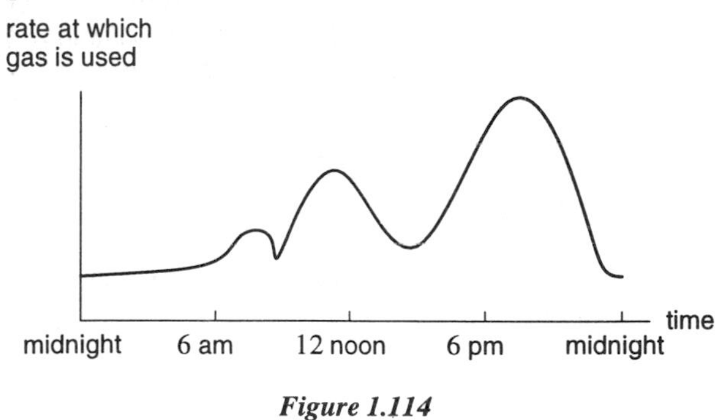

Figure 1.114

4. Sketch a possible graph for a function that is decreasing everywhere, concave up for negative x and concave down for positive x.

5. Consider the graph in Figure 1.115.

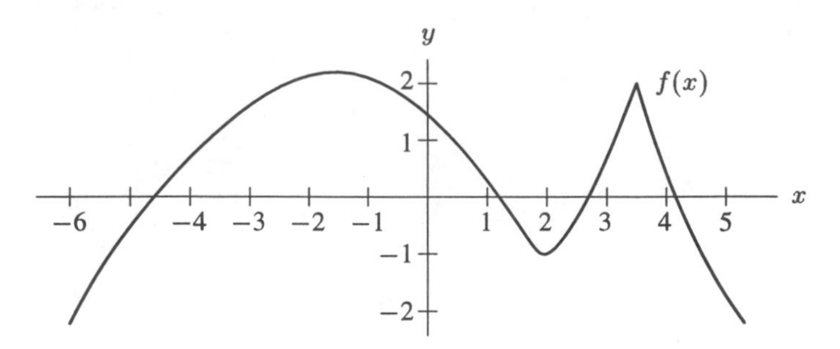

Figure 1.115

(a) How many zeros does this function have? Approximately where are they?
(b) Give approximate values for $f(2)$ and $f(4)$.
(c) Is the function increasing or decreasing near $x = -1$? How about near $x = 3$?
(d) Is the graph concave up or concave down near $x = 2$? How about near $x = -4$?
(e) List all intervals (approximately) on which the function is increasing.

6. Table 1.40 gives the average temperature in Wallingford, Connecticut, for the first 10 days in March 1990.

(a) Over which intervals was the average temperature increasing? Decreasing?
(b) Find a pair of consecutive intervals over which the average temperature was increasing at a decreasing rate. Find another pair of consecutive intervals over which the average temperature was increasing at an increasing rate.

TABLE 1.40

Date in March	1	2	3	4	5	6	7	8	9	10
Average temperature (°F)	42°	42°	34°	25°	22°	34°	38°	40°	49°	49°

7. For tax purposes, you may have to report the value of your assets, such as cars or refrigerators. The value you report depreciates, or drops, with time. The idea is that a car you originally paid $10,000 for may be worth only $5000 a few years later. The simplest way to calculate the value of your asset is using "straight-line depreciation," which assumes that the value is a linear function of time. If a $950 refrigerator depreciates completely in seven years, find a formula for its value as a function of time.

8. An airplane uses a fixed amount of fuel for takeoff, a (different) fixed amount for landing, and a fixed amount per mile when it is in the air. How does the total quantity of fuel required depend on the length of the trip? Write a formula for the function involved. Explain the meaning of the constants in your formula.

9. Sketch reasonable graphs for the following. Pay particular attention to the concavity of the graphs, and explain your reasoning.

(a) The total revenue generated by a car rental business, plotted against the amount spent on advertising.
(b) The temperature of a cup of hot coffee standing in a room, plotted as a function of time.

10. If $f(x) = x^2 + 1$, find:
(a) $f(t+1)$ (b) $f(t^2+1)$ (c) $f(2)$ (d) $2f(t)$ (e) $[f(t)]^2 + 1$

11. For $g(x) = x^2 + 2x + 3$, determine:
(a) $g(2+h)$ (b) $g(2)$ (c) $g(2+h) - g(2)$

12. For $f(n) = 3n^2 - 2$ and $g(n) = n + 1$, determine:
(a) $f(n) + g(n)$
(b) $f(n)g(n)$
(c) The domain of $f(n)/g(n)$.
(d) $f(g(n))$
(e) $g(f(n))$

13. Sketch the graph of a function defined for $x \geq 0$ with all of the following properties. (There are lots of possible answers.)

(a) $f(0) = 2$.
(b) $f(x)$ is increasing for $0 \leq x < 1$.
(c) $f(x)$ is decreasing for $1 < x < 3$.
(d) $f(x)$ is increasing for $x > 3$.
(e) $f(x) \to 5$ as $x \to \infty$.

14. When a new product is advertised, more and more people try it. However, the rate at which new people try it slows as time goes on.

(a) Sketch a graph of the total number of people who have tried such a product against time.
(b) What do you know about the concavity of the graph?

15. An amusement park operator has fixed costs of $5000 per day and variable costs averaging $2 per customer. The amusement park fee is $7.

(a) How many customers does the park need in order to make a profit?
(b) Find the cost and revenue functions and graph them on the same axes. Mark the break-even point on the graph.

16. Values of the functions $F(t)$, $G(t)$, and $H(t)$ are listed in the following table. Determine which one is concave up, which one is concave down, and which one is linear.

t	$F(t)$	$G(t)$	$H(t)$
10	15	15	15
20	22	18	17
30	28	21	20
40	33	24	24
50	37	27	29
60	40	30	35

17. Suppose y_1, y_2, and y_3 are functions of x such that one of them is proportional to x, one is inversely proportional to x, and one is proportional to x^2. Using the table below, write y_1, y_2, and y_3 as functions of x and find the constants of proportionality.

x	y_1	y_2	y_3
5	600	50	1.25
10	300	200	2.50
15	200	450	3.75
20	150	800	5.00
25	120	1250	6.25

18. Table 1.41 gives the world's population for three different years. If the world's population increased exponentially from 1950 through 1980 and continued to increase according to this pattern between 1980 and 1991, what would the world's population have been in 1991? How does this compare to the actual data and what conclusions, if any, can you draw?

TABLE 1.41 *Population of the world*

Year	Population (in billions)
1950	2.564
1980	4.478
1991	5.423

19. The profit function for a skateboard company is given by $\pi(p) = -p^2 + 70p - 125$, where p is the price charged by the company for a skateboard.

(a) Sketch a graph of this function and find the price that will maximize profits.
(b) For what prices will the company make a profit?

20. A company produces and sells customized shirts. The fixed costs for the company are \$7000 and the variable costs are \$5 per shirt.

(a) Assume the company sells the shirts for \$12 each. Find the cost and revenue functions, as functions of the quantity of shirts, q.
(b) The company is considering changing the selling price of the shirts. Suppose the demand equation is $q = 2000 - 40p$, where p is the price charged by the company for a shirt and q is the number of shirts sold at that price. What quantity is sold at the current price of \$12? What profit is realized at this price?
(c) Use the demand equation to write cost and revenue as functions of the price, p. Then use the fact that profit = revenue − cost to write profit as a function of price.
(d) Graph the profit function against price, and use the graph to find the price that will maximize profits. What is this price?

Convert the functions in Problems 21–22 into the form $P = P_0a^t$.

21. $P = 2.91e^{0.55t}$

22. $P = (5 \cdot 10^{-3})e^{-1.9 \cdot 10^{-2}t}$

23. (a) Use the data from Table 1.42 to determine a formula of the form

$$Q = Q_0e^{rt}$$

which would give the number of rabbits, Q, at time t (in months).

(b) What is the approximate doubling time for this population of rabbits?
(c) Use your equation to predict when the rabbit population will reach 1000.

TABLE 1.42

t	0	1	2	3	4	5
Q	25	43	75	130	226	391

24. If you need \$20,000 in your bank account in 6 years, how much must be deposited now? (Assume an annual interest rate of 10%, compounded continuously.)

25. What nominal annual interest rate has an effective annual yield of 5% under continuous compounding?

26. What is the effective annual yield, under continuous compounding, for a nominal annual interest rate of 8%?

27. Different kinds of the same element (called different *isotopes*) can have very different half-lives. The decay of plutonium-240 is described by the formula

$$Q = Q_0e^{-0.00011t}$$

whereas the decay of plutonium-242 is described by

$$Q = Q_0e^{-0.0000018t}.$$

Find the half-lives of plutonium-240 and plutonium-242.

28. An animal skull still has 20% of the carbon-14 that was present when the animal died. The half-life of carbon-14 is 5730 years. Find the approximate age of the skull.

29. Suppose prices are increasing by 0.1% a day.

(a) By what percent do prices increase a year?
(b) Looking at your answer to part (a), guess the approximate doubling time of prices increasing at this rate. Check your guess.

30. (a) Consider the functions shown in Figure 1.116(a). Find the coordinates of C.
(b) Consider the functions shown in Figure 1.116(b). Find the coordinates of C in terms of b.

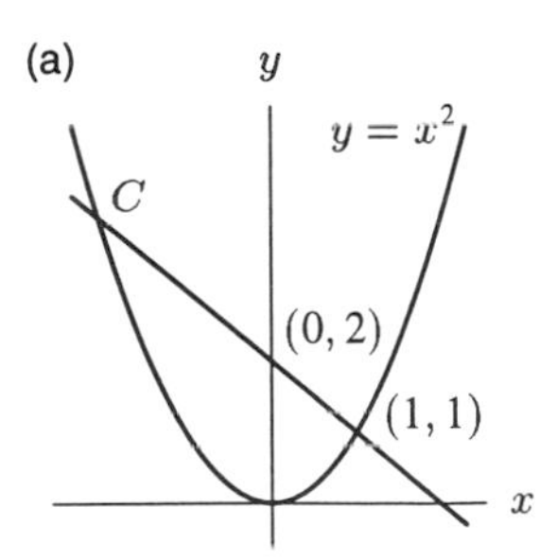

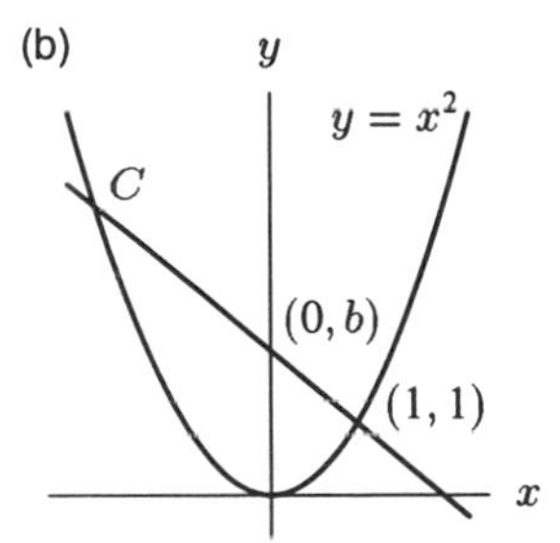

Figure 1.116

31. Match the following formulas with the graphs in Figure 1.117:

(a) $y = 1 - 2^{-x}$
(b) $y = x^2 + 4x + 5$
(c) $y = 2\cos x$
(d) $y = 1 - x^2$
(e) $y = 2 + e^x$
(f) $y = x^3 - x^2 - x + 1$
(g) $y = -2\ln x$
(h) $y = 1 + \cos x$
(i) $y = \frac{1}{x}$

(I)

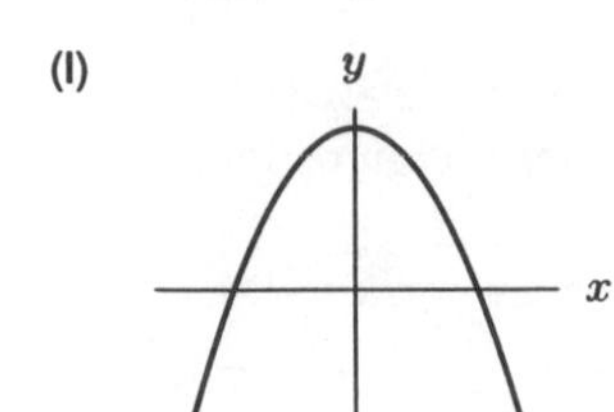

(II)

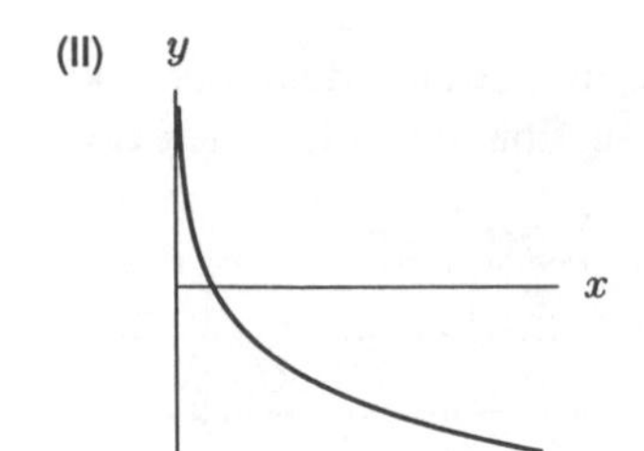

(III)

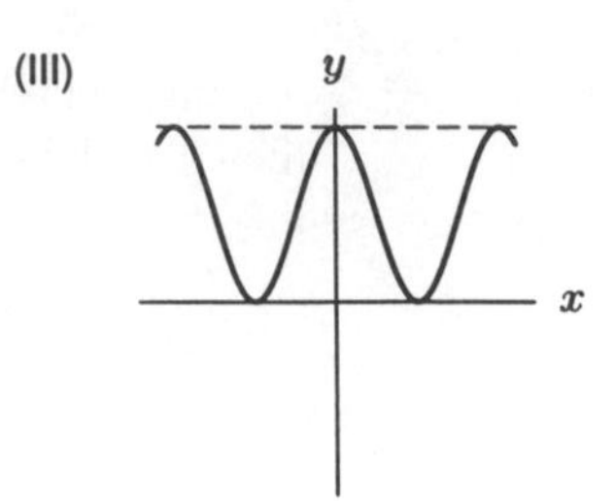

(IV)

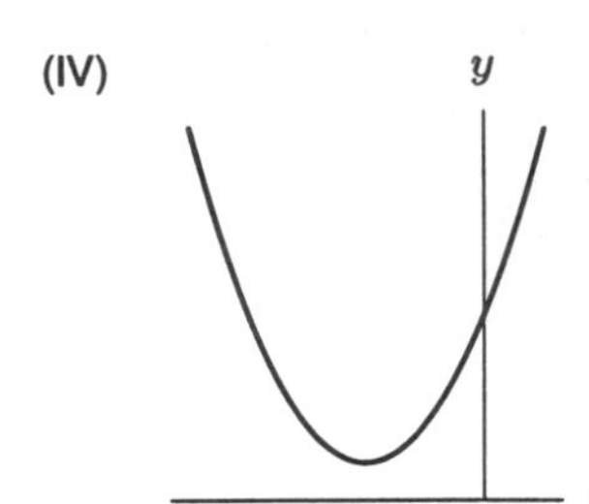

(V)

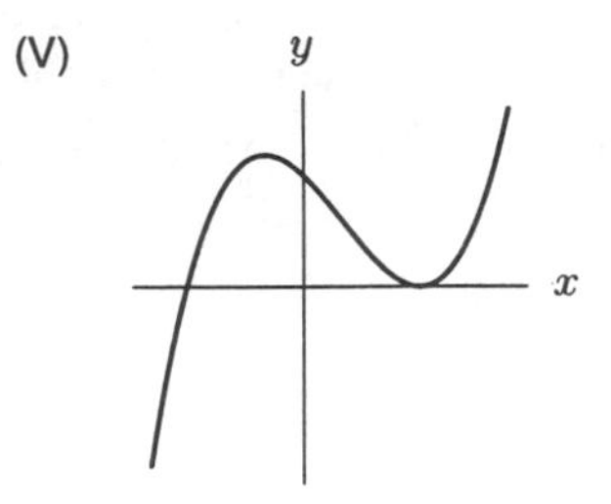

(VI)

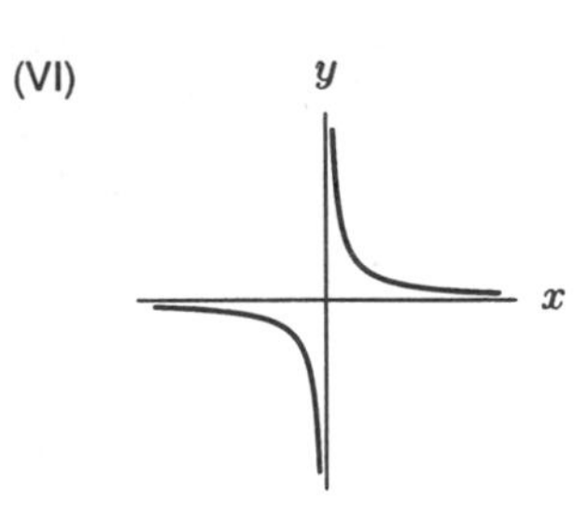

(VII)

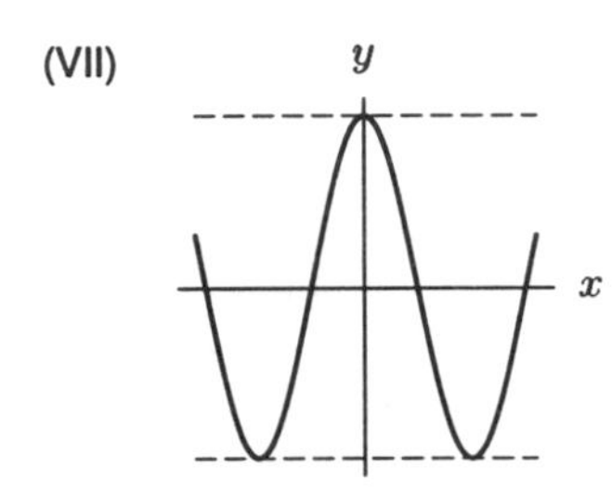

(VIII)

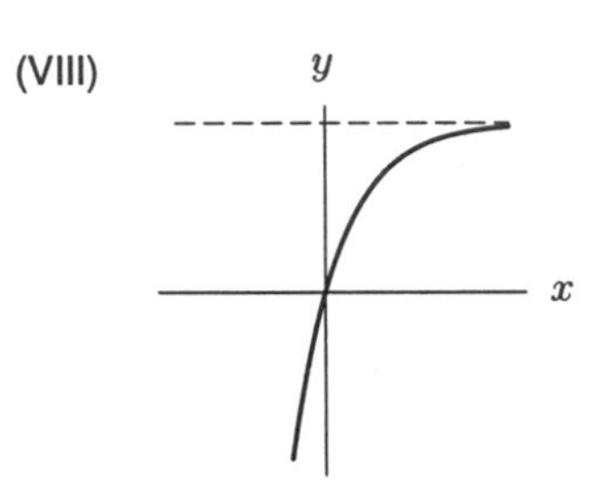

(IX)

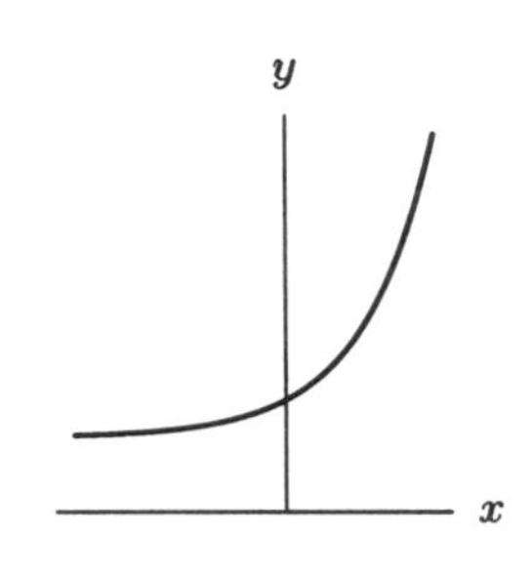

Figure 1.117

Find possible formulas for the functions graphed in Problems 32–40.

32.

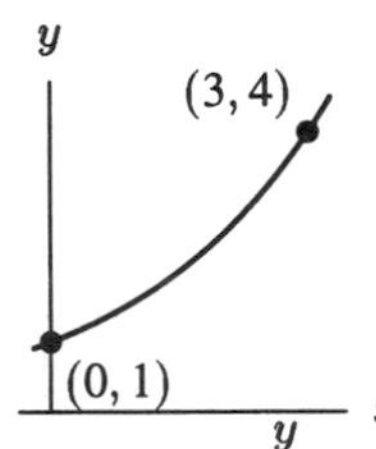

33.

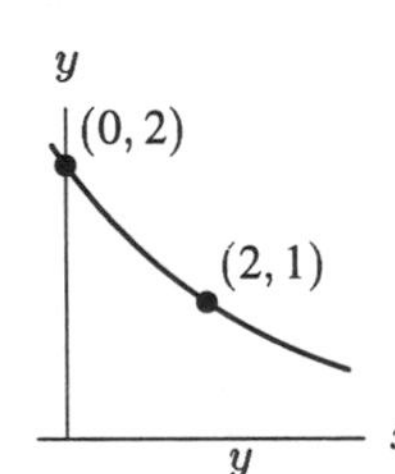

34.

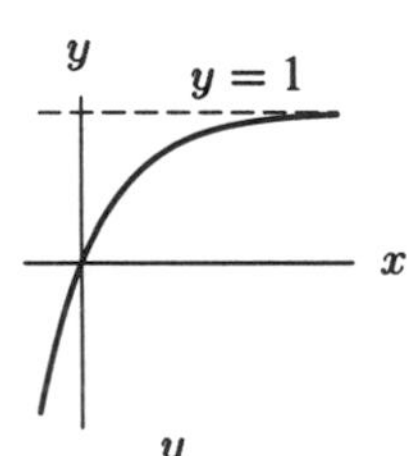

35.

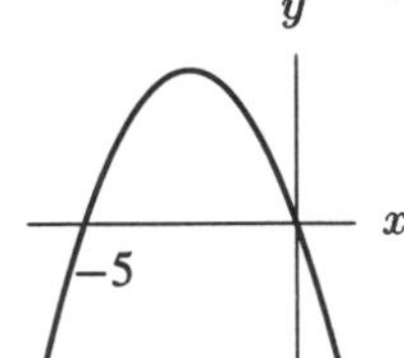

36.

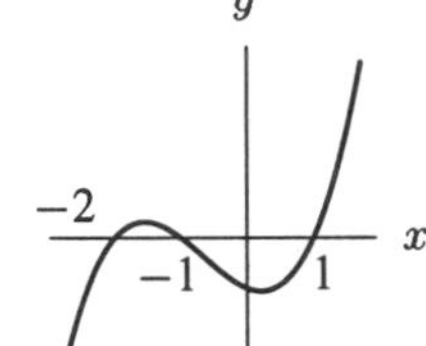

37.

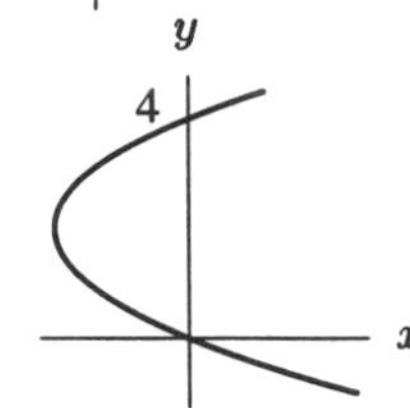

38.

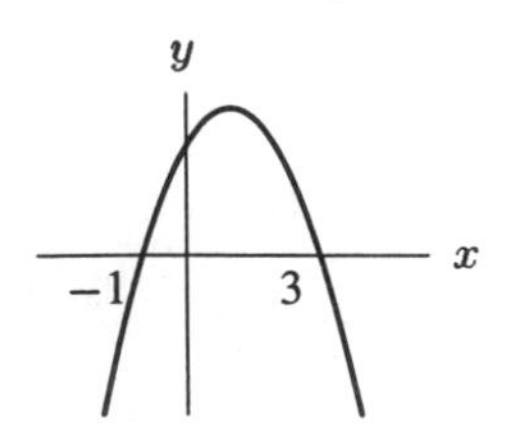

39.

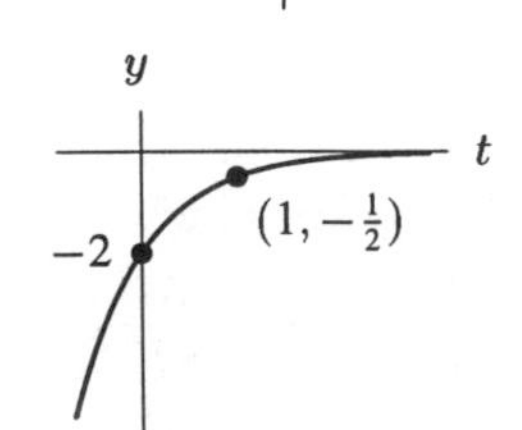

40.

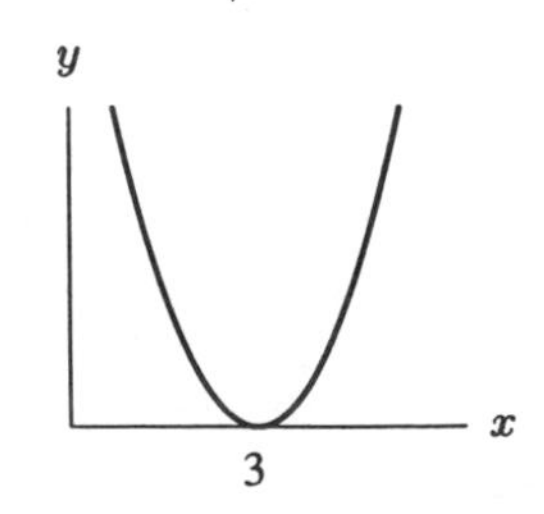

41. Each of the functions described by the data in Table 1.43 is increasing over its domain, but each increases in a different way. Which of the graphs in Figure 1.118 best fits each function?

TABLE 1.43

x	$f(x)$	x	$g(x)$	x	$h(x)$
1	1	3.0	1	10	1
2	2	3.2	2	20	2
4	3	3.4	3	28	3
7	4	3.6	4	34	4
11	5	3.8	5	39	5
16	6	4.0	6	43	6
22	7	4.2	7	46.5	7
29	8	4.4	8	49	8
37	9	4.6	9	51	9
47	10	4.8	10	52	10

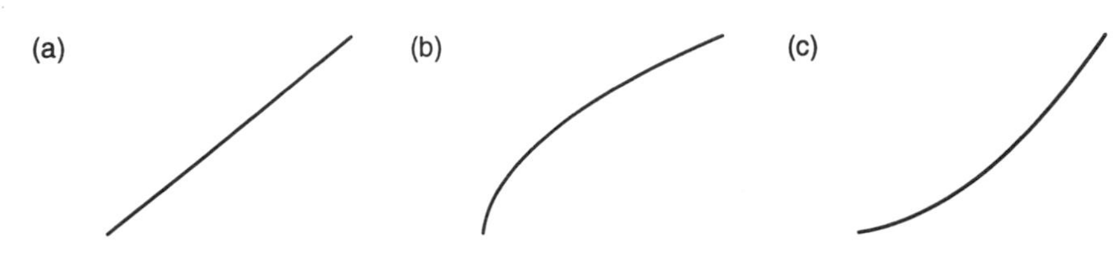

Figure 1.118

42. (a) What effect does the transformation

$$y = p(x) \quad \text{to} \quad y = p(1+x)$$

have on the graph of $p(x)$?

(b) If p is a polynomial of degree ≤ 2 such that for all x

$$p(x) = p(1+x),$$

what can you say about p?

43. For each of the following two conditions, find all polynomials, p, of degree ≤ 2 which satisfy the condition for all x.

(a) $p(x) = p(-x)$
(b) $p(2x) = 2p(x)$

44. A *catalyst* in a chemical reaction is a substance which speeds up the reaction but which does not itself change. If the product of a reaction is itself a catalyst, the reaction is said to be *autocatalytic*. Suppose the rate, r, of a particular autocatalytic reaction is proportional to the quantity of the original material remaining times the quantity of product, p, produced. If the initial quantity of the original material is A and the amount remaining is $A - p$:

(a) Express r as a function of p.
(b) What is the value of p when the reaction is proceeding fastest?

45. Glucose is fed by intravenous injection at a constant rate, k, into a patient's bloodstream. Once there, the glucose is removed at a rate proportional to the amount of glucose present. If R is the net rate at which the quantity, G, of glucose in the blood is increasing:

(a) Write a formula giving R as a function of G.
(b) Sketch a graph of R against G.

46. In the early 1920s, Germany had tremendously high inflation, called hyperinflation. Photographs of the time show people going to the store with wheelbarrows full of money. If a loaf of bread cost $1/4$ RM in 1919 and 2,400,000 RM in 1922, what was the average yearly inflation rate between 1919 and 1922?

47. A fish population is reproducing at an annual rate equal to 5% of the current population, P. Meanwhile, fish are being caught by fishermen at a constant rate, Y (measured in fish per year).

(a) Write a formula for the rate, R, at which the fish population is increasing as a function of P.
(b) Sketch a graph of R against P.

48. Find a formula for the function in the graph that follows:

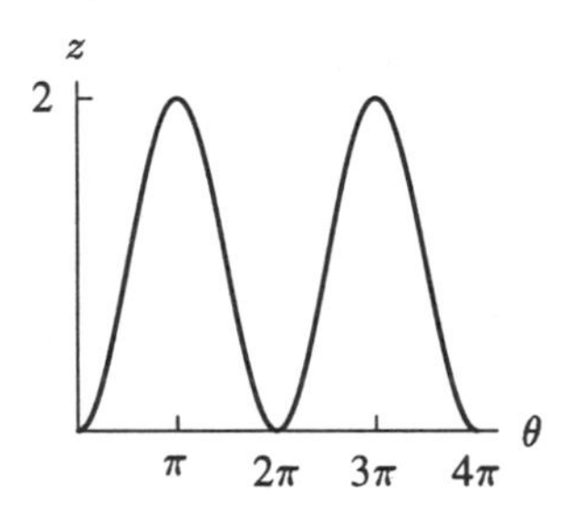

49. The depth of water in a tank oscillates once every 6 hours around an average depth of 7 feet. If the smallest depth is 5.5 feet and the largest depth is 8.5 feet, find a formula for the depth in terms of time, measured in hours. (There are many possible answers.)

50. Use a graphing calculator or computer to find the period of $2\sin 4x + 3\cos 2x$.

CHAPTER TWO

KEY CONCEPT: THE DERIVATIVE

We begin this chapter by investigating the problem of speed: how can we measure the speed of a moving object at a given instant in time? Or, more fundamentally, what do we mean by the term *speed*? We'll come up with a definition of speed that has wide-ranging implications — not just for the speed problem, but for measuring any rate of change. Our journey will lead us to the key concept of *derivative*, which forms the basis for our study of calculus.

The derivative can be interpreted geometrically as the slope of a curve, and physically as a rate of change. Because derivatives can be used to represent everything from fluctuations in interest rates to the rates at which fish are dying and gas molecules are moving, they have many applications.

2.1 AVERAGE RATE OF CHANGE

Average Velocity

Suppose that you drive 200 miles and the trip takes 4 hours. What is your average velocity for the trip? Since velocity equals distance divided by time, your average speed is $200/4 = 50$ miles per hour. The concept of average velocity can be extended to any kind of motion. Consider the motion of a grapefruit thrown in the air. The behavior of the grapefruit is not surprising: It goes up, slows down, reverses direction, falls down, and finally, "*splat!*" Informally, we can say that the grapefruit leaves the thrower's hand at high speed, slows down until it reaches its maximum height, and then gradually speeds up in the downward direction. (See Figure 2.1.)

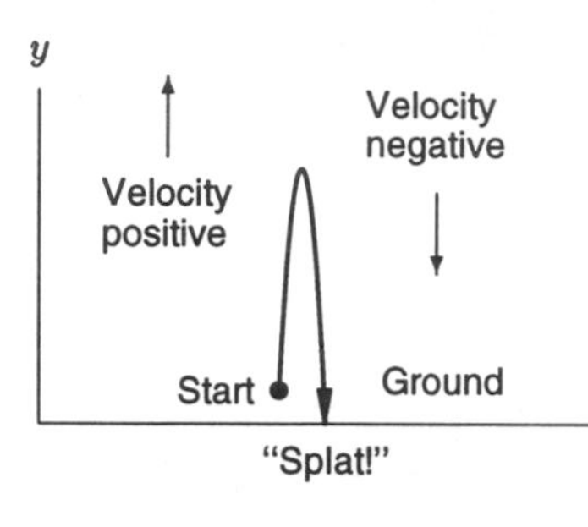

Figure 2.1: The grapefruit's path

TABLE 2.1 *Height of the grapefruit above the ground*

t (sec)	0	1	2	3	4	5	6
y (feet)	6	90	142	162	150	106	30

How can we determine the average velocity of the grapefruit over any part of its trip? We'll assume that we can measure the height of the grapefruit above the ground at any time t: we'll think of the height, $y = s(t)$, as a function of time t. (See Table 2.1.) "*Splat!*" comes sometime between 6 and 7 seconds. The numbers show the behavior noted above: in the first second the grapefruit travels $90 - 6 = 84$ feet, and in the second second it travels only $142 - 90 = 52$ feet. Hence the grapefruit traveled faster over the first interval, $0 \leq t \leq 1$, than the second interval, $1 \leq t \leq 2$. Since speed = distance/time, we say that the average speed over the interval $0 \leq t \leq 1$ is 84 ft/sec, and the average speed over the next interval is 52 ft/sec.

Velocity versus Speed

From now on, we will make a distinction between velocity and speed. Suppose an object moves along a line. If we pick one direction to be positive, the *velocity* is positive if it is in the same direction, and negative if it is in the opposite direction. For the grapefruit, upward is positive and downward is negative. (See Figure 2.1.) *Speed* is the magnitude of the velocity and so is always positive or zero.

The **average velocity** of an object over the interval $a \leq t \leq b$ is the net change in position during the interval divided by the change in time. If position is given by $y = s(t)$, we have

$$\text{Average velocity} = \frac{\text{Change in position}}{\text{Change in time}} = \frac{s(b) - s(a)}{b - a}.$$

Example 1 Compute the average velocity of the grapefruit over the interval $4 \leq t \leq 5$. What is the significance of the sign of your answer?

Solution During this interval, the grapefruit moves $(106 - 150) = -44$ feet. Therefore the average velocity is -44 ft/sec. The negative sign means the height is decreasing and the grapefruit is moving downward.

Example 2 Compute the average velocity of the grapefruit over the interval $1 \leq t \leq 3$.

Solution Average velocity $= (162 - 90)/(3 - 1) = 72/2 = 36$ ft/sec.

The average velocity is a useful concept since it gives a rough idea of the behavior of the grapefruit: if two identical grapefruits are hurled into the air, and one has an average velocity of 10 ft/sec over the interval $0 \leq t \leq 1$ while the second has an average velocity of 100 ft/sec over the same interval, clearly the second one was thrown harder.

Visualizing the Average Velocity

We have just seen how to compute average velocity. Now we will see how to visualize average velocity using a graph of height above ground as a function of time. Let's go back to the grapefruit. Suppose that Figure 2.2 shows the height of the grapefruit plotted against time. (Note that this is not a picture of the grapefruit's path, which is straight up and down.)

How can we visualize the average velocity on this graph? Suppose $y = s(t)$. Let's consider the interval $1 \leq t \leq 2$ and the expression

$$\text{Average velocity} = \frac{\text{Distance moved}}{\text{Time elapsed}} = \frac{s(2) - s(1)}{2 - 1} = \frac{142 - 90}{1} = 52 \text{ ft/sec.}$$

Now $s(2) - s(1)$ is the change in height over the interval, or the distance moved, and it is marked vertically in Figure 2.2. The 1 in the denominator is the time elapsed and is marked horizontally in Figure 2.2. Therefore,

$$\text{Average velocity} = \frac{\text{Distance moved}}{\text{Time elapsed}} = \text{Slope of line joining } BC.$$

(See Figure 2.2.) A similar argument shows the following:

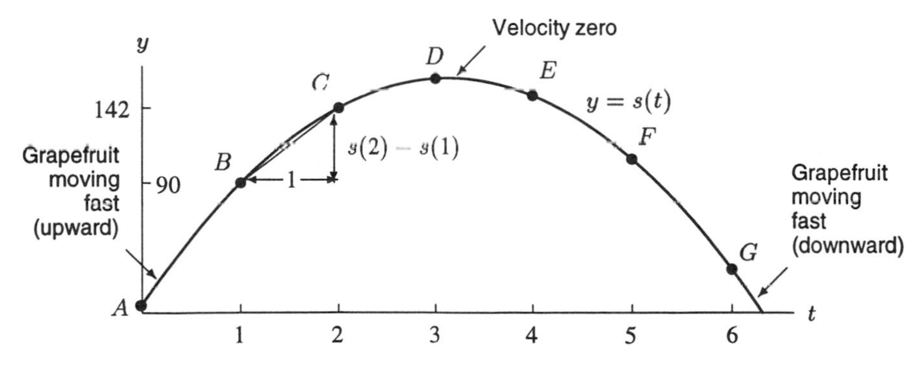

Figure 2.2: The height, y, of the grapefruit at time t

> The **average velocity** over any interval is the slope of the line joining the points on the graph of $s(t)$ corresponding to the endpoints of the interval.

Example 3 A car travels away from home in a straight line and its distance from home at time t is shown in Figure 2.3. Is the average velocity of the car greater during the first hour or second hour?

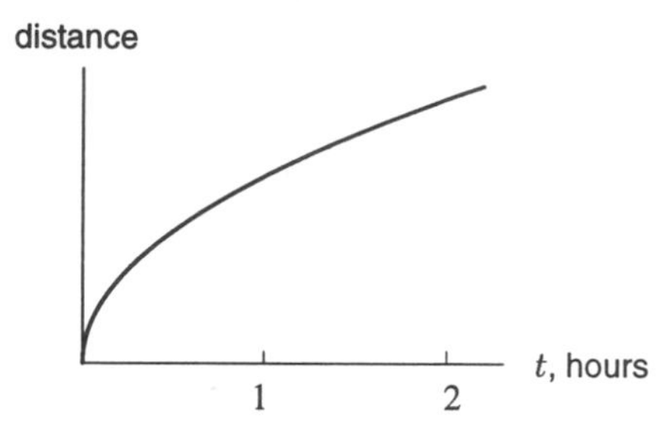

Figure 2.3: Distance of car from home

distance

slope=average velocity between $t = 1$ and $t = 2$

slope=average velocity between $t = 0$ and $t = 1$

t, hours

1 2

Figure 2.4: Visualizing average velocity of a car

Solution Average velocity is represented graphically by the slope of a secant line. We see that the secant line between $t = 0$ and $t = 1$ is much steeper than the secant line between $t = 1$ and $t = 2$. (See Figure 2.4.) Thus, the average velocity is greater during the first hour.

Average Rate of Change

Table 2.2 gives the public debt of the United States for the years 1980 to 1993. The total change in debt during this 13-year period was $4351.2 - 907.7 = 3443.5$ billion dollars. Often, it is more helpful to look at the rate of change than the total change. In this case, the rate of change tells us the rate at which the debt has been increasing, in billions of dollars *per year.* The United States public debt increased 3443.5 billion dollars during the 13-year period from 1980 to 1993, so the average rate of change was $3443.5/13 = 264.88$ billion dollars per year.

TABLE 2.2 *Public Debt of the United States [The World Almanac, 1995]*

Year	Debt (billions)	Year	Debt (billions)
1980	907.7	1987	2350.3
1981	997.9	1988	2602.3
1982	1142.0	1989	2857.4
1983	1377.2	1990	3233.3
1984	1572.3	1991	3665.3
1985	1823.1	1992	4064.6
1986	2125.3	1993	4351.2

Example 4 Find the average rate of change of the US public debt between 1980 and 1985, and between 1985 and 1993.

Solution Between 1980 and 1985

$$\text{Average rate of change} = \frac{1823.1 - 907.7}{1985 - 1980} = \frac{915.4}{5} = 183.1 \text{ billion dollars per year.}$$

Between 1985 and 1993

$$\text{Average rate of change} = \frac{4351.2 - 1823.1}{1993 - 1985} = \frac{2528.1}{8} = 316.0 \text{ billion dollars per year.}$$

We can extend this concept of average rate of change to find the average rate of change of any function $y = f(x)$. As in the case of position as a function of time, or public debt as a function of time, we are interested in the ratio of the change in the y-variable to the change in the x-variable. We have

Average rate of change of f over the interval from a to b $= \dfrac{f(b) - f(a)}{b - a}$

To emphasize the independent variable, we talk about the average rate of change of f with respect to x. Notice that the units of rate of change are the units of the dependent variable divided by the units of the independent variable, such as miles per (divided by) hour, feet per second, or billion dollars per year.

The quantity

$$\frac{f(b) - f(a)}{b - a}$$

is called a *difference quotient* since it is the quotient of the difference in the dependent variable divided by the difference in the independent variable.

How do we visualize average rate of change? Since the difference quotient is the same as the slope formula, we see that, as with average velocity, the average rate of change is represented graphically as the slope of the secant line.

Average rate of change of f between a and b = Slope of the secant line to f between a and b

Example 5 Figure 2.5 shows a graph of the number of farms (in millions) in the United States between 1940 and 1993. [The World Almanac, 1995] Estimate the average rate at which the number of farms is changing between 1950 and 1970. Interpret your answer.

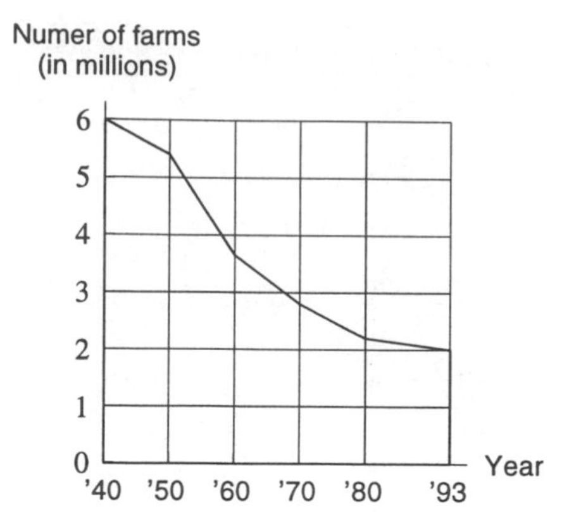

Figure 2.5: Number of farms in the US

Solution We see from Figure 2.5 that the number of farms in the US is approximately 5.4 million in 1950 and approximately 2.8 million in 1970. We have

$$\text{Average rate of change} = \frac{2.8 - 5.4}{1970 - 1950} = -0.13 \text{ million farms per year.}$$

The average rate of change is negative since the number of farms is decreasing. We see that during this 20 year period, the number of farms in the US went down at an average rate of 0.13 million farms a year, or an average decrease of 130,000 farms per year.

Example 6

a) Find the average rate of change of $f(x) = 7x - x^2$ between $x = 1$ and $x = 4$.
b) Sketch a graph of $f(x)$ and represent the average rate of change graphically as the slope of a line.

Solution

a) Since $f(x) = 7x - x^2$, we see that the average rate of change between $x = 1$ and $x = 4$ is

$$\frac{f(4) - f(1)}{4 - 1} = \frac{12 - 6}{3} = 2.$$

b) A graph of $f(x) = 7x - x^2$ is given in Figure 2.6. The average rate of change is represented graphically as the slope of the secant line between $x = 1$ and $x = 4$.

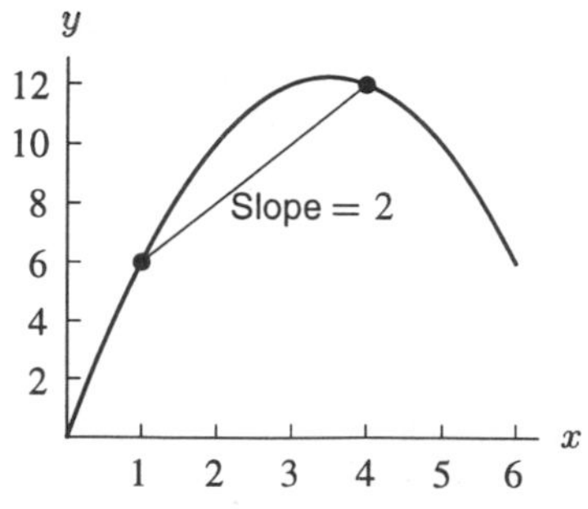

Figure 2.6: Average rate of change = Slope of secant line

Problems for Section 2.1

1. The position, s, of a car is given in Table 2.3.
   (a) Find the average velocity of the car between $t = 0$ and $t = 15$. Give units with your answer.
   (b) Find the average velocity of the car between $t = 10$ and $t = 30$. Give units with your answer.
   (c) Find the distance traveled by the car between $t = 10$ and $t = 30$. Give units with your answer.

TABLE 2.3

t (sec)	0	5	10	15	20	25	30
s (ft)	0	30	55	105	180	260	410

2. Figure 2.7 shows the position of an object at time t.
   (a) Is average velocity greater between $t = 0$ and $t = 3$ or between $t = 3$ and $t = 6$?
   (b) Is average velocity positive or negative between $t = 6$ and $t = 9$?
   (c) On a sketch similar to Figure 2.7, represent the average velocity between $t = 2$ and $t = 8$ as the slope of a line.

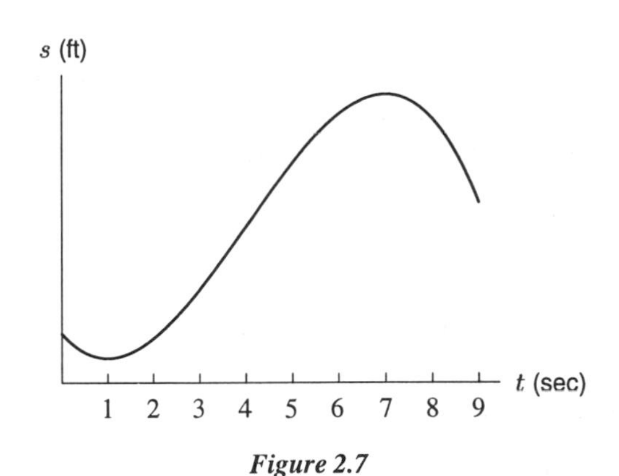

Figure 2.7

3. Draw a graph of distance against time satisfying the following properties: the average velocity is positive for all intervals and the average velocity for the first half of the trip is less than the average velocity for the second half of the trip.
4. The quantity of a substance present at time t (in minutes) is given by $Q(t) = 10e^{-0.3t}$ milligrams.
   (a) Find the average rate of change of the quantity between $t = 0$ and $t = 4$. Give units with your answer.
   (b) Is your answer to part (a) positive or negative? What does this tell you about how the quantity is changing?
   (c) Sketch a graph of $Q(t)$ against t and represent your answer to part (a) on it graphically.

5. Table 2.4 shows the total amount (in billions of dollars) spent by consumers on tobacco products in the US between 1987 and 1993.
   (a) What is the average rate of change in the amount spent on tobacco between 1987 and 1993? Give units with your answer and interpret it in terms of money spent on tobacco.
   (b) During this 6 year period, is there any interval during which the average rate of change is negative? If so when?

TABLE 2.4

Year	1987	1988	1989	1990	1991	1992	1993
Billions of dollars	35.6	36.2	40.5	43.4	45.4	50.9	50.5

6. Table 2.5 shows the total labor force (in thousands of workers) in the US between 1930 and 1990. [The World Almanac, 1995] Find the average rate of change between 1930 and 1950. Between 1950 and 1970. Give units with yours answers and interpret your answers in terms of the labor force. Is the average rate of change increasing or decreasing over time?

TABLE 2.5

Year	1930	1940	1950	1960	1970	1980	1990
Labor force (in thousands)	29,424	32,376	45,222	54,234	70,920	90,564	103,905

7. Total sales for a company are shown in Figure 2.8.
   (a) Is the average rate of change of sales positive or negative between $t = 0$ and $t = 5$?
   (b) Is the average rate of change of sales positive or negative between $t = 0$ and $t = 10$?
   (c) Is the average rate of change of sales positive or negative between $t = 0$ and $t = 15$?
   (d) Is the average rate of change of sales positive or negative between $t = 0$ and $t = 20$?
   (e) During which time interval is the average rate of change larger: $0 \leq t \leq 5$ or $0 \leq t \leq 10$?
   (f) During which time interval is the average rate of change larger: $0 \leq t \leq 10$ or $0 \leq t \leq 20$?
   (g) Estimate the average rate of change between $t = 0$ and $t = 10$ and interpret your answer in terms of sales.

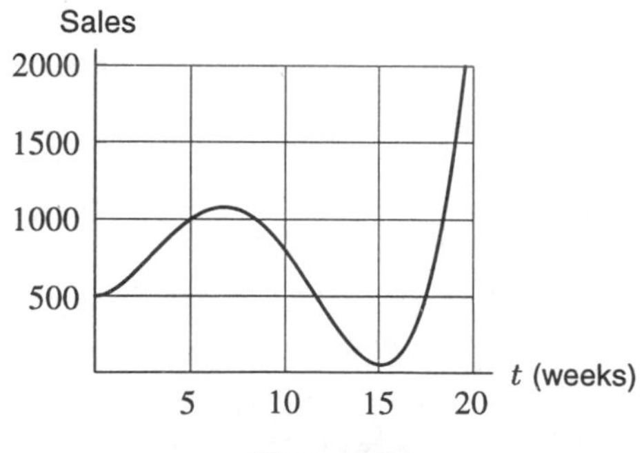

Figure 2.8

8. Find the average rate of change of $f(x) = 3x^2 + 4$, from $x = -2$ to $x = 1$, and illustrate your answer graphically.

2.2 INSTANTANEOUS RATE OF CHANGE: THE DERIVATIVE

In the previous section, we looked at the average rate of change of a function over an interval. In this section, we consider the rate of change of a function at a single point. As in the previous section, we begin by considering position functions. In this case, rate of change is the same as velocity.

Instantaneous Velocity

The notion of speed—and, in particular, the speed of an object at an instant in time—is surprisingly subtle and difficult to define precisely. Consider the statement "At the instant it crossed the finish line, the horse was traveling at 42 mph." How can such a claim be substantiated? A photograph taken at that instant will show the horse motionless—it is no help at all. There is some paradox in trying to quantify the property of motion at a particular instant in time, since by focusing on a single instant you stop the motion!

A similar difficulty arises whenever we attempt to measure the rate of change of anything—for example, oil leaking out of a damaged tanker. The statement "One hour after the ship's hull ruptured, oil was leaking at a rate of 200 barrels per second" seems not to make sense. You could argue that at any given instant *no* oil is leaking.

Problems of motion were of central concern to Zeno and other philosophers as early as the fifth century B.C. The modern approach, made famous by Newton's calculus, is to stop looking for a simple notion of speed at an instant, and instead to look at speed over small intervals containing the instant. This method sidesteps the philosophical problems mentioned earlier but brings new ones of its own.

We shall illustrate the ideas discussed above by an idealized example, called a thought experiment. It is idealized in the sense that we assume that we can make measurements of distance and time as accurately as we wish. In fact, the numbers we shall use come from a mathematical formula, not from real measurements, but that doesn't matter for our purposes.

Velocity of a Grapefruit at an Instant

We return to the grapefruit that we threw in the air in the last section. Table 2.1 on page 122 gives the height of the grapefruit at time t. Suppose we would like to determine the velocity of the grapefruit at, say, $t = 1$. How fast is the grapefruit going exactly one second after we let go? We can use average velocities to help us estimate this quantity.

We saw in the previous section that the average velocity on the interval $0 \leq t \leq 1$ is 84 ft/sec, and the average velocity on the interval $1 \leq t \leq 2$ is 52 ft/sec. Notice that the average velocity before $t = 1$ is more than the average velocity after $t = 1$ since the grapefruit is slowing down. We would expect to define the velocity *at* $t = 1$ to be between these two average velocities. How can we find a more accurate measure of the velocity at *exactly* $t = 1$? We'll have to look at what happens near $t = 1$ in more detail. Suppose that we could measure the height of the grapefruit at any instant. Then we could find the average velocities on either side of $t = 1$ over smaller and smaller intervals. This is done in Figure 2.9.

We would expect to define the instantaneous velocity at $t = 1$ to be between average velocities on either side of $t = 1$. We see in Figure 2.9 that, as the size of the interval shrinks, the values of the velocity before $t = 1$ and the velocity after $t = 1$ get closer together. By the smallest interval in Figure 2.9, both velocities are 68.0 ft/sec (to one decimal place), so we will define the velocity at $t = 1$ to be 68.0 ft/sec (to one decimal place).

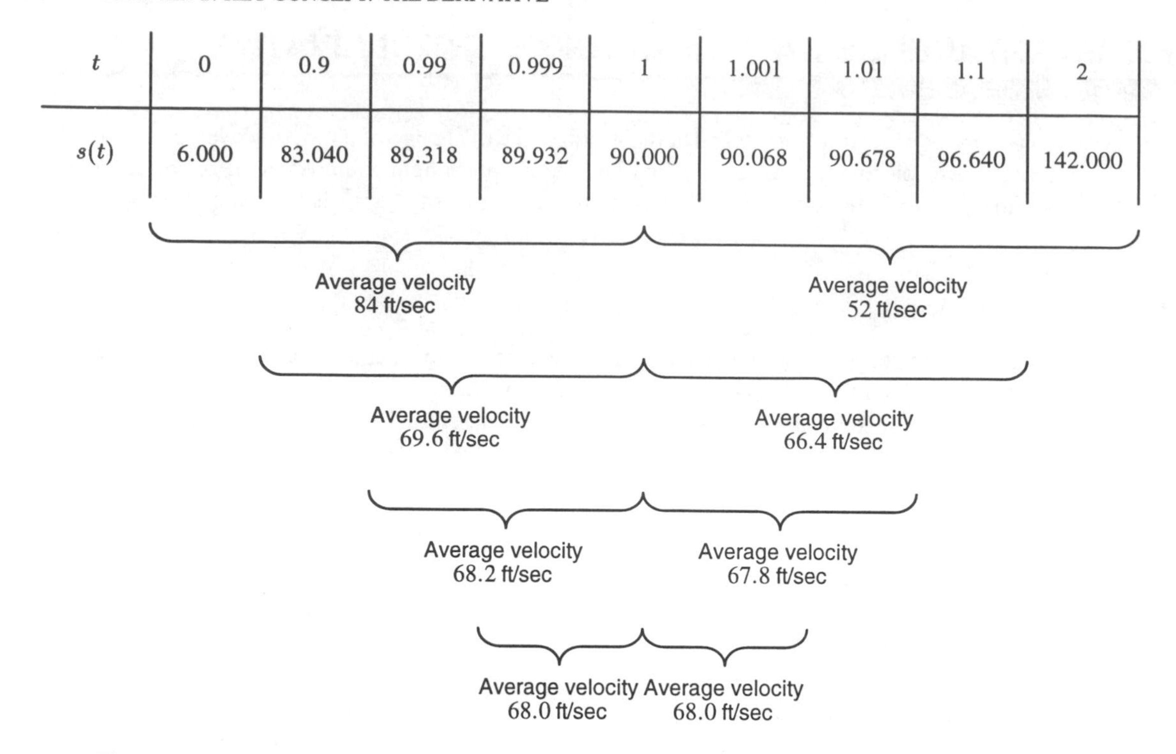

t	0	0.9	0.99	0.999	1	1.001	1.01	1.1	2
$s(t)$	6.000	83.040	89.318	89.932	90.000	90.068	90.678	96.640	142.000

Figure 2.9: Average velocities over intervals on either side of $t = 1$: showing successively smaller intervals

Of course, if we showed more decimal places, average velocities before and after $t = 1$ would no longer agree even in the smallest interval. To calculate the velocity at $t = 1$ to more decimal places, we would have to take smaller and smaller intervals on either side of $t = 1$ until the average velocities agree to the number of decimal places we wanted. The velocity at $t = 1$ is then defined to be this common average velocity.

Defining Instantaneous Velocity Using the Idea of a Limit

When you take smaller and smaller intervals, it turns out that the average velocities are always just above or just below 68 ft/sec. It seems natural, then, to define velocity at the instant $t = 1$ to be 68 ft/sec. This is called the *instantaneous velocity* at this point, and its definition depends on our being adequately convinced that smaller and smaller intervals will provide average speeds that come arbitrarily close to 68. Modern mathematics has a name for this process: it is called *taking the limit.*

> The **instantaneous velocity** of an object at time t is given by the limit of the average velocity over the interval as the interval shrinks around t.

Make sure that you see how we have replaced the original difficulty of computing velocity at a point by a search for an argument to convince ourselves that the average velocities do approach a number as the time intervals shrink in size. In a sense, we have traded one hard question for another, since we don't yet have any idea how to be certain what number the average velocities are approaching. In our thought experiment, the number seems to be exactly 68, but what if it were 68.000001? How can we be sure that we have taken small enough intervals?

For most practical purposes, it is not likely to be important whether the velocity is exactly 68 or 68.000001. Showing that the limit is exactly 68 requires more precise knowledge of how the velocities were calculated and of the limiting process; we will see this in Chapter 4.

Visualizing Velocity: Slope of Curve

We saw in the previous section that average velocity is represented graphically as the slope of a secant line over the interval. The next question is how to visualize the velocity at an instant. Let's think about how we found the instantaneous velocity. We took average velocities across smaller and smaller intervals ending at the point $t = 1$. Two such velocities are represented by the slopes of the lines in Figure 2.10. As the length of the interval shrinks, the slope of the line gets closer to the slope of the curve at $t = 1$.

The cornerstone of the idea is the fact that, on a very small scale, most functions look almost like straight lines. Imagine taking the graph of a function near a point and "zooming in" to get a close-up view. (See Figure 2.11.) The more you zoom in, the more the curve will appear to be a straight line. In other words, if you repeatedly zoom in on a section of the curve centered at a point of interest, the section of curve will eventually look like a straight line. We call the slope of this line the *slope of the curve* at the point. Therefore, the slope of the magnified line is the instantaneous velocity. Thus, we can say:

The **instantaneous velocity** is the slope of the curve at a point.

Curve
More linear
Almost completely linear
P
Slope of line = Slope of curve at P

Figure 2.11: Finding the slope of the curve at the point by "zooming in"

Look back at the graph of the grapefruit's height as a function of time in Figure 2.2. If you think of the velocity at any point as the slope of the curve there, you can see how the grapefruit's velocity

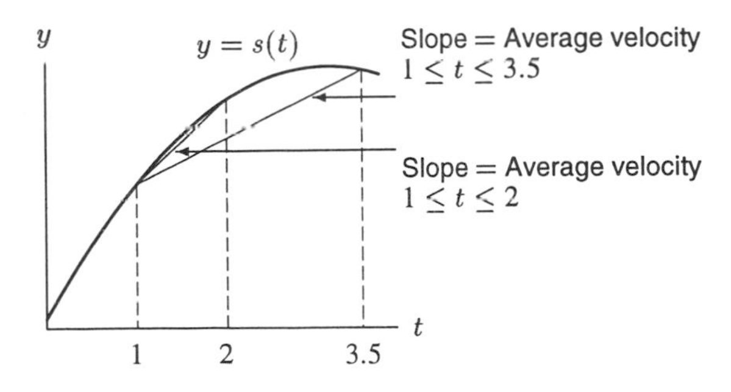

Figure 2.10: Average velocities over small intervals

varies during its journey. At points A and B the curve has a large positive slope, indicating that the grapefruit is traveling up rapidly. Point D is almost at the top: the grapefruit is slowing as it reaches the peak. At the peak, the slope of the curve is zero: the fruit has slowed to a stop in preparation for its return to earth. At point E the curve has a small negative slope, indicating a slow velocity of descent. Finally, the slope of the curve at point G is large and negative, indicating a large downward velocity that is responsible for the *"Splat."*

Example 1 At the top of the grapefruit's path, its velocity is zero. Does this mean the grapefruit has stopped?

Solution At the top of its path the grapefruit changes direction—from moving upward to moving downward—and at the moment it makes this change, the grapefruit's instantaneous velocity is zero. Deciding whether the grapefruit has stopped depends on what you think "stopped" means. Most people think an object has stopped only if it has zero velocity for some interval of time; with this definition, our grapefruit has not stopped, but only had zero instantaneous velocity for one instant of time.

Instantaneous Rate of Change

We can define the *instantaneous rate of change* of any function $y = f(x)$ at a point $x = a$. We simply mimic what we did for velocity, namely, look at the average rate of change over smaller and smaller intervals. As with velocity, the instantaneous rate of change is the slope of the curve at that point.

> The **instantaneous rate of change** of f at a is given by the average rates of change over the interval as the interval shrinks around a. Graphically, the instantaneous rate of change is the slope of the curve at a.

Example 2 The quantity (in mg) of a drug in the blood at time t (in minutes) is given by

$$Q(t) = 25e^{-0.2t}.$$

Estimate the rate of change of the quantity at $t = 3$ and interpret your answer.

Solution We estimate the rate of change at $t = 3$ by computing the average rate of change over intervals near $t = 3$. If we use the intervals $2 \leq t \leq 3$ and $3 \leq t \leq 4$ we have:

t	2	3	4
$Q(t)$	16.758	13.720	11.233

Average Rate of change $= \dfrac{13.720 - 16.758}{3 - 2} = -3.038.$

Average Rate of change $= \dfrac{11.233 - 13.720}{4 - 3} = -2.487.$

We estimate that the rate of change at $t = 3$ is between -3.038 and -2.487. We can make our estimate as accurate as we like by choosing our intervals small enough. Let's look at the average rate of change over the intervals $2.99 \leq t \leq 3$ and $3 \leq t \leq 3.01$:

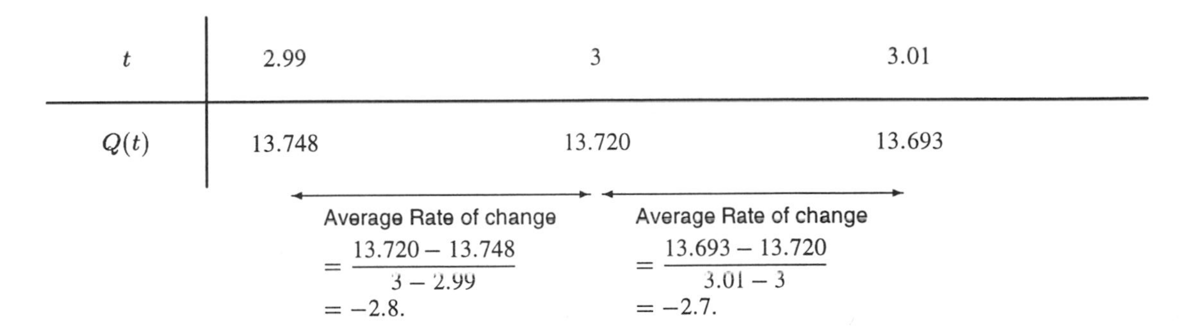

t	2.99	3	3.01
$Q(t)$	13.748	13.720	13.693

Average Rate of change $= \dfrac{13.720 - 13.748}{3 - 2.99} = -2.8.$

Average Rate of change $= \dfrac{13.693 - 13.720}{3.01 - 3} = -2.7.$

A reasonable estimate for the rate of change of the quantity at $t = 3$ is -2.75. The units are $Q(t)$ units over t units, or mg/minute. Since the rate of change is negative, the quantity of the drug is decreasing. At $t = 3$, the quantity of the drug in the body is changing at a rate of about -2.75 mg/minute.

Example 3 A car starts out slowly and then goes faster and faster. Sketch a graph of the distance the car has traveled as a function of time.

Solution Since the car is moving, the distance it has traveled is an increasing function. The speed of the car is the slope of the distance curve. Therefore, the slope of the curve is small at first and then gets steeper and steeper. One possible graph is shown in Figure 2.12.

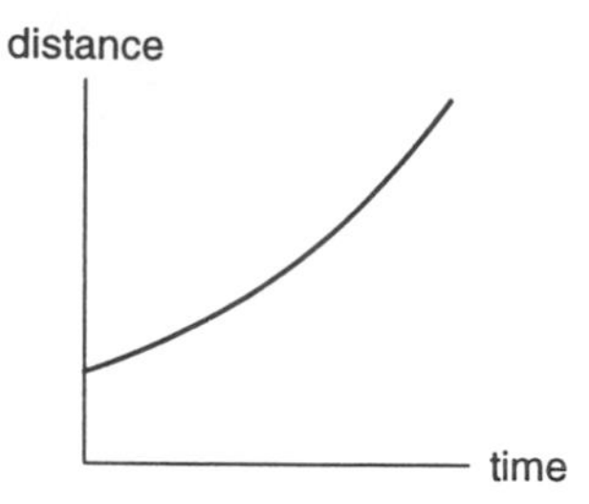

Figure 2.12: This car is speeding up

The Derivative

The rate of change of f at a point a is so important that it is given its own name, the *derivative of f at a*, denoted $f'(a)$. We will give a more formal definition of the derivative in the next section.

> The **derivative of f at a** $= f'(a) =$ the rate of change of f at a.

If we want to emphasize that $f'(a)$ is the rate of change of $f(x)$ as the variable x increases, we call $f'(a)$ the derivative of f *with respect to x* at $x = a$. Notice that the derivative is just another name for the rate of change of a function at a point.

Example 4 Estimate $f'(2)$ if $f(x) = x^3$.

Solution We know that $f'(2)$ is the derivative of $f(x) = x^3$ at 2. This is the same as the rate of change of x^3 at 2. We estimate this by looking at the average rate of change over intervals near 2. If we use the intervals $1.999 \le x \le 2$ and $2 \le x \le 2.001$, we see that

$$\begin{aligned} \begin{array}{l}\text{Average rate of change}\\ \text{on } 1.999 \le x \le 2\end{array} &= \frac{2^3 - (1.999)^3}{2 - 1.999} = \frac{8 - 7.988}{0.001} = 12.0 \\ \begin{array}{l}\text{Average rate of change}\\ \text{on } 2 \le x \le 2.001\end{array} &= \frac{(2.001)^3 - 2^3}{2.001 - 2} = \frac{8.012 - 8}{0.001} = 12.0 \end{aligned}$$

It appears that the rate of change of $f(x)$ at $x = 2$ is 12, so we estimate $f'(2) = 12$.

Visualizing the Derivative: Slope of Curve and Slope of Tangent

We have already seen that the rate of change (the derivative) is equal to the slope of the curve. Since the derivative is found by taking the average rate of change over smaller and smaller intervals, we can also think of the derivative as the slope of the tangent line to the curve at that point. See Figures 2.13 and 2.14.

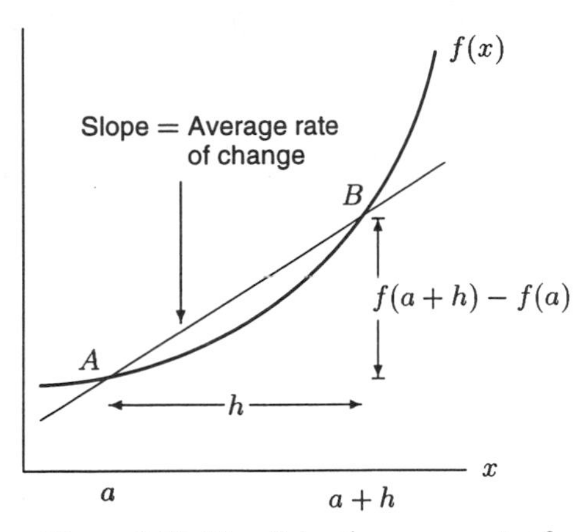

Figure 2.13: Visualizing the average rate of change of f

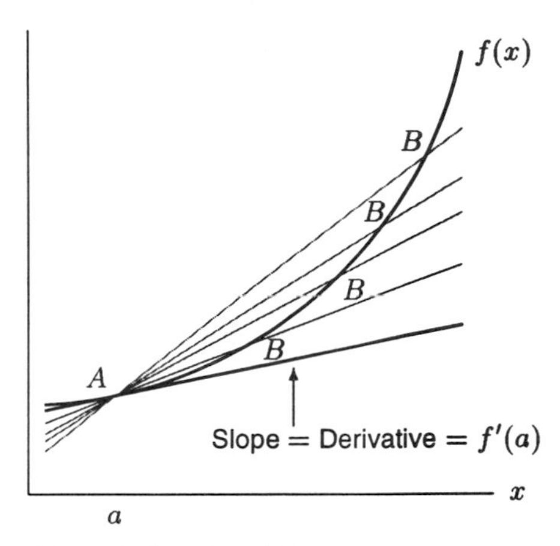

Figure 2.14: Visualizing the instantaneous rate of change of f

> Therefore, the derivative at point A can be thought of as:
> - The slope of the curve at A.
> - The slope of the tangent line to the curve at A.

The slope interpretation is often useful in gaining rough information about the derivative, as the following examples show.

Example 5 Is the derivative of $\ln(x+1)$ at $x=0$ positive or negative?

Solution Looking at a graph of $\ln(x+1)$ in Figure 2.15, we see that a tangent line drawn at $x=0$ has positive slope, so the derivative at this point is positive.

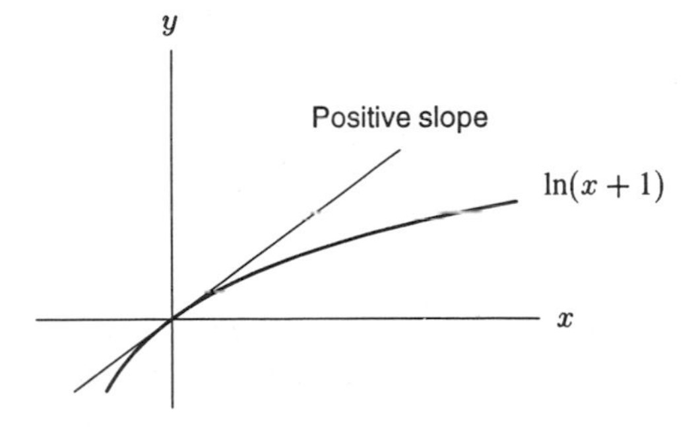

Figure 2.15: Tangent line to $\ln(x+1)$ at $x=0$

Remember that if you zoom in on the graph of a function $y = f(x)$ at the point where $x = a$, you will usually find that the graph looks more and more like a straight line with slope $f'(a)$.

Example 6 By zooming in on the point $(0, 0)$ on the graph of $y = \ln(x + 1)$, estimate the value of the derivative of this function at $x = 0$.

Solution Figure 2.16 shows successive graphs of $\ln(x + 1)$, with smaller and smaller scales. On the interval $-0.1 \leq x \leq 0.1$, the graph looks like a straight line of slope 1. Thus, the derivative of $\ln(x + 1)$ at $x = 0$ is about 1.

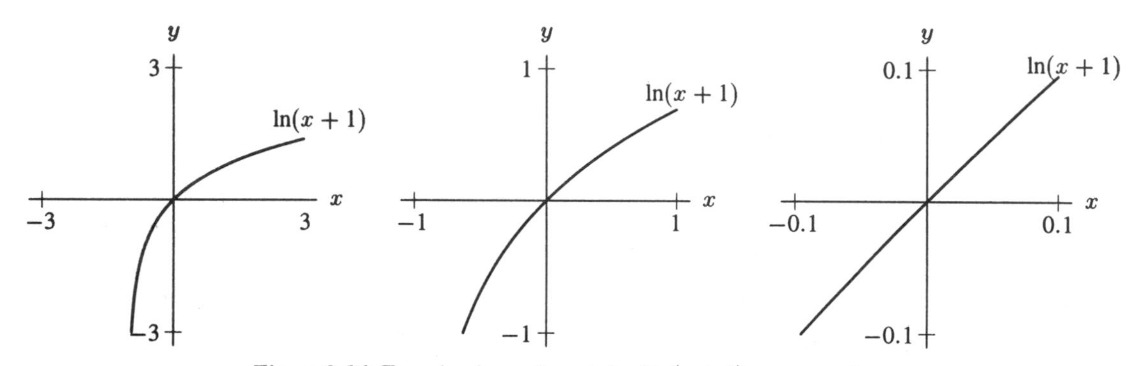

Figure 2.16: Zooming in on the graph of $\ln(x + 1)$ near $x = 0$

Later we will show that the derivative of $\ln(x + 1)$ at $x = 0$ is exactly 1. For the next example, we will assume that this is so.

Example 7 Use the tangent line at $x = 0$ to estimate values of $\ln(x + 1)$ near $x = 0$.

Solution The previous example shows that near $x = 0$, the graph of $y = \ln(x + 1)$ looks like the graph of the straight line $y = x$; we can use this to estimate values of $\ln(x + 1)$ when x is close to 0. For example, the point on the straight line $y = x$ with x coordinate 0.13 is $(0.13, 0.13)$. Since the line is close to the graph of $y = \ln(x + 1)$, we estimate that $\ln 1.13 \approx 0.13$. (See Figure 2.17.) Checking on the calculator, we find that $\ln 1.13 = 0.1222$, so our estimate is quite close. Notice that the graph leads you to expect that the real value of $\ln 1.13$ is slightly less than 0.13.

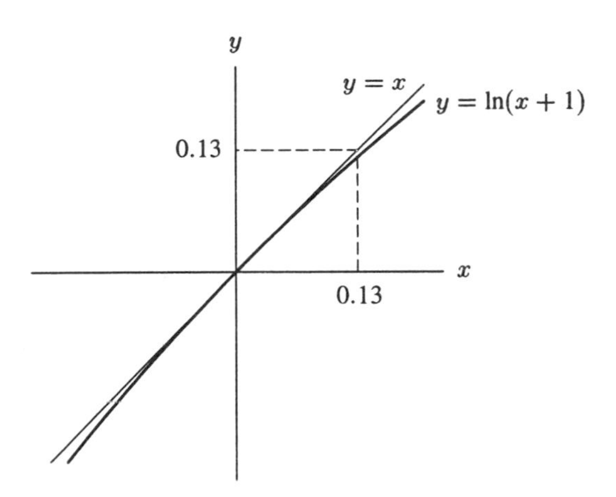

Figure 2.17: Approximating $y = \ln(x+1)$ by $y = x$

Examples: Computing the Derivative at a Point

Example 8 Estimate the value of the derivative of $f(x) = 2^x$ at $x = 0$ graphically and numerically.

Solution Graphically: If you draw a tangent line at $x = 0$ to the exponential curve in Figure 2.18, you will see that it has a positive slope. Since the slope of the line BA is $(2^0 - 2^{-1})/(0-(-1)) = 1/2$ and the slope of the line AC is $(2^1 - 2^0)/(1-0) = 1$, we know that the derivative is between $1/2$ and 1.

Numerically: To find the derivative at $x = 0$, we compute the average rate of change on intervals near 0.

$$\begin{array}{l}\text{Average rate of change} \\ \text{on } -0.0001 \le x \le 0\end{array} = \frac{2^0 - 2^{-0.0001}}{0-(-0.0001)} = \frac{1 - 0.999930688}{0.0001} = 0.69312$$

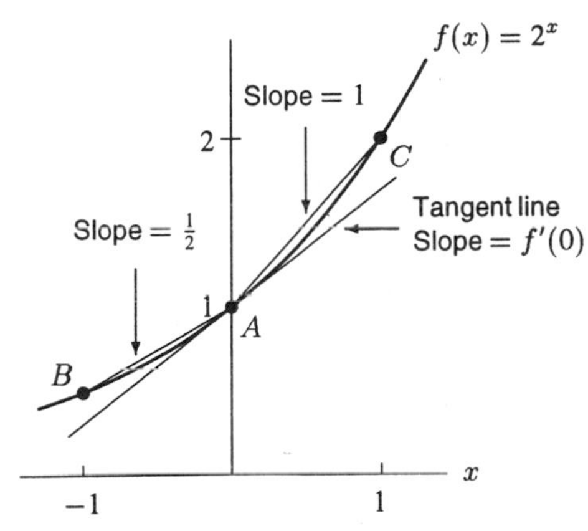

Figure 2.18: Graph of $y = 2^x$ showing the derivative at $x = 0$

$$\begin{array}{l}\text{Average rate of change} \\ \text{on } 0 \leq x \leq 0.0001\end{array} = \frac{2^{0.0001} - 2^0}{0.0001 - 0} = \frac{1.000069317 - 1}{0.0001} = 0.69317$$

It appears that the derivative is between 0.69312 and 0.69317. To three decimal places, $f'(0) = 0.693$.

Example 9 Find an approximate equation for the tangent line to $f(x) = 2^x$ at $x = 0$.

Solution From the previous example, we know the slope of the tangent line is about 0.693. Since we also know the line has y intercept 1, its equation is approximately

$$y = 0.693x + 1.$$

Example 10 Worldwide, the amount of land per person that is used for farming has been decreasing over time. Table 2.6 shows crop land (in acres per person) as a function of year[1].

TABLE 2.6 *Amount of land on earth, per person, used for farming*

Year	1800	1900	1950	1970
Crop land (acres/person)	3.51	2.03	1.24	0.84

(a) What is the average annual rate of change in world crop land between 1800 and 1970?
(b) Estimate the rate, in acres per person per year, at which crop land is decreasing in the year 1950.

Solution (a) Between 1800 and 1970,

$$\text{Average rate of change} = \frac{0.84 - 3.51}{1970 - 1800} = \frac{-2.67}{170} = -0.0157.$$

The number of acres per person used for farming between 1800 and 1970 has been decreasing at an average rate of 0.0157 acres per person per year.

(b) We want to estimate the instantaneous rate of change (the derivative) at the year 1950. We will use the interval from 1950 to 1970 to estimate the instantaneous rate of change at 1950:

$$\text{Rate of change in } 1950 \approx \frac{0.84 - 1.24}{1970 - 1950} = \frac{-0.4}{20} = -0.02.$$

In 1950, the amount of farmland per person worldwide was decreasing at a rate of approximately 0.02 acres per person per year.

---

[1]From *An Introduction to Population, Environment, Society*, by Lawrence Schaefer, (Hamden, CT, E-P Education Services, 1972, p.30.)

Example 11 The graph of a function $y = f(x)$ is shown in Figure 2.19. Indicate whether each of the following is positive or negative, and illustrate your answers graphically.

(a) $f'(1)$ (b) $\dfrac{f(3) - f(1)}{3 - 1}$ (c) $f(4) - f(2)$

(d) The average rate of change of $f(x)$ between $x = 3$ and $x = 7$.

(e) The instantaneous rate of change of $f(x)$ at $x = 3$.

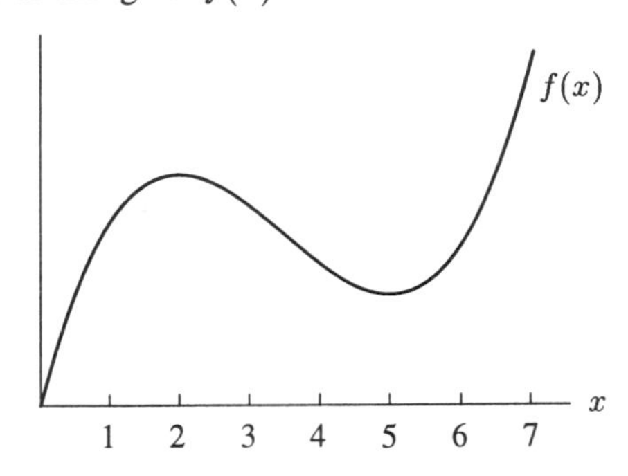

Figure 2.19: A graph of a function

Solution (a) Since $f'(1)$ is the slope of the curve at $x = 1$, we see in Figure 2.20 that $f'(1)$ is positive.

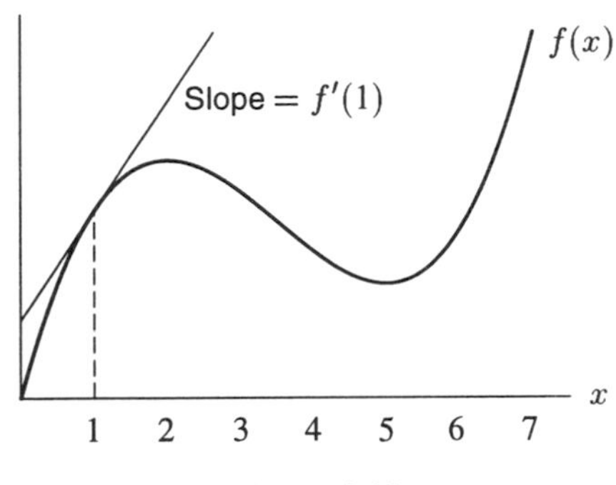

Figure 2.20

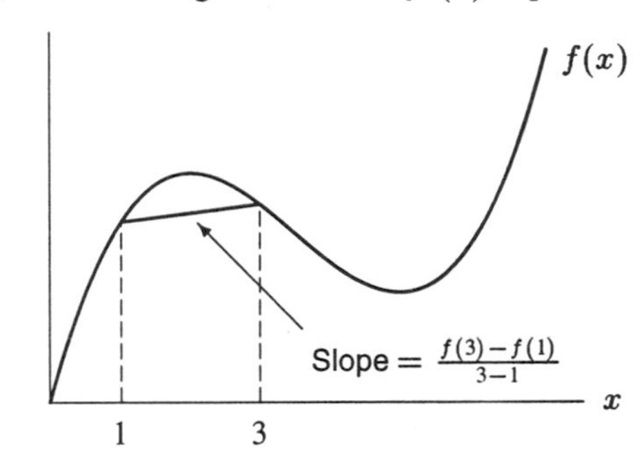

Figure 2.21

(b) This expression is the slope of the secant line between $x = 1$ and $x = 3$ and we see in Figure 2.21 that this slope is positive.

(c) Since $f(4)$ is the value of the function at $x = 4$ and $f(2)$ is the value of the function at $x = 2$, the expression $f(4) - f(2)$ is the change in the value of the function between $x = 4$ and $x = 2$. Since $f(4)$ lies below $f(2)$, this change is negative. See Figure 2.22.

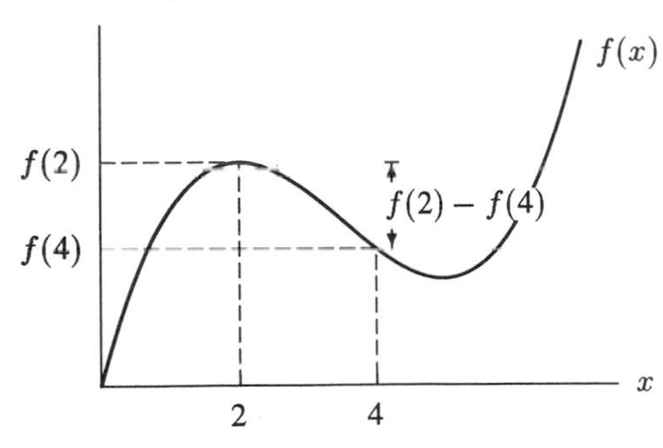

Figure 2.22

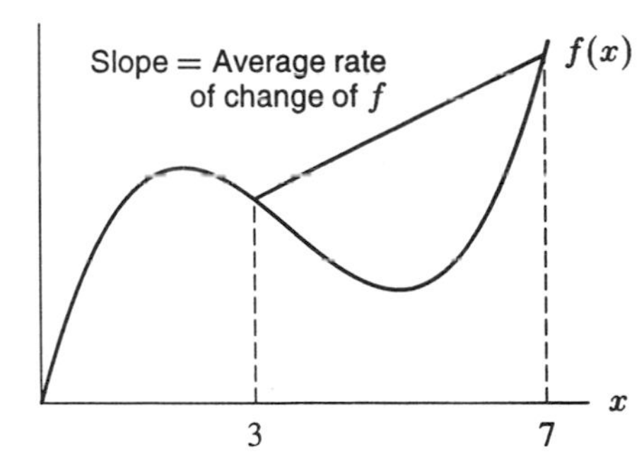

Figure 2.23

(d) The average rate of change is the slope of the secant line in Figure 2.23, which shows that this slope is positive.
(e) The instantaneous rate of change is the slope of the curve at $x = 3$, which we see from Figure 2.24 is negative.

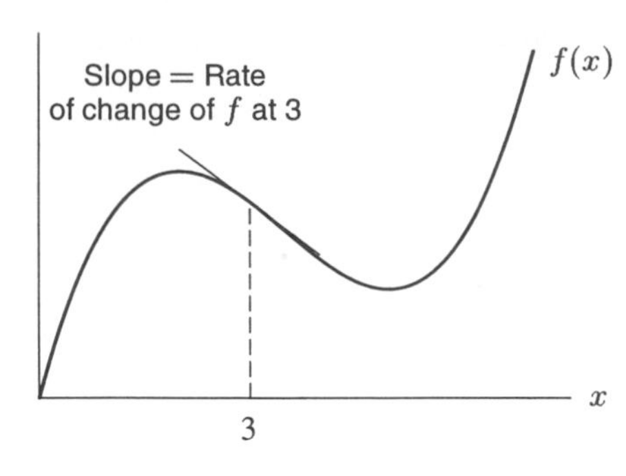

Figure 2.24

Problems for Section 2.2

1. A car is driven at a constant speed. Sketch a graph of the distance the car has traveled as a function of time.
2. A car is driven at an increasing speed. Sketch a graph of the distance the car has traveled as a function of time.
3. A car starts at a high speed, and its speed then decreases slowly. Sketch a graph of the distance the car has traveled as a function of time.
4. A bicyclist pedals at a fairly constant rate, with evenly spaced intervals of coasting. Sketch a graph of the distance she has traveled as a function of time.
5. Find the average velocity over the interval $0 \leq t \leq 0.2$, and estimate the velocity at $t = 0.2$ of a car whose position, s, is given by

t (sec)	0	0.2	0.4	0.6	0.8	1.0
s (ft)	0	0.5	1.8	3.8	6.5	9.6

6. Suppose the distance (in feet) of an object from a point is given by $s(t) = t^2$, where t is measured in seconds.
   (a) What is the average velocity of the object between $t = 3$ and $t = 5$?
   (b) By using smaller and smaller intervals around 3, estimate the instantaneous velocity at time $t = 3$.

7. A ball is tossed into the air from a bridge, and its height, y (in feet), above the ground t seconds after it is thrown is given by

$$y = f(t) = -16t^2 + 50t + 36.$$

(a) How high above the ground is the bridge?
(b) What is the average velocity of the ball for the first second?
(c) Approximate the velocity of the ball at $t = 1$ second.
(d) Graph the function f, and determine the maximum height the ball will reach. What should the velocity be at the time the ball is at the peak?
(e) Use the graph to decide at what time, t, the ball reaches its maximum height.

8. Match the points labeled on the curve in Figure 2.25 with the given slopes.

Slope	Point
-3	
-1	
0	
$1/2$	
1	
2	

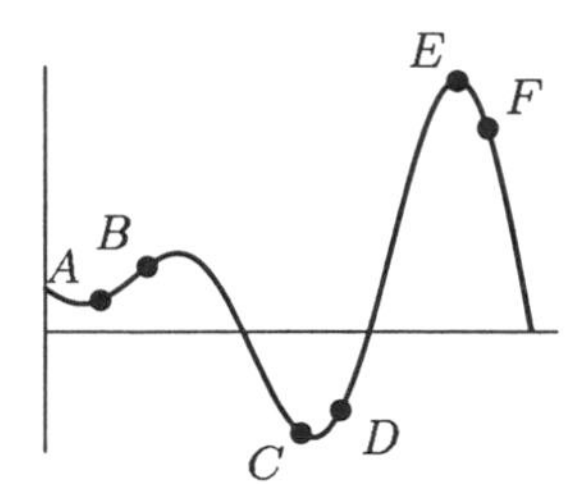

Figure 2.25

9. For the function shown in Figure 2.26, at what labeled points is the slope of the curve positive? Negative? Which labeled point has the greatest (i.e., most positive) slope? The least slope (i.e., negative and with the largest magnitude)?

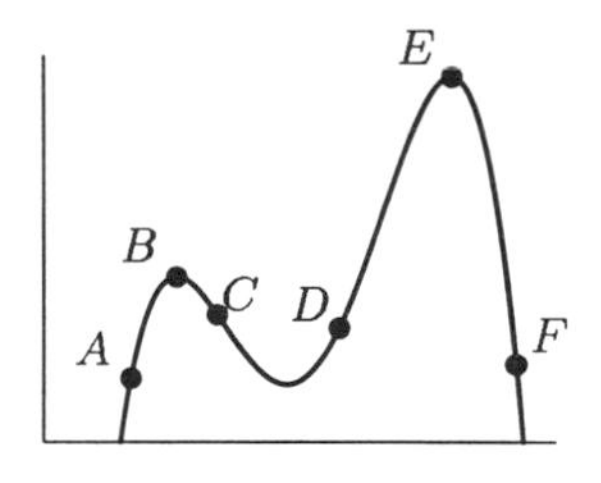

Figure 2.26

10. For the graph $y = f(x)$ shown in Figure 2.27, arrange the following numbers in ascending (i.e., smallest to largest) order:

- The slope of the curve at A.
- The slope of the curve at B.
- The slope of the curve at C.
- The slope of the line AB.
- The number 0.
- The number 1.

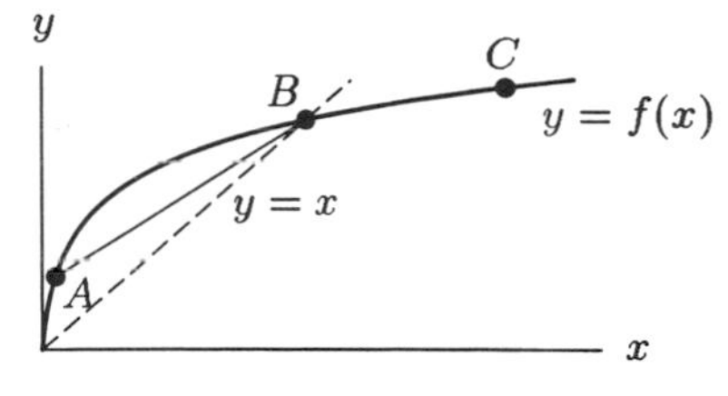

Figure 2.27

11. Suppose a particle is moving at varying velocity along a straight line and that $s = f(t)$ represents the distance of the particle from a point as a function of time, t. Sketch a possible graph for f if the average velocity of the particle between $t = 2$ and $t = 6$ is the same as the instantaneous velocity at $t = 5$.

12. Table 2.7 gives the percent of the US population living in urban areas as a function of year[2].

TABLE 2.7 *Percent of US population living in urban areas*

Year	1800	1830	1860	1890	1920	1950	1960	1970	1980
Percent urban	6	9	20	35	51	64	69.9	73.5	73.7

(a) Find the average rate of change in the percent of the US population living in urban areas between 1860 and 1960.
(b) Estimate the rate at which this percent is increasing at the year 1960.
(c) Estimate the derivative of this function for the year 1830 and explain what it is telling you.
(d) Estimate the derivative of this function for the year 1970 and explain what it is telling you.
(e) Is the derivative of this function getting larger or smaller over time? What does this mean about how the population is changing?
(f) Is this function increasing or decreasing? Concave up or concave down?

13. Sketch a rough graph of $f(x) = e^x$, and use the graph to decide whether the derivative of $f(x)$ at $x = 1$ is positive or negative. Give reasons for your decision.

14. Show how you can represent the following on a sketch similar to that in Figure 2.28.

(a) $f(4)$ (b) $f(4) - f(2)$ (c) $\dfrac{f(5) - f(2)}{5 - 2}$

15. Consider the function $y = f(x)$ shown in Figure 2.28. For each of the following pairs of numbers, decide which is larger. Explain your answer.

(a) $f(3)$ or $f(4)$? (b) $f(3) - f(2)$ or $f(2) - f(1)$?

(c) $\dfrac{f(2) - f(1)}{2 - 1}$ or $\dfrac{f(3) - f(1)}{3 - 1}$?

16. Suppose $y = f(x)$ graphed in Figure 2.28 represents the cost of manufacturing x kilograms of a chemical. Then $f(x)/x$ represents the average cost of producing 1 kilogram when x kilograms are made. This problem asks you to visualize these averages graphically.

(a) Show how to represent $f(4)/4$ as the slope of a line.
(b) Which is larger, $f(3)/3$ or $f(4)/4$?

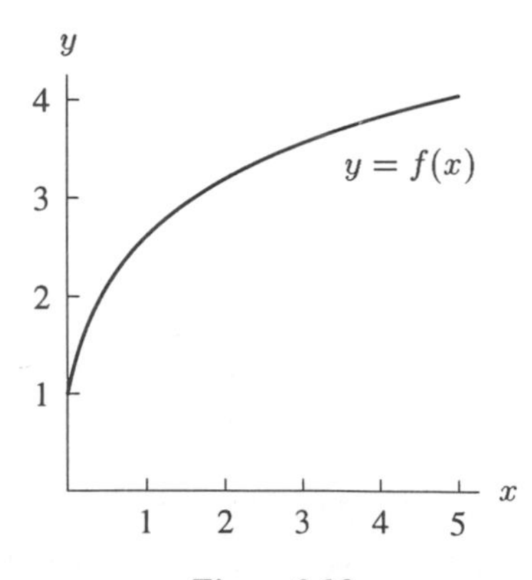

Figure 2.28

[2] *Statistical Abstracts of the US*, 1985, US Department of Commerce, Bureau of the Census, p.22

17. With the function f given by Figure 2.28, arrange the following quantities in ascending order:

$$0, \quad 1, \quad f'(2), \quad f'(3), \quad f(3) - f(2)$$

18. Refer to the graph of the function k in Figure 2.29:
    (a) Between which pair of consecutive points is the average rate of change of k greatest?
    (b) Between which pair of consecutive points is the average rate of change of k closest to zero?
    (c) Between which two pairs of consecutive points are the average rates of change of k closest?

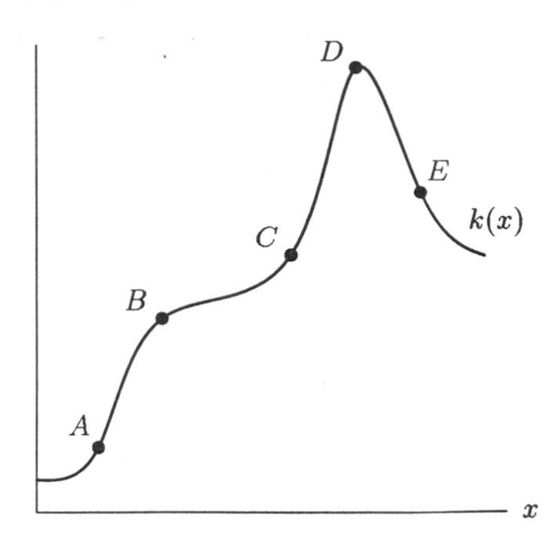

Figure 2.29

19. Estimate $f'(0)$ if $f(x) = e^{2x}$.
20. Estimate the derivative of $f(x) = x^x$ at $x = 2$.
21. For $y = f(x) = 3x^{3/2} - x$, use your calculator to construct a graph of $y = f(x)$, for $0 \leq x \leq 2$. From your graph, estimate $f'(0)$ and $f'(1)$.
22. Let $f(x) = \ln x$. Estimate $f'(2)$ by two different methods.
    (a) Using a graph.
    (b) By using smaller and smaller intervals around 2. Give your answer accurate to one decimal place.
23. Let $f(x) = x + e^x + e^{-x}$.
    (a) Use the graph of $f(x)$ to decide if the derivative is positive, negative, or zero at the points $x = -1$, $x = 0$, and $x = 1$.
    (b) By zooming in on the graph of $f(x)$, estimate the value of $f'(0)$.
24. The population, P, of China, in billions, can be approximated by the function

$$P = 1.15(1.014)^t$$

where t is the number of years since the start of 1993. According to this model, how fast is the population growing at the start of 1993 and at the start of 1995? Give your answers in millions of people per year.

25. (a) Sketch graphs of the functions $f(x) = \frac{1}{2}x^2$ and $g(x) = f(x) + 3$ on the same set of axes. What can you say about the slopes of the tangent lines to the two graphs at the point $x = 0$? $x = 2$? $x = x_0$?
    (b) Show that adding a constant value, C, to any function does not change the value of the slope of its graph at any point. [Hint: Let $g(x) = f(x) + C$, and calculate the difference quotients for f and g.]
26. Let $f(x) = \ln(\cos x)$. Use your calculator to approximate the instantaneous rate of change of f at the point $x = 1$. Do the same thing for $x = \pi/4$. (Note: Be sure that your calculator is set in radians.)

2.3 THE DEFINITION OF THE DERIVATIVE

In the last section, we saw how to estimate a derivative as the rate of change at a point by using the average rate of change over smaller and smaller intervals. In this section, we will see how to use the definition to find a derivative. We begin by returning to average rate of change. Recall that if $y = f(x)$, the average rate of change on an interval $a \le x \le b$ is given by

$$\text{Average rate of change} = \frac{\text{change in } y}{\text{change in } x} = \frac{f(b) - f(a)}{b - a}.$$

To find the derivative at $x = a$, we want to look at smaller and smaller intervals near $x = a$. We will consider intervals of the form $a \le x \le a + h$, where h is the length of the interval. Then, over the interval $a \le x \le a + h$,

$$\text{Average rate of change} = \frac{f(a+h) - f(a)}{(a+h) - a} = \frac{f(a+h) - f(a)}{h}.$$

Figure 2.30 illustrates the case of $h > 0$.

In the previous section we agreed that the instantaneous rate of change is the number that the average rates of change approach as the intervals decrease in size, that is, as h becomes smaller. Thus, the instantaneous rate of change at $x = a$ is

$$\text{The limit, as } h \text{ approaches 0, of } \frac{f(a+h) - f(a)}{h}.$$

One final small change is for the sake of economy: Instead of writing the phrase "limit, as h approaches 0, of," we will just use the notation

$$\lim_{h \to 0}.$$

We have

$$\begin{array}{c}\text{Instantaneous rate of} \\ \text{change at } x = a\end{array} = \lim_{h \to 0} \frac{f(a+h) - f(a)}{h}.$$

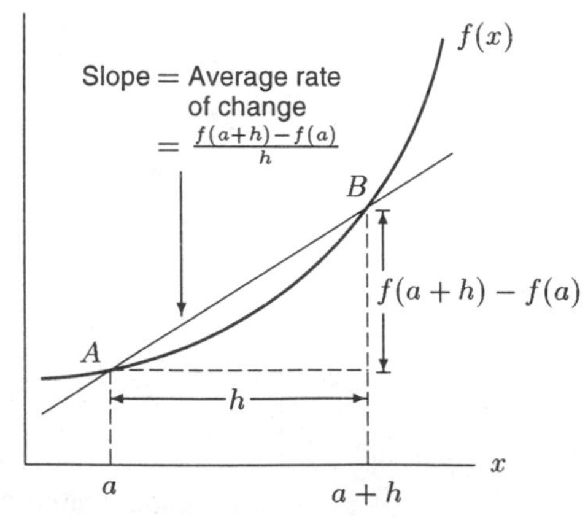

Figure 2.30: Visualizing the average rate of change of f

Since the derivative is just another name for the instantaneous rate of change, we can now exhibit the definition of the derivative in all its finery:

$$\textbf{The derivative of } f \textbf{ at } a = \lim_{h\to 0} \frac{f(a+h) - f(a)}{h}.$$

This expression forms the foundation of the rest of calculus. Be sure that you are not confused by the notation and recognize it for what it is: the number that the average rates of change approach as the intervals shrink. To find that number, the limit, we look at intervals of smaller and smaller, but never zero, length. You should realize that we haven't introduced any new ideas in this definition; we have simply introduced a compact way to write the ideas developed previously.

What a Limit Is and How to Find One

Let's look at what it means to take a limit in more detail. Suppose, for example, that $a = 5$ and $f(t) = t^2$, and that we want to calculate the limit, L:

$$L = \lim_{h\to 0} \frac{f(5+h) - f(5)}{h}.$$

Then L is the *number* obtained by evaluating the expression $(f(5+h) - f(5))/h$ at values of h getting closer and closer to 0, and observing to what number these values converge. Let's pick a value of h close to 0, say $h = 0.1$, and see what value we get for this expression:

$$\frac{f(5+h) - f(5)}{h} = \frac{f(5.1) - f(5)}{0.1} = \frac{(5.1)^2 - (5)^2}{0.1} = \frac{26.01 - 25}{0.1} = \frac{1.01}{0.1} = 10.1$$

Notice that we have just calculated the average rate of change between 5 and 5.1. We see that the average rate of change is 10.1 when $h = 0.1$. To find the instantaneous rate of change, we calculate the average rate of change as $h \to 0$, so we take values of h closer and closer to 0 (both positive and negative). The results are shown in Table 2.8.

TABLE 2.8 *Finding a limit numerically*

h	0.1	0.01	0.001	0.0001	-0.1	-0.01	-0.001	-0.0001
$(f(5+h) - f(5))/h$	10.1	10.01	10.001	10.0001	9.9	9.99	9.999	9.9999

The values of $(f(5+h) - f(5))/h$ in Table 2.8 seem to be converging to the number 10 as $h \to 0$. So it is a good guess that

$$L = \lim_{h\to 0} \frac{f(5+h) - f(5)}{h} = 10.$$

However, from Table 2.8 we can't be sure that the limit isn't, for example, 10.0000001 or 9.99999. Showing that the limit is *exactly* 10 requires algebra techniques not developed in this text.

Notice that by taking a small value of h, we get a good approximation to L, though we don't get L exactly. We generally estimate limits numerically by the method below.

If
$$L = \lim_{h\to 0} \frac{f(a+h)-f(a)}{h},$$
then we often approximate L by taking a small value of h, giving
$$L \approx \frac{f(a+h)-f(a)}{h}.$$

Using the Definition of the Derivative

Example 1 Find the derivative of the function $f(x) = x^2$ at the point $x = 1$.

Solution We need to look at
$$f'(1) = \lim_{h\to 0} \frac{f(1+h)-f(1)}{h}.$$

This is the same as
$$\lim_{h\to 0} \frac{(1+h)^2 - 1^2}{h} = \lim_{h\to 0} \frac{(1+2h+h^2)-1}{h} = \lim_{h\to 0} \frac{2h+h^2}{h}.$$

Now choose several small values for h (e.g. $0.1, 0.01, 0.001$), compute $(2h+h^2)/h$ for these, and see if you can guess the limit. An alternative approach is to divide by h in the expression $(2h+h^2)/h$, a valid operation since the limit only examines values of h close to, but not equal to, zero. We get
$$\lim_{h\to 0} \frac{h(2+h)}{h} = \lim_{h\to 0}(2+h).$$

You can see that this limit is 2, so $f'(1) = 2$. Thus, at $x = 1$ the rate of change of x^2 is 2.

Graphically, what does it mean that the derivative of $f(x) = x^2$ at the point $x = 1$ is 2? Since the derivative is the rate of change, it means that for small changes in x, near $x = 1$, the change in $f(x) = x^2$ is about twice as big as the change in x. As an example, if x changes from 1 to 1.1, a net change of 0.1, $f(x)$ should change by about 0.2. Figure 2.31 shows this geometrically.

Table 2.9 shows the derivative of $f(x) = x^2$ numerically. Notice that near $x = 1$, every time the value of x increases by 0.001, the value of x^2 increases by approximately 0.002. Thus near $x = 1$ the graph is approximately linear with slope $0.002/0.001 = 2$.

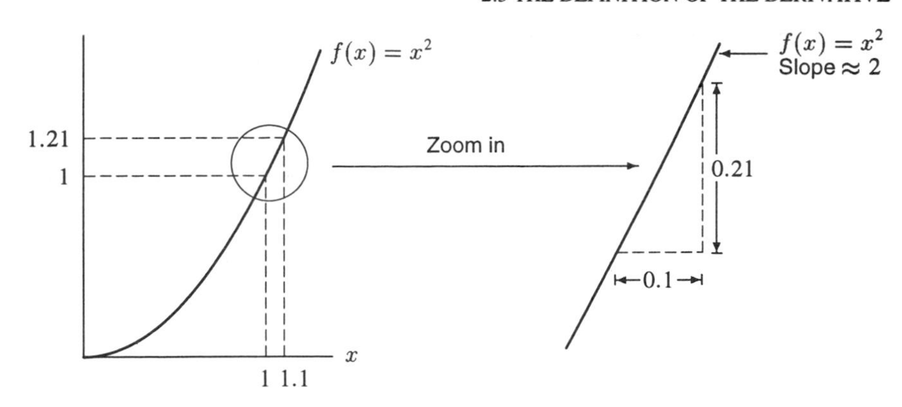

Figure 2.31: Graph of $f(x) = x^2$ near $x = 1$ has slope ≈ 2

TABLE 2.9 *Values of $f(x) = x^2$ near $x = 1$*

x	x^2	Difference in successive x^2 values
0.998	0.996004	
		0.001997
0.999	0.998001	
		0.001999
1.000	1.000000	
		0.002001
1.001	1.002001	
		0.002003
1.002	1.004004	
↑ x increments 0.001		↑ all approximately 0.002

Example 2 Find the value of the derivative of $f(x) = x^2 + 1$ at $x = 3$ algebraically. Find the equation of the tangent line to f at $x = 3$.

Solution We need to look at the difference quotient and take the limit as h approaches zero. The difference quotient is

$$\frac{f(3+h)-f(3)}{h} = \frac{(3+h)^2+1-10}{h} = \frac{9+6h+h^2-9}{h} = \frac{6h+h^2}{h} = \frac{h(6+h)}{h}.$$

Since $h \neq 0$, we can divide by h in the last expression to get $6+h$. Now the limit as h goes to 0 of $6+h$ is clearly 6, so,

$$f'(3) = \lim_{h\to 0} \frac{6h+h^2}{h} = \lim_{h\to 0}(6+h) = 6.$$

Thus, we know that the slope of the tangent line at $x = 3$ is 6. Since $f(3) = 10$, the tangent line passes through $(3, 10)$, so the equation of the tangent line is

$$y - 10 = 6(x-3) \quad \text{or} \quad y = 6x - 8.$$

Example 3 Table 2.10 shows values of $f(x) = x^3$ near $x = 2$. Use it to estimate $f'(2)$.

TABLE 2.10 *Values of x^3 (to three decimal places)*

x	1.998	1.999	2.000	2.001	2.002
x^3	7.976	7.988	8.000	8.012	8.024

Solution The derivative, $f'(2)$, is the rate of change of x^3 at $x = 2$. Notice that each time x changes by 0.001 in the table, the value of x^3 changes by 0.012. Therefore, we estimate

$$f'(2) = \begin{array}{c}\text{Rate of change}\\ \text{of } f \text{ at } x = 2\end{array} \approx \frac{0.012}{0.001} = 12.$$

The function values in the table look exactly linear because they have been rounded. For example, the exact value of x^3 when $x = 2.001$ is 8.012006001, not 8.012. Thus, the table can tell us only that the derivative is approximately 12. Showing that the derivative is exactly 12 (which it is) requires algebra and will be done in Chapter 4.

Problems for Section 2.3

1. On a sketch of $y = f(x)$ similar to that in Figure 2.32, mark lengths that represent the quantities in parts (a) – (d). (Pick any convenient x, and assume $h > 0$.)
(a) $f(x)$ (b) $f(x+h)$ (c) $f(x+h) - f(x)$ (d) h
(e) Using your answers to parts (a)–(d), show how the quantity $\dfrac{f(x+h) - f(x)}{h}$ can be represented as the slope of a line on the graph.

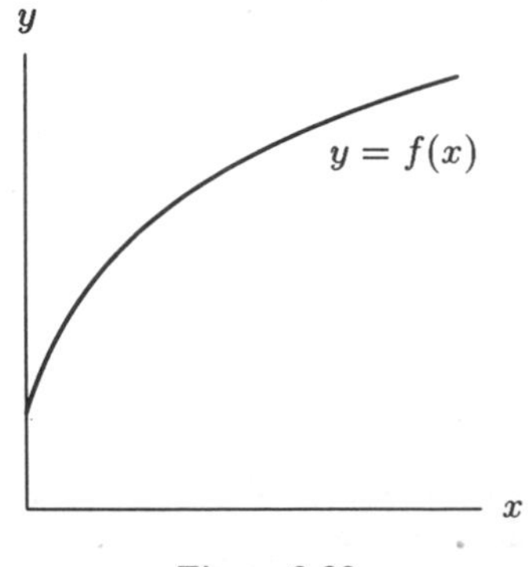

Figure 2.32

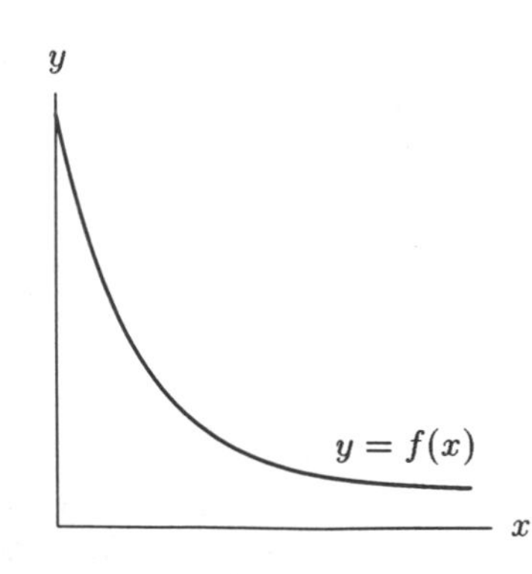

Figure 2.33

2. On a sketch of $y = f(x)$ similar to that in Figure 2.33, mark lengths that represent the quantities in parts (a) – (d). (Pick any convenient x, and assume $h > 0$.)
(a) $f(x)$ (b) $f(x+h)$ (c) $f(x+h) - f(x)$ (d) h
(e) Using your answers to parts (a)–(d), show how the quantity $\dfrac{f(x+h)-f(x)}{h}$ can be represented as the slope of a line on the graph.

Estimate the limits in Problems 3–6 by substituting smaller and smaller values of h. Give your answers to one decimal place.

3. $\displaystyle\lim_{h\to 0} \frac{(3+h)^2 - 9}{h}$

4. $\displaystyle\lim_{h\to 0} \frac{7^h - 1}{h}$

5. $\displaystyle\lim_{h\to 0} \frac{\ln(2+h) - \ln(2)}{h}$

6. $\displaystyle\lim_{h\to 0} \frac{e^{1+h} - e}{h}$

7. Find the derivative of $f(x) = x^3 + 5$ at $x = 1$ algebraically.

8. Find the derivative of $g(x) = x^2 - 3x$ at $x = 2$ algebraically.

9. If $g(t) = 3t^2 + 5t$, find $g'(-1)$ algebraically.

10. If $g(z) = z^2 + 4z - 5$, find $g'(2)$ algebraically.

11. Find the equation of the tangent line to $f(x) = x^2 + x$ at $x = 3$. Sketch a graph of the function and this tangent line.

12. Find the equation of the tangent line to $f(x) = 4 - x^2$ at the point $(1, 3)$. Sketch a graph of the function and this tangent line.

13. If $f(x) = x^3 + 4x$, estimate $f'(3)$ using a table similar to that in Example 3 on page 148.

14. For $g(x) = x^5$, use tables similar to that in Example 3 on page 148 to estimate $g'(2)$ and $g'(-2)$. What relationship do you notice between $g'(2)$ and $g'(-2)$? Explain geometrically why this must occur.

2.4 THE DERIVATIVE FUNCTION

In the last section we looked at the derivative of a function at a fixed point. Now we'll consider what happens at a variety of points and see that, in general, the derivative takes on different values at different points and is itself a function.

First, remember that the derivative of a function at a point tells you the rate at which the value of the function is changing at that point. Geometrically, if you "zoom in" on a point in the graph until it looks like a straight line, the slope of that line is the derivative at that point. Equivalently, you can think of the derivative as the slope of the tangent line at the point, because as you "zoom in," the curve and the tangent line become indistinguishable.

Example 1 Estimate the derivative of the function given by the graph in Figure 2.34 at $x = -2, -1, 0, 1, 2, 3, 4, 5$.

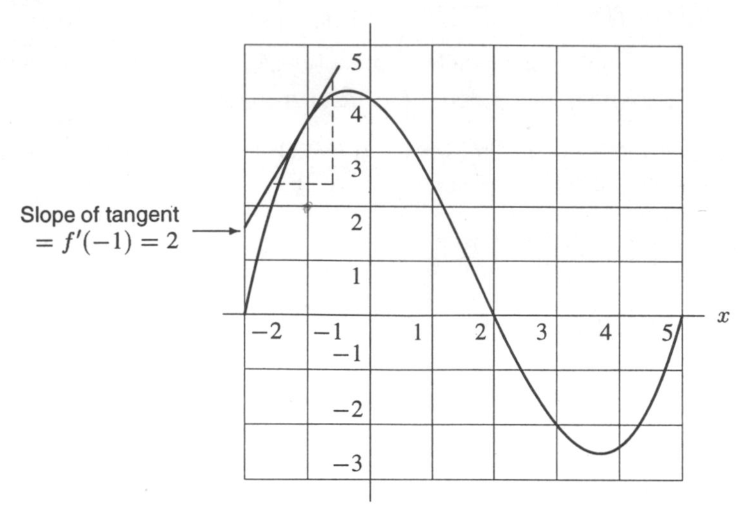

Figure 2.34: Estimating the derivative graphically as the slope of the tangent line

Solution From the graph you can estimate the derivative at any point by placing a straight edge so that it forms the tangent line at that point, and then using the grid squares to estimate the slope of the straight edge. (For example, the tangent at $x = -1$ is drawn in Figure 2.34, and has a slope of about 2, so $f'(-1) \approx 2$.) Notice that the slope at $x = -2$ is positive and fairly large; the slope at $x = -1$ is positive but smaller. At $x = 0$, the slope is negative, by $x = 1$ it has become more negative, and so on. Some estimates of the derivative are listed in Table 2.11. You should check these values yourself. Are they reasonable? Is the derivative positive where you expect? Negative?

TABLE 2.11 *Estimated values of derivative of function in Figure 2.34*

x	−2	−1	0	1	2	3	4	5
Derivative at x	6	2	−1	−2	−2	−1	1	4

The important point to notice is that for every x value, there's a corresponding value of the derivative. The derivative, therefore, is itself a function of x.

For any function f, we define the **derivative function**, f', by

$$f'(x) = \text{Rate of change of } f \text{ at } x = \lim_{h \to 0} \frac{f(x+h) - f(x)}{h}.$$

For every x value for which this limit exists, we say f is *differentiable* at that x value. If the limit exists for all x in the domain of f, we say f is *differentiable everywhere*. The functions we shall deal with will be differentiable at every point in their domain, except perhaps for a few isolated points.

Finding the Derivative of a Function Given Graphically

Example 2 Sketch the graph of the derivative of the function shown in Figure 2.34.

Solution Table 2.11 gives some values of this derivative which we can plot. However, it is a good idea first to identify some of the key features of the derivative graph from the graph of the original function. For example, we can see from Figure 2.34 (repeated in Figure 2.35) that the function is increasing from $x = -2$ to about $x = -0.5$. Thus, the derivative is positive in this interval, and so we must draw the graph of f' above the x-axis from $x = -2$ to about $x = -0.4$. Between $x = -0.4$ and about $x = 3.7$ the function is decreasing, so the derivative is negative and its graph must be below the x-axis. Beyond $x = 3.7$ the function is increasing, so the derivative is positive again and the graph of f' is above the axis. Somewhere in the region where the derivative is negative, it is going to reach its lowest point; this will be at the point where the graph of the original function is decreasing most steeply. From Figure 2.34 we see that this occurs a little before $x = 2$, where the slope is slightly steeper than -2. Thus, our derivative graph should have a minimum value of slightly below -2 occurring slightly to the left of $x = 2$. With this in mind, and using the data in Table 2.11, we obtain Figure 2.35, which shows a graph of the derivative, along with the original function.

You should check for yourself that this graph of f' makes sense. Notice that at the points where f has large upward slope, such as $x = -2$, the graph of the derivative is far above the x-axis, as it should be, since the value of the derivative should be large there. On the other hand, at points where the slope is gentle, the graph of f' is close to the x-axis, since the derivative is small.

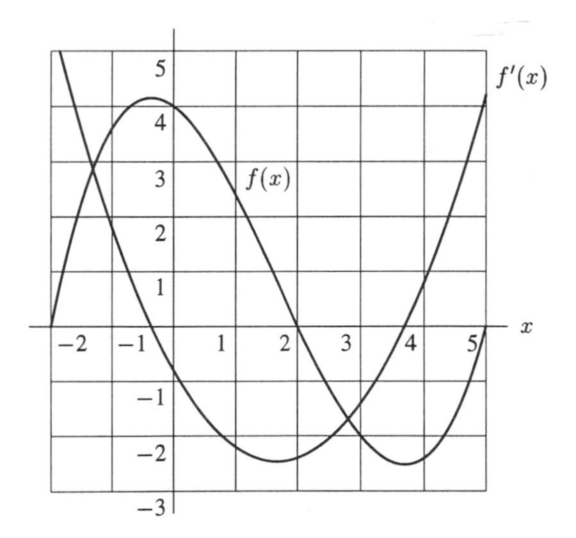

Figure 2.35: Function and derivative from Example 2

Finding the Derivative of a Function Given Numerically

If we are given a table of function values instead of a graph of the function, we can estimate values of the derivative.

Example 3 Suppose Table 2.12 gives values of $c(t)$, the concentration (mg/cc) of a drug in the bloodstream at time t (min). Construct a table of estimated values for $c'(t)$, the rate of change of $c(t)$ with respect to time.

TABLE 2.12 *Concentration as a function of time*

t (min)	0	0.1	0.2	0.3	0.4	0.5	0.6	0.7	0.8	0.9	1.0
$c(t)$ (mg/cc)	0.84	0.89	0.94	0.98	1.00	1.00	0.97	0.90	0.79	0.63	0.41

Solution We want to estimate the derivative of c using the values in the table. To do this, we have to assume that the data points are close enough together that the concentration doesn't change wildly between them. From the table, we can see that the concentration is increasing between $t = 0$ and $t = 0.4$, so we'd expect a positive derivative there. However, the increase is quite slow, so we would expect the derivative to be small. The concentration doesn't change between 0.4 and 0.5, so we expect the derivative to be 0 there. From $t = 0.5$ to $t = 1.0$, the concentration starts to decrease, and the rate of decrease gets larger and larger, so we would expect the derivative to be negative and of greater and greater magnitude.

Using the data in the table, we can estimate the derivative using the difference quotient:

$$c'(t) \approx \frac{c(t+h) - c(t)}{h}$$

with $h = 0.1$. For example,

$$c'(0) \approx \frac{c(0.1) - c(0)}{0.1} = \frac{0.89 - 0.84}{0.1} = 0.5 \text{ mg/cc/min.}$$

Thus, we estimate that

$$c'(0) \approx 0.5.$$

Similarly, we get the estimates

$$c'(0.1) \approx \frac{c(0.2) - c(0.1)}{0.1} = \frac{0.94 - 0.89}{0.1} = 0.5$$
$$c'(0.2) \approx \frac{c(0.3) - c(0.2)}{0.1} = \frac{0.98 - 0.94}{0.1} = 0.4$$
$$c'(0.3) \approx \frac{c(0.4) - c(0.3)}{0.1} = \frac{1.00 - 0.98}{0.1} = 0.2$$
$$c'(0.4) \approx \frac{c(0.5) - c(0.4)}{0.1} = \frac{1.00 - 1.00}{0.1} = 0.0$$

and so on. These values are tabulated in Table 2.13. Notice that the derivative has small positive values up until $t = 0.4$, and then it gets more and more negative, as we expected. The slopes are shown on the graph of $c(t)$ in Figure 2.36.

TABLE 2.13
Derivative of concentration

t	$c'(t)$
0	0.5
0.1	0.5
0.2	0.4
0.3	0.2
0.4	0.0
0.5	−0.3
0.6	−0.7
0.7	−1.1
0.8	−1.6
0.9	−2.2

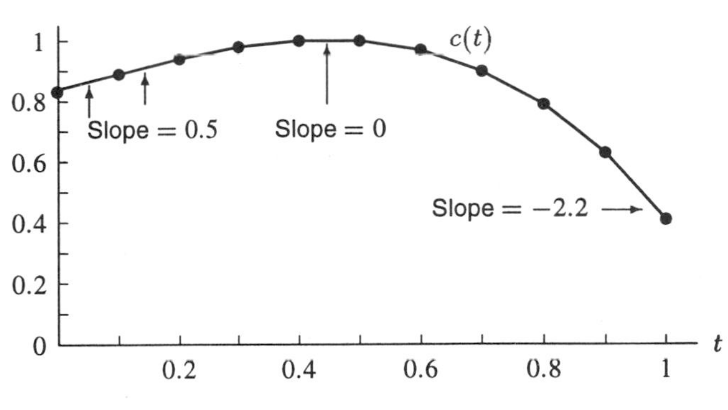

Figure 2.36: Graph of concentration as a function of time

Other Ways We Could Calculate the Derivative Numerically

In the previous example, our estimate for the derivative at 0.2 used the point to the right. We found the average rate of change between $t = 0.2$ and $t = 0.3$. However, we could equally well have gone to the left and used the rate of change between $t = 0.1$ and $t = 0.2$ to approximate the derivative at 0.2. For a more accurate result, we could average these slopes and say

$$c'(0.2) \approx \frac{1}{2}\left(\begin{matrix}\text{slope to left}\\ \text{of } 0.2\end{matrix} + \begin{matrix}\text{slope to right}\\ \text{of } 0.2\end{matrix}\right) = \frac{0.5 + 0.4}{2} = 0.45.$$

Each of these methods of approximating the derivative gives a reasonable answer. For convenience, unless there is a reason to do otherwise, we will estimate the derivative by going to the right.

Finding the Derivative of a Function Given by a Formula

If we are given a formula for f, can we come up with a formula for f'? Using the definition of the derivative, we often can, as shown in the next example. Indeed, much of the power of calculus depends on our ability to find formulas for the derivatives of all the functions in our library. This is done in detail in Chapter 4.

Example 4 Find a formula for the derivative of $f(x) = x^2$.

Solution Before computing the formula for $f'(x)$ algebraically, let's try to guess the formula by looking for a pattern in the values of $f'(x)$. Table 2.14 contains values of $f(x) = x^2$ (rounded to three decimals), which we can use to estimate the values of $f'(1)$, $f'(2)$, and $f'(3)$.

TABLE 2.14 *Values of $f(x) = x^2$ near $x = 1$, $x = 2$, $x = 3$ (rounded to three decimals)*

x	x^2 (approx)
0.999	0.998
1.000	1.000
1.001	1.002
1.002	1.004

x	x^2 (approx)
1.999	3.996
2.000	4.000
2.001	4.004
2.002	4.008

x	x^2 (approx)
2.999	8.994
3.000	9.000
3.001	9.006
3.002	9.012

Near $x = 1$, x^2 increases by about 0.002 each time x increases by 0.001, so

$$f'(1) \approx \frac{0.002}{0.001} = 2.$$

Similarly,

$$f'(2) \approx \frac{0.004}{0.001} = 4$$
$$f'(3) \approx \frac{0.006}{0.001} = 6.$$

Knowing the value of f' at specific points can never tell us the formula for f', but it certainly can be suggestive: knowing $f'(1) \approx 2$, $f'(2) \approx 4$, $f'(3) \approx 6$ certainly suggests that $f'(x) = 2x$.

Now we'll show that $f'(x) = 2x$ is the correct formula algebraically. The derivative is calculated by forming the difference quotient and taking the limit as h goes to zero. The difference quotient is

$$\frac{f(x+h) - f(x)}{h} = \frac{(x+h)^2 - x^2}{h} = \frac{x^2 + 2xh + h^2 - x^2}{h} = \frac{2xh + h^2}{h}.$$

Since h never actually reaches zero, we can divide it out in the last expression to get $2x + h$. The limit of this as h goes to zero is $2x$, so

$$f'(x) = \lim_{h\to 0}(2x + h) = 2x.$$

What Does the Derivative Tell Us Graphically?

When f' is positive, the tangent line is sloping up; when f' is negative, the tangent line is sloping down. If $f' = 0$ everywhere, then the tangent line is horizontal everywhere and so f is constant. Thus, the sign of f' tells us whether f is increasing or decreasing.

> If $f' > 0$ on an interval, then f is *increasing* over that interval.
> If $f' < 0$ on an interval, then f is *decreasing* over that interval.
> If $f' = 0$ on an interval, then f is *constant* over that interval.

Moreover, the magnitude of the derivative gives us the magnitude of the rate of change; so if f' is large (positive or negative), then the graph of f will be steep (up or down), whereas if f' is small the graph of f will slope gently. With this in mind, you can deduce a lot about the behavior of a function from the behavior of its derivative.

Example 5 Suppose the derivative of f is the spike function illustrated in Figure 2.37. What can you say about the graph of f itself?

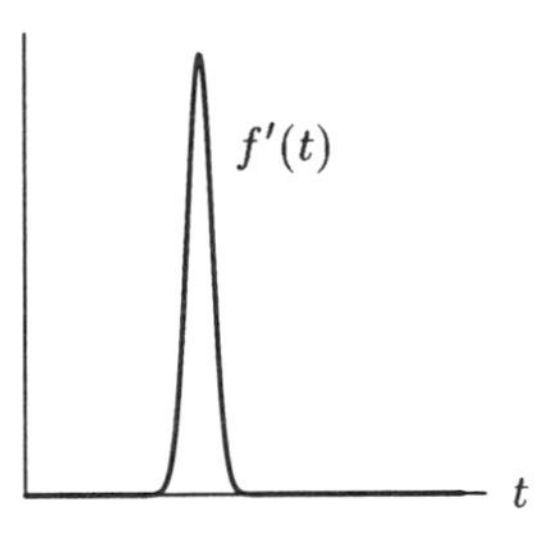

Figure 2.37: Spike derivative function

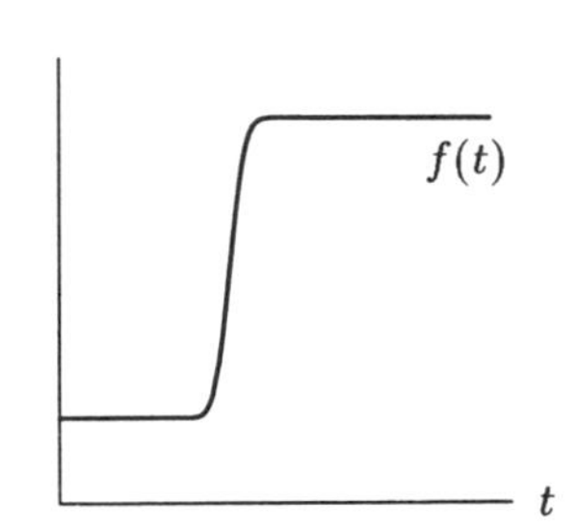

Figure 2.38: Step function

Solution On intervals where $f' = 0$, f is not changing at all, and is therefore constant. On the small interval where $f' > 0$, f is increasing; at the point where f' hits the top of its spike, f is increasing quite sharply. So f should be constant for a while, have a sudden increase, and then be constant again. A possible graph for f is shown in Figure 2.38.

Problems for Section 2.4

For Problems 1–6, sketch a graph of the derivative function of each of the given functions.

1.

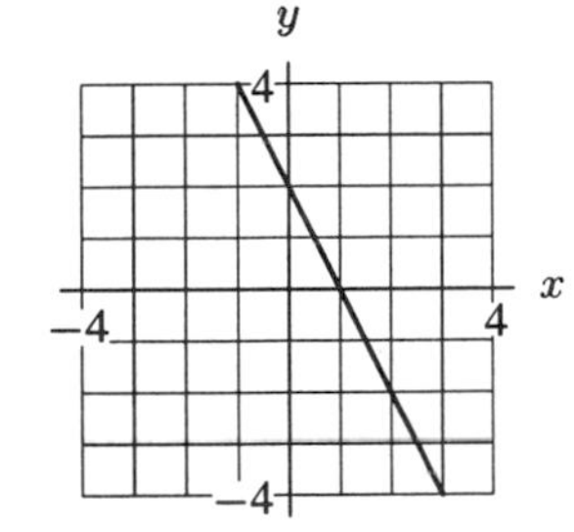

2.

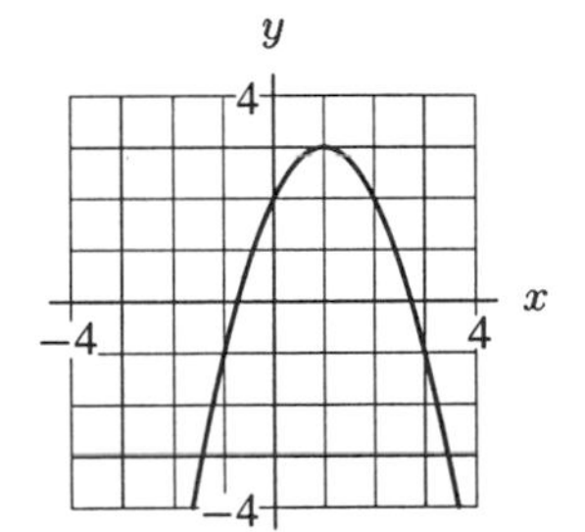

3.

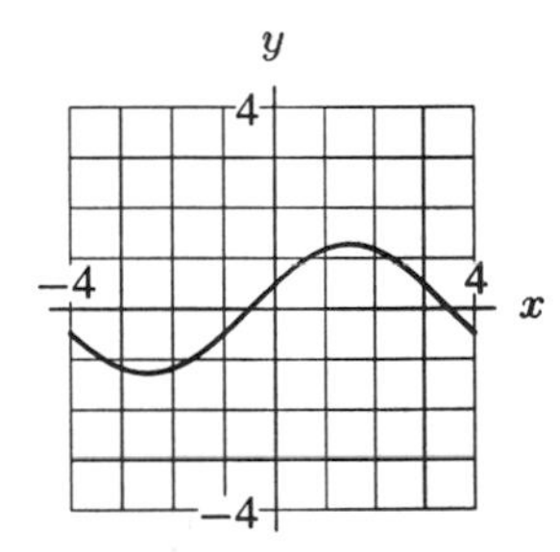

4.

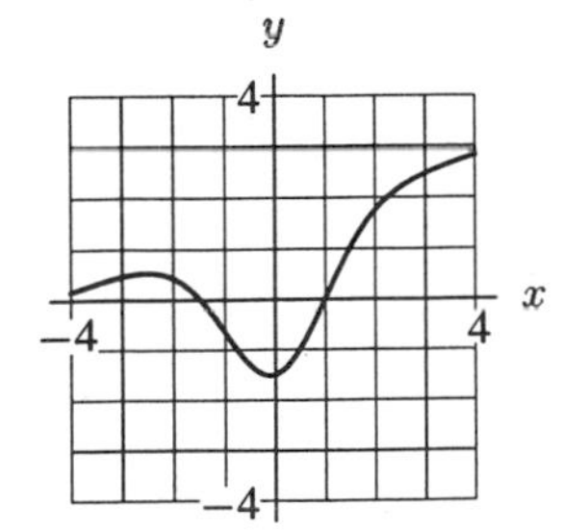

5.

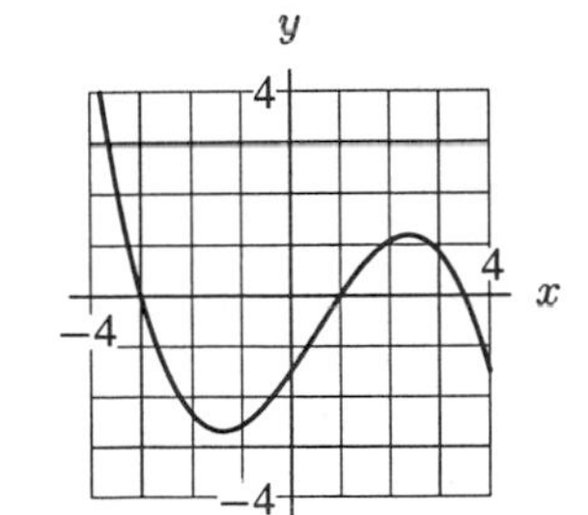

6.

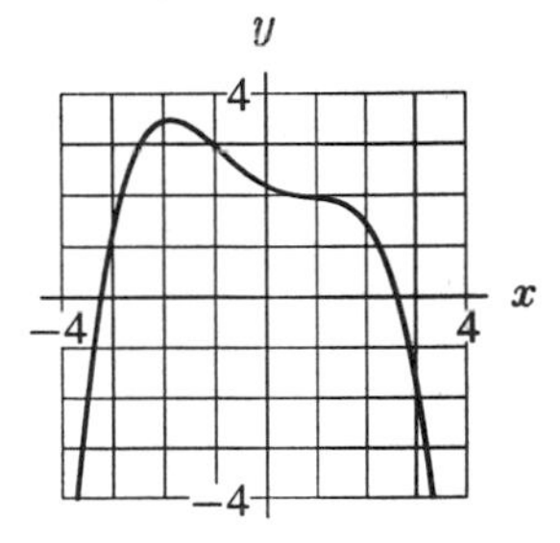

7. Sketch the derivative of the downward step function in Figure 2.39.

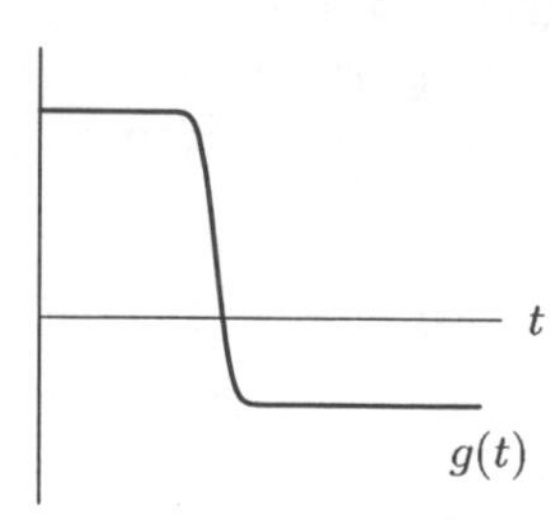

Figure 2.39: Downward step function

8. (a) Sketch a smooth curve whose slope is everywhere positive and increasing gradually.
(b) Sketch a smooth curve whose slope is everywhere positive and decreasing gradually.
(c) Sketch a smooth curve whose slope is everywhere negative and increasing gradually (i.e., becoming less and less negative).
(d) Sketch a smooth curve whose slope is everywhere negative and decreasing gradually (i.e., becoming more and more negative).

9. Given the numerical values shown, find approximate values for the derivative of $f(x)$ at each of the x values given. Where is the rate of change of $f(x)$ positive? Where is it negative? Where does the rate of change of $f(x)$ seem to be greatest?

x	0	1	2	3	4	5	6	7	8
$f(x)$	18	13	10	9	9	11	15	21	30

10. Suppose $f(x) = \frac{1}{3}x^3$. Make tables similar to Table 2.14 on page 153 to estimate $f'(2)$, $f'(3)$, and $f'(4)$. What do you notice? Can you guess a formula for $f'(x)$?

11. If $g(t) = t^2 + t$, use tables similar to Table 2.14 on page 153 to estimate $g'(1)$, $g'(2)$, and $g'(3)$. Use these to guess a formula for $g'(t)$.

Find a formula for the derivatives of the functions in Problems 12–15 algebraically.

12. $f(x) = x^3$ 13. $g(x) = 2x^2 - 3$ 14. $k(x) = 1/x$ 15. $l(x) = x^2 - 4x$

16. Draw a possible graph of $y = f(x)$ given the following information about its derivative.

- $f'(x) > 0$ on $1 < x < 3$
- $f'(x) < 0$ for $x < 1$ and $x > 3$
- $f'(x) = 0$ at $x = 1$ and $x = 3$

17. Draw a possible graph of $y = f(x)$ given the following information about its derivative.

- $f'(x) > 0$ for $x < -1$
- $f'(x) < 0$ for $x > -1$
- $f'(x) = 0$ at $x = -1$

18. In the graph of f in Figure 2.40, at which of the labeled x values is

(a) $f(x)$ greatest? (b) $f(x)$ least?
(c) $f'(x)$ greatest? (d) $f'(x)$ least?

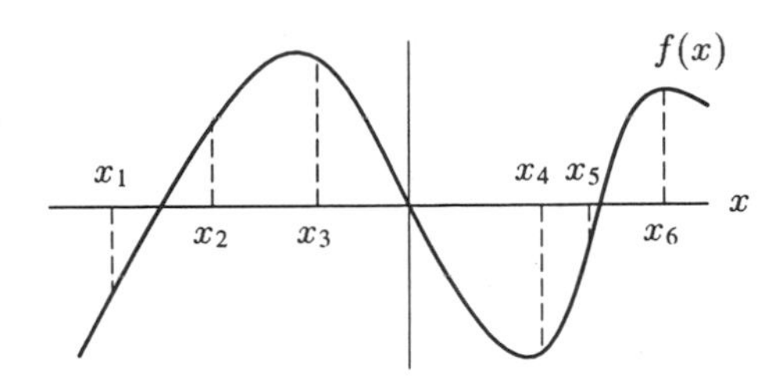

Figure 2.40

19. The solid waste generated each year by our cities has been increasing. Values for yearly solid waste (measured in millions of tons) as a function of year are given in Table 2.15[3].
    (a) Estimate the derivative of this function for the years 1960, 1965, 1970, 1975, and 1980.
    (b) Interpret these derivatives in terms of the amount of solid waste being generated.
    (c) Are the derivatives getting larger or smaller over time? What does this mean in terms of solid waste?

TABLE 2.15 *Municipal solid waste in the US*

Year	1960	1965	1970	1975	1980	1984
Waste (millions of tons)	82.3	98.3	118.3	122.7	139.1	148.1

For Problems 20–25, sketch the graph of $f(x)$, and use this graph to sketch the graph of $f'(x)$.

20. $f(x) = x(x-1)$
21. $f(x) = 5x$
22. $f(x) = x^3$
23. $f(x) = e^x$
24. $f(x) = \ln x$
25. $f(x) = \dfrac{1}{x^3}$

For Problems 26–31, sketch the graph of $y = f'(x)$ for the function given.

26.
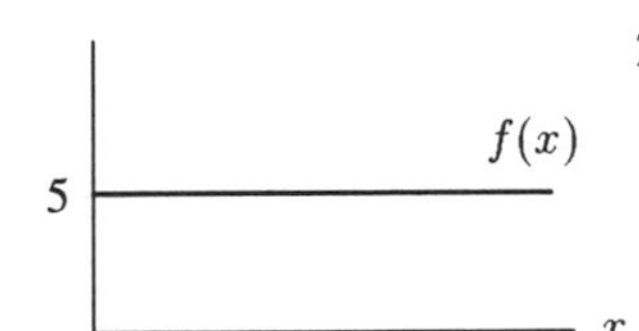

27.
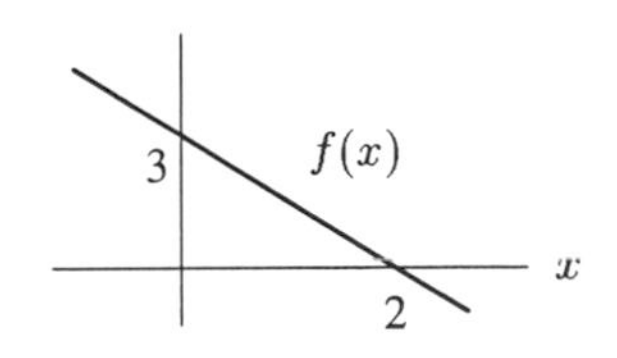

28.
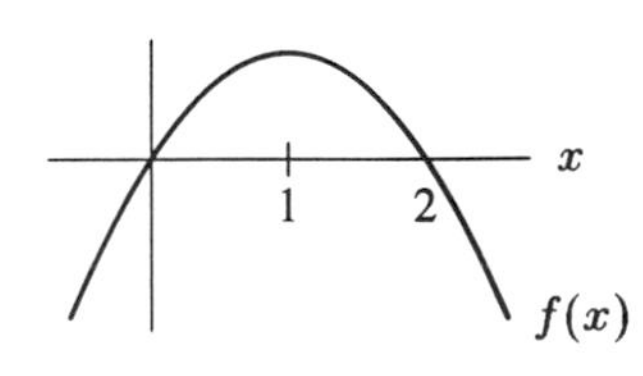

29.
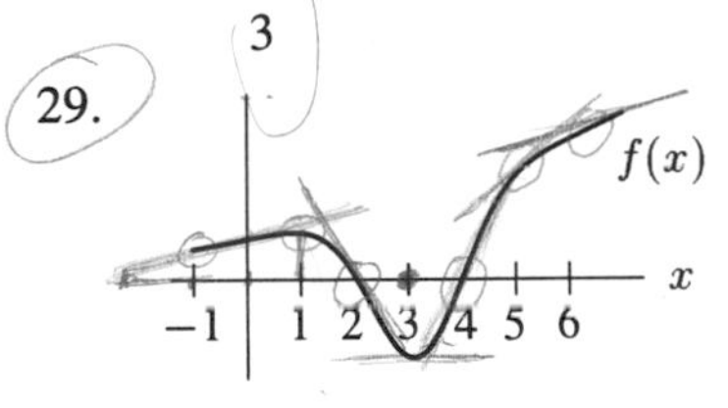

30.

31.
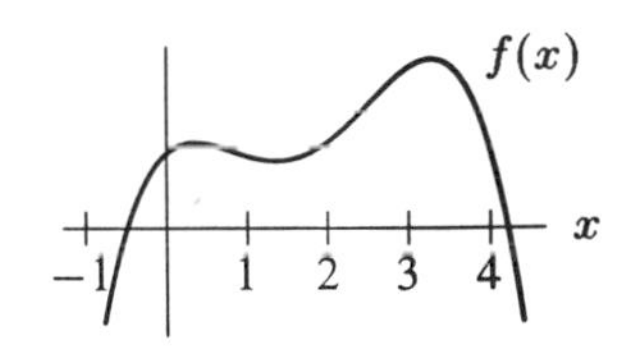

32. The amount of a certain bacteria in the blood (measured in number of bacteria per cubic centimeter) can be modeled by
$$f(t) = 100te^{-0.2t},$$

[3] *Statistical Abstracts of the US*, 1988, p193, Table 333

where t is measured in days since the person first became ill.

(a) How does the amount of bacteria vary with time? Sketch a graph of $f(t)$ from the time the person first becomes ill until three weeks later.
(b) Use the graph to decide when the population is a maximum. What is that maximum? Is there a minimum after the person first becomes ill? If so, when?
(c) Use the graph to decide when the population of bacteria is growing fastest. When is it decreasing fastest?
(d) Estimate roughly how fast the population is changing one week after exposure.

33. Show that if $f(x)$ is an even function, then $f'(x)$ is odd.
34. Show that if $g(x)$ is an odd function, then $g'(x)$ is even.
35. To give a patient an antibiotic slowly, the drug is injected into the muscle. (For example, penicillin for venereal disease is administered this way.) The quantity of the drug in the bloodstream starts out at zero, increases to a maximum, and then decays to zero again.
    (a) Sketch a possible graph of the quantity of the drug in the bloodstream as a function of time. Mark the time at which the drug is at a maximum by t_0.
    (b) Describe in words how the rate at which the drug is entering or leaving the blood changes over time. Sketch a graph of this rate against time, marking t_0 on the time axis.
36. The population of a herd of deer is modeled by

$$P(t) = 4000 + 500 \sin\left(2\pi t - \frac{\pi}{2}\right)$$

where t is measured in years.

(a) How does this population vary with time? Sketch a graph of $P(t)$ for one year.
(b) Use the graph to decide when in the year the population is a maximum. What is that maximum? Is there a minimum? If so, when?
(c) Use the graph to decide when the population is growing fastest. When is it decreasing fastest?
(d) Estimate roughly how fast the population is changing on the first of July.

2.5 INTERPRETATIONS OF THE DERIVATIVE

We have already seen how the derivative can be interpreted as a slope and as a rate of change. In this section, you will see examples of other interpretations. The point of these examples is not to make a catalog of interpretations but to illustrate the process of obtaining them.

Acceleration

We started this chapter by showing how velocity could be calculated as a rate of change of position with respect to time. If $s(t)$ measures the distance an object has moved from a reference point along a straight line, and $v(t)$ is its velocity at time t, then:

$$\text{Instantaneous velocity} = s'(t) = \lim_{h \to 0} \frac{s(t+h) - s(t)}{h}.$$

Now, *acceleration*, $a(t)$, is the rate of change of velocity with respect to time, so we define

$$\begin{array}{c}\text{Average acceleration} \\ \text{from } t \text{ to } t+h\end{array} = \frac{v(t+h) - v(t)}{h}$$

and

$$\text{Instantaneous acceleration} = v'(t) = \lim_{h \to 0} \frac{v(t+h) - v(t)}{h}.$$

If the term velocity or acceleration is used alone, it is assumed to be instantaneous.

Example 1 An accelerating sports car goes from 0 mph to 60 mph in five seconds. Its velocity is given in Table 2.16, converted from miles per hour to feet per second, so that all time measurements are in seconds. (Note: 1 mph is 22/15 ft/sec.) Find the average acceleration of the car over each of the first two seconds.

TABLE 2.16 *Velocity of sports car*

Time, t (sec)	0	1	2	3	4	5
Velocity, v (ft/sec)	0	30	52	68	80	88

Solution To measure the average acceleration over an interval, we calculate the average rate of change of velocity over the interval. The units of acceleration are ft/sec per second, or (ft/sec)/sec, written ft/sec^2.

$$\begin{array}{c}\text{Average acceleration} \\ \text{for } 0 \le t \le 1\end{array} = \frac{\text{Change in velocity}}{\text{Time}} = \frac{30-0}{1} = 30 \text{ ft/sec}^2.$$

$$\begin{array}{c}\text{Average acceleration} \\ \text{for } 1 \le t \le 2\end{array} = \frac{52-30}{2-1} = 22 \text{ ft/sec}^2.$$

Example 2 The graph in Figure 2.41 shows the velocity of a racing car as a function of time. Estimate its acceleration when $t = 1$.

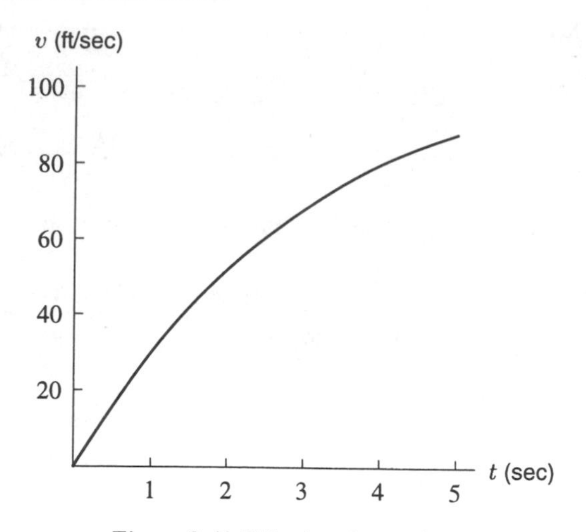

Figure 2.41: Velocity of racing car

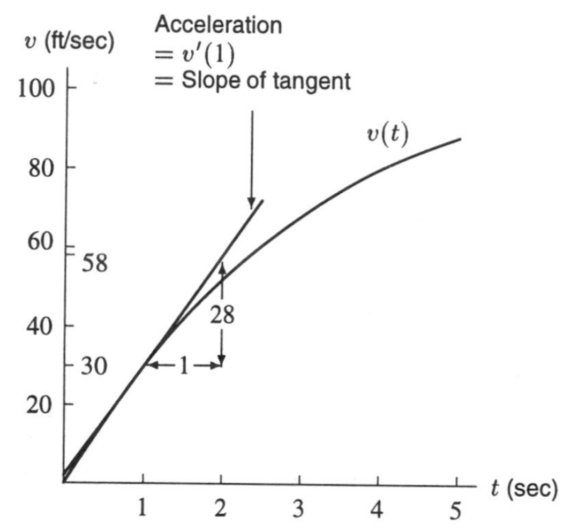

Figure 2.42: Estimating racing car's acceleration from velocity graph

Solution The acceleration at $t = 1$ is the derivative $v'(1)$, the slope of the tangent to the velocity curve at $t = 1$. (See Figure 2.42.) Thus, we estimate

$$\begin{array}{c}\text{Acceleration} \\ \text{at } t = 1\end{array} = v'(1) \approx \frac{28}{1} = 28 \text{ ft/sec}^2.$$

An Alternative Notation for the Derivative

So far we have used the notation f' to stand for the derivative of the function f. An alternative notation for derivatives was introduced by the German mathematician Gottfried Wilhelm Leibniz

(1646–1716) when calculus was first being developed in the seventeenth century. If the variable y depends on the variable x, that is, $y = f(x)$, then we let dy/dx stand for the derivative $f'(x)$. In other words, if

$$y = f(x),$$

then we write

$$\frac{dy}{dx} = f'(x).$$

Leibniz's notation is quite suggestive, especially if you think of the letter d in dy/dx as standing for "small difference in . . . ." The notation dy/dx reminds us that the derivative is a limit of ratios of the form

$$\frac{\text{Difference in } y \text{ values}}{\text{Difference in } x \text{ values}}.$$

It is always a good idea to have a mathematical symbol let us know where it came from and what it means: dy/dx does this and $f'(x)$ does not. The notation dy/dx is useful for determining the units for the derivative: the units for dy/dx are the units for y divided by (or "per") the units for x. The d/dx notation can also be very convenient. For example, it is a lot easier to say

"$\frac{d}{dx}(x^2 + 3x) = 2x + 3$" than it is to say "if $f(x) = x^2 + 3x$, then $f'(x) = 2x + 3$."

The separate entities dy and dx officially have no independent meaning: they are all part of one notation. In fact, a good formal way to view the notation dy/dx is to think of d/dx as a single symbol meaning "the derivative with respect to x of . . .". Thus dy/dx could be viewed as

$$\frac{d}{dx}(y), \quad \text{meaning "the derivative with respect to } x \text{ of } y\text{."}$$

On the other hand, many scientists and mathematicians really do think of dy and dx as separate entities representing "infinitesimally" small differences in y and x, even though it is difficult to say exactly how small "infinitesimal" is. It may not be formally correct, but it is very helpful to the intuition to think of dy/dx as a very small change in y divided by a very small change in x.

For example, recall that if $s = f(t)$ is the position of a moving object at time t, then $v = f'(t)$ is the velocity of the object at time t. Writing

$$v = \frac{ds}{dt}$$

directly reminds you of this fact, since it suggests a distance, ds, over a time, dt, and we know that distance over time is velocity. Similarly, we recognize

$$\frac{dy}{dx} = f'(x)$$

as the slope of the graph of $y = f(x)$ by remembering that slope is vertical rise, dy, over horizontal run, dx.

The disadvantage of the Leibniz's notation is that it is rather awkward if you want to specify the x value at which you are evaluating the derivative. To specify $f'(2)$, for example, we have to write

$$\left.\frac{dy}{dx}\right|_{x=2}.$$

Using Units to Interpret the Derivative

The following examples illustrate the fact that if you want to interpret a derivative in practical terms, it often helps to think about units of measurement.

For example, suppose $s = f(t)$ gives the position in meters of a body from a fixed point as a function of time, t, in seconds. Then knowing that

$$\frac{ds}{dt} = f'(2) = 10 \text{ meters/sec}$$

tells us that when $t = 2$ sec, the body is moving at a velocity of 10 meters/sec. This is an instantaneous velocity, meaning that if the body continued to move at this speed for a whole second, it would cover 10 meters. In practice, however, the velocity of the body is probably changing and so doesn't remain 10 meters/sec for long. Notice that the units of instantaneous velocity and of average velocity are the same.

Example 3 The cost C (in dollars) of building a house A square feet in area is given by the function $C = f(A)$. What is the practical interpretation of the function $f'(A)$?

Solution In the alternative notation,

$$f'(A) = \frac{dC}{dA}.$$

This is a cost divided by an area, so it is measured in dollars per square foot. You can think of dC as the extra cost of building an extra dA square feet of house. Thus, dC/dA is the additional cost per square foot. So if you are planning to build a house roughly A square feet in area, $f'(A)$ is the cost per square foot of the *extra* area involved in building a slightly larger house, and is called the *marginal cost*. The marginal cost is not necessarily the same thing as the average cost per square foot for the entire house, since once you are already set up to build a large house, the cost of adding a few square feet could be comparatively small.

Example 4 The cost of extracting T tons of ore from a copper mine is $C = f(T)$ dollars. What does it mean to say that $f'(2000) = 100$?

Solution In the alternative notation,

$$f'(2000) = \left.\frac{dC}{dT}\right|_{T=2000}.$$

Since C is measured in dollars and T is measured in tons, dC/dT must be measured in dollars per ton. So the statement

$$\left.\frac{dC}{dT}\right|_{T=2000} = 100$$

says that when 2000 tons of ore have already been extracted from the mine, the cost of extracting the next ton is approximately \$100. In other words, after 2000 tons have been removed, extraction costs are \$100 per ton. Another way of saying this is that it costs about \$100 to extract ton number 2000 or 2001. Note that this may well be different from the cost of extracting the tenth ton, which is likely to be more accessible.

Example 5 If $q = f(p)$ gives the number of pounds of sugar produced when the price per unit is p dollars, then what are the units and the meaning of

$$\left.\frac{dq}{dp}\right|_{p=3} = f'(3) = 50?$$

Solution Since $f'(3)$ is the limit as $h \to 0$ of

$$\frac{f(3+h) - f(3)}{h}$$

and $f(3 + h) - f(3)$ is in pounds, while h is in dollars, the units of the difference quotient is pounds/dollar. Since $f'(3)$ is the limit of the difference quotient, its units are also pounds/dollar. The statement

$$\frac{dq}{dp} = f'(3) = 50 \text{ pounds/dollar}$$

tells us that the rate of change of q with respect to p is 50 when $p = 3$. Rephrasing, this means that when the price is \$3, the quantity produced is increasing at 50 pounds/dollar. This is an instantaneous rate of change, meaning that, if the rate were to remain 50 pounds/dollar, and if the price were to increase by a whole dollar, the quantity produced would increase by 50 pounds. In fact, the rate probably doesn't remain constant and so the quantity produced would probably not be exactly 50 pounds. Notice, however, that the units of the derivative and of the average rate of change are again the same. This is because the units of the instantaneous and the average rate of change are *always* the same.

Example 6 You are told that water is flowing through a pipe at a rate of 10 cubic feet per second. Interpret this rate as the derivative of some function.

Solution You might think at first that the statement has something to do with the velocity of the water, but in fact a flow rate of 10 cubic feet per second could be achieved either with very slowly moving water through a large pipe, or with very rapidly moving water through a narrow pipe. If we look at the units — cubic feet per second — we realize that we are being given the rate of change of a quantity measured in cubic feet. But a cubic foot is a measure of volume, so we are being told the rate of change of a volume. If you imagine all the water that is flowing through ending up in a tank somewhere and let $V(t)$ be the volume of the tank at time t, then we are being told that the rate of change of $V(t)$ is 10, or

$$V'(t) = \frac{dV}{dt} = 10$$

Problems for Section 2.5

1. Consider the graph shown in Figure 2.43.
   (a) If $f(t)$ gives the position of a particle at time t, list the points at which the particle has zero velocity.
   (b) If we now suppose instead that $f(t)$ is the *velocity* of a particle at time t, what is the significance of the points listed in your answer to part (a)?

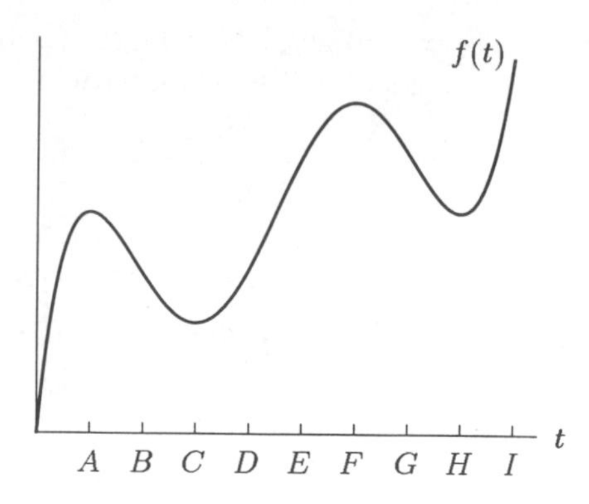

Figure 2.43

2. If $\lim_{x\to\infty} f(x) = 50$ and $f'(x)$ is positive for all x, what is $\lim_{x\to\infty} f'(x)$? (Assume this limit exists.) Explain your answer with a picture.
3. Let $f(x)$ be the elevation in feet of the Mississippi river x miles from its source. What are the units of $f'(x)$? What can you say about the sign of $f'(x)$? (Assume that $0 \le x \le$ length of the river.)
4. Let $g(t)$ be the height, in inches, of Amelia Earhart (one of the first woman airplane pilots) t years after her birth. What are the units of $g'(t)$? What can you say about the signs of $g'(10)$ and $g'(30)$? (Assume that $0 \le t < 39$, the age at which Amelia Earhart's plane disappeared.)
5. Suppose $C(r)$ is the total cost of paying off a car loan borrowed at an annual interest rate of r%. What are the units of $C'(r)$? What is the practical meaning of $C'(r)$? What is its sign?
6. Suppose $P(t)$ is the monthly payment on a mortgage which will take t years to pay off. What are the units of $P'(t)$? What is the practical meaning of $P'(t)$? What is its sign?
7. After investing \$1000 at an annual interest rate of 7% compounded continuously for t years, your balance is \$$B$, where $B = f(t)$. What are the units of dB/dt? What is the financial interpretation of dB/dt?
8. An economist is interested in how the price of a certain commodity affects its sales. Suppose that at a price of \$$p$, a quantity, q, of the commodity is sold. If $q = f(p)$, explain in economic terms the meaning of the statements $f(10) = 240{,}000$ and $f'(10) = -29{,}000$.
9. The temperature, T, in degrees Fahrenheit, of a cold yam placed in a hot oven is given by $T = f(t)$, where t is the time in minutes since the yam was put in the oven.
   (a) What is the sign of $f'(t)$? Why?
   (b) What are the units of $f'(20)$? What is the practical meaning of the statement $f'(20) = 2$?
10. Investing \$1000 at an annual interest rate of r%, compounded continuously, for 10 years gives you a balance of \$$B$, where $B = g(r)$. What is a financial interpretation of the statements
    (a) $g(5) \approx 1649$?
    (b) $g'(5) \approx 165$? What are the units of $g'(5)$?
11. If $g(v)$ is the fuel efficiency of a car going at v miles per hour (i.e., $g(v)$ = the number of miles per gallon at v mph), what are the units of $g'(55)$? What is the practical meaning of the statement $g'(55) = -0.54$?
12. Let P be the total petroleum reservoir on earth in the year t. (In other words, P represents the total quantity of petroleum, including what's not yet discovered, on earth at time t.) Assume that no new petroleum is being made and that P is measured in barrels. What are the units of

dP/dt? What is the meaning of dP/dt? What is its sign? How would you set about estimating this derivative in practice? What would you need to know to make such an estimate?

13. (a) If you jump out of an airplane without a parachute, you will fall faster and faster until wind resistance causes you to approach a steady velocity, called a *terminal* velocity. Sketch a graph of your velocity against time.
    (b) Explain the concavity of your graph.
    (c) Assuming wind resistance to be negligible at $t = 0$, what natural phenomenon is represented by the slope of the graph at $t = 0$?

14. A company's revenue from car sales, C (measured in thousands of dollars), is a function of advertising expenditure, a, also measured in thousands of dollars. Suppose $C = f(a)$.
    (a) What does the company hope is true about the sign of f'?
    (b) What does the statement $f'(100) = 2$ mean in practical terms? How about $f'(100) = 0.5$?
    (c) Suppose the company plans to spend about $100,000 on advertising. If $f'(100) = 2$, should the company spend slightly more or slightly less than $100,000 on advertising? What if $f'(100) = 0.5$?

15. Let $P(x)$ = the number of people in the US of height $\leq x$ inches. What is the meaning of $P'(66)$? What are its units? Estimate $P'(66)$ (using common sense). Is $P'(x)$ ever negative? [Hint: You may want to approximate $P'(66)$ by a difference quotient, using $h = 1$. Also, you may use the fact that the US population is about 250 million, and that 66 inches = 5 feet 6 inches.]

TABLE 2.17 *US population (in millions), 1790–1990*

Year	Population	Year	Population	Year	Population	Year	Population
1790	3.9	1850	23.1	1910	92.0	1970	205.0
1800	5.3	1860	31.4	1920	105.7	1980	226.5
1810	7.2	1870	38.6	1930	122.8	1990	248.7
1820	9.6	1880	50.2	1940	131.7		
1830	12.9	1890	62.9	1950	150.7		
1840	17.1	1900	76.0	1960	179.0		

16. Table 2.17 gives the US population figures between 1790 and 1990.
    (a) Estimate the rate of change of the population for the years 1900, 1945, and 1990.
    (b) When, approximately, was the rate of change of the population greatest?
    (c) Estimate the US population in 1956.
    (d) Based on the data from the table, what would you predict for the census in the year 2000?

17. (a) In Problem 16, we thought of the US population as a smooth function of time. To what extent is this justified? What happens if we zoom in at a point of the graph? What about events such as the Louisiana Purchase? Or the moment of your birth?
    (b) What do we in fact mean by the rate of change of the population for a particular time t?
    (c) Give another example of a real-world function which is not smooth but is usually treated as such.

2.6 THE SECOND DERIVATIVE

Since the derivative is itself a function, we can often calculate its derivative. For a function f, the derivative of its derivative is called the *second derivative*, and written f'' (read "f double-prime"). If $y = f(x)$, the second derivative can also be written as $\frac{d^2y}{dx^2}$, which means $\frac{d}{dx}\left(\frac{dy}{dx}\right)$, the derivative of $\frac{dy}{dx}$.

What Does the Second Derivative Tell Us?

Recall that the derivative of a function tells you whether a function is increasing or decreasing:

> If $f' > 0$ on an interval, then f is *increasing* over that interval.
> If $f' < 0$ on an interval, then f is *decreasing* over that interval.

Since f'' is the derivative of f',

> If $f'' > 0$ on an interval, then f' is *increasing* over that interval.
> If $f'' < 0$ on an interval, then f' is *decreasing* over that interval.

So the question becomes: what does it mean for f' to be increasing or decreasing? The case in which f' is increasing is shown in Figure 2.44, where the curve is bending upward, or is *concave up*.

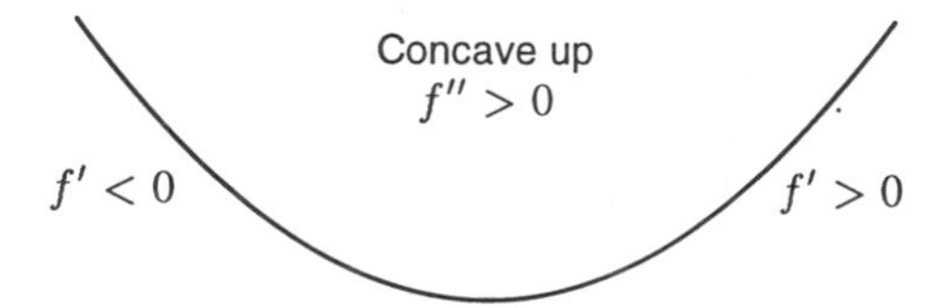

Figure 2.44: Meaning of f'': The slope increases from left to right, so f'' is positive and f is concave up

In the case when f' is decreasing, shown in Figure 2.45, the graph is bending downward, or is *concave down*.

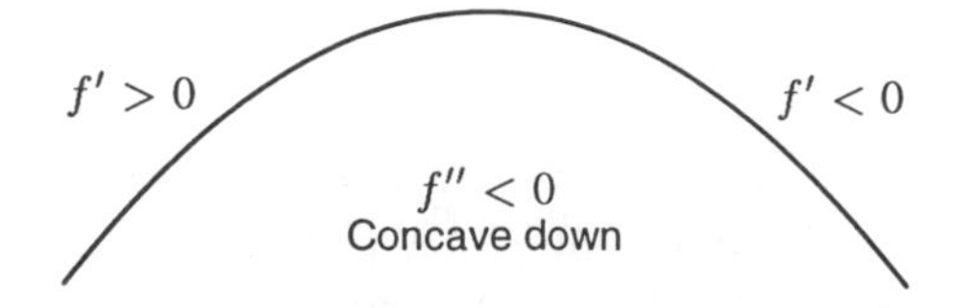

Figure 2.45: Meaning of f'': The slope decreases from left to right, so f'' is negative and f is concave down

$f'' > 0$	means f' is increasing, so the graph of f is concave up on that interval,
$f'' < 0$	means f' is decreasing, so the graph of f is concave down on that interval.

Example 1 For the functions whose graphs are given in Figure 2.46, decide where their second derivatives are positive and where they are negative.

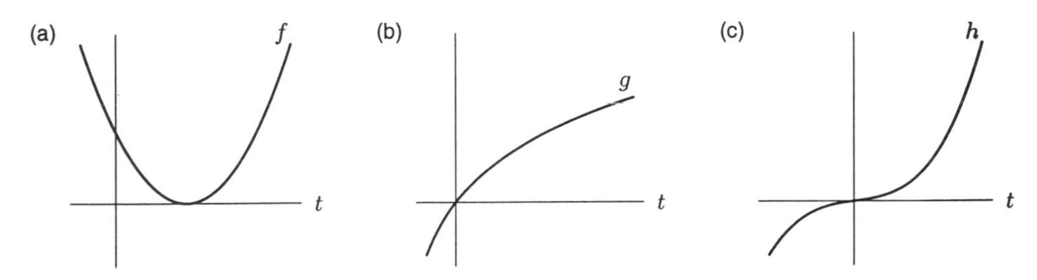

Figure 2.46: What signs do the second derivatives have?

Solution From the graphs it appears that:

(a) $f'' > 0$ everywhere, because the graph of f is concave up everywhere.
(b) $g'' < 0$ everywhere, because the graph is concave down everywhere.
(c) $h'' > 0$ for $t > 0$, because the graph of h is concave up there; $h'' < 0$ for $t < 0$, because the graph of h is concave down there.

Interpretation of the Second Derivative as a Rate of Change

If we think of the derivative as a rate of change, then the second derivative is a rate of change of a rate of change. If the second derivative is positive, the rate of change is increasing; if the second derivative is negative, the rate of change is decreasing.

The second derivative is often a matter of practical concern. In 1985 a newspaper headline reported the Secretary of Defense as saying that Congress and the Senate had cut the defense budget. As his opponents pointed out, however, Congress had merely cut the rate at which the defense budget was increasing.[4] In other words, the derivative of the defense budget was still positive (the budget was increasing), but the second derivative was negative (the budget's rate of increase had slowed).

Example 2 A population, P, growing in a confined environment often follows a *logistic* growth curve, like the graph shown in Figure 2.47. Describe how the rate at which the population is increasing changes over time. What is the practical interpretation of t_0 and L?

[4] In the *Boston Globe*, March 13, 1985, Representative William Gray (D–Pa.) was reported as saying: "It's confusing to the American people to imply that Congress threatens national security with reductions when you're really talking about a reduction in the increase."

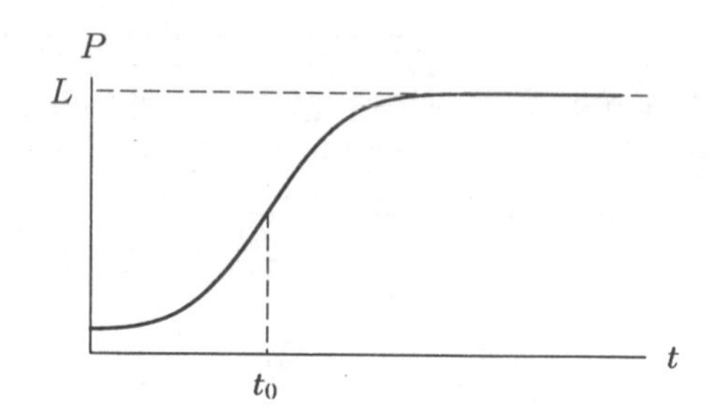

Figure 2.47: Logistic growth curve

Solution Initially, the population is increasing, and at an increasing rate. Thus, initially dP/dt is increasing and so $d^2P/dt^2 > 0$. At t_0, the rate at which the population is increasing is a maximum. Thus at time t_0 the population is growing fastest. Beyond t_0, the rate at which the population is growing is decreasing, and so $d^2P/dt^2 < 0$. At t_0, the concavity changes from positive to negative, and $d^2P/dt^2 = 0$.

The quantity L represents the limiting value of the population which is approached as t tends to infinity. L is called the *carrying capacity* of the environment and represents the maximum population that the environment can support.

Example 3 Table 2.18 shows the number of abortions per year, A, performed in the US in year t (as reported to the Center for Disease Control and Prevention). Suppose these data points lie on a smooth curve $A = f(t)$.

TABLE 2.18 *Abortions reported in the US (1972–1985)*

Year, t	1972	1976	1980	1985
Number of abortions reported, A	586,760	988,267	1,297,606	1,328,570

(a) Estimate dA/dt for the time intervals shown between 1972 and 1985.
(b) What can you say about the sign of d^2A/dt^2 during the period 1972–1985?

Solution (a) For each time interval we can calculate the average rate of change of the number of abortions per year over this interval. For example, between 1972 and 1976

$$\frac{dA}{dt} \approx \text{Average rate of change} = \frac{988{,}267 - 586{,}760}{1976 - 1972} \approx 100{,}377.$$

Values of dA/dt are listed in Table 2.19:

TABLE 2.19 *Rate of change of number of abortions reported*

Time	1972–1976	1976–1980	1980–1985
Average rate of change, dA/dt	100,377	77,335	6,193

(b) Since the values of dA/dt are decreasing for 1976–1985, d^2A/dt^2 is negative for this period. For 1972–1976, the sign of d^2A/dt^2 is less clear; abortion data from 1968 would help. The graph of A against t in Figure 2.48 confirms this; the graph is concave down for 1976–1985, but straight for 1972–1976. The fact that dA/dt is positive tells us that the number of abortions reported has increased during the period 1972–1985. The fact that d^2A/dt^2 is negative for 1976–1985 tells us that the rate of increase has slowed over this period.

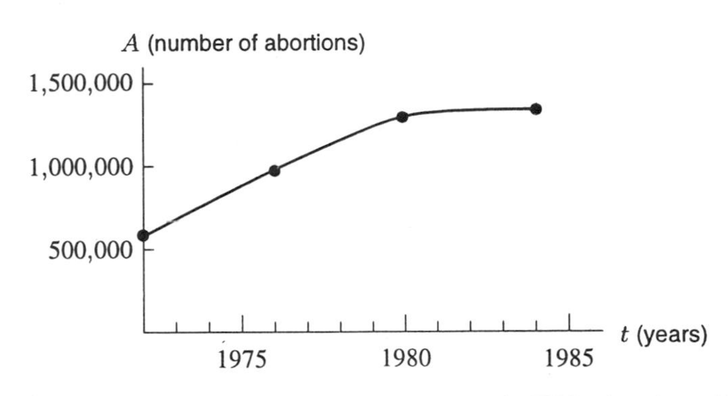

Figure 2.48: How the number of reported abortions in the US is changing with time

Velocity and Acceleration

The velocity, v, of a moving body is the rate at which the position, s, of the body is changing with respect to time:

$$v = \frac{ds}{dt}.$$

The acceleration of a moving body describes how fast the velocity is changing with time, so

$$a = \frac{dv}{dt}$$

or, since the velocity is itself a derivative,

$$a = \frac{dv}{dt} = \frac{d^2s}{dt^2}.$$

Example 4 A particle is moving along a straight line. If its distance, s, to the right of a fixed point is given by Figure 2.49, estimate:

(a) When the particle is moving to the right and when it is moving to the left.
(b) When the particle has positive acceleration and when it has negative acceleration.

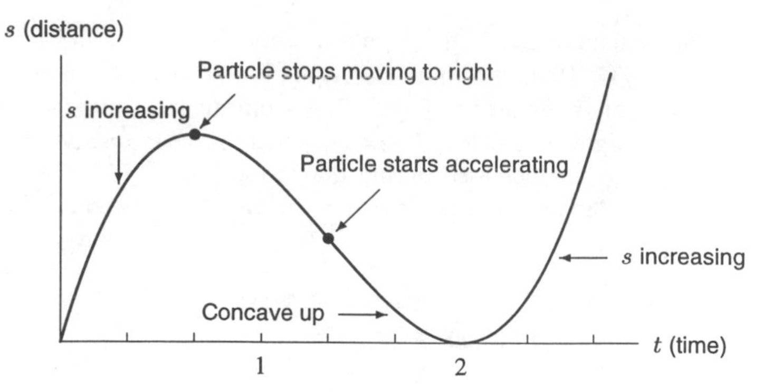

Figure 2.49: Distance of particle to right of fixed point

Solution (a) The particle is moving to the right whenever s is increasing. From the graph, this appears to be for $0 < t < \frac{2}{3}$ and for $t > 2$. For $\frac{2}{3} < t < 2$, the value of s is decreasing, so the particle is moving to the left.

(b) The particle has positive acceleration whenever the curve is concave up, which appears to be for $t > \frac{4}{3}$. The particle has negative acceleration when the curve is concave down, for $t < \frac{4}{3}$.

Problems for Section 2.6

1. (a) If f'' is positive on an interval, then f' is __________ on that interval, and f is __________ on that interval.
   (b) If f'' is negative on an interval, then f' is __________ on that interval, and f is __________ on that interval.
2. (a) Sketch a curve whose first and second derivatives are everywhere positive.
   (b) Sketch a curve whose second derivative is everywhere negative but whose first derivative is everywhere positive.
   (c) Sketch a curve whose second derivative is everywhere positive but whose first derivative is everywhere negative.
   (d) Sketch a curve whose first and second derivatives are everywhere negative.
3. (a) Sketch a smooth curve whose slope is both positive and increasing at first but later is positive and decreasing.
   (b) Sketch the graph of the first derivative of the curve in part (a).
   (c) Sketch the graph of the second derivative of the curve in part (a).
4. "Winning the war on poverty" has been described cynically as slowing the rate at which people are slipping below the poverty line. Assuming that this is happening:
   (a) Sketch a graph of the total number of people in poverty against time.
   (b) If N is the number of people below the poverty line at time t, what are the signs of dN/dt and d^2N/dt^2?
5. In economics "total utility" refers to the total satisfaction from consuming some commodity. According to the economist Samuelson[5]:

[5]From Paul A. Samuelson, *Economics*, 11th edition (New York: McGraw-Hill, 1981).

As you consume more of the same good, the total (psychological) utility increases. However, . . . with successive new units of the good, your total utility will grow at a slower and slower rate because of a fundamental tendency for your psychological ability to appreciate more of the good to become less keen.

(a) Sketch the total utility as a function of the number of units consumed.
(b) In terms of derivatives, what is Samuelson telling us?

6. Let $P(t)$ represent the price of a share of stock of a corporation at time t. What does each of the following statements tell us about the signs of the first and second derivatives of $P(t)$?
   (a) "The price of the stock is rising faster and faster."
   (b) "The price of the stock is close to bottoming out."

7. IBM-Peru uses second derivatives to assess the relative success of various advertising campaigns. They assume that all campaigns produce some increase in sales. If a graph of sales against time shows a positive second derivative during a new advertising campaign, what does this suggest to IBM management? Why? What does a negative second derivative during a new campaign suggest?

8. Given the following data:

x	0	0.2	0.4	0.6	0.8	1.0
$f(x)$	3.7	3.5	3.5	3.9	4.0	3.9

   (a) Estimate $f'(0.6)$ and $f'(0.5)$. (b) Estimate $f''(0.6)$.
   (c) Where do you think the maximum and minimum values of f occur in the interval $0 \leq x \leq 1$?

9. An industry is being charged by the Environmental Protection Agency (EPA) with dumping unacceptable levels of toxic pollutants in a lake. Over a several month period, an engineering firm makes daily measurements of the rate at which pollutants are being discharged into the lake.

   Suppose the engineers produce a graph similar to either Figure 2.50(a) or Figure 2.50(b). For each case, give an idea of what argument the EPA might make in court against the industry and of the industry's defense.

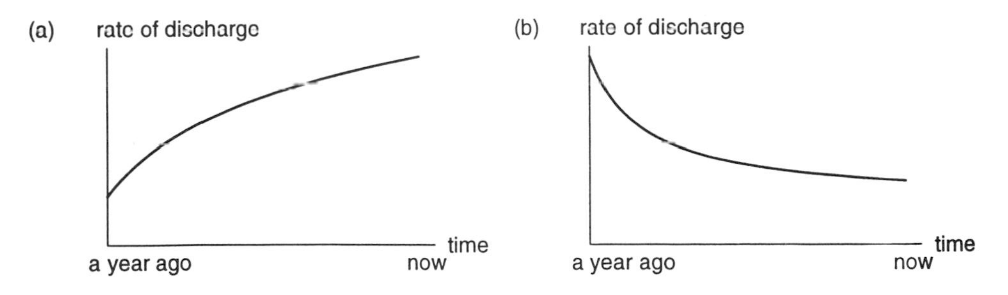

Figure 2.50: Discharge of pollutants

10. A high school principal is concerned about the drop in the percentage of high school students who graduate from her school, shown in the table below.

Year entering high school, t	Percent graduating, P
1977	62.4
1980	54.1
1983	48.0
1986	43.5
1989	41.8

(a) Estimate dP/dt for each of the three-year intervals between 1977 and 1989.
(b) Does d^2P/dt^2 appear to be positive or negative between 1977 and 1989?
(c) Explain why the values of P and dP/dt are troublesome to the principal.
(d) Explain why the sign of d^2P/dt^2 and the magnitude of dP/dt in the year 1986 may give the principal some cause for optimism.

11. The graph of f' (not f) is given in Figure 2.51. At which of the marked values of x is
(a) $f(x)$ greatest? (b) $f(x)$ least? (c) $f'(x)$ greatest? (d) $f'(x)$ least?
(e) $f''(x)$ greatest? (f) $f''(x)$ least?

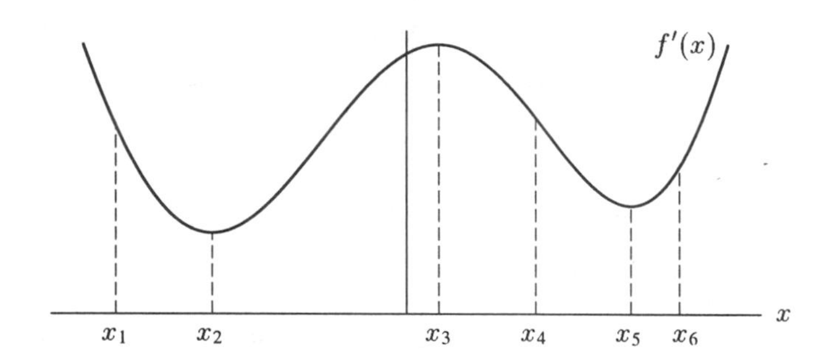

Figure 2.51: Note that this is a graph of f' against x

12. Which of the points labeled by letters in the graph of f in Figure 2.52 have
(a) f' and f'' nonzero and of the same sign?
(b) At least two of f, f', and f'' equal to zero?

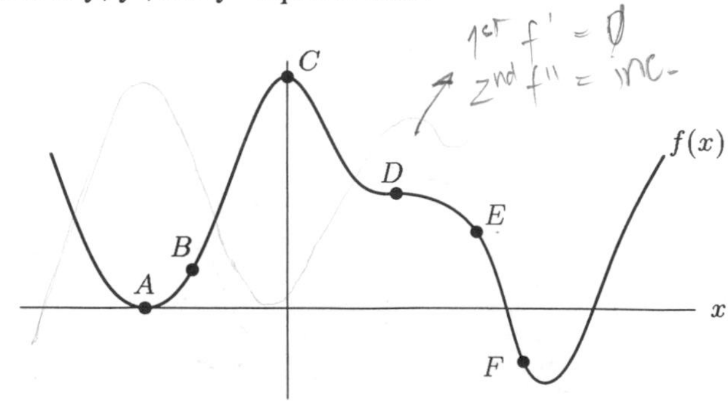

Figure 2.52

2.7 MARGINAL COST AND REVENUE

Management decisions within a particular firm or industry usually depend on the costs and revenues involved. In this section we will look at applications of the derivative to the cost and revenue functions.

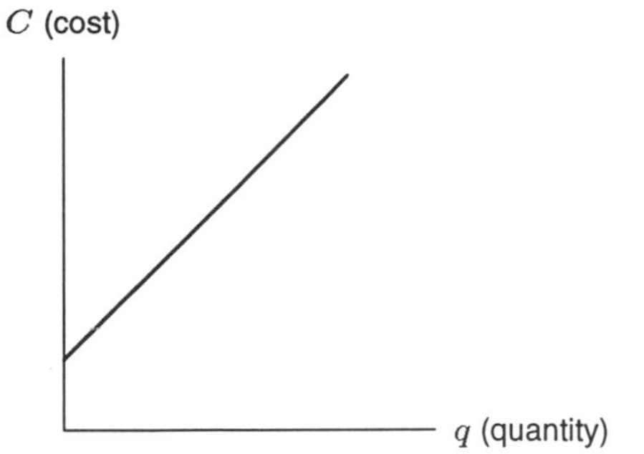

Figure 2.53: A linear cost function

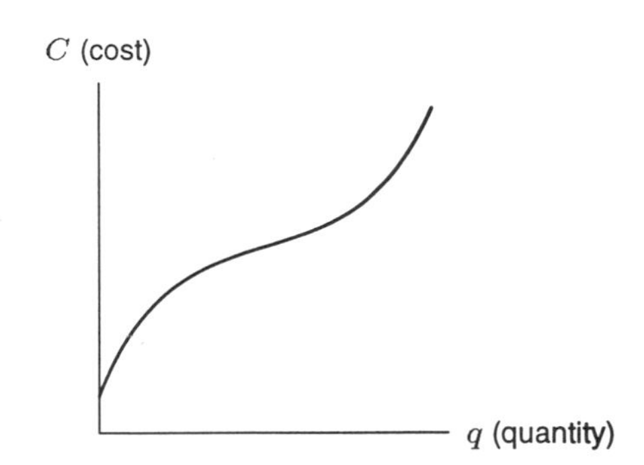

Figure 2.54: A nonlinear cost function

In Chapter 1, we considered primarily linear cost functions (See Figure 2.53.) In practice, however, cost functions usually have the general shape shown in Figure 2.54. Recall that the intercept on the C-axis represents the fixed costs, which are incurred even if nothing is produced. (This includes, for instance, the machinery needed to begin production.) The cost function increases quickly at first and then more slowly because producing larger quantities of a good is usually more efficient than producing smaller quantities—this is called economy of scale. At still higher production levels, the cost function starts to increase faster again as resources become scarce, and sharp increases may occur when new factories have to be built. Thus, $C(q)$ starts out concave down and becomes concave up later on.

Recall that if the price, p, is a constant, then the revenue function is $R = pq$ and the graph of R against q is a straight line through the origin with slope equal to the price. See Figure 2.55. In practice, for large values of q, the market may become glutted, causing the price to drop, giving R the shape in Figure 2.56.

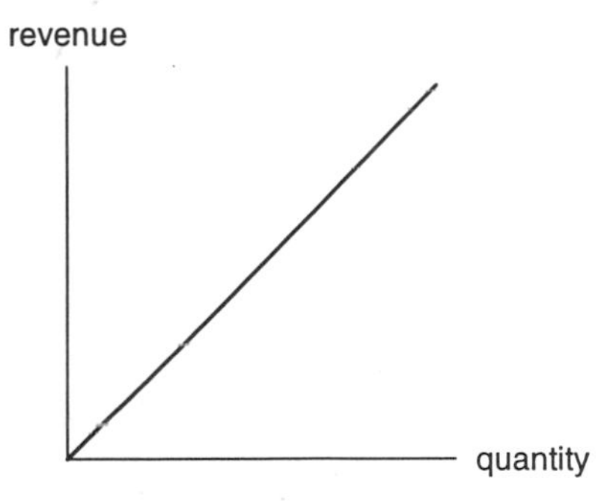

Figure 2.55: Revenue: Constant price

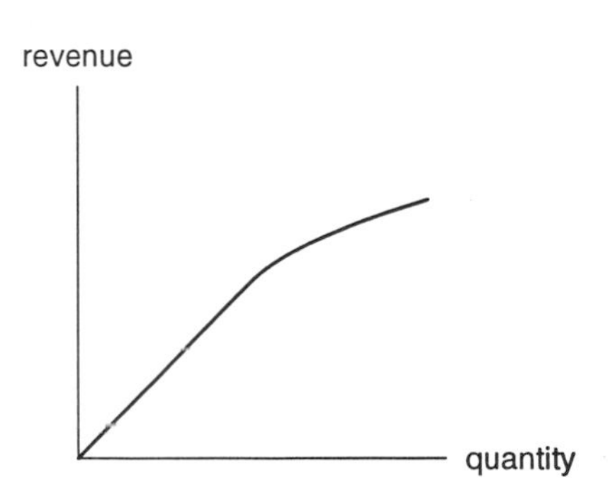

Figure 2.56: Revenue: Decreasing price

Example 1 If $C(q)$ and $R(q)$ are given by the graphs in Figure 2.57, for what values of q does the firm make a profit?

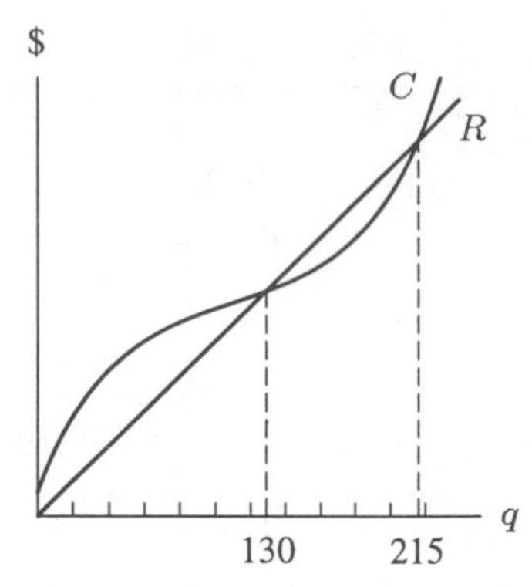

Figure 2.57: Costs and revenues for Example 1

Solution The firm makes a profit whenever revenues are greater than costs, that is, when $R > C$. The graph of R is above the graph of C approximately when $130 < q < 215$, so production between $q = 130$ and $q = 215$ will generate a profit.

Marginal Analysis

Many economic decisions are based on an analysis of the costs and revenues "at the margin." Let's look at this idea through an example.

Suppose you are running an airline and you are trying to decide whether to offer an additional flight. How should you decide? We'll assume that the decision is to be made purely on financial grounds: if the flight will make money for the company, it should be added. Obviously you need to consider the costs and revenues involved. Since the choice is between adding this flight and leaving things the way they are, the crucial question is whether the *additional costs* incurred are greater or smaller than the *additional revenues* generated by the flight. These additional costs and revenues are called the *marginal costs* and *marginal revenues*.

Suppose $C(q)$ is the function giving the total cost of running q flights. If the airline had originally planned to run 100 flights, its costs would be $C(100)$. With the additional flight, its costs would be $C(101)$. Therefore,

$$\text{Marginal cost} = C(101) - C(100).$$

Now

$$C(101) - C(100) = \frac{C(101) - C(100)}{1},$$

and this quantity is the average rate of change of cost between 100 and 101 flights. In Figure 2.58 the average rate of change is the slope of the secant line. If the cost function is not curving too fast near the point, the slope of the secant line will be close to the slope of the tangent line there. Therefore, the average rate of change is close to the instantaneous rate of change. Since these rates of change are not very different, many economists choose to define marginal cost, MC, as the instantaneous rate of change of cost with quantity:

$$\boxed{\text{Marginal cost} = MC = C'(q).}$$

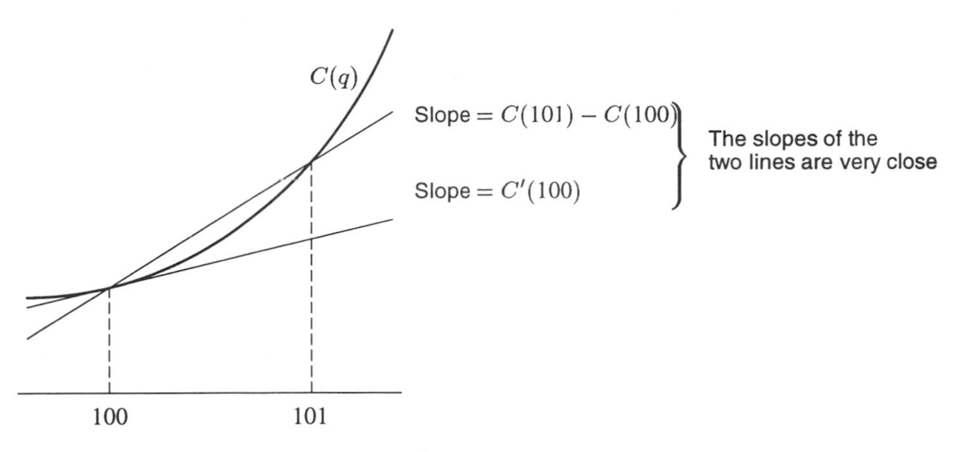

Figure 2.58: Marginal cost: Slope of one of these lines

Similarly if the revenue generated by q flights is $R(q)$, the additional revenue generated by increasing the number of flights from 100 to 101 is

$$\text{Marginal revenue} = R(101) - R(100).$$

Now $R(101) - R(100)$ is the average rate of change of revenue between 100 and 101 flights. As before, the average rate of change is usually almost equal to the instantaneous rate of change, so economists often define:

$$\text{Marginal revenue} = MR = R'(q).$$

Example 2 If $C(q)$ and $R(q)$ for the airline are given in Figure 2.59, should the company add the 101st flight?

Solution The marginal revenue is the slope of the revenue line. The marginal cost is the slope of the graph of C at the point 100. From Figure 2.59, you can see that the slope at the point A is smaller than the slope at B, so $MC < MR$. This means that the airline will make more in extra revenue than it will spend in extra costs if it runs another flight, so it should go ahead and run the 101st flight.

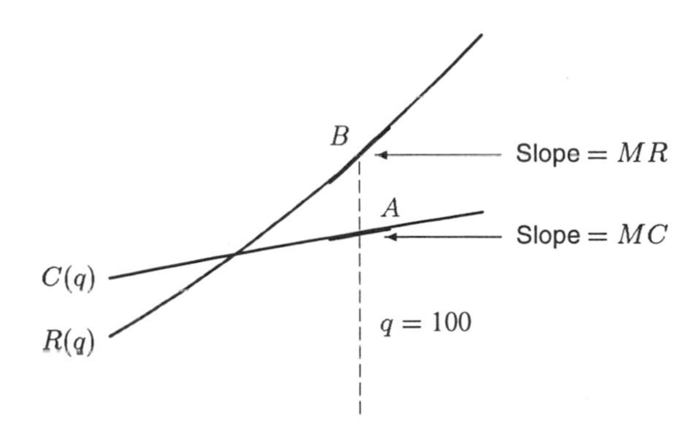

Figure 2.59: Cost and revenue for Example 2

Since MC and MR are derivative functions, they can be estimated from the graphs of total cost and total revenue.

Example 3 If R and C are given by the graphs in Figure 2.60, sketch graphs of $MR = R'(q)$ and $MC = C'(q)$.

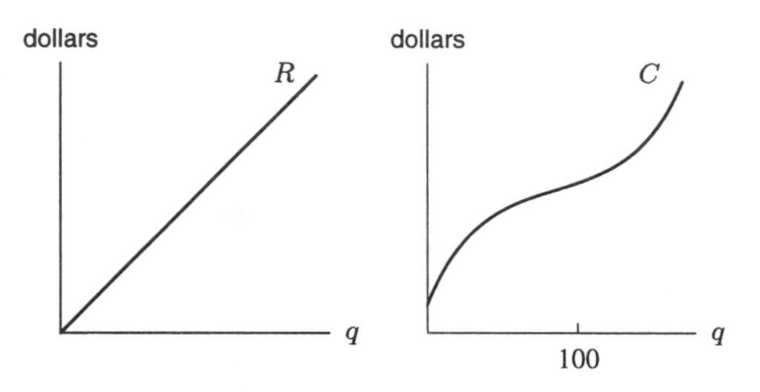

Figure 2.60: Total revenue and total cost for Example 3

Solution The revenue graph is a line through the origin, with equation

$$R = pq$$

where p which represents the constant price, so the slope is p and

$$MR = R'(q) = p.$$

The total cost is increasing, so the marginal cost is always positive. For small q values, the total cost curve is concave down, so the marginal cost is decreasing. For larger q, say $q > 100$, the total cost is concave up and the marginal cost is increasing. Thus the marginal cost has a minimum at about $q = 100$. (See Figure 2.61.)

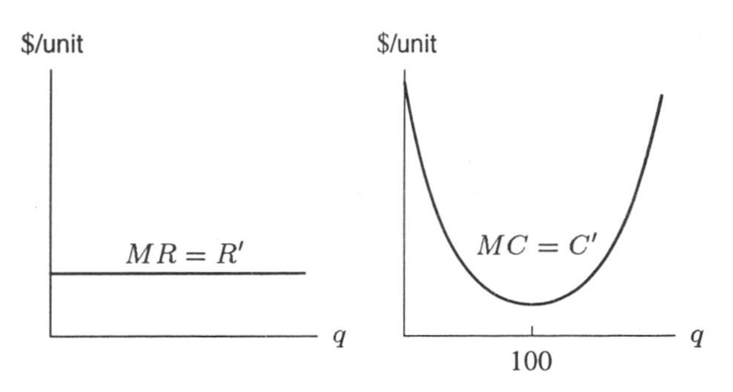

Figure 2.61: Marginal revenue and costs for Example 3

Maximizing Profit

Now let's look at how to maximize total profit, given functions for total revenue and cost. This is clearly a fundamental issue for any producer of goods. The next example looks at this problem.

Example 4 Find the maximum profit if the total revenue and total cost are given by the curves R and C, respectively, in Figure 2.62.

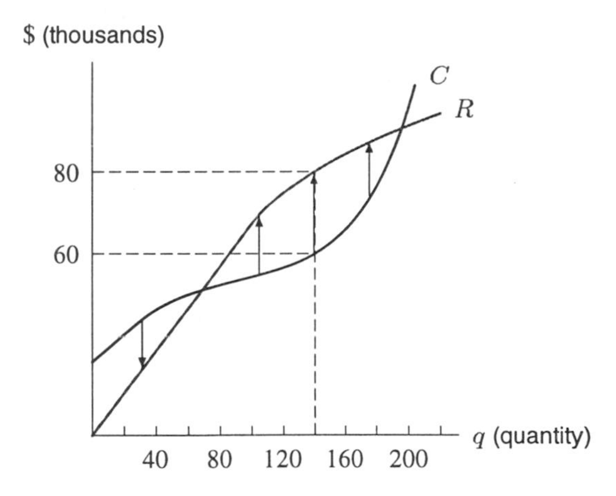

Figure 2.62: Maximum profit at $q = 140$

Solution The profit is represented by the vertical difference between the curves and is marked by the vertical arrows on the graph. When revenue is below cost, the company is taking a loss; when revenue is above cost, the company is making a profit. You can see that the profit is largest at about $q = 140$, so this is the production level we're looking for.

You should observe that at $q = 140$ the tangents to the two curves are parallel. To the left of $q = 140$, the revenue curve has a larger slope than the cost curve, and the profit increases as q increases. To the right of $q = 140$, the slope of the revenue curve is less than the slope of the cost curve, and the profit is decreasing. At the point where the slopes are equal, the profit appears to be at a maximum. Since the slopes are equal, we see that the marginal revenue equals marginal cost. We will see in Chapter 5 that this is a result which is valid generally.

Thus, we have found the quantity (namely, $q = 140$) which gives the maximum profit. If we wanted the actual profit, we would have to estimate the vertical distance between the curves at $q = 140$. This tells us that the maximum profit = \$80,000 − \$60,000 = \$20,000.

Average Cost

Example 5 A regional yogurt company has cost function $C(q) = 0.01q^3 - 0.6q^2 + 13q + 1000$ (in dollars), where q is the number of cases of yogurt produced. Find the average cost per case if 100 cases are produced.

Solution The total cost of producing the 100 cases is given by

$$C(100) = 0.01(100^3) - 0.6(100^2) + 13(100) + 1000 = 6300.$$

The total cost of producing the 100 cases is \$6300. We can find the average cost per case by dividing by 100, the number of cases produced.

$$\text{Average cost per case} = \frac{6300}{100} = 63.$$

If 100 cases of yogurt are produced, the average cost per case is \$63.

The *average cost* is the cost per unit of producing a certain quantity; it is the total cost divided by the number produced.

> The **average cost**, $a(q)$, is given by $a(q) = C(q)/q$.

Be careful not to confuse the average cost (the cost per unit of producing a certain quantity) with the marginal cost (the cost of producing the next one).

Example 6 If the cost function, in dollars, is $C(q) = 1000 + 20q$, where q is the number of items produced, find the marginal cost to produce the 100th item, and find the average cost of producing 100 items.

Solution This is a linear cost function with constant variable costs of \$20 per unit, so the marginal cost for each item is \$20. (Another way to see this is that the graph of the cost function has a constant slope of 20, and so the derivative of the cost function is 20 at every point. Since marginal cost is equal to the derivative of the cost function, the marginal cost is \$20 per unit.) After 99 units have been produced, it costs \$20 to produce the next one.

The average cost of producing 100 units is given by

$$a(100) = \frac{C(100)}{100} = \frac{3000}{100} = 30.$$

The average cost of producing 100 units is \$30 per unit. Notice that the average cost includes the fixed costs spread over the entire production, whereas marginal cost does not, so average cost is greater than marginal cost for this example.

How can we visualize the average cost on a graph? Since $(q, C(q))$ represents a point on the graph of the cost function $C(q)$,

$$\text{Average cost} = \frac{C(q)}{q} = \frac{C(q)-0}{q-0} = \text{Slope of a line from the origin to the point } (q, C(q)).$$

Example 7 The graph of a cost function is given in Figure 2.63. Mark on the graph the quantity at which the average cost is minimized.

Solution The average cost of producing a quantity q is given by the slope of a line through the origin to the graph of $C(q)$ at q. In Figure 2.64, several of these lines have been drawn at points q_1, q_2, q_3, and q_4. The slopes of these lines are steep for small q, become less steep as q increases, and then get steeper again as q continues to increase. Thus, as q increases, the average costs decrease and then increase, and so there is a minimum value. This minimum occurs at the point q_0 in Figure 2.64.

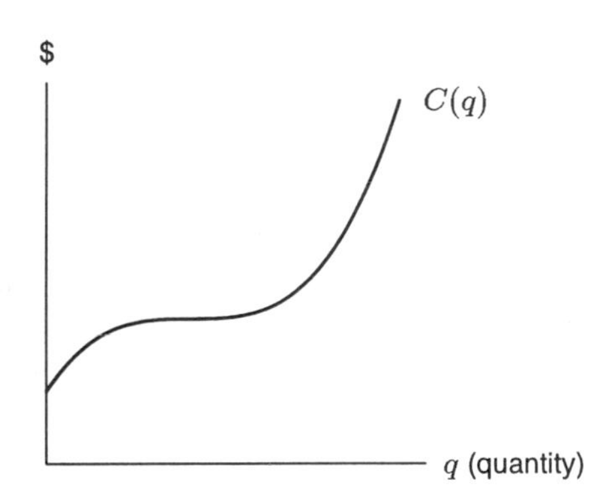

Figure 2.63: A cost function

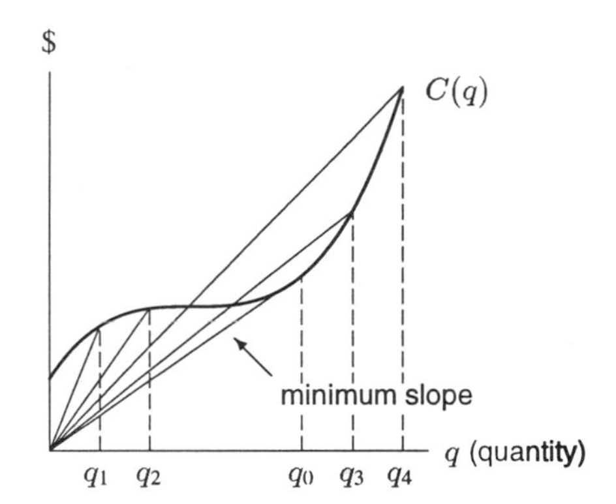

Figure 2.64: Average cost = slope of a line through the origin to the point $(q, C(q))$

Problems for Section 2.7

1. The cost and revenue functions for a charter bus company are shown in Figure 2.65. Should the company add a 50th bus? How about a 100th? Explain your answers using marginal revenue and cost.
2. The graph of a cost function is given in Figure 2.66.
   (a) Estimate the marginal cost at $q = 5$, at $q = 20$, and at $q = 40$.
   (b) Estimate the average cost at $q = 30$.

3. Suppose a cost function, in dollars, is given by $C(q) = 2500 + 12q$, where q is the number of items produced.
   (a) What is the marginal cost of producing the 100th item? the 1000th item?
   (b) What is the average cost of producing 100 items? 1000 items?

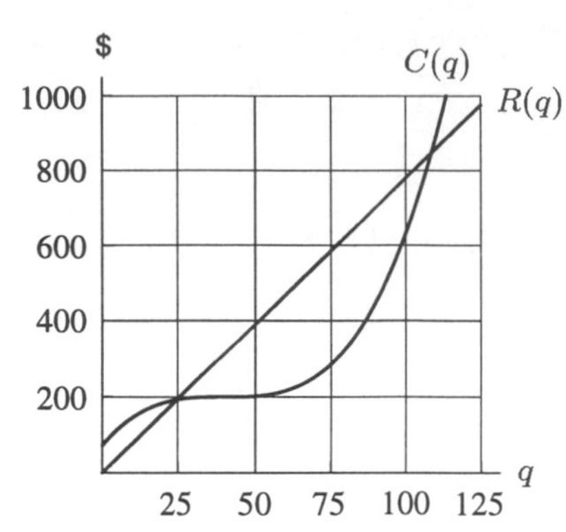

Figure 2.65: Cost and revenue for a charter bus company

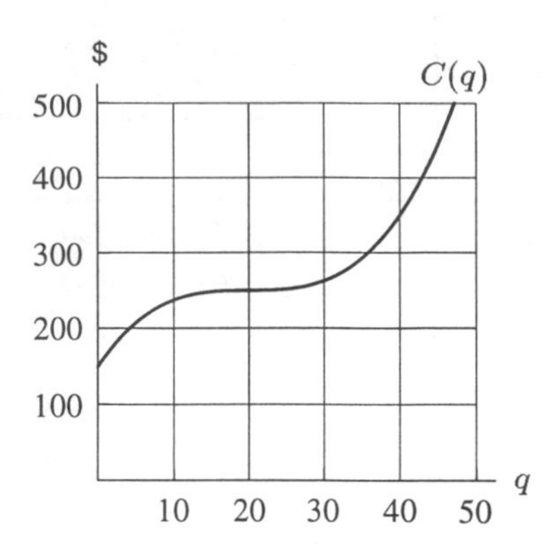

Figure 2.66: A cost function

4. An agricultural worker in Uganda is interested in planting clover to increase the number of bees making their home in the region. There are 100 bees in the region naturally, and for every acre put under clover, 20 more bees are found in the region.
   (a) Draw a graph of the total number, $N(x)$, of bees as a function of x, the number of acres devoted to clover.
   (b) Explain, both geometrically and algebraically, the shape of the graph of:
      (i) The marginal rate of increase of the number of bees with acres of clover, $N'(x)$.
      (ii) The average number of bees per acre of clover, $N(x)/x$.

5. A manufacturer reports the total cost and total revenue functions shown in Figure 2.67. Sketch graphs, as a function of quantity, of:
   (a) Total profit, (b) Marginal cost, (c) Marginal revenue.
   Label the points q_1 and q_2 on your graphs.

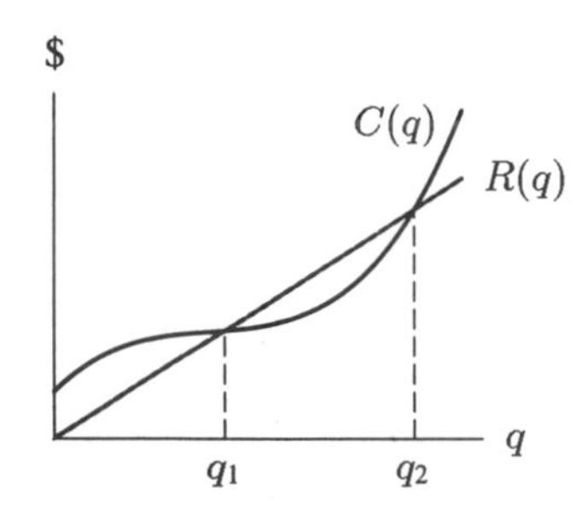

Figure 2.67: Cost and revenue for a good

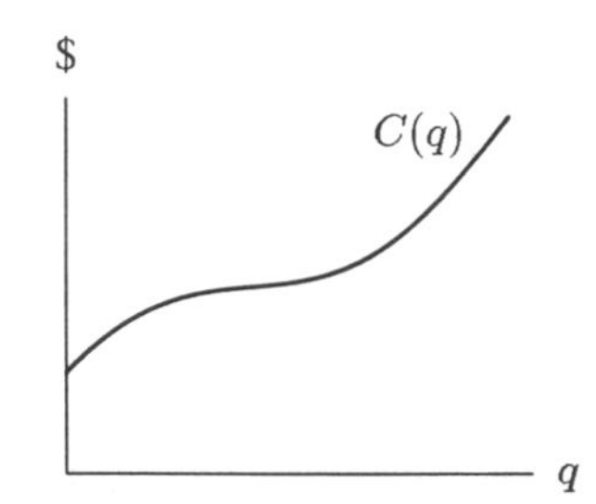

Figure 2.68: Cost of producing a good

6. Let $C(q)$ be the total cost of producing a quantity q of a certain good. (See Figure 2.68.)
   (a) What is the meaning of $C(0)$?

(b) Describe in words how the marginal cost changes as the quantity produced increases.
(c) Explain the concavity of the graph (in terms of economics).
(d) Explain the economic significance (in terms of marginal cost) of the point at which the concavity changes.
(e) Do you think the graph of $C(q)$ looks like this for all types of goods?

7. Assume that the demand equation for a product shows that the price of an item is related to the quantity sold by the equation

$$p = 100e^{-0.02q},$$

where q is the quantity sold at price p.

(a) Write revenue, R, as a function of quantity q.
(b) Sketch a graph of revenue, R, against quantity, q.
(c) Use the graph to estimate the marginal revenue when $q = 40$, when $q = 50$, and when $q = 60$. Interpret your answers in terms of revenue.

8. The cost function of a paper recycling plant is given in the following table. Estimate the marginal cost at $q = 2000$, and give units with your answer. Interpret your answer in terms of cost. At approximately what value of q does marginal cost appear smallest?

q (tons of recycled paper)	1000	1500	2000	2500	3000	3500
$C(q)$ (dollars)	2500	3200	3640	3825	3900	4400

9. The graphs of the cost and revenue functions for a certain company are given in Figure 2.69. Approximately what quantity should be produced if the company wants to maximize profits? Explain your answer graphically.

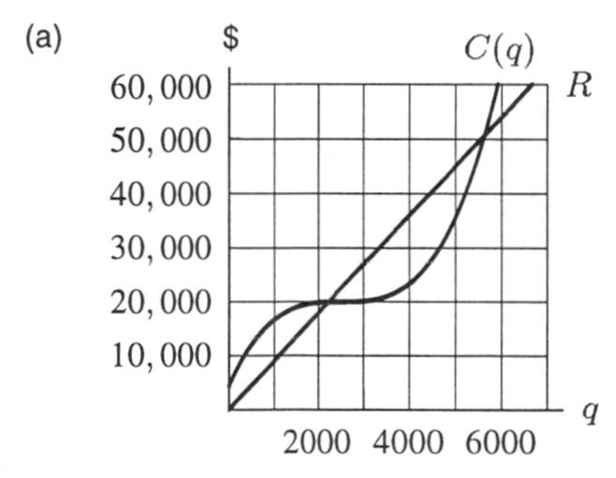

Figure 2.69: What quantity maximizes profits?

10. Two different possible cost functions are given in Figure 2.70.

(a) The graph labeled (a) is concave down. For this cost function, is there a value of q at which average cost is minimized?
(b) The graph labeled (b) is concave up. For this cost function, is there a value of q at which average cost is minimized?

Explain your answers graphically.

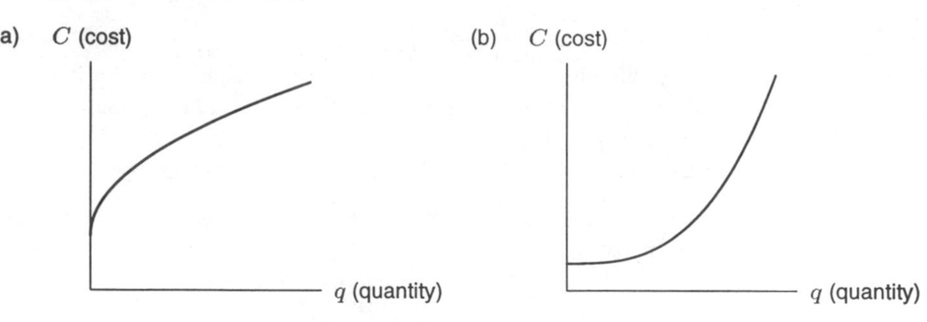

Figure 2.70: Cost functions with different concavity

11. Consider the cost function whose graph is given in Figure 2.54. Sketch a graph of the corresponding average cost function.

12. Let $C(q) = 0.04q^3 - 3q^2 + 75q + 96$ be the total cost of producing q items.
    (a) Find the average cost per item as a function of q.
    (b) Use a graphing calculator or computer to sketch a graph of average cost against q.
    (c) For what values of q is the average cost per item decreasing?
    (d) For what values of q is the average cost per item increasing?
    (e) For what value of q is the average cost per item smallest and what is the average cost per item at that point?

13. Let $C(q)$ represent the total cost of producing q items. Then $C'(q)$ gives the marginal cost in dollars per item. Suppose a company determines that $C(15) = 2300$ and $C'(15) = 108$.
    (a) Estimate the total cost of producing 16 items.
    (b) Estimate the total cost of producing 14 items.

14. Let $C(q)$ represent the total cost, $R(q)$ the total revenue and $\pi(q)$ the total profit, where q is the number of items produced. Assume that $C(q)$, $R(q)$, and $\pi(q)$ are all measured in dollars.
    (a) If $C'(50) = 75$ and $R'(50) = 84$, approximately how much profit will be earned by the 51st item?
    (b) If $C'(90) = 71$ and $R'(90) = 68$, approximately how much profit will be earned by the 91st item?
    (c) If $\pi(q)$ reaches its maximum value when $q = 78$, how do you think $C'(78)$ and $R'(78)$ will compare? Explain.

15. A developer has recently purchased a laundry and an adjacent factory. For years, the laundry has taken pains to keep the smoke from the factory from soiling the air used by its clothes dryers. Now that the developer owns both the laundry and the factory, she could install filters within the factory's smokestacks to reduce the emission of smoke directly, instead of merely protecting the laundry from it, at a cost which would depend on the number of filters used. The developer faces the schedule of factory's cost of filters versus laundry's cost of protecting against dirt shown in Table 2.20.

TABLE 2.20

# of Filters	Total Expense for Filters	Total Costs of Protecting Laundry from Smoke
0	$0	$127
1	$5	$63
2	$11	$31
3	$18	$15
4	$26	$6
5	$35	$3
6	$45	$0
7	$56	$0

(a) Draw up a table which shows, for each possible number of filters (0 through 7,) the marginal cost of the filter, the average cost of the filters, and the marginal savings protecting the laundry from smoke.

(b) Since the developer wishes to minimize the total costs to both her businesses, what should she do? Use the table from part (a) to explain your answer.

(c) What should the developer do if, in addition to the cost of the filters, the filters must be mounted on a rack which costs $100?

(d) What should the developer do if the rack costs $50?

16. In Figure 2.71, find the marginal cost when the level of production is 10,000 units and interpret it.

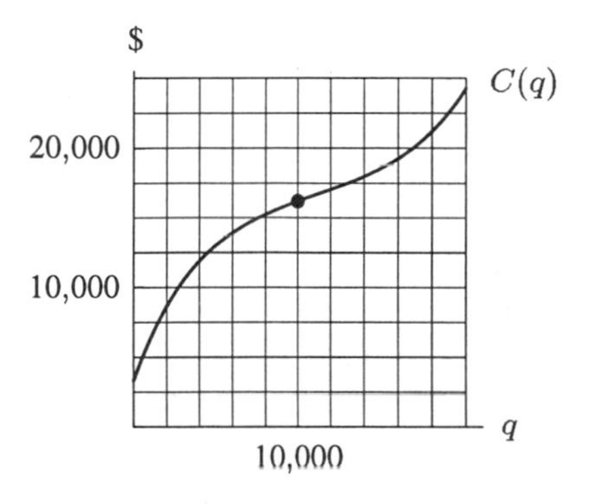

Figure 2.71

17. In Figure 2.72, find the marginal revenue when the level of production is 600 units and interpret it.

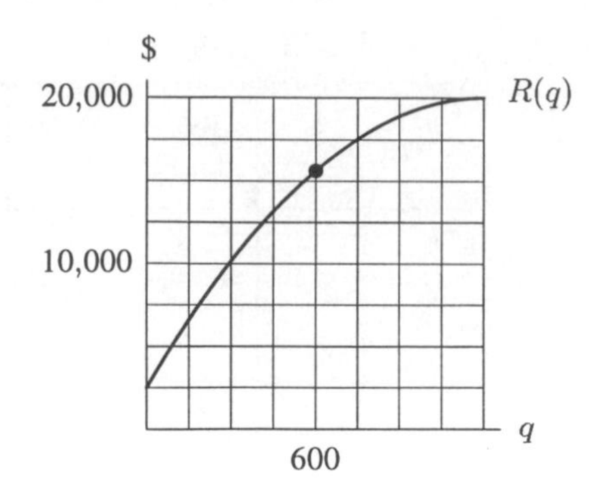

Figure 2.72

REVIEW PROBLEMS FOR CHAPTER TWO

1. For $-3 \leq x \leq 7$, use a graphing calculator or computer to graph

$$f(x) = (x^3 - 6x^2 + 8x)(2 - e^x).$$

   (a) How many zeros does f have in this interval?
   (b) Is f increasing or decreasing at $x = 0$? At $x = 2$? At $x = 4$?
   (c) On which interval is the average rate of change of f greater: $-1 \leq x \leq 0$ or $2 \leq x \leq 3$?
   (d) Is the instantaneous rate of change of f greater at $x = 0$ or $x = 2$?

2. For the function $f(x) = \ln x$, estimate $f'(1)$. From the graph of $f(x)$, would you expect your estimate to be greater than or less than $f'(1)$?

3. Let $f(x) = x^2$. In Section 2.3 we calculated an estimate of $f'(1)$ by tabulating values of x^2 near $x = 1$ (see Table 2.9 on page 147). Do the same for $x = -1$ and for $x = 5$. In each case estimate the derivative at that point. Recalling from Example 4 of Section 2.3 that $f'(2) \approx 4$ and $f'(3) \approx 6$, guess a general formula for the derivative of $f(x) = x^2$.

4. Construct tables of values, rounded to three decimals, for $f(x) = x^3$ near $x = 1$, near $x = 3$ and near $x = 5$, as in Problem 3. In each case estimate the derivatives $f'(1)$, $f'(3)$, and $f'(5)$. Then guess a general formula for $f'(x)$.

5. For $f(x) = \ln x$, construct tables, rounded to four decimals, near $x = 1$, $x = 2$, $x = 5$, and $x = 10$ as in Problem 3. Use the tables to estimate $f'(1)$, $f'(2)$, $f'(5)$, and $f'(10)$. Then guess a general formula for $f'(x)$.

6. (a) If f is even and $f'(10) = 6$, what must $f'(-10)$ equal?
   (b) If f is any even function and $f'(0)$ exists, what must $f'(0)$ equal?

7. If g is an odd function and $g'(4) = 5$, what must $g'(-4)$ equal?

For Problems 8–9, sketch a graph of the derivative function of the given functions.

8.

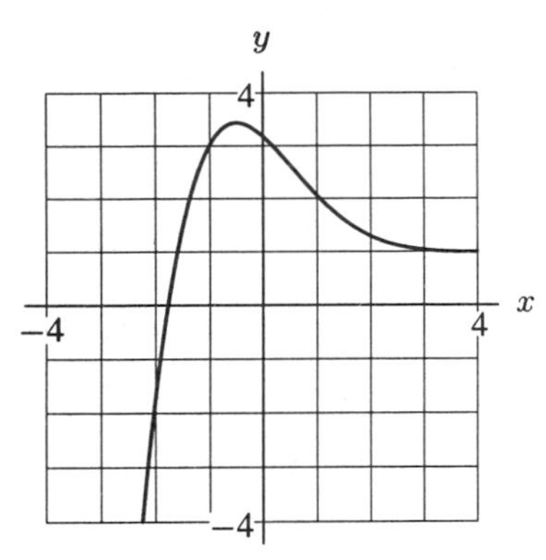

9.

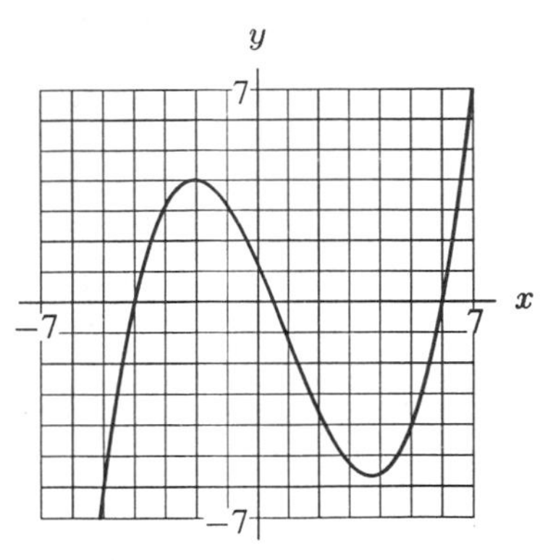

Sketch the graphs of the derivatives of the functions shown in Problems 10–14. Be sure your sketches are consistent with the important features of the original functions.

10.

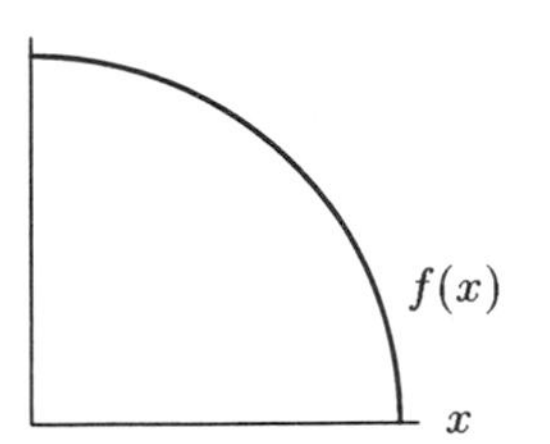

11.

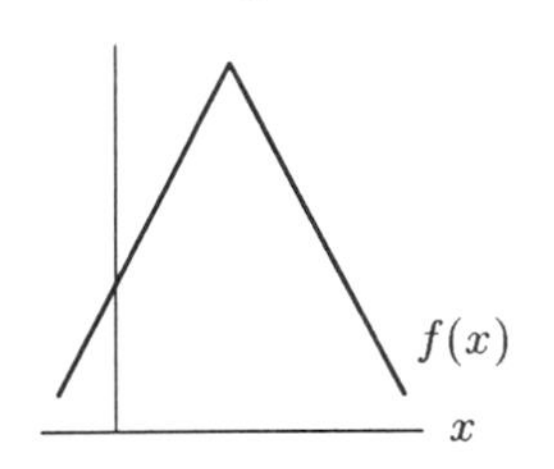

12.

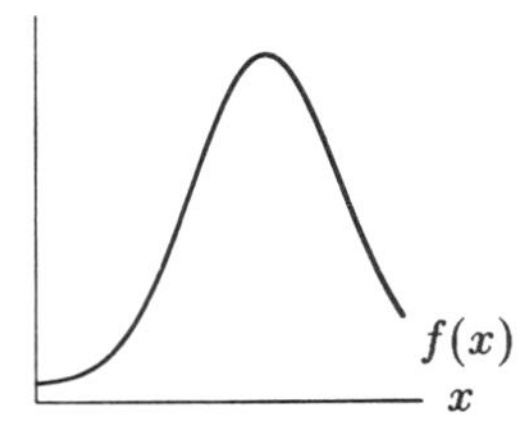

13.

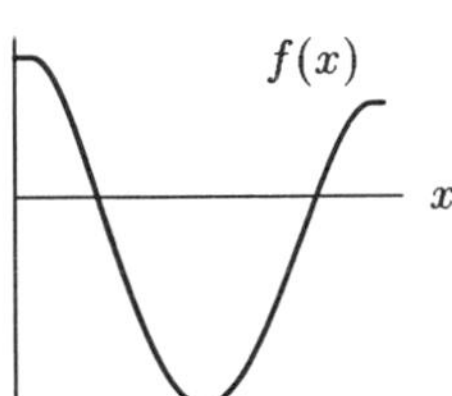

14.

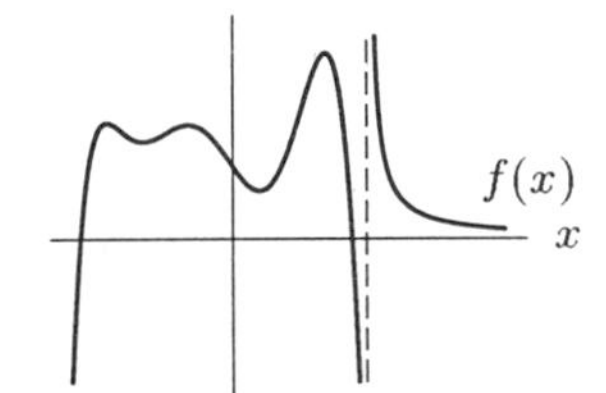

15. Draw a possible graph of $y = f(x)$ given the following information about its derivative:

- For $x < -2$, $f'(x) > 0$ and the derivative is increasing.
- For $-2 < x < 1$, $f'(x) > 0$ and the derivative is decreasing.
- At $x = 1$, $f'(x) = 0$.
- For $x > 1$, $f'(x) < 0$ and the derivative is decreasing (getting more and more negative).

16. Given the following data:

x	0	0.2	0.4	0.6	0.8	1.0
$f(x)$	3.7	3.5	3.5	3.9	4.0	3.9

(a) Estimate an equation of the tangent line to $y = f(x)$ at $x = 0.6$.

(b) Using this equation, estimate $f(0.7)$, $f(1.2)$, and $f(1.4)$. Which of these estimates do you feel most confident about? Why?

17. Given the following data about a function, f:

x	6.5	7.0	7.5	8.0	8.5	9.0
$f(x)$	10.3	8.2	6.5	5.2	4.1	3.2

(a) Estimate $f'(7.0)$, $f'(8.5)$ and $f'(6.75)$.
(b) Estimate the rate of change of f' at $x = 7$.
(c) Find, approximately, an equation of the tangent line to the graph of f at $x = 7$.
(d) Estimate $f(6.8)$.

18. Suppose you put a yam in a hot oven, maintained at a constant temperature of 200°C. As the yam picks up heat from the oven, its temperature rises.[6]

(a) Draw a possible graph of the temperature T of the yam against time t (minutes) since it is put into the oven. Explain any interesting features of the graph, and in particular explain its concavity.
(b) Suppose that, at $t = 30$, the temperature T of the yam is 120° and increasing at the (instantaneous) rate of 2°/min. Using this information, plus what you know about the shape of the T graph, estimate the temperature at time $t = 40$.
(c) Suppose in addition you are told that at $t = 60$, the temperature of the yam is 165°. Can you improve your estimate of the temperature at $t = 40$?
(d) Assuming all the data given so far, estimate the time at which the temperature of the yam is 150°.

19. The graphs of the revenue and cost functions for a certain company are given in Figure 2.73.

(a) Estimate the marginal cost at $q = 400$.
(b) Should the company produce the 500th item? Explain.
(c) Estimate the quantity which maximizes profit.

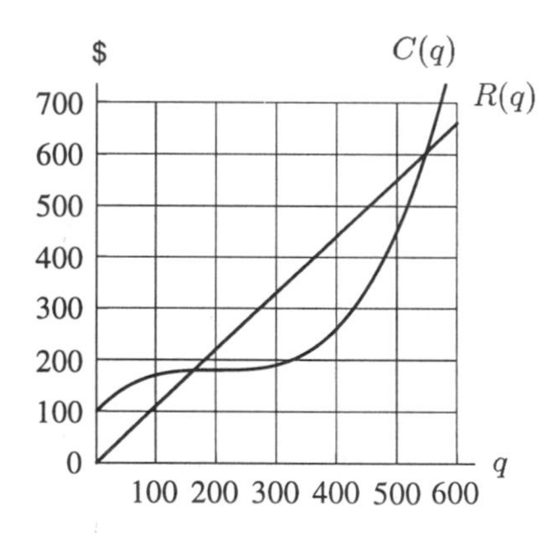

Figure 2.73: Cost and revenue functions

20. A software firm is trying to determine how much money to invest each year in research and development. An extensive study indicates that if they invest x million dollars each year in research and development then they can expect to make a profit, $\pi(x)$, given in thousands of dollars, by

$$\pi(x) = -0.01x^4 + 0.33x^3 + 2.69x^2 + 4.35x - 500.$$

(a) Use a graphing calculator or computer to sketch the graph of $\pi(x)$.
(b) For what values of x is the company going to make a profit?
(c) For what values of x is $\pi(x)$ increasing and for what values of x is $\pi(x)$ decreasing?
(d) For what values of x is $\pi'(x)$ increasing and for what values of x is $\pi'(x)$ decreasing?

[6] From Peter D. Taylor, *Calculus: The Analysis of Functions* (Toronto: Wall & Emerson, Inc., 1992)

(e) Using the fact that $\pi'(x) = -0.04x^3 + 0.99x^2 + 5.38x + 4.35$, graph $\pi'(x)$ and check your answer to part (d).
(f) Which value of x maximizes the rate of change of $\pi(x)$ with respect to x?

21. Students were asked to evaluate $f'(4)$ from the following table which shows the values of the function f:

x	1	2	3	4	5	6
$f(x)$	4.2	4.1	4.2	4.5	5.0	5.7

- Student A estimated the derivative as $f'(4) \approx \dfrac{f(5) - f(4)}{5 - 4} = 0.5$.
- Student B estimated the derivative as $f'(4) \approx \dfrac{f(4) - f(3)}{4 - 3} = 0.3$.
- Student C suggested that they should split the difference and estimate the average of these two results, that is, $f'(4) \approx \frac{1}{2}(0.5 + 0.3) = 0.4$.

(a) Sketch the graph of f, and indicate how the three estimates are represented on the graph.
(b) Explain which answer is likely to be best.
(c) Use Student C's method to find an algebraic formula which approximates $f'(x)$ using increments of size h.

22. Each of the graphs in Figure 2.74 shows the position of a particle moving along the x-axis as a function of time, $0 \le t \le 5$. The vertical scales of the graphs are the same. During this time interval, which particle has

(a) Constant velocity?
(b) The greatest initial velocity?
(c) The greatest average velocity?
(d) Zero average velocity?
(e) Zero acceleration?
(f) Positive acceleration throughout?

(I)
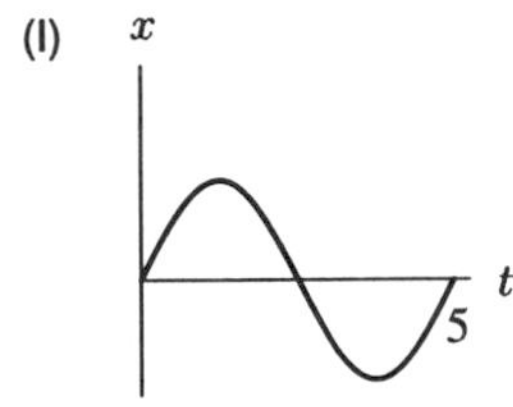

(II)
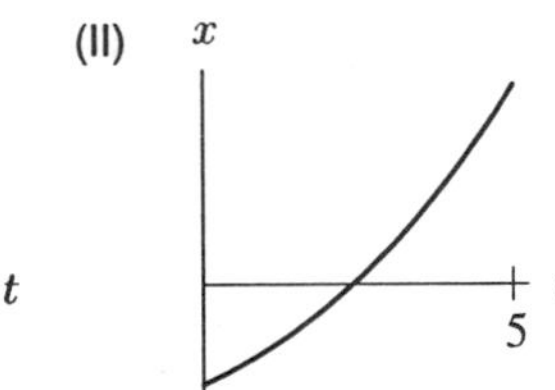

(III)
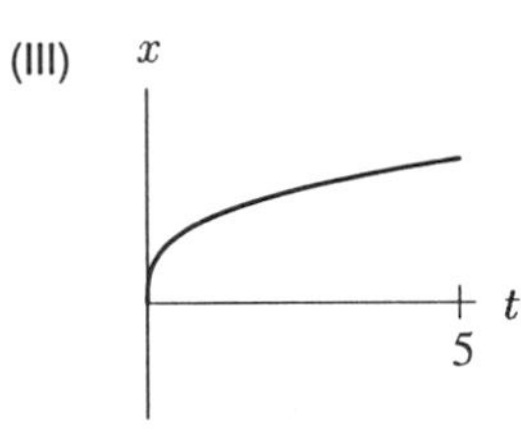

(IV)
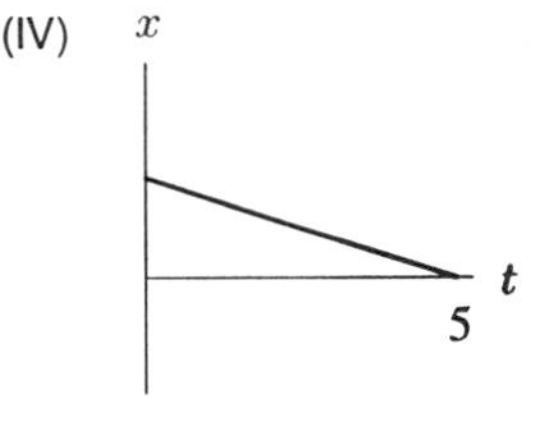

Figure 2.74

23. Let $g(x) = \sqrt{x}$ and $f(x) = kx^2$, where k is a constant.

(a) Find the slope of the tangent line to the graph of g at the point $(4, 2)$.
(b) Find the equation of this tangent line.
(c) If the graph of f contains the point $(4, 2)$, find k.
(d) Where, other than point $(4, 2)$, does the graph of f intersect the tangent line?

24. A circle with center at the origin and radius of length $\sqrt{19}$ has equation $x^2 + y^2 = 19$. Graph the circle.

(a) Just from looking at the graph, what can you say about the slope of the line tangent to the circle at the point $(0, \sqrt{19})$? What about the slope of the tangent at $(\sqrt{19}, 0)$?
(b) Estimate the slope of the tangent to the circle at the point $(2, -\sqrt{15})$ by graphing the tangent carefully at that point.
(c) Use the result of part (b) and the symmetry of the circle to find slopes of the tangents drawn to the circle at $(-2, \sqrt{15})$, $(-2, -\sqrt{15})$, and $(2, \sqrt{15})$.

25. A person with a certain liver disease first exhibits larger and larger concentrations of certain enzymes (called SGOT and SGPT) in the blood. As the disease progresses, the concentration of these enzymes drops, first to the predisease level and eventually to zero (when almost all of the liver cells have died). Monitoring the levels of these enzymes allows doctors to track the progress of a patient with this disease. If $C = f(t)$ is the concentration of the enzymes in the blood as a function of time:

(a) Sketch a possible graph of $C = f(t)$.
(b) Mark on the graph the intervals where $f' > 0$ and where $f' < 0$.
(c) What does $f'(t)$ represent, in practical terms?

26. A continuous function defined for all x has the following properties:
- f is increasing.
- f is concave down.
- $f(5) = 2$.
- $f'(5) = \frac{1}{2}$.

(a) Sketch a possible graph for f.
(b) How many zeros does f have?
(c) What can you say about the location of the zeros?
(d) What is $\lim_{x \to -\infty} f(x)$?
(e) Is it possible that $f'(1) = 1$?
(f) Is it possible that $f'(1) = \frac{1}{4}$?

27. Roughly sketch the shape of the graph of a quadratic polynomial, f, if it is known that:
- $(1, 3)$ is on the graph of f.
- $f'(0) = 3$, $f'(2) = 1$, $f'(3) = 0$.

28. Roughly sketch the shape of the graph of a cubic polynomial, f, if it is known that:
- $(0, 3)$ is on the graph of f.
- $f'(0) = 4$, $f'(1) = 0$, $f'(2) = -\frac{4}{3}$, $f'(4) = 4$.

29. The population of a herd of deer is modeled by

$$P(t) = 4000 + 400 \sin\left(\frac{\pi}{6}t\right) + 180 \sin\left(\frac{\pi}{3}t\right)$$

where t is measured in months from the first of April.

(a) Use a calculator or computer to sketch a graph showing how this population varies with time.

Use the graph to answer the following questions.

(b) When is the herd largest? How many deer are in it at that time?
(c) When is the herd smallest? How many deer are in it then?
(d) When is the herd growing the fastest? When is it shrinking the fastest?
(e) How fast is the herd growing on April 1?

CHAPTER THREE

KEY CONCEPT: THE DEFINITE INTEGRAL

We started Chapter 2 by calculating the velocity from the distance traveled. This led us to the notion of the derivative, or rate of change, of a function. Now we will consider the reverse problem: given the velocity, how can we calculate the distance traveled? This will lead us to our second key concept, the *definite integral,* which computes the total change in a function from its rate of change. We will then discover that the definite integral can be used to calculate not only distance but also other quantities, such as the area under a curve and the average value of a function.

We end this chapter with the Fundamental Theorem of Calculus, which tells us that we can use a definite integral to get information about a function from its derivative. Calculating derivatives and calculating definite integrals are, in a sense, reverse processes.

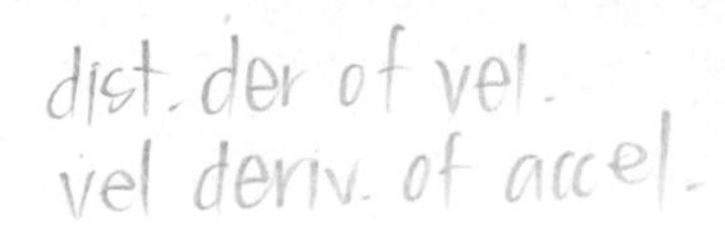

3.1 HOW DO WE MEASURE DISTANCE TRAVELED?

If the velocity is a constant, we can find the distance traveled using the formula

$$\text{Distance} = \text{Velocity} \times \text{Time}.$$

In this section we will see how to estimate the distance when the velocity is not a constant.

A Thought Experiment: How Far Did the Car Go?

One Second Velocity Data

Suppose a car is moving with increasing velocity. First, let's suppose we are given the velocity every second, as in Table 3.1:

TABLE 3.1 *Velocity of car every second*

Time (sec)	0	1	2	3	4	5
Velocity (ft/sec)	20	30	38	44	48	50

How far has the car moved? Since we don't know how fast the car is moving at every moment, we can't figure the distance out exactly, but we can make an estimate. Since the velocity is increasing, the car goes at least 20 feet during the first second. Likewise, it goes at least 30 feet during the next second, at least 38 feet during the third second, at least 44 feet during the fourth second, and at least 48 feet during the last second.

During the five-second period it goes at least

$$20 + 30 + 38 + 44 + 48 = 180 \text{ feet}.$$

Thus, 180 feet is an underestimate of the total distance moved during the five seconds.

To get an overestimate, we can reason this way: In the first second the car moved at most 30 feet, in the next second it moved at most 38 feet, in the third second it moved at most 44 feet, and so on. Therefore, altogether it moved at most

$$30 + 38 + 44 + 48 + 50 = 210 \text{ feet}.$$

Thus, the total distance moved is between 180 and 210 feet:

$$180 \text{ feet} \leq \text{Total distance traveled} \leq 210 \text{ feet}.$$

There is a difference of 30 feet between our upper and lower estimates.

Half Second Velocity Data

What if we wanted a more accurate estimate? Then we should ask for more frequent velocity measurements, say every 0.5 seconds. See Table 3.2.

TABLE 3.2 *Velocity of car every half second*

Time (sec)	0	0.5	1.0	1.5	2.0	2.5	3.0	3.5	4.0	4.5	5.0
Velocity (ft/sec)	20	26	30	35	38	42	44	46	48	49	50

As before, we get a lower estimate for each half second by using the velocity at the beginning of that half second. During the first half second the velocity is at least 20 ft/sec, and so the car travels at least $(20)(0.5) = 10$ feet. During the next half second the car moves at least $(26)(0.5) = 13$ feet, and so on. So now we can say

$$\begin{aligned}\text{Lower estimate} &= (20)(0.5) + (26)(0.5) + (30)(0.5) + (35)(0.5) + (38)(0.5)\\ &\quad + (42)(0.5) + (44)(0.5) + (46)(0.5) + (48)(0.5) + (49)(0.5)\\ &= 189 \text{ feet.}\end{aligned}$$

Notice that this is higher than our old lower estimate of 180 feet.

We get a new upper estimate by considering the velocity at the end of each half second. During the first half second the velocity is at most 26 ft/sec, and so the car moves at most $(26)(0.5) = 13$ feet; in the next half second it moves at most $(30)(0.5) = 15$ feet, and so on. Thus,

$$\begin{aligned}\text{Upper estimate} &= (26)(0.5) + (30)(0.5) + (35)(0.5) + (38)(0.5) + (42)(0.5)\\ &\quad + (44)(0.5) + (46)(0.5) + (48)(0.5) + (49)(0.5) + (50)(0.5)\\ &= 204 \text{ feet.}\end{aligned}$$

This is lower than our old upper estimate of 210 feet. Now we know that

$$189 \text{ feet} \leq \text{Total distance traveled} \leq 204 \text{ feet.}$$

Notice that the difference between our new upper and lower estimates is now 15 feet, half of what it was before. By halving our interval of measurement, we have halved the difference between the upper and lower estimates.

Visualizing Distance on the Velocity Graph: One Second Data

We can represent both upper and lower estimates on a graph of the velocity, as well as see how changing the interval of measurement changes the accuracy of our estimates.

The velocity can be graphed by plotting the data in Table 3.1 and drawing a smooth curve through them. (See Figure 3.1.)

The area of the first dark rectangle is $(20)(1) = 20$, the lower estimate of the distance moved in the first second. The area of the second dark rectangle is $(30)(1) = 30$, the lower estimate for the distance moved in the next second. Therefore, the total area of the dark rectangles represents the lower estimate for the total distance moved during the five seconds.

If the dark and light rectangles are considered together, the first area is $(30)(1) = 30$, the upper estimate for the distance moved in the first second. Continuing this calculation shows that the upper estimate for the total distance is represented by the area of the dark and light rectangles together. Therefore, the area of the light rectangles alone represents the difference between the two estimates.

To calculate the difference between the two estimates, look at Figure 3.1 and imagine the light rectangles all pushed to the right and stacked on top of each other; this gives a rectangle of width 1 and height 30. Notice that the height, 30, is just the difference between the initial and final values of the velocity, $30 = 50 - 20$; and the width, 1, is the time interval between velocity measurements.

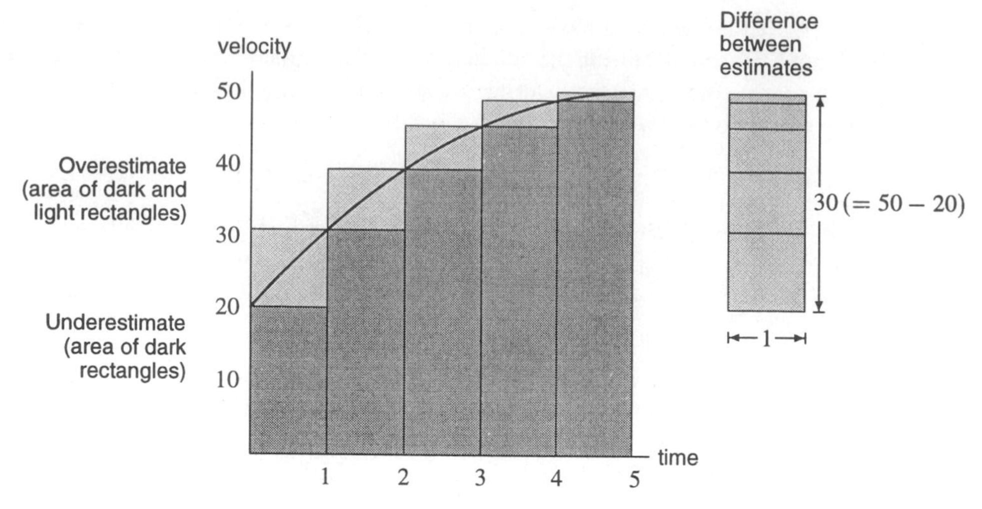

Figure 3.1: Velocity measured each second

Visualizing Distance on the Velocity Graph: Half Second Data

The data for the velocities measured each half second is in Figure 3.2.

The area of the dark rectangles again represents the lower estimate, and the area of the dark and light rectangles together represents the upper estimate. As before, the difference between the two estimates is represented by the area of the light rectangles. This difference can be calculated by stacking the light rectangles vertically, giving a rectangle of the same height as before but of half the width. Its area is therefore half what it was before. Again, the height of this stack is $50 - 20 = 30$, and its width is the time interval 0.5.

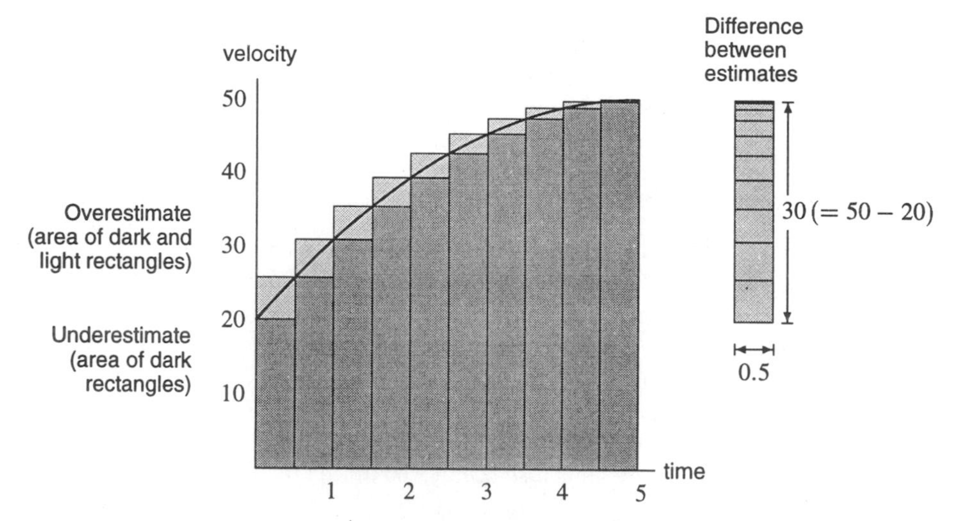

Figure 3.2: Velocity measured each half second

Determining Distance Precisely

In this section we will obtain a precise expression for the distance traveled. We express the exact distance traveled as a limit of estimates in much the same way as we expressed velocity as a limit of average velocities.

Suppose we want to know the distance traveled by a moving object over the time interval $a \leq t \leq b$. Let the velocity at time t be given by the function $v = f(t)$. Suppose we take measurements of $f(t)$ at equally spaced times $t_0, t_1, t_2, \ldots, t_n$ with time $t_0 = a$ and time $t_n = b$. The time interval between any two consecutive measurements is

$$\Delta t = \frac{b-a}{n}$$

where Δt (read "delta t") means the change, or increment, in t.

During the first time interval, the velocity can be approximated by $f(t_0)$, so the distance traveled is approximately

$$f(t_0)\Delta t.$$

During the second time interval, the velocity is about $f(t_1)$, so the distance traveled is about

$$f(t_1)\Delta t.$$

Continuing in this way and adding up all the estimates, we get an estimate for the total distance. In the last interval, the velocity is approximately $f(t_{n-1})$, so the last term is $f(t_{n-1})\Delta t$:

$$\text{Total distance traveled between } a \text{ and } b \approx f(t_0)\Delta t + f(t_1)\Delta t + f(t_2)\Delta t + \cdots + f(t_{n-1})\Delta t.$$

This is called a *left-hand sum* because we used the value of velocity from the left end of each time interval. It is represented by the sum of the areas of the rectangles in Figure 3.3.

We can also calculate a *right-hand sum* by using the value of the velocity at the right end of each time interval. In that case the estimate for the first interval is $f(t_1)\Delta t$, for the second interval it is $f(t_2)\Delta t$, and so on. The estimate for the last interval is now $f(t_n)\Delta t$, so

$$\text{Total distance traveled between } a \text{ and } b \approx f(t_1)\Delta t + f(t_2)\Delta t + f(t_3)\Delta t + \cdots + f(t_n)\Delta t.$$

The right-hand sum is represented by the area of the rectangles in Figure 3.4.

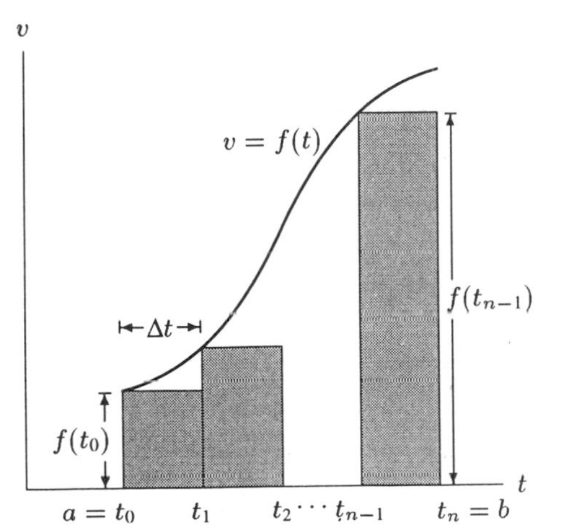

Figure 3.3: Left-hand sums

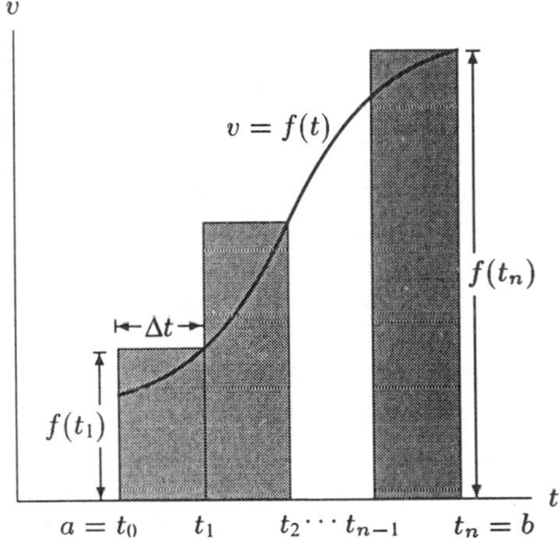

Figure 3.4: Right-hand sums

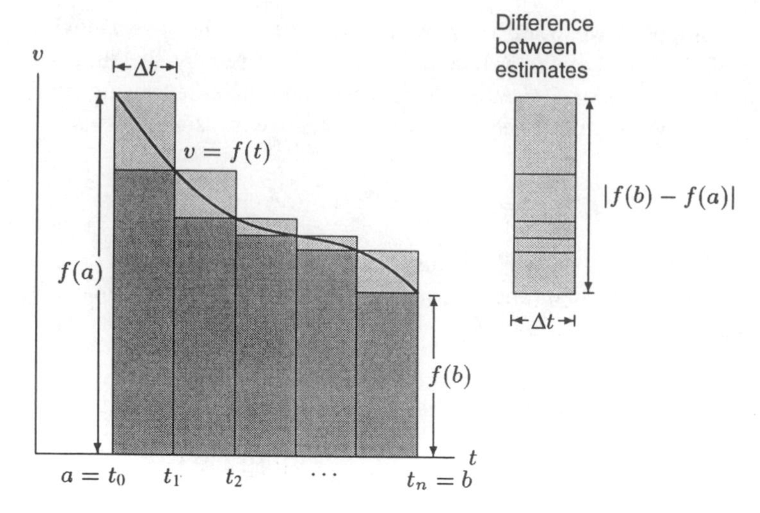

Figure 3.5: Left and right sums if f is decreasing

If f is an increasing function, the left-hand sum will be an underestimate of the total distance traveled, since for each time interval we use the velocity at the start of that interval to compute the distance traveled, whereas in fact the velocity continues to increase after that measurement. We get an overestimate by using the velocity at the right end of each time interval. If f is decreasing, as in Figure 3.5, then the roles of the two sums are reversed: the left-hand sum is an overestimate, and the right-hand sum is an underestimate.

For either increasing or decreasing functions, the exact value of the distance traveled lies somewhere between the two estimates. Thus the accuracy of our estimate depends on how close these two sums are. For a function which is increasing throughout or decreasing throughout the interval $[a, b]$:

$$\left|\begin{array}{c}\text{Difference between}\\ \text{upper and lower estimates}\end{array}\right| = \left|\begin{array}{c}\text{Difference between}\\ f(a) \text{ and } f(b)\end{array}\right| \times \Delta t = |f(b) - f(a)| \cdot \Delta t.$$

(Absolute values are used to make the difference non-negative.) By making the measurements close enough together, we can make Δt as small as we like; therefore, we can make the difference between our lower and upper estimates as small as we like.

To find exactly the total distance traveled between a and b, we take the limit of the sums, as n, the number of subdivisions of the interval $[a, b]$, goes to infinity. The sum of the areas of the rectangles approaches the area under the curve between $t = a$ and $t = b$. So we conclude that

$$\begin{aligned}\begin{array}{l}\text{Total distance traveled}\\ \text{between } t = a \text{ and } t = b\end{array} &= \lim_{n\to\infty} (\text{Left-hand sum})\\ &= \lim_{n\to\infty} \left[f(t_0)\Delta t + f(t_1)\Delta t + \cdots + f(t_{n-1})\Delta t\right]\\ &= \text{Area under curve } f(t) \text{ from } t = a \text{ to } t = b\end{aligned}$$

and

$$\begin{aligned}\begin{array}{l}\text{Total distance traveled}\\ \text{between } t = a \text{ and } t = b\end{array} &= \lim_{n\to\infty} (\text{Right-hand sum})\\ &= \lim_{n\to\infty} \left[f(t_1)\Delta t + f(t_2)\Delta t + \cdots + f(t_n)\Delta t\right]\\ &= \text{Area under curve } f(t) \text{ from } t = a \text{ to } t = b.\end{aligned}$$

Thus, if n is large enough, both the left-hand and the right-hand sums are accurate estimates of the distance traveled. This method of calculating the distance by taking the limit of a sum works even if the velocity is not increasing throughout, or decreasing throughout, the interval.

Distance Traveled as Area

We have seen graphically that the left-hand and right-hand sums used to estimate distance traveled are represented graphically by the sums of areas of rectangles. As n gets larger, the estimate of distance traveled gets more accurate, and the estimate of the area under the velocity curve becomes more accurate. We see that distance traveled is represented graphically by the area under the velocity curve.

> If $v(t)$ represents velocity and $v(t) > 0$ between $t = a$ and $t = b$, then
>
> Distance traveled between $t = a$ and $t = b$ = Area under the graph of $v(t)$ between $t = a$ and $t = b$.

Estimating distance traveled is the same as estimating the area under the velocity curve.

Example 1 The graph of a function $y = f(x)$ is given in Figure 3.6. Estimate the area of the region bounded by this graph, the x-axis, the y-axis, and the vertical line $x = 6$.

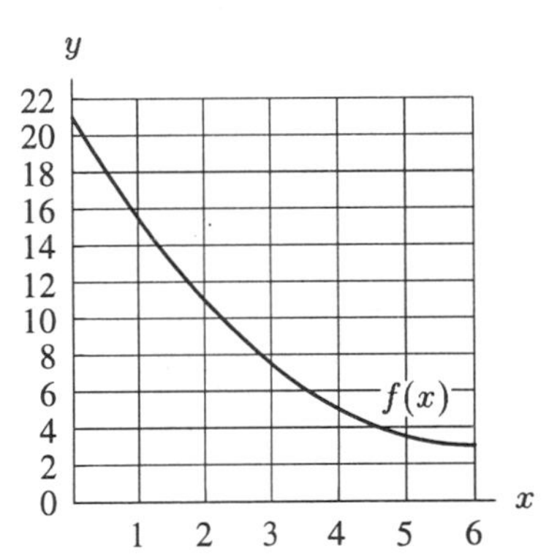

Figure 3.6: Estimate the area under this function

Solution As we saw before, to find the area under this curve, we go through the same process used to determine distance traveled from velocity. We can make n as large as we like depending on how accurate we want the estimate to be. Let's start with $n = 3$. Since $0 \le x \le 6$, we have $\Delta x = 2$, and we are partitioning the area into three parts: the part over $0 \le x \le 2$, the part over $2 \le x \le 4$, and the part over $4 \le x \le 6$. We will compute the left-hand sum first. Figure 3.7 shows the three rectangles. The height of the first rectangle is $f(0) = 21$ and the width is 2, so the area of the first rectangle is $(21)(2) = 42$.

$$\begin{aligned}\text{Left-hand estimate} &= (f(0))(2) + (f(2))(2) + (f(4))(2)\\ &= (21)(2) + (11)(2) + (5)(2)\\ &= 42 + 22 + 10\\ &= 74 \text{ square units.}\end{aligned}$$

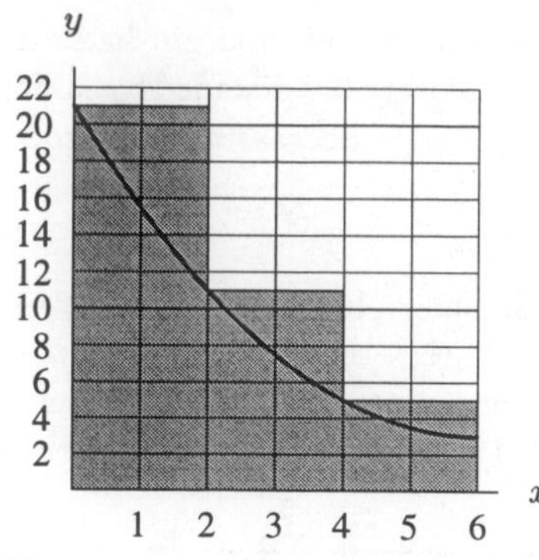

Figure 3.7: Left-hand estimate, with $n = 3$

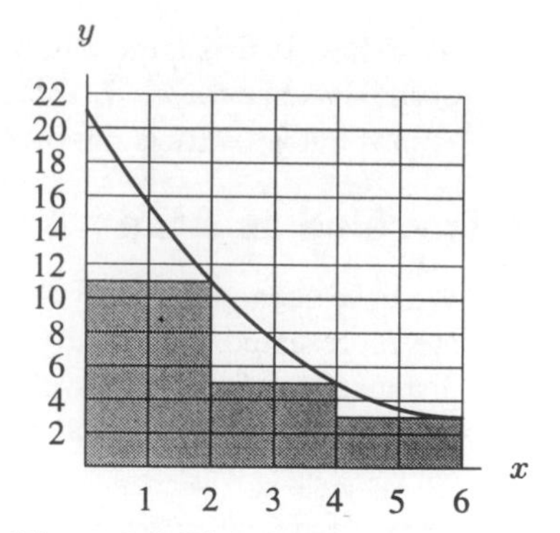

Figure 3.8: Right-hand estimate, with $n = 3$

Similarly, we compute a right-hand estimate. (See Figure 3.8.)

$$\begin{aligned}\text{Right-hand estimate} &= (f(2))(2) + (f(4))(2) + (f(6))(2)\\ &= (11)(2) + (5)(2) + (3)(2)\\ &= 22 + 10 + 6\\ &= 38 \text{ square units.}\end{aligned}$$

We can see from Figure 3.7 and Figure 3.8 that the left-hand sum is an overestimate and the right-hand sum is an underestimate, so the true area is between 38 and 74 square units.

To improve the accuracy, we use a larger number of rectangles. If $n = 6$, the width of each rectangle is 1, and we obtain the following estimates:

$$\text{Left-hand estimate} = (21)(1) + (15.5)(1) + (11)(1) + (7.5)(1) + (5)(1) + (3.5)(1) = 63.5,$$

and

$$\text{Right-hand estimate} = (15.5)(1) + (11)(1) + (7.5)(1) + (5)(1) + (3.5)(1) + (3)(1) = 45.5.$$

We have improved the estimate. Now we know that the true value of the area is between 45.5 and 63.5 square units. For even greater accuracy, we would take an even larger value of n.

Example 2 Suppose that the velocity of an object (in m/sec) is given by $v(t) = 10 + 8t - t^2$. Estimate the distance traveled by the object during the first 5 seconds (that is, between $t = 0$ and $t = 5$) using $n = 5$.

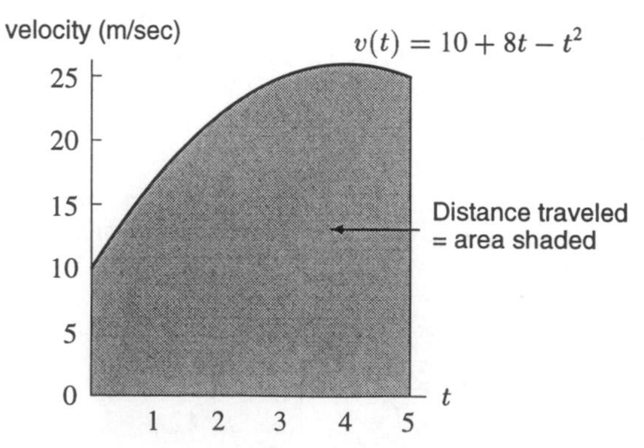

Figure 3.9: A velocity function

Solution A graph of this velocity function is given in Figure 3.9. Finding the distance traveled is equivalent to finding the area under this curve between $t = 0$ and $t = 5$. We estimate that this area is about 100 since the average height appears to be about 20 and the width is 5. To find a more accurate estimate we find right-hand and left-hand sums and average them. If we use $n = 5$, the width of each rectangle is 1 and, using $v(t) = 10 + 8t - t^2$, we compute the following estimates:

$$\begin{aligned}\text{Left-hand estimate} &= (v(0))(1) + (v(1))(1) + (v(2))(1) + (v(3))(1) + (v(4))(1)\\ &= (10)(1) + (17)(1) + (22)(1) + (25)(1) + (26)(1)\\ &= 100 \text{ meters.}\end{aligned}$$

$$\begin{aligned}\text{Right-hand estimate} &= (v(1))(1) + (v(2))(1) + (v(3))(1) + (v(4))(1) + (v(5))(1)\\ &= (17)(1) + (22)(1) + (25)(1) + (26)(1) + (25)(1)\\ &= 115 \text{ meters.}\end{aligned}$$

We average these to obtain a better estimate:

$$\text{Distance traveled} \approx \frac{100 + 115}{2} = 107.5 \text{ meters.}$$

Problems for Section 3.1

1. A car comes to a stop five seconds after the driver slams on the brakes. While the brakes are on, the following velocities are recorded:

Time since brakes applied (sec)	0	1	2	3	4	5
Velocity (ft/sec)	88	60	40	25	10	0

   (a) Give lower and upper estimates of the distance the car traveled after the brakes were applied.
   (b) On a sketch of velocity against time, show the lower and upper estimates and the difference between them.

2. Roger decides to run a marathon. Roger's friend Jeff rides behind him on a bicycle and clocks his pace every 15 minutes. Roger starts out strong, but after an hour and a half he is so exhausted that he has to stop. The data Jeff collected is summarized below:

Time spent running (min)	0	15	30	45	60	75	90
Speed (mph)	12	11	10	10	8	7	0

   (a) Assuming that Roger's speed is always decreasing, give upper and lower estimates for the distance Roger ran during the first half hour.
   (b) Give upper and lower estimates for the distance Roger ran in total.

3. Coal gas is produced at a gasworks. Pollutants in the gas are removed by scrubbers, which become less and less efficient as time goes on. Measurements made at the start of each month showing the rate at which pollutants are escaping in the gas are as follows:

Time (months)	0	1	2	3	4	5	6
Rate pollutants are escaping (tons/month)	5	7	8	10	13	16	20

   (a) Make an overestimate and an underestimate of the total quantity of pollutants that escaped during the first month.
   (b) Make an overestimate and an underestimate of the total quantity of pollutants that escaped during the first six months.

4. A graph of the velocity function $v(t) = 10 + 8t - t^2$ is given in Figure 3.9. In Example 2, we calculated left- and right-hand estimates for distance traveled between $t = 0$ and $t = 5$. On a graph similar to Figure 3.9, shade in rectangles to represent each of these estimates.

5. In Figure 3.10, use the grid to estimate the area of the region bounded by the curve, the horizontal axis and the lines $x = \pm 3$. Explain what you are doing.

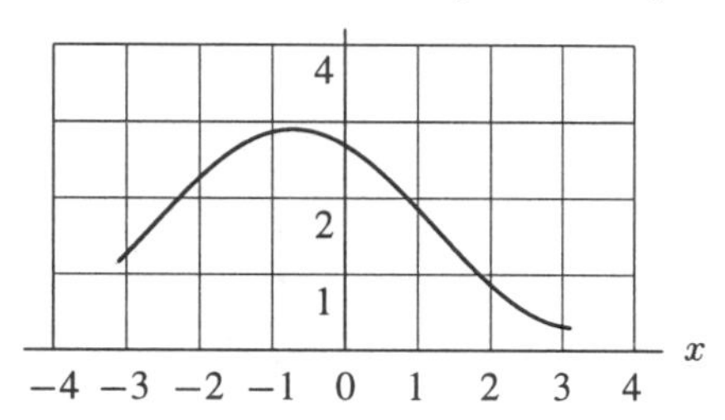

Figure 3.10

6. Figure 3.11 shows the graph of the velocity, v, of an object (in ft/sec).
   (a) Estimate the distance the object traveled between $t = 0$ and $t = 8$, using a right-hand sum with $n = 4$.
   (b) On a graph similar to Figure 3.11, shade in rectangles to represent your estimate in part (a).
   (c) Is the estimate in part (a) an underestimate or an overestimate of the actual distance traveled? Explain.

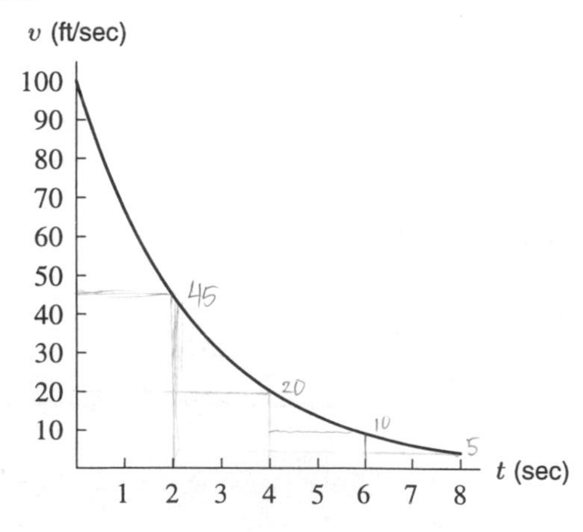

Figure 3.11

7. Figure 3.12 shows the graph of the velocity, v, of an object (in m/sec). Estimate the total distance the object traveled between $t = 0$ and $t = 6$.

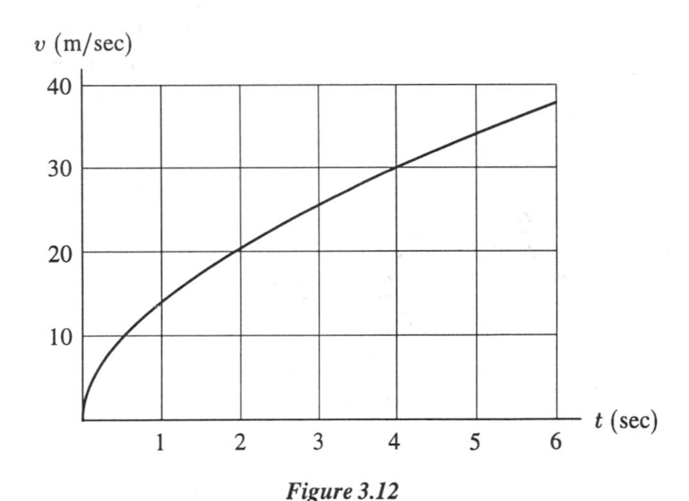

Figure 3.12

8. Your velocity is given by $v(t) = t^2 + 1$ in m/sec, with t in seconds. Estimate the distance traveled between $t = 0$ and $t = 5$. Explain how you arrive at your estimate.

9. Your velocity is given by $v(t) = 10e^{-0.5t}$ in m/min, with t in minutes.
   (a) Sketch a graph of this velocity function, and estimate the distance traveled between $t = 0$ and $t = 3$, using $n = 3$.
   (b) Do you travel farther during the first second (between $t = 0$ and $t = 1$) or between $t = 2$ and $t = 3$? Justify your answer.

10. You jump out of an airplane. Before your parachute opens you fall faster and faster, but your acceleration decreases as you fall because of air resistance. The table below gives your acceleration, a (in m/sec^2), after t seconds.

t	0	1	2	3	4	5
a	9.81	8.03	6.53	5.38	4.41	3.61

   (a) Give upper and lower estimates of your speed at $t = 5$.
   (b) Get a new estimate by taking the average of your upper and lower estimates. What does the concavity of the graph of acceleration tell you about your new estimate?

11. An object has zero initial velocity and a constant acceleration of 32 ft/sec^2. Find a formula for its velocity as a function of time. Then use left and right sums to find upper and lower bounds on the distance that the object travels in four seconds. How can the precise distance be found? [Hint: Find the area under a curve.]

3.2 THE DEFINITE INTEGRAL

In Section 3.1 we saw how distance traveled can be approximated by sums and expressed exactly as the limit of a sum. This process of forming sums and taking their limits has many interpretations. This section will be devoted to looking at this process in more detail. In particular, we will show how these sums can be defined for any function f, whether or not it represents a velocity. This is similar to what we did for the derivative: first we introduced the limit of difference quotients as the solution to the velocity problem, and then we studied the limit in its own right.

Suppose we have a function $f(t)$ which is continuous for $a \leq t \leq b$, except perhaps at a few points, and bounded everywhere. We divide the interval from $a \leq t \leq b$ into n equal subdivisions, and we call the width of an individual subdivision Δt. Thus,

$$\Delta t = \frac{b-a}{n}.$$

We will let $t_0, t_1, t_2, \ldots, t_n$ be endpoints of the subdivisions, as in Figures 3.13 and 3.14. As before, we construct two sums:

$$\text{Left-hand sum} = f(t_0)\Delta t + f(t_1)\Delta t + \cdots + f(t_{n-1})\Delta t$$

and

$$\text{Right-hand sum} = f(t_1)\Delta t + f(t_2)\Delta t + \cdots + f(t_n)\Delta t.$$

These sums may be represented graphically as the areas in Figures 3.13 and 3.14, provided $f(t) \geq 0$. In Figure 3.13, the first rectangle has width Δt and height $f(t_0)$, since the top of its left edge just touches the curve, and hence it has area $f(t_0)\Delta t$; the second rectangle has width Δt and height $f(t_1)$ and hence has area $f(t_1)\Delta t$; and so on. The sum of all these areas is the left-hand sum. The right-hand sum, shown in Figure 3.14, is constructed in the same way, except that each rectangle touches the curve on its right edge instead of its left.

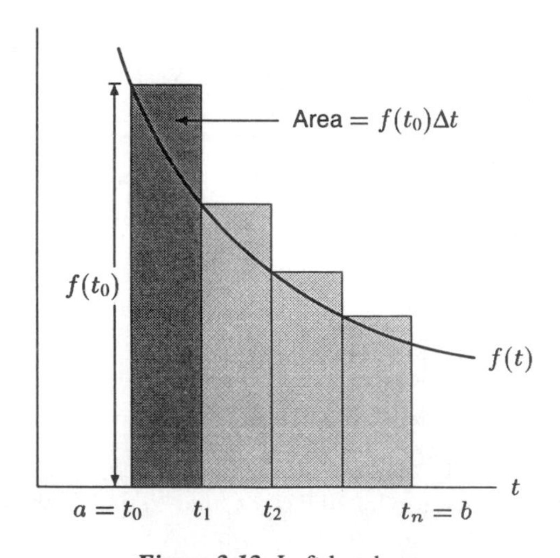

Figure 3.13: Left-hand sum

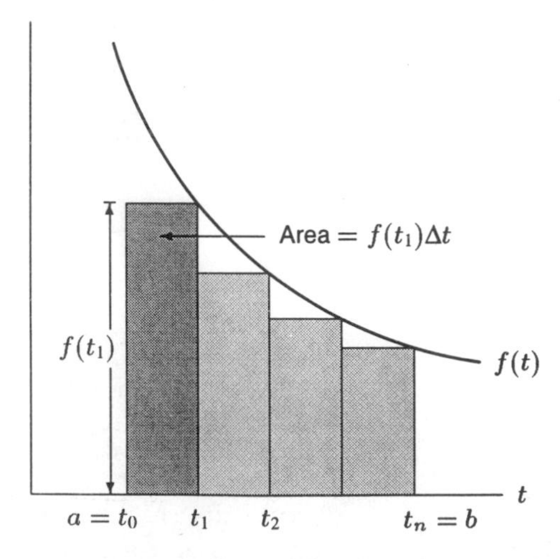

Figure 3.14: Right-hand sum

Writing Left and Right Sums Using Sigma Notation

Both the left-hand and right-hand sums can be written more compactly using *sigma*, or summation, notation. The symbol $\sum$ is a capital sigma, or Greek letter "S." We write

$$\text{Right-hand sum} = \sum_{i=1}^{n} f(t_i)\Delta t = f(t_1)\Delta t + f(t_2)\Delta t + \cdots + f(t_n)\Delta t$$

where the $\sum$ tells us to add terms of the form $f(t_i)\Delta t$. The "$i = 1$" at the base of the sigma sign tells us to start at $i = 1$, and the "n" at the top tells us to stop at $i = n$.

In the left-hand sum we start at $i = 0$ and stop at $i = n - 1$, so we write

$$\text{Left-hand sum} = \sum_{i=0}^{n-1} f(t_i)\Delta t = f(t_0)\Delta t + f(t_1)\Delta t + \cdots + f(t_{n-1})\Delta t.$$

Taking the Limit to Define the Definite Integral

In the previous section we took the limit of these sums as n went to infinity, and we do the same here. Whenever the integrand is continuous for $a \leq x \leq b$, and in fact for any function you are ever likely to meet, the limits of the left- and right-hand sums will exist and be equal. In such cases, we define the *definite integral* to be the limit of these sums. When $f(t) \geq 0$, the definite integral represents the area between the graph of $f(t)$ and the t-axis, from $t = a$ to $t = b$. The definite integral is well approximated by a left- or right-hand sum if n is large enough.

The **definite integral** of f from a to b, written

$$\int_a^b f(t)\,dt,$$

is the limit of the left-hand or right-hand sums with n subdivisions as n gets arbitrarily large. In other words,

$$\int_a^b f(t)\,dt = \lim_{n\to\infty} (\text{Left-hand sum}) = \lim_{n\to\infty} \left(\sum_{i=0}^{n-1} f(t_i)\Delta t \right)$$

and

$$\int_a^b f(t)\,dt = \lim_{n\to\infty} (\text{Right-hand sum}) = \lim_{n\to\infty} \left(\sum_{i=1}^{n} f(t_i)\Delta t \right).$$

Each of these sums is called a *Riemann sum*, f is called the *integrand*, and a and b are called the *limits of integration*.

The "$\int$" notation comes from an old-fashioned "S," which stands for "sum" in the same way that $\sum$ does. The "dt" in the integral comes from the factor Δt. Notice that the limits on the $\sum$ sign are 0 and $n - 1$ for the left-hand sum, or 1 and n for the right-hand sum, whereas the limits on the $\int$ sign are a and b.

Example 1 Find the left-hand and right-hand sums with $n = 2$ and $n = 10$ for $\int_1^2 \frac{1}{t}\,dt$. How do the values of these sums compare with the true value of the integral?

Solution Here $a = 1$ and $b = 2$, so for $n = 2$, $\Delta t = (2-1)/2 = 0.5$. Therefore, $t_0 = 1$, $t_1 = 1.5$ and $t_2 = 2$. (See Figure 3.15.) Thus,

$$\begin{aligned}\text{Left-hand sum} &= f(1)\Delta t + f(1.5)\Delta t\\ &= 1(0.5) + \frac{1}{1.5}(0.5)\\ &\approx 0.8333,\end{aligned}$$

and

$$\begin{aligned}\text{Right-hand sum} &= f(1.5)\Delta t + f(2)\Delta t\\ &= \frac{1}{1.5}(0.5) + \frac{1}{2}(0.5)\\ &\approx 0.5833.\end{aligned}$$

From Figure 3.15 you can see that the left-hand sum is bigger than the area under the curve and the right-hand sum is smaller, so the area under the curve $f(t) = 1/t$ from $t = 1$ to $t = 2$ is between 0.5833 and 0.8333. Thus,

$$0.5833 < \int_1^2 \frac{1}{t}\,dt < 0.8333.$$

When $n = 10$, $\Delta t = 0.1$ (see Figure 3.16), so

$$\begin{aligned}\text{Left-hand sum} &= f(1)\Delta t + f(1.1)\Delta t + \cdots + f(1.9)\Delta t\\ &= \left(1 + \frac{1}{1.1} + \cdots + \frac{1}{1.9}\right)0.1\\ &\approx 0.7188,\\ \text{Right-hand sum} &= f(1.1)\Delta t + f(1.2)\Delta t + \cdots + f(2)\Delta t\\ &= \left(\frac{1}{1.1} + \frac{1}{1.2} + \cdots + \frac{1}{2}\right)0.1\\ &\approx 0.6688.\end{aligned}$$

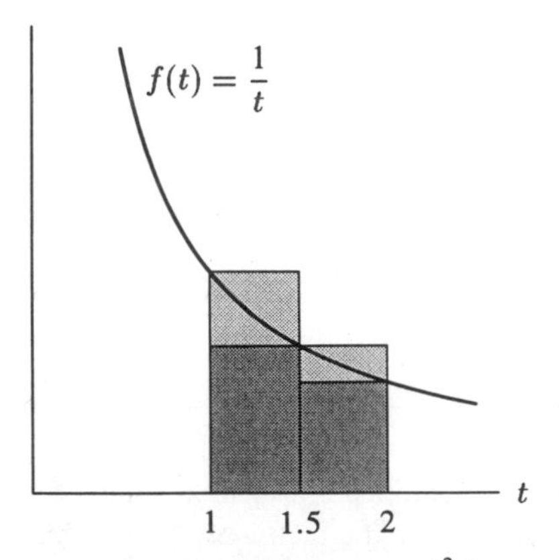

Figure 3.15: Approximating $\int_1^2 \frac{1}{t}\,dt$ with $n = 2$

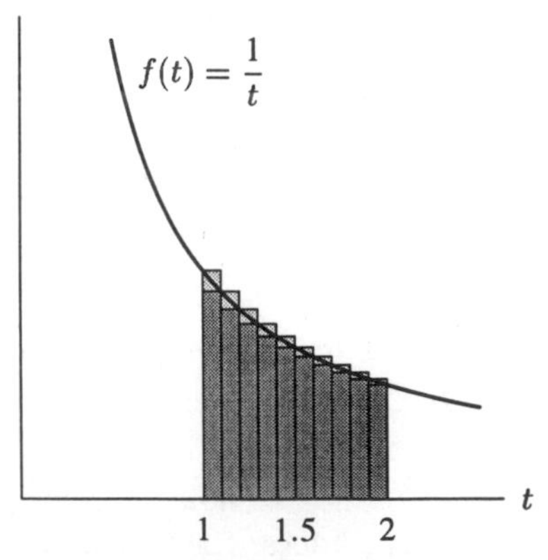

Figure 3.16: Approximation of $\int_1^2 \frac{1}{t}\,dt$ with $n = 10$

From Figure 3.16 you can see that the left-hand sum is larger than the area under the curve, and the right-hand sum smaller, so

$$0.6688 < \int_1^2 \frac{1}{t}\,dt < 0.7188.$$

Notice that the left- and right-hand sums trap the true value of the integral between them. As the subdivisions become finer, the left- and right-hand sums get closer together.

Convergence of Left and Right Sums

A function which is either increasing throughout an interval or decreasing throughout that interval is said to be *monotonic* there. It is usually easy to tell whether or not a function is monotonic just by looking at its graph.

When f Is Monotonic

If f is monotonic, the left- and right-hand sums trap the exact value of the integral between them. Let us continue the example

$$\int_1^2 \frac{1}{t}\,dt.$$

The left- and right-hand sums for $n = 2$, 10, 50, and 250 are listed in Table 3.3.

TABLE 3.3 *Left- and right-hand sums for* $\int_1^2 \frac{1}{t}\,dt$

n	Left-hand sum	Right-hand sum
2	0.8333	0.5833
10	0.7188	0.6688
50	0.6982	0.6882
250	0.6941	0.6921

Because the function $f(t) = 1/t$ is decreasing, the left-hand sums converge down on the integral from above, and the right-hand sums converge up from below. From the last row of the table we can deduce that

$$0.6921 < \int_1^2 \frac{1}{t}\,dt < 0.6941$$

so $\int_1^2 \frac{1}{t}\,dt \approx 0.69$ to two decimal places. To approximate definite integrals accurately, we evaluate left- and right-hand sums with a large number of subdivisions, using a computer or calculator.

When f Is Not Monotonic

If f is not monotonic, the definite integral is not always bracketed between the left- and right-hand sums, but in practice you can still see the convergence quite easily. For example, Table 3.4 gives sums for the integral $\int_0^{2.5} \sin(t^2)\,dt$.

TABLE 3.4 *Left- and right-hand sums for $\int_0^{2.5} \sin(t^2)\,dt$*

n	Left-hand sum	Right-hand sum
2	1.2500	1.2085
10	0.4614	0.4531
50	0.4324	0.4307
250	0.4307	0.4304
1000	0.4306	0.4305

sin (t²) ≠ monotonic

* Although $\sin(t^2)$ is certainly not monotonic, by the time we get to $n = 250$, it is pretty clear that $\int_0^{2.5} \sin(t^2)\,dt \approx 0.43$ to two decimal places. As you can see, the sums wander around a bit at the beginning, and it's not certain that they won't wander later on. Nonetheless, it is a good rule of thumb that when the digits begin to stabilize, or remain the same as n increases, you are getting close to the true value of the integral. This rule of thumb is not foolproof, but it works well in practice. For $\int_0^{2.5} \sin(t^2)\,dt$, we do not have a guarantee of the accuracy, as we do for monotonic functions. Notice that 0.43 does not lie between 1.2500 and 1.2085, the left- and right-hand sums for $n = 2$, or even between 0.4614 and 0.4531, the two sums for $n = 10$. If the integrand is not monotonic, the left- and right-hand sums may both be larger (or smaller) than the integral. (See Problems 17 and 16, page 207, to see how this can happen.)

If f is not monotonic, it is often possible to construct upper and lower estimates for an integral by first splitting the interval $[a, b]$ into subintervals on which f is monotonic.

Using a Calculator or Computer to Evaluate Definite Integrals

In Example 1, we approximated the integral $\int_1^2 \frac{1}{t}dt$ using $n = 10$. Although 10 is a relatively small value for n, this calculation was a bit tedious. The approximations given in Table 3.3 and Table 3.4 would be very cumbersome to do by hand. In practice, we usually use a calculator or computer to compute Riemann sums.

We can make the approximation more accurate by increasing the value of n, until n is so large that round-off error comes into play. Of course, as we make n larger, the computation will take longer.

Example 2 Compute $\int_1^3 t^2\,dt$ to one decimal place.

Solution Since $f(t) = t^2$ is an increasing function on the interval from 1 to 3, the left-hand sum will underestimate the integral and the right-hand sum will overestimate it, as shown in Figure 3.17.

How large should n be? The problem asks us to find a value for the integral to one decimal place, that is, a value which we are certain is within 0.05 of the true value. [1] (Note: This is *not* telling us that Δt is 0.05.) One way is to use a computer or calculator to calculate left- and right-hand sums for various values of n.

For $n = 100$, the left-hand sum is 8.5868 and the right-hand sum is 8.7468, so

$$8.5868 < \int_1^3 t^2\,dt < 8.7468.$$

[1] See Appendix A for a discussion of decimal place accuracy.

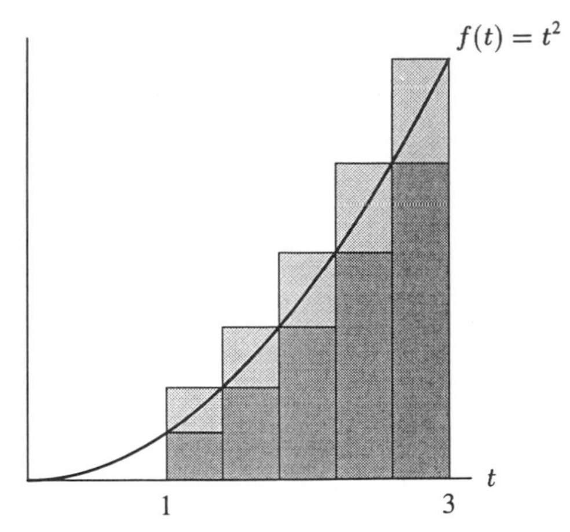

Figure 3.17: Approximation of $\int_1^3 t^2\,dt$

The left- and right-hand sums differ by 0.16, and averaging them gives

$$\int_1^3 t^2\,dt \approx 8.6668,$$

which differs from the true value by at most $0.16/2 = 0.08$. However, we want more accuracy, so we need a larger value of n. Taking $n = 5000$ makes the left-hand sum 8.6650 (rounding down) and the right-hand sum 8.6683 (rounding up), so

$$8.6650 < \int_1^3 t^2\,dt < 8.6683.$$

Averaging the left and right sums gives

$$\int_1^3 t^2\,dt \approx 8.6667.$$

Since the left and right sums differ by 0.0033, the average must be within 0.00165 of the true value.

The Definite Integral as Distance Traveled

The left- and right-hand sums that we use to compute the definite integral are the same sums used in the previous section to compute distance traveled from velocity. Thus, we know that:

> If the velocity of a particle is given by $v(t)$, where t is time, then
>
> $$\int_a^b v(t)dt = \text{Total distance traveled between the times } t = a \text{ and } t = b.$$

Example 3 When you jump out of an airplane and your parachute fails to open, your downward velocity (in meters/second) t seconds after the jump is approximated by

$$v(t) = 49(1 - e^{-0.2t}).$$

(a) Write a definite integral to represent the distance you travel between time $t = 0$ and time $t = 5$.
(b) Graph $v(t)$ and represent this distance graphically.
(c) Use a computer or graphing calculator to estimate the distance you fall during the first five seconds.

Solution (a) Distance is the definite integral of velocity, so between time $t = 0$ and time $t = 5$,

$$\text{Distance traveled} = \int_0^5 49(1 - e^{-0.2t})\, dt.$$

(b) A graph of this velocity is given in Figure 3.18, and the distance traveled is represented by the shaded area under this velocity curve.
(c) We use a computer or calculator to see that

$$\int_0^5 49(1 - e^{-0.2t})\, dt = 90.14.$$

Thus, if you fall out of an airplane without a parachute, you fall about 90 meters during the first five seconds.

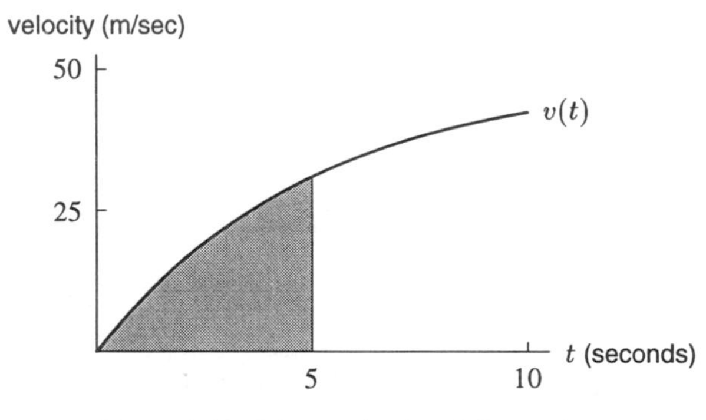

Figure 3.18: Velocity of a free-falling object

Problems for Section 3.2

For Problems 1–2, estimate the value of the definite integral by computing left-hand and right-hand sums with $n = 2$ and with $n = 4$. Draw a graph of the integrand $f(x)$ in each case, and illustrate (using rectangles) each of your approximations. Is each approximation an underestimate or an overestimate?

1. $\displaystyle\int_0^2 (x^2 + 1)\, dx$

2. $\displaystyle\int_1^5 \left(\frac{2}{x}\right)\, dx$

For Problems 3–6, using a computer or calculator, construct a table of left- and right-hand sums with 2, 10, 50, and 250 subdivisions. Observe the limit to which your sums are tending as the number of subdivisions gets larger, and estimate the value of the definite integral.

3. $\int_0^1 x^3\,dx$

4. $\int_1^3 \ln x\,dx$

5. $\int_0^1 e^{t^2}\,dt$

6. $\int_1^2 x^x\,dx$

In Problems 7–13, compute the definite integrals to one decimal place. In each case, say how many subdivisions you are using. [Note that, except for Problem 13, each function is monotonic over the given interval.]

7. $\int_0^5 x^2\,dx$

8. $\int_1^4 \frac{1}{\sqrt{1+x^2}}\,dx$

9. $\int_1^5 (\ln x)^2\,dx$

10. $\int_{1.1}^{1.7} e^t \ln t\,dt$

11. $\int_1^2 2^x\,dx$

12. $\int_1^2 (1.03)^t\,dt$

13. $\int_{-3}^{3} e^{-t^2}\,dt$

14. The velocity of a car (in miles per hour) is given by $v(t) = 40t - 10t^2$ where t is measured in hours.
    (a) Write a definite integral for the distance the car travels during the first three hours.
    (b) Sketch a graph of velocity against time and represent the distance traveled during the first three hours as an area on your graph.
    (c) Use a computer or calculator to find this distance.

15. As in Example 3, if you jump out of an airplane without a parachute, your downward velocity (in m/sec) t seconds after you jump is given by

$$v(t) = 49(1 - e^{-0.2t}).$$

In Example 3, we learned that you fall approximately 90 meters during the first five seconds. Write a definite integral to represent the distance you fall during the first ten seconds and then find this distance. How far do you fall between $t = 5$ and $t = 10$?

16. Sketch the graph of a function f (you do not need to give a formula for f) on an interval $[a, b]$ with the property that with $n = 2$ subdivisions,

$$\int_a^b f(x)\,dx < \text{Lcft-hand sum} < \text{Right-hand sum}.$$

17. Using the graph of $2 + \cos x$, for $0 \le x \le 4\pi$, list the following quantities in increasing order: the value of the integral $\int_0^{4\pi} (2 + \cos x)\,dx$, the left-hand and right-hand sum with $n = 2$ subdivisions.

18. (a) In Figure 3.19 estimate the shaded area with an error of at most 0.1.
    (b) How can you approximate this shaded area to any desired degree of accuracy?

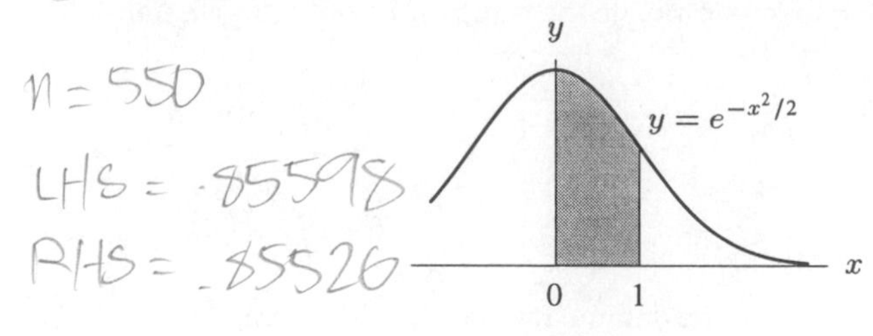

Figure 3.19

19. Your velocity is given by $v(t) = e^{-t^2}$ for $0 \le t \le 1.1$. Estimate the distance traveled during this time (to one decimal place).

20. For $0 \le t \le 1$, a bug is crawling at a velocity, v, determined by the formula

$$v = \frac{1}{1+t},$$

where t is in hours and v is in meters/hour. Estimate the distance that the bug crawls during this hour. Your estimate should be within 0.1 meter of the true answer.

3.3 THE DEFINITE INTEGRAL AS AREA

The Definite Integral as an Area, when $f(x)$ is Positive

If $f(x)$ is positive we can interpret each term $f(x_0)\Delta x, f(x_1)\Delta x, \ldots$ in a left- or right-hand Riemann sum as the area of a rectangle. As the width Δx of the rectangles approaches zero, the tops of the rectangles fit the curve of the graph more exactly, and the sum of their areas gets closer and closer to the area under the curve, shaded in Figure 3.20. Thus, we know that:

When $f(x)$ is positive and $a < b$:

$$\begin{matrix}\text{Area under graph of } f \\ \text{between } a \text{ and } b\end{matrix} = \int_a^b f(x)\,dx.$$

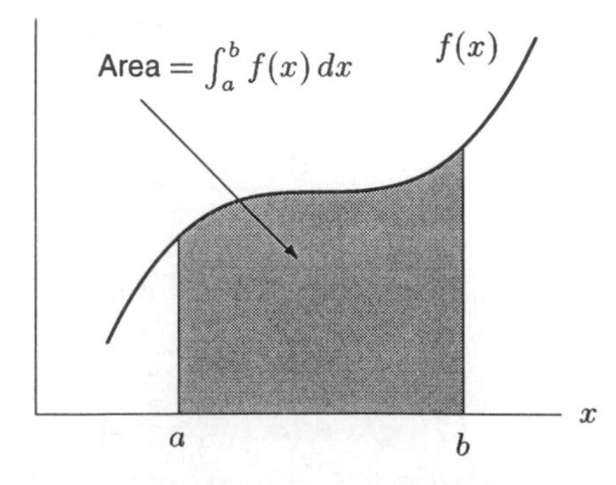

Figure 3.20: The definite integral $\int_a^b f(x)\,dx$

Example 1 A graph of $y = f(x)$ is shown in Fig 3.21. Estimate $\int_0^3 f(x)dx$.

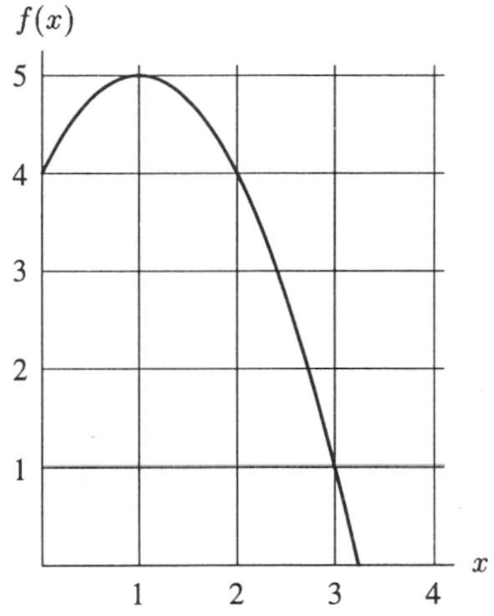

Figure 3.21: Estimate $\int_0^3 f(x)dx$

Solution We know that $\int_0^3 f(x)dx$ is equal to the area under the curve between $x = 0$ and $x = 3$, so we must estimate this area. (See Figure 3.22.) A rough estimate is about 12 square units, since the average height is about 4, and the width is 3. To get a better estimate, we can use left-hand and right-hand sums, choosing n depending on how ambitious we feel! Another excellent way to estimate area when a grid is given is simply to estimate the number of boxes in the area we are estimating. If we count boxes here, it appears that the shaded area in Figure 3.22 includes about 11.5 boxes. Since each box has area 1, we estimate that the shaded area is about 11.5 square units. We have

$$\int_0^3 f(x)dx \approx 11.5.$$

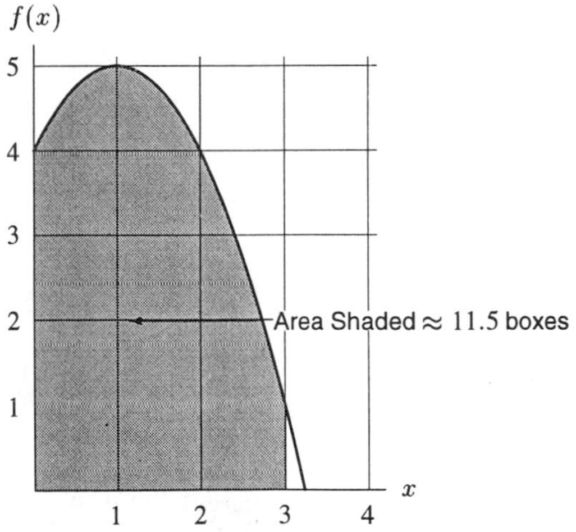

Figure 3.22: Area shaded $= \int_0^3 f(x)dx$

Example 2 A table of values for a function $P = f(t)$, where t is time measured in minutes, is given in Table 3.5. Estimate $\int_{20}^{30} f(t)dt$.

TABLE 3.5 *Estimate* $\int_{20}^{30} f(t)dt$

t	20	22	24	26	28	30
P	5	7	11	18	29	45

Solution Since we only have a table of values, we must use Riemann sums to approximate the integral. We are given information at 6 times which divide the interval $20 \leq t \leq 30$ into 5 time intervals, so we use $n = 5$. We see that measurements are taken every 2 minutes, so $\Delta t = 2$. Our best estimate is obtained by calculating the left-hand and right-hand sums, and averaging the two.

$$\begin{aligned}\text{Left-hand sum} &= f(20)2 + f(22)2 + f(24)2 + f(26)2 + f(28)2 \\ &= 5 \cdot 2 + 7 \cdot 2 + 11 \cdot 2 + 18 \cdot 2 + 29 \cdot 2 \\ &= 10 + 14 + 22 + 36 + 58 \\ &= 140.\end{aligned}$$

$$\begin{aligned}\text{Right-hand sum} &= f(22)2 + f(24)2 + f(26)2 + f(28)2 + f(30)2 \\ &= 7 \cdot 2 + 11 \cdot 2 + 18 \cdot 2 + 29 \cdot 2 + 45 \cdot 2 \\ &= 14 + 22 + 36 + 58 + 90 \\ &= 220.\end{aligned}$$

We average the two to obtain our best estimate of the integral:

$$\int_{20}^{30} f(t)dt \approx \frac{140 + 220}{2} = 180.$$

Example 3 Find the area under the graph of $y = 10xe^{-x}$ between $x = 0$ and $x = 3$.

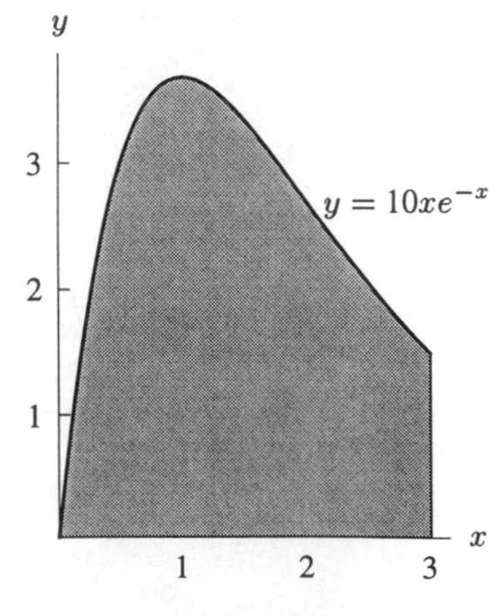

Figure 3.23

Solution A graph of this function is shown in Figure 3.23. We see that a rough estimate of the area is about 9. To find the area as accurately as we wish, we realize that

$$\text{Area shaded} = \int_0^3 10xe^{-x}\,dx,$$

and we use technology to evaluate the integral. Using a left-hand sum with $n = 100$, we obtain

$$\int_0^3 10xe^{-x}\,dx \approx 7.985.$$

If we use $n = 500$, we obtain

$$\int_0^3 10xe^{-x}\,dx \approx 8.004.$$

The shaded area is about 8.0 square units.

Example 4 Consider the integral $\int_{-1}^{1} \sqrt{1-x^2}\,dx$.

(a) Interpret the integral as an area, and find its exact value.

(b) Estimate the integral using a left-hand or right-hand sum with 100 subdivisions. Compare your answer to the exact value.

Solution (a) The integral is the area under the graph of $y = \sqrt{1-x^2}$ between -1 and 1. This is a semicircle of radius 1 and area $\frac{1}{2}\pi r^2 = \pi/2$ (see Figure 3.24).

(b) With $n = 100$, the left-hand and right-hand sums are 1.569. For comparison, $\pi/2 = 1.5707\ldots$

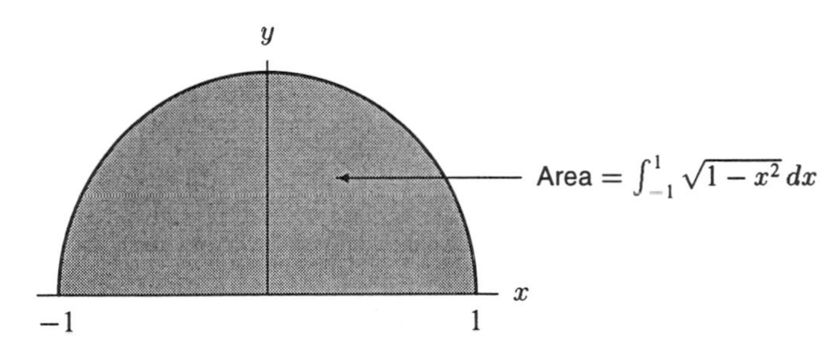

Figure 3.24: Area interpretation of $\int_{-1}^{1} \sqrt{1-x^2}\,dx$

Interpreting the Definite Integral as Area, when $f(x)$ is Not Positive

We have assumed in drawing Figure 3.20 that the graph of $f(x)$ lies above the x-axis. If the graph lies below the x-axis, then each value of $f(x)$ is negative, so each $f(x)\Delta x$ is negative, and the area gets counted negatively. In that case, the definite integral is the negative of the area.

Example 5 What is the relation between the definite integral $\int_{-1}^{1} (x^2 - 1)\, dx$, and the area between the parabola $y = x^2 - 1$ and the x-axis?

Solution The parabola lies below the axis between $x = -1$ and $x = 1$. (See Figure 3.25.) So,

$$\int_{-1}^{1} (x^2 - 1)\, dx = -\text{Area} \approx -1.33.$$

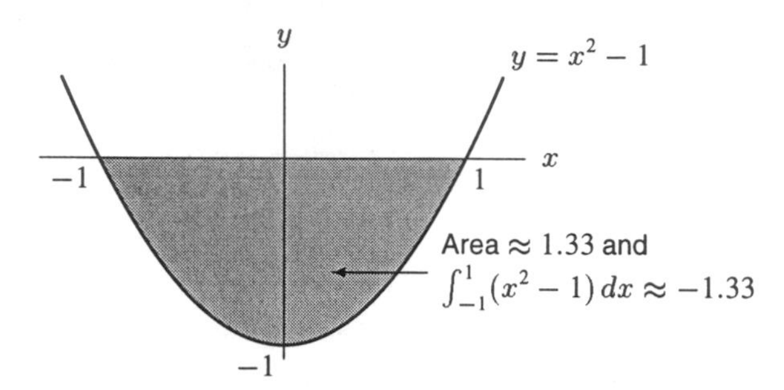

Figure 3.25: Integral $\int_{-1}^{1} (x^2 - 1)\, dx$ is negative of shaded area

---

When $f(x)$ is positive for some x values and negative for others, and $a < b$:
$\int_{a}^{b} f(x)\, dx$ is the sum of the areas above the x-axis, counted positively, and areas below the x-axis, counted negatively.

Example 6 Interpret the definite integral $\int_{0}^{4} (x^3 - 7x^2 + 11x)\, dx$ in terms of areas.

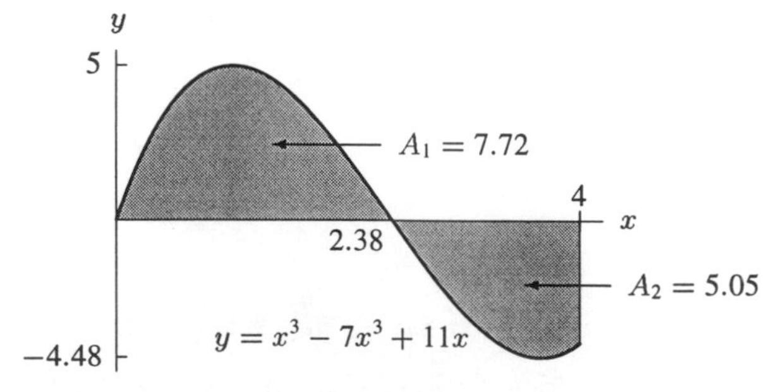

Figure 3.26: Integral $\int_{0}^{4} (x^3 - 7x^2 + 11x)\, dx = A_1 - A_2$

Solution The graph of $f(x) = x^3 - 7x^2 + 11x$ is shown in Figure 3.26. The graph crosses the x-axis at approximately $x = 2.38$. We see in Figure 3.26 that $f(x)$ is positive for $0 < x < 2.38$ and negative for $2.38 < x < 4$. The integral is the area above the x-axis, A_1, minus the area below the x-axis, A_2. Approximating the integral with $n = 1000$ shows

$$\int_0^4 (x^3 - 7x^2 + 11x)\,dx \approx 2.67.$$

Breaking the integral into two parts and calculating each one separately gives

$$\int_0^{2.38} (x^3 - 7x^2 + 11x)\,dx \approx 7.72 \quad \text{and} \quad \int_{2.38}^4 (x^3 - 7x^2 + 11x)\,dx \approx -5.05$$

so $A_1 \approx 7.72$ and $A_2 \approx 5.05$. Then, as we would expect,

$$\int_0^4 (x^3 - 7x^2 + 11x)\,dx = A_1 - A_2 \approx 7.72 - 5.05 = 2.67.$$

Example 7 Find the total area of the shaded regions in Figure 3.26.

Solution We saw in Example 6 that $A_1 \approx 7.72$ and $A_2 \approx 5.05$. Thus we have

$$\text{Total shaded area} = A_1 + A_2 \approx 7.72 + 5.05 = 12.77.$$

Example 8 Represent the definite integral $\int_0^5 4x^2e^{-x}\,dx$ as the area under a curve, and explain why we can conclude that

$$0 < \int_0^5 4x^2e^{-x}\,dx < 15.$$

Solution The graph of $f(x) = 4x^2e^{-x}$ is shown in Figure 3.27, and the value of the integral is given by the shaded area.

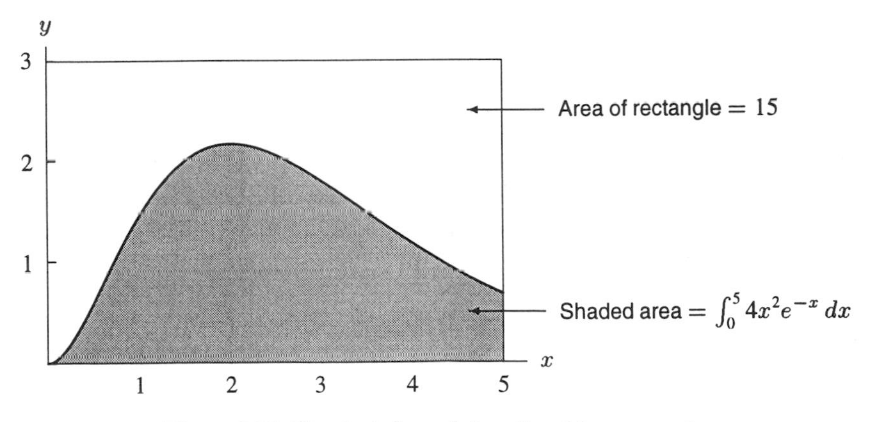

Figure 3.27: The shaded area is less than 15 square units

Since $f(x) \geq 0$ on the interval from 0 to 5, the integral represents the area under the curve and is positive. The maximum value of $f(x)$ is less than 3, and so the entire area under the curve fits inside the rectangle shown in Figure 3.27. Since this rectangle has area 15, the area under the curve is less than 15. Therefore, we have

$$0 < \int_0^5 4x^2e^{-x} < 15.$$

Example 9 Graphs of several functions are shown in Figure 3.28. In each case, indicate whether $\int_0^5 f(x)dx$ is positive, negative or approximately zero. Explain your answers.

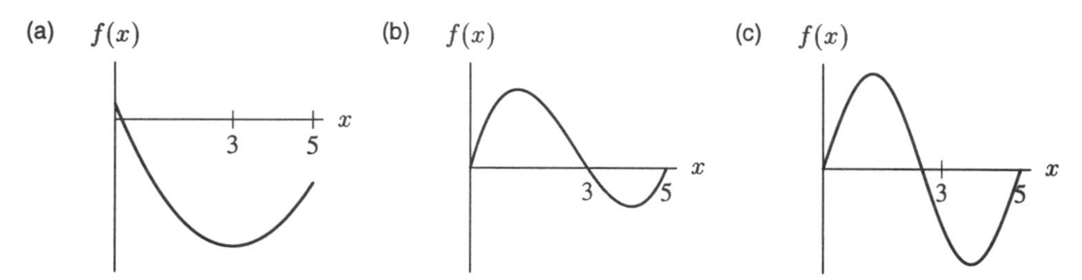

Figure 3.28: Is $\int_0^5 f(x)dx$ positive, negative or zero?

Solution

a) The graph lies almost entirely below the x-axis, so the integral is negative.
b) The graph lies partly below the x-axis and partly above the x-axis. However, we see that the area above the x-axis is larger than the area below the x-axis, and so the integral is positive.
c) The graph lies partly below the x-axis and partly above the x-axis. Since the areas above and below the x-axis appear to be approximately equal, they will cancel each other out. Therefore, the integral is approximately zero.

Example 10 Find the area between the graph of $y = x^2 - 2$ and the x-axis between $x = 0$ and $x = 3$.

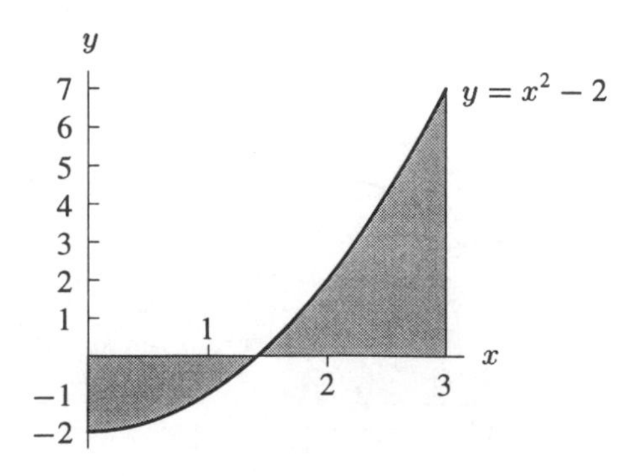

Figure 3.29

Solution The graph of this function is shown in Figure 3.29, and the relevant area is shaded. If you compute the integral $\int_0^3 (x^2 - 2)dx$, you will find that

$$\int_0^3 (x^2 - 2)dx = 3.0.$$

However, since part of the area lies below the x-axis and part lies above the x-axis, this computation does not help us at all. (In fact, it is clear from the graph that the shaded area is more than 3.) We have to find the area above the x-axis and the area below the x-axis separately. We find that the root of the function is at $x = 1.414$, and we compute the two areas separately:

$$\int_0^{1.414} (x^2 - 2)dx = -1.885 \quad \text{and} \quad \int_{1.414}^3 (x^2 - 2)dx = 4.885.$$

As we expect, we see that the integral between 0 and 1.414 is negative and the integral between 1.414 and 3 is positive. The total area shaded is the sum of the absolute values of the two areas:

$$\text{Area shaded} = 1.885 + 4.885 = 6.77.$$

The shaded area is 6.77 square units. (Notice that, as we saw earlier, if we integrate the function between 0 and 3, we obtain $-1.885 + 4.885 = 3$.)

Area Between Two Curves

We can extend the use of rectangles to approximate the area between two curves. If $f(x) \geq g(x)$, as in Figure 3.30, we see that the height of a rectangle is given by $f(x) - g(x)$. The area of the rectangle is $(f(x) - g(x))\Delta x$, so the area between the two curves is given as follows.

> If $g(x) \leq f(x)$ for $a \leq x \leq b$:
>
> $$\begin{array}{c}\text{Area between graphs of } f(x) \text{ and } g(x)\\ \text{between } a \text{ and } b\end{array} = \int_a^b (f(x) - g(x))\,dx.$$

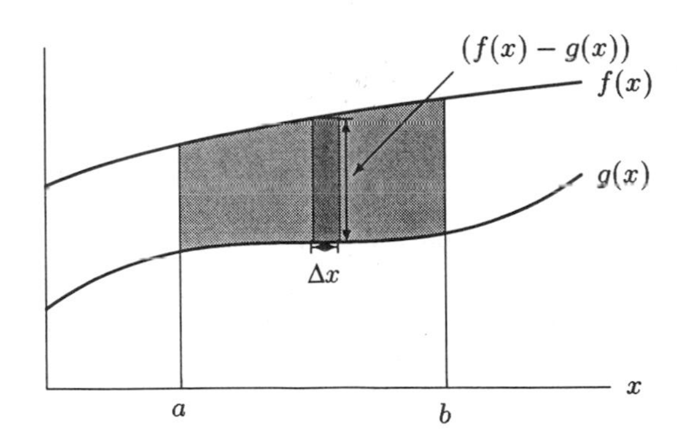

Figure 3.30: Area between two curves

Example 11 Sketch the graphs of $f(x) = 4x - x^2$ and $g(x) = \ln(x+1)$ for $x \geq 0$. Write a definite integral representing the area between the graphs of these two functions and use it to estimate the area.

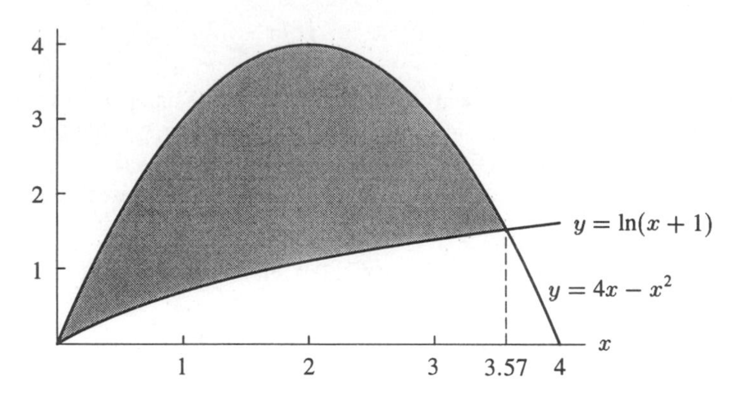

Figure 3.31: Area between $y = 4x - x^2$ and $y = \ln(x+1)$

Solution The graphs of the two functions are shown in Figure 3.31; the area between them is shaded. We see that the two graphs cross at $x = 0$ and at $x \approx 3.57$. Between these values, the graph of $y = 4x - x^2$ lies above the graph of $y = \ln(x+1)$. Thus, the area between the two functions is given by

$$\text{Area} = \int_0^{3.57} ((4x - x^2) - (\ln(x+1)))\, dx.$$

Counting squares on the graph shows that the shaded area is about 7 square units. Alternatively, we can use a calculator or computer to estimate the value of the definite integral to any desired degree of accuracy. A left-hand estimate with $n = 250$ gives an area of 6.949.

Problems for Section 3.3

1. If the graph of f is in Figure 3.32, what is $\int_1^6 f(x)\, dx$?

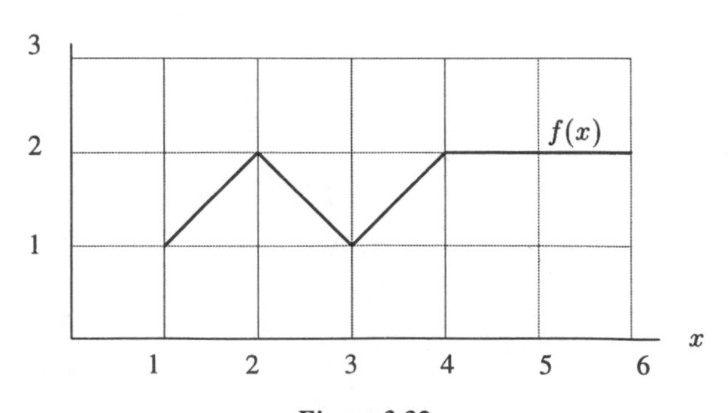

Figure 3.32

2. A graph of $y = f(x)$ is given in Figure 3.33. Estimate $\int_0^3 f(x)\,dx$.

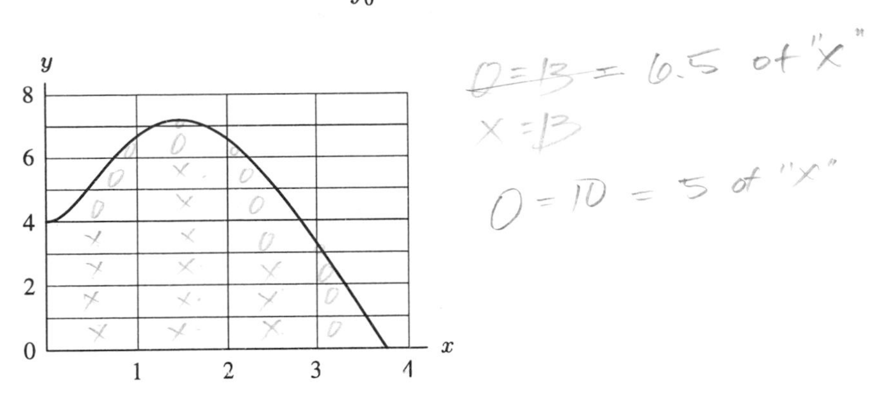

Figure 3.33

3. Find the area under $y = x^3 + 2$ between $x = 0$ and $x = 2$.
4. Find the area under $P = 100e^{-0.5t}$ between $t = 0$ and $t = 8$.
5. A table of values for the function $y = f(x)$ is given in Table 3.6. Estimate $\int_0^{25} f(x)dx$.

TABLE 3.6

x	0	5	10	15	10	25
$f(x)$	100	82	69	60	53	49

6. The graph of f is shown in Figure 3.34.
   (a) Estimate (by counting the squares) the total area shaded in Figure 3.34.
   (b) Estimate

$$\int_0^8 f(x)\,dx.$$

   (c) Why are your answers to parts (a) and (b) different?

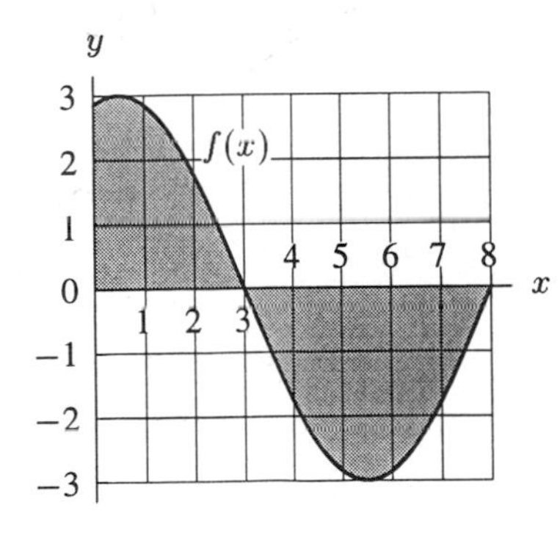

Figure 3.34

Without computing the integrals in Problems 7–10, decide if each is positive or negative, and explain your decision. [Hint: Sketch the graph of each. Compare areas above and below the x-axis.]

7. $\int_0^3 (x^3 - 4x^2 + 2)\,dx$

8. $\int_0^2 (x^2 - x)\,dx$

9. $\int_0^4 (2 - x)e^x\,dx$

10. $\int_2^5 \frac{4 - x}{\ln x}\,dx$

11. Find the area between the parabola $y = 4 - x^2$ and the x-axis.
12. Find the area between $y = x^2 - 9$ and the x-axis.
13. Find the area between $y = 3x$ and $y = x^2$.
14. Find the area between $y = x$ and $y = \sqrt{x}$.
15. (a) Sketch a graph of $f(x) = x(x + 2)(x - 1)$.
    (b) Find the total area between the graph and the x-axis between $x = -2$ and $x = 1$.
    (c) Find $\int_{-2}^1 f(x)\,dx$ and interpret it in terms of areas.
16. Compute the definite integral $\int_1^4 (x - 3\ln x)\,dx$ and interpret the result in terms of areas.
17. Is $\int_{-1}^1 e^{x^2}\,dx$ positive, negative, or zero? Explain. [Hint: Sketch a graph of e^{x^2}.]
18. Explain why $0 < \int_0^1 e^{x^2}\,dx < 3$.
19. (a) Sketch a graph of $x^3 - 5x^2 + 4x$ and mark on it points where $x = 1, 2, 3, 4, 5$.
    (b) Use your graph and area interpretation of the definite integral to decide which of the five numbers

$$I_n = \int_0^n (x^3 - 5x^2 + 4x)\,dx \quad \text{for } n = 1, 2, 3, 4, 5$$

is largest. Which is smallest? How many of the numbers are positive? (You do not need to calculate these integrals.)

20. For the function f graphed in Figure 3.35:
    (a) Suppose you know $\int_0^2 f(x)\,dx$. What is $\int_{-2}^2 f(x)\,dx$?
    (b) Suppose you know $\int_0^5 f(x)\,dx$ and $\int_2^5 f(x)\,dx$. What is $\int_0^2 f(x)\,dx$?
    (c) Suppose you know $\int_{-2}^5 f(x)\,dx$ and $\int_{-2}^2 f(x)\,dx$. What is $\int_0^5 f(x)\,dx$?

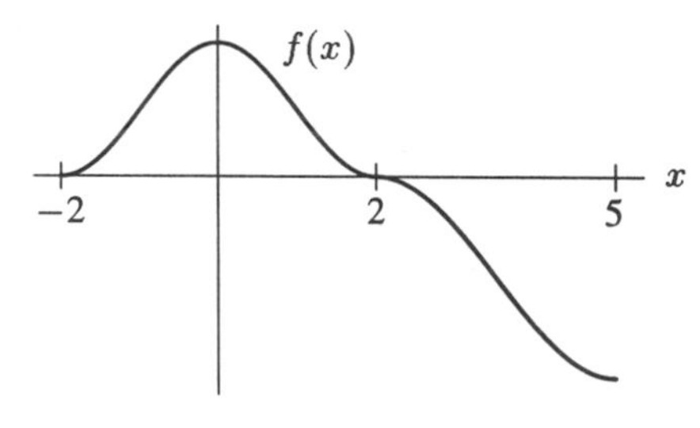

Figure 3.35

21. For the function f graphed in Figure 3.35:
    (a) Suppose you know $\int_{-2}^2 f(x)\,dx$ and $\int_0^5 f(x)\,dx$. What is $\int_2^5 f(x)\,dx$?
    (b) Suppose you know $\int_{-2}^5 f(x)\,dx$ and $\int_{-2}^0 f(x)\,dx$. What is $\int_2^5 f(x)\,dx$?
    (c) Suppose you know $\int_2^5 f(x)\,dx$ and $\int_{-2}^5 f(x)\,dx$. What is $\int_0^2 f(x)\,dx$.

3.4 THE DEFINITE INTEGRAL AS AVERAGE VALUE

The Average Value of a Function

We know how to find the average of n numbers: add them and divide by n. But how do we find the average value of a continuously varying function? Let us consider an example. Suppose $C = f(t)$ is the temperature at time t, measured in hours since midnight, and that we want to calculate the average temperature over a 24-hour period. One way to start would be to average the temperatures at n times, $t_1, t_2, \ldots, t_n$, during the day.

$$\text{Average temperature} \approx \frac{f(t_1) + f(t_2) + \cdots + f(t_n)}{n}.$$

The larger we make n, the better the approximation. We can rewrite this expression as a Riemann sum over the interval $0 \le t \le 24$ if we use the fact that $\Delta t = 24/n$, so $n = 24/\Delta t$:

$$\begin{aligned}\text{Average temperature} &\approx \frac{f(t_1) + f(t_2) + \cdots + f(t_n)}{24/\Delta t} \\ &= \frac{f(t_1)\Delta t + f(t_2)\Delta t + \cdots + f(t_n)\Delta t}{24} \\ &= \frac{1}{24}\sum_{i=1}^{n} f(t_i)\Delta t.\end{aligned}$$

As $n \to \infty$, the Riemann sum tends towards an integral and also approximates the average temperature better. Thus, in the limit

$$\begin{aligned}\text{Average temperature} &= \lim_{n\to\infty} \frac{1}{24}\sum_{i=1}^{n} f(t_i)\Delta t \\ &= \frac{1}{24}\int_0^{24} f(t)\, dt.\end{aligned}$$

Thus we have found a way of expressing the average temperature in terms of an integral. Generalizing for any function f, we define

$$\text{Average value of } f \text{ from } a \text{ to } b = \frac{1}{b-a}\int_a^b f(x)\, dx$$

How to Visualize the Average on a Graph

The definition of average value tells us that

$$(\text{Average value of } f)\cdot(b-a) = \int_a^b f(x)\, dx.$$

Thus, if we interpret the integral as the area under the graph of f, then we can think of the average value of f as the height of the rectangle with the same area that is on the same base, $(b-a)$. (See Figure 3.36.)

If we think of the graph of the function as a wave in a fish tank, the average value is the height of the water when the wave settles down to a horizontal line.

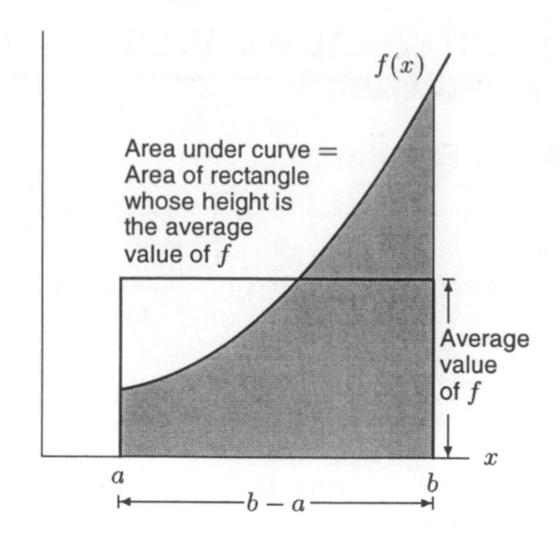

Figure 3.36: Area and average value

Example 1 Find the average value of $y = 5 + 6x - x^2$ on the interval $x = 0$ to $x = 5$.

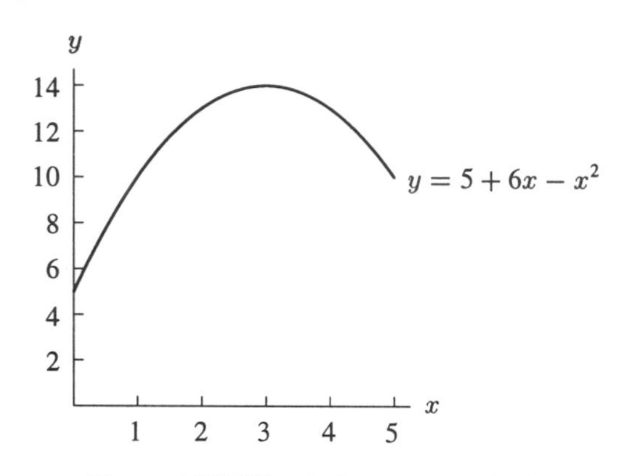

Figure 3.37: What is the average value?

Solution A graph of this function is shown in Figure 3.37. We see that on this interval, all the y values are between 5 and 14. The average y-value must therefore be between 5 and 14. It appears in Figure 3.37 that the average value will be about 11. We use the formula for average value to calculate it exactly:

$$\begin{aligned}
\text{Average value} &= \frac{1}{b-a}\int_a^b f(x)dx \\
&= \frac{1}{5-0}\int_0^5 (5 + 6x - x^2)dx \\
&= \frac{1}{5}(58.333) \\
&= 11.667.
\end{aligned}$$

The average value of y is 11.667. This matches what we see in the graph.

Example 2 The market research division of a fast food restaurant plans to track the effect of an extensive marketing campaign on business. In Figure 3.38, the number of customers per week, $N(t)$, is graphed against time, t (in weeks), for a period of six months (26 weeks). The marketing campaign starts at time $t = 0$ and lasts for 10 weeks.

(a) Describe the effect of the marketing campaign on the number of customers per week during the six month period shown on the graph.
(b) From the graph, estimate visually the average number of customers per week over the six month period.
(c) Set up a definite integral representing the average number of customers per week over the six month period.
(d) Estimate the total number of customers to visit the store during this six month period.

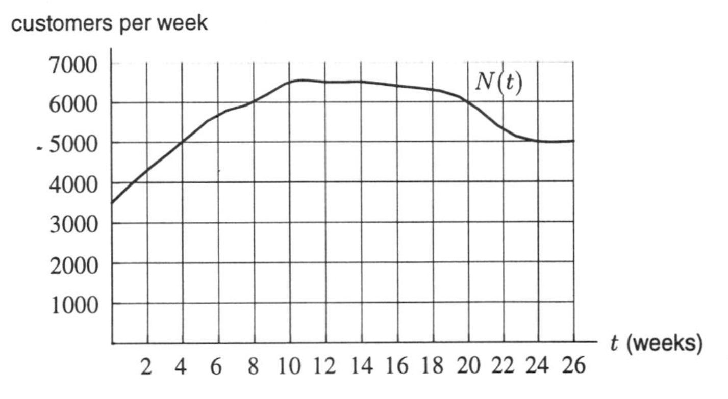

Figure 3.38: Number of customers at a fast food restaurant

Solution (a) While the marketing campaign lasts, the number of customers per week increases from about 3500 at the start of the campaign ($t = 0$) to about 6500 at the end of the campaign ($t = 10$). For the next 8 weeks, the number of customers stays pretty steady at about 6500. After this, the number of customers drops off again, and appears to be stabilizing at about 5000 customers per week.

(b) To approximate the average number of customers per week, we want to draw a horizontal line so that the area under this line is the same as the area under the curve. In other words, the area between the line and the curve that is below the line should be approximately equal to the area between the line and the curve that is above the line. We see in Figure 3.39 that a horizontal line at about 5500 satisfies this condition, so the average number of customers per week is about 5500.

(c)
$$\text{Average number of customers per week} = \frac{1}{26-0}\int_0^{26} N(t)\,dt$$

(d) The total number of customers is represented by the total area under the curve, so we estimate this area. Each grid square has an area of $(1000)(2) = 2000$, and there are about 74 squares under the curve, so we have

$$\text{Total number of customers} \approx (74)(2000) = 148,000.$$

Alternatively, we can estimate the total number of customers by multiplying the average number of customers per week (5500) by the number of weeks (26). We have

$$\text{Total number of customers} = (5500)(26) = 143,000.$$

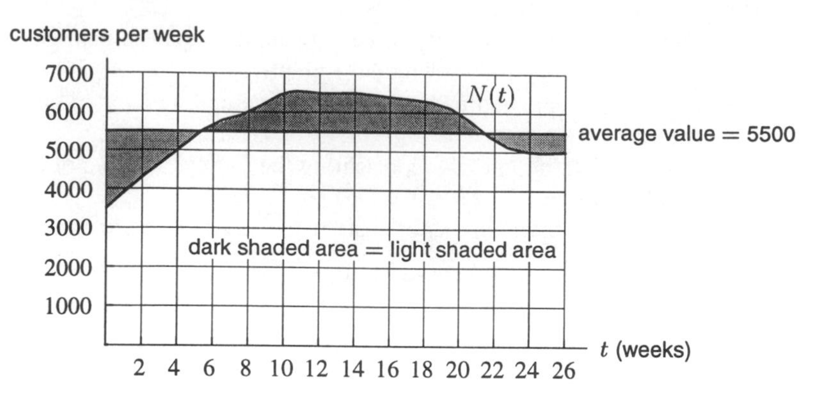

Figure 3.39: Average number of customers per week

Problems for Section 3.4

1. The graph of f is in figure 3.40. You calculated $\int_1^6 f(x)dx$ in the previous section. What is the average value of f on the interval $[1, 6]$?

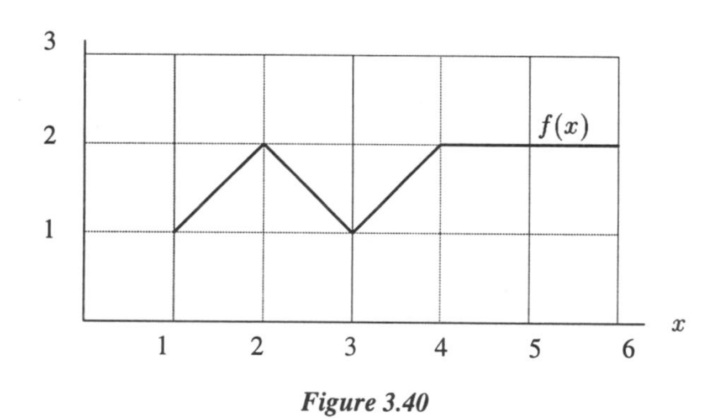

Figure 3.40

2. The graph of f is shown in Figure 3.41.
   (a) Estimate (by counting the squares)
   $$\int_0^5 f(x)\,dx.$$
   (b) Estimate the average value of f between $x = 0$ and $x = 5$ by estimating visually the average height.

(c) Estimate the average value of f between $x = 0$ and $x = 5$ by using your answer to part (a) and the formula for average value. (Your answers to parts (a) and (b) should be about the same.)

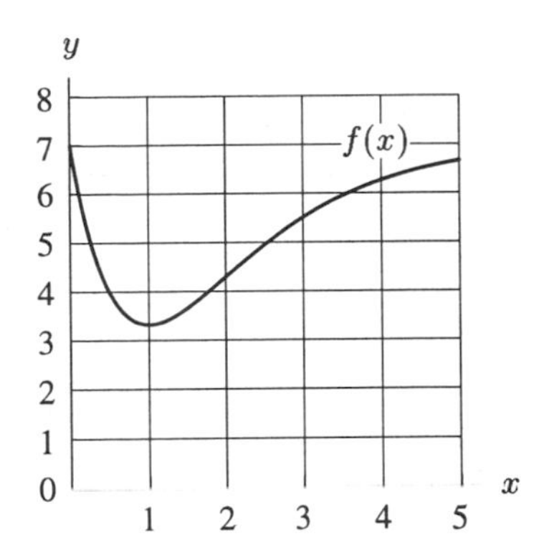

Figure 3.41

3. Find the average value of $g(t) = 1 + t$ over the interval $[0, 2]$.
4. Find the average value of $g(t) = e^t$ over the interval $[0, 10]$.
5. Graphs of two functions are shown in Figure 3.42. In each case, give a rough estimate of the average value of the function between $x = 0$ and $x = 7$, and explain how you arrived at your answer.

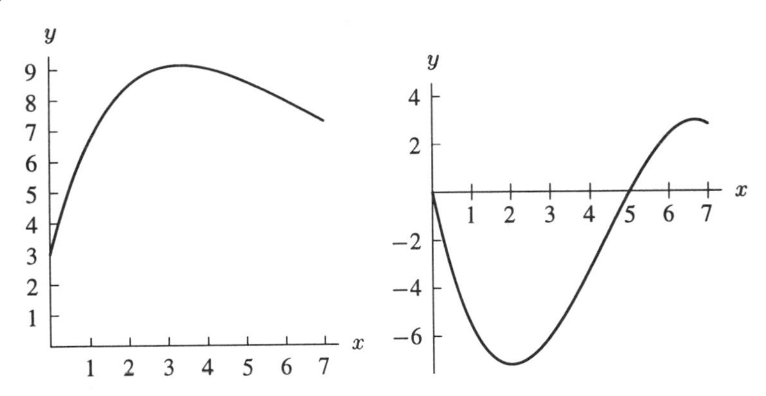

Figure 3.42: Estimate the average value

6. (a) What is the average value of $f(x) = \sqrt{1 - x^2}$ over the interval $0 \leq x \leq 1$?
   (b) How can you tell whether this average value is more or less than 0.5 without doing any calculations?
7. The warehouse for a mail-order company receives shipment for a certain item on June 1, and the stock of the item is steadily depleted over the next few months. Suppose the inventory function for this item is given by
$$I(t) = 5000e^{-0.1t},$$
where t is measured in days since June 1.
   (a) Find the average inventory in the warehouse during the 90 days after June 1.
   (b) Graph the function $I(t)$ and illustrate the average inventory graphically.

8. The value, V, of a Tiffany lamp, worth \$225 in 1965, increases at 15% per year. Its value in dollars t years after 1965 is given by

$$V = 225(1.15)^t.$$

Find the average value of the lamp over the period 1965–2000.

9. A bar of metal is cooling from 1000°C to room temperature, 20°C. The temperature, H, of the bar t minutes after it starts cooling is given, in °C, by

$$H = 20 + 980e^{-0.1t}.$$

(a) Find the temperature of the bar at the end of one hour.
(b) Find the average value of the temperature over the first hour.
(c) Is your answer to part (b) greater or smaller than the average of the temperatures at the beginning and the end of the hour? Explain this in terms of the concavity of the graph of H.

10. For the function f in Figure 3.43, write an expression involving one or more definite integrals that denote(s):

(a) The average value of f for $0 \leq x \leq 5$.
(b) The average value of $|f|$ for $0 \leq x \leq 5$.

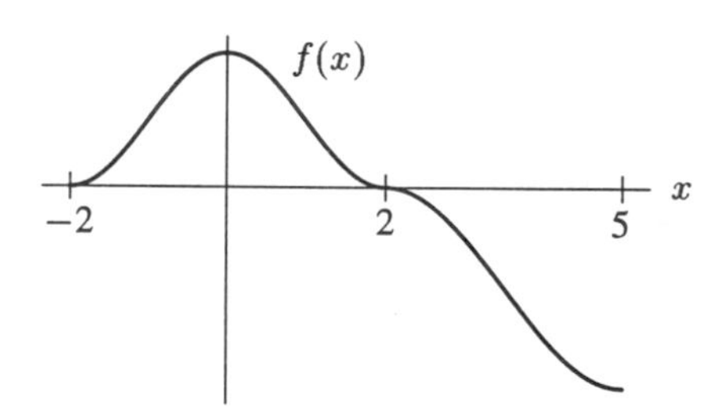

Figure 3.43

11. For the function f in Figure 3.43, consider the average value of f over the following intervals:
(I) $0 \leq x \leq 1$ (II) $0 \leq x \leq 2$ (III) $0 \leq x \leq 5$ (IV) $-2 \leq x \leq 2$

(a) For which interval is the average value of f least?
(b) For which interval is the average value of f greatest?
(c) For which pair of intervals are the average values equal?

12. The number of hours, H, of daylight in Madrid as a function of date is approximated by the formula

$$H = 12 + 2.4\sin[0.0172(t - 80)],$$

where t is the number of days since the start of the year. Find the average number of hours of daylight in Madrid:

(a) in January (b) in June (c) over a whole year
(d) Comment on the relative magnitudes of your answers to parts (a), (b), and (c). Why are they reasonable?

13. The quantity of a certain radioactive substance at time t is given by

$$Q = 4e^{-0.036t} \text{ grams.}$$

(a) Find $Q(10)$ and $Q(20)$.
(b) Find the average of $Q(10)$ and $Q(20)$.
(c) Find the average value of Q over the interval $10 \leq t \leq 20$. (Use numerical methods to get a good estimate.)
(d) Use what you know about the graph of Q to explain the relative sizes of your answers in parts (b) and (c).

14. Given the graph of f in Figure 3.44, arrange the following numbers in order from least to greatest:
(a) $f'(1)$
(b) the average value of f on $[0, 4]$
(c) $\int_0^1 f(x)dx$

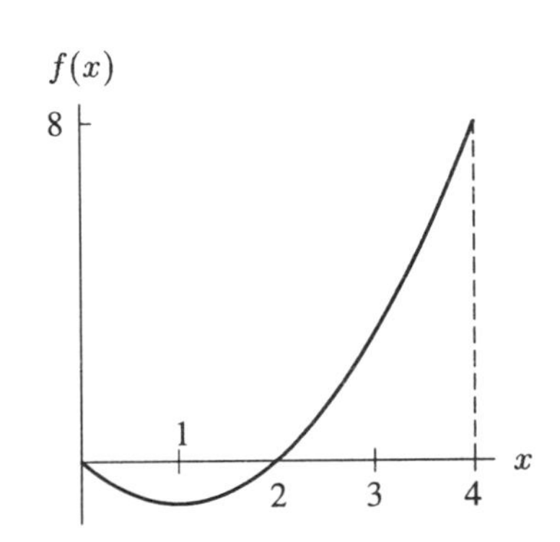

Figure 3.44

3.5 INTERPRETATIONS OF THE DEFINITE INTEGRAL

The Notation and Units for the Definite Integral

Just as the Leibniz notation dy/dx for the derivative reminds us that the derivative is the limit of a ratio of differences, the notation for the definite integral helps us recall the meaning of the integral. The symbol

$$\int_a^b f(x)\,dx$$

reminds us that an integral is a limit of sums (the integral sign is a misshapen S) of terms of the form "$f(x)$ times a small difference of x." Officially, dx is not a separate entity, but a part of the whole integral symbol. Thus, just as one thinks of d/dx as a single symbol meaning "the derivative with respect to x of. . . ," one can think of $\int_a^b \ldots dx$ as a single symbol meaning "the integral of . . . with respect to x."

However, most scientists and mathematicians informally think of dx as an "infinitesimally" small bit of x which in this context is multiplied by a function value $f(x)$. This viewpoint is often the key to interpreting the meaning of a definite integral. For example, if $f(t)$ is the velocity of a moving particle at time t, then $f(t)dt$ may by thought of informally as velocity $\times$ time, giving the distance traveled by the particle during a small bit of time dt. The integral $\int_a^b f(t)\,dt$ may then be thought of as the sum of all these small distances, giving us the net change in position of the particle between $t = a$ and $t = b$.

The notation for the integral also helps us determine what units should be used for the numerical value of the integral. Since the terms being added up are products of the form "$f(x)$ times a difference in x," the unit of measurement for $\int_a^b f(x)\,dx$ is the product of the units for x and the units for $f(x)$. Thus if $f(t)$ is velocity measured in meters/second and t is time measured in seconds, then

$$\int_a^b f(t)\,dt$$

has for units (meters/sec)$\times$(sec)=meters, which is what we expect since the value of the integral represents change in position. Similarly, if we graph $y = f(x)$ with the same units of measurement of length along the x and y axes, then $f(x)$ and x are measured in the same units so

$$\int_a^b f(x)\,dx$$

is measured in square units, say cm$\times$cm=cm^2. Again this is what we would expect since in this context the integral represents an area. Finally for the average value,

$$\frac{1}{b-a}\int_a^b f(x)\,dx,$$

the units for dx inside the integral are canceled by the units for $1/(b-a)$ outside the integral, leaving only the units for $f(x)$. This is as it should be since the average value of f should be measured in the same units as $f(x)$.

Example 1 Suppose that $C(t)$ represents the cost per day to heat your house measured in dollars per day, where t is time measured in days and $t = 0$ corresponds to January 1, 1996. Interpret $\displaystyle\int_0^{90} C(t)\,dt$ and $\displaystyle\frac{1}{90-0}\int_0^{90} C(t)\,dt$.

Solution The units for the integral $\displaystyle\int_0^{90} C(t)\,dt$ are (dollars/day)$\times$(days)=dollars. The integral represents the total cost in dollars to heat your house for the first 90 days of 1996, namely the months of January, February, and March. The second expression is measured in (1/days)(dollars) or dollars per day, the same units as $C(t)$. It represents the average cost per day to heat your house during the first 90 days of 1996.

The Definite Integral of a Rate Gives Total Change

In Chapter 2, we learned how to use the derivative to find the rate of change from total change. Here we see how to go in the other direction. If we know the rate of change, how can we determine the total change?

Example 2 A new video game has been put on the market, and the rate of sales (in games per week) is recorded over a 20-week period. The results are shown in Table 3.7, with t in weeks since the game went on the market. Assuming that the rate of sales increased throughout the 20-week period, estimate the total sales of the game during this period.

TABLE 3.7 *Weekly sales of a video game*

Time (weeks)	0	5	10	15	20
Rate of sales (games per week)	0	585	892	1875	2350

Solution How many games were sold during the first five weeks? During this time, sales went from 0 (the moment the game went on the market) to 585 games per week during the fifth week. If we assume that 585 games were sold every week during the first five weeks, we will clearly have an overestimate for total sales of the video game during this period. Since

$$\text{Total sales} = \text{Rate of sales per week} \times \text{Number of weeks},$$

an overestimate for the first five weeks is $(585)(5) = 2925$. We can use similar overestimates for each of the five week periods to calculate an overestimate for the entire twenty-week period

$$\text{Overestimate for total sales} = (585)(5) + (892)(5) + (1875)(5) + (2350)(5) = 28,510 \text{ games.}$$

We can find an underestimate for total sales by taking the lower value for rate of sales during each of the five week periods. The underestimate is given by

$$\text{Underestimate for total sales} = (0)(5) + (585)(5) + (892)(5) + (1875)(5) = 16,760 \text{ games.}$$

Thus, the total sales of the game during the twenty-week period is between 16,760 and 28,510 games. Our best single estimate of total sales is the average of these two numbers:

$$\text{total sales} = \frac{16760 + 28510}{2} = 22,635 \text{ games.}$$

The sums computed in Example 2 are the same sums we use to evaluate a definite integral. We will generalize this result to show that the integral of the rate of change of any quantity will give the total change in that quantity. Suppose $f(t)$ is the rate of change of some quantity $F(t)$ with respect to time, and that we are interested in the total change in $F(t)$ between $t = a$ and $t = b$. We divide the interval $a \le t \le b$ into n subintervals, each of length Δt. For each small interval, we will calculate the change in $F(t)$, written ΔF, and then add all these up. For each subinterval we assume the rate of change of $F(t)$ is approximately constant, so that we can say

$$\Delta F \approx \text{Rate of change of } F \times \text{Time}.$$

For the first subinterval, from t_0 to t_1, the rate of change of $F(t)$ is approximately $f(t_0)$, so

$$\Delta F \approx f(t_0)\Delta t.$$

Similarly, for the second interval

$$\Delta F \approx f(t_1)\Delta t.$$

Summing over all the subintervals,

$$\text{Total change in } F = \sum \Delta F \approx \sum_{i=0}^{n-1} f(t_i)\,\Delta t.$$

Thus, we have approximated the change in $F(t)$ as a left-hand sum. By a similar argument, we can approximate the change in $F(t)$ by a right-hand sum:

$$\text{Total change in } F = \sum \Delta F \approx \sum_{i=1}^{n} f(t_i)\,\Delta t.$$

The total change in $F(t)$ between the times $t = a$ and $t = b$ can be written as $F(b) - F(a)$. Thus, taking the limit as n goes to infinity, we get the following result:

> If $f(t)$ gives the rate of change of some quantity:
>
> $$\begin{array}{c}\text{Total change} \\ \text{between } t = a \text{ and } t = b\end{array} = \int_a^b f(x)\,dx.$$

Example 3 A bacteria colony has a population of 14 million bacteria at time $t = 0$. Suppose that the bacteria population is growing at a rate of $f(t) = 2^t$ million bacteria per hour.

(a) Give a definite integral which represents the total change in the bacteria population during the two hours from $t = 0$ to $t = 2$.

(b) Find the population at time $t = 2$.

Solution (a) Since $f(t) = 2^t$ gives the rate of change of population, the total change between $t = 0$ and $t = 2$ is given by

$$\int_0^2 2^t\,dt.$$

(b) Using a computer or graphing calculator, we find $\int_0^2 2^t\,dt = 4.328$. The total change in the bacteria population during these two hours is 4.328 million bacteria. The bacteria population was 14 million at time $t = 0$, and increased 4.328 million between $t = 0$ and $t = 2$. Therefore, the population at time $t = 2$ is $14 + 4.238 = 18.328$ million bacteria.

Example 4 The graph of rate of change of profit with respect to time is shown in Figure 3.45, with time t given in months. Did the company make money or lose money during the 5 months shown?

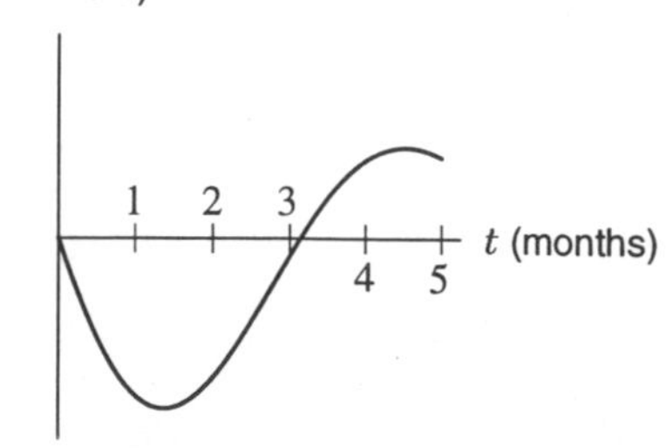

Figure 3.45: Did the company make money?

Solution We want to find total profit between $t = 0$ and $t = 5$. Since total profit is the integral of the rate of change of profit, we are looking for

$$\int_0^5 \pi'(t)dt,$$

where $\pi'(t)$ represents the rate of change of profit, shown in Figure 3.45. The integral equals the area above the t-axis minus the area below the t-axis. Since the area below the axis is greater than the area above the axis, the integral is negative. Total profit during this time is negative, so the company lost money. (We see in the graph that profits are increasing at the 5-month mark, however, and so the company might be in the black soon.)

Example 5 A man takes a trip in a car, and his velocity (in mph) is given in Figure 3.46. At time $t = 0$, the man is 50 miles from his house. Positive velocities take him toward his house and negative velocities take him away from his house.

a) When is the man closest to his house, and approximately how far away is he then?
b) When is the man farthest from his house, and how far away is he then?

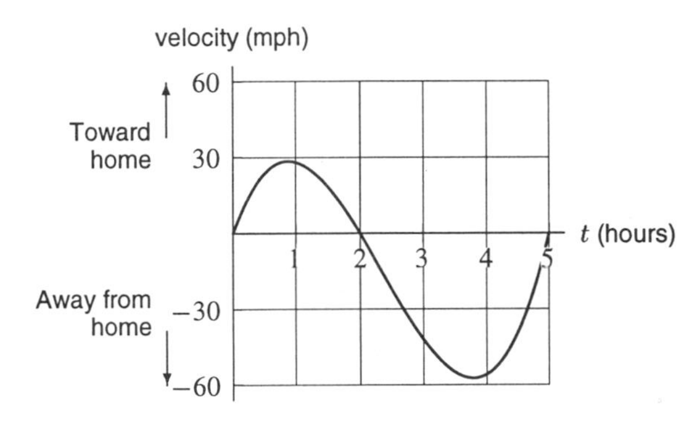

Figure 3.46: Describe this trip

Solution What happens on this trip? The velocity function is positive the first two hours and negative between $t = 2$ and $t = 5$. Thus the man moves toward his house during the first two hours, then turns around at $t = 2$ and moves away from his house. The distance he travels is represented by area; since the area below the axis is greater than the area above the axis, we see that he ends up farther away from home than when he started. Thus he is closest to home at $t = 2$ and farthest from home at $t = 5$. We can estimate how far he went in each direction by estimating the area.

a) The man starts out 50 miles from home. During the first two hours, the distance the man travels is the area under the curve between $t = 0$ and $t = 2$. This area corresponds to about one box. Since each box has area $(30)(1) = 30$, the man travels about 30 miles toward home. He is closest to home after 2 hours, and he is about 20 miles away at that time.
b) Between $t = 2$ and $t = 5$, the man moves away from his house. Since this area is equal to about 4 boxes, which is $(4)(30) = 120$ miles, he has moved 120 miles farther from home. He was already 20 miles from home so at $t = 5$, he is about 140 miles from home. He is farthest from home at $t = 5$, and he is about 140 miles away at that time.

The Definite Integral of Marginal Cost Gives Total Cost

Recall that the derivative of the cost function is the marginal cost. A marginal cost function $C'(q)$ is given in Figure 3.47, where q is the quantity of items produced. We see, for example, that $C'(100) = 6$, which means that the additional cost to produce the 100th item is about \$6. How do we interpret

$$\int_a^b C'(q)\, dq?$$

Since marginal cost $C'(q)$ is the rate of change of the cost function, this definite integral represents the total change in the cost function between $q = a$ and $q = b$. In other words, this is the amount it will cost to increase production from a units to b units. Recall that the cost of producing 0 units is the fixed cost. The area under the marginal cost curve between 0 and q is the total increase in cost between a production of 0 and a production of q. This is what we called the total variable cost. If we add this to the fixed cost, we obtain the total cost to produce q units. In general,

> If $C'(q)$ is the marginal cost function,
>
> $$\int_a^b C'(q)\, dq = \text{Amount it will cost to increase production from } a \text{ units to } b \text{ units.}$$
>
> $$\int_0^b C'(q)\, dq = \text{Total variable cost to produce } b \text{ units.}$$
>
> To obtain the total cost of producing b units, add the fixed cost to the total variable cost.

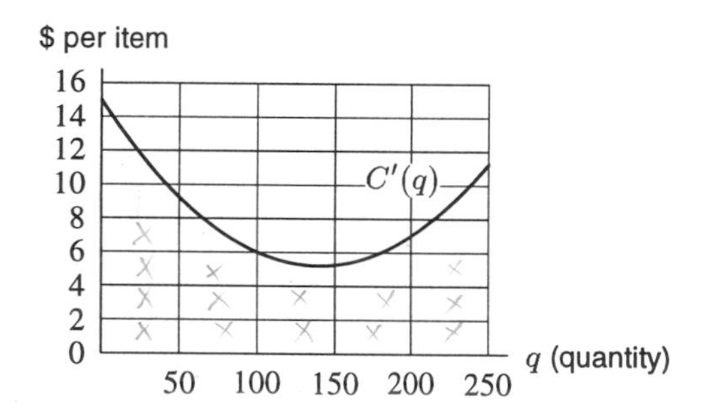

Figure 3.47: A marginal cost curve

Example 6 The marginal cost curve for a certain product is given in Figure 3.47. If the fixed costs for this production are \$1000, estimate the total cost in producing 200 items.

Solution The total cost of production = fixed cost + variable cost. We know that the fixed cost is \$1000 and the variable cost in producing 200 items is represented by the area under the marginal cost curve in Figure 3.47 between $q = 0$ and $q = 200$. Approximating this area as about 20 boxes each of area $2(50) = 100$, we get

$$\text{Total variable cost} = \int_0^{200} C'(q)\, dq \approx 2000.$$

We now find the total cost to produce 200 items:

$$\begin{aligned}\text{Total cost} &= \text{Fixed cost} + \text{Variable cost}\\ &= \$1000 + \$2000\\ &= \$3000\end{aligned}$$

A Pondwater Example

Example 7 Biological activity in a pond is reflected in the rate at which carbon dioxide, CO_2, is added to or withdrawn from the water. Plants take CO_2 out of the water during the day for photosynthesis and put CO_2 into the water at night. Animals put CO_2 into the water all the time as they breathe. Biologists are interested in how the net rate at which CO_2 enters or leaves a pond varies during the day. Figure 3.48 shows this rate as a function of time of day.[2] The rate is measured in millimoles (mmol) of CO_2 per liter of water per hour; time is measured in hours past dawn. At dawn, there were 2.600 mmol of CO_2 per liter of water.

(a) What can be concluded from the fact that the rate is negative during the day and positive at night?

(b) Some scientists have suggested that plants and animals respire (breathe) at a constant rate at night, and that plants photosynthesize at a constant rate during the day. Does Figure 3.48 support this view?

(c) When was the CO_2 content of the water at its lowest? How low did it go?

(d) How much CO_2 was released into the water during the 12 hours of darkness? Compare this quantity with the amount of CO_2 withdrawn from the water during the 12 hours of daylight. How can you tell by looking at the graph whether the CO_2 in the pond is in equilibrium?

(e) Estimate the CO_2 content of the water at three hour intervals throughout the day. Use your estimates to plot a graph of CO_2 content throughout the day.

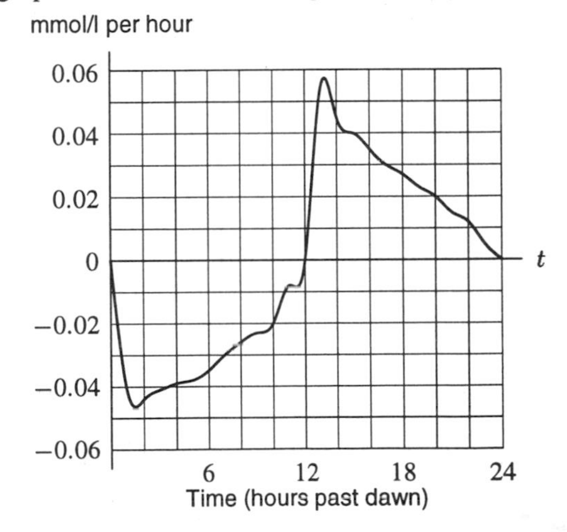

Figure 3.48: Rate at which CO_2 is entering or leaving water, $f'(t)$

[2]Data from R. J. Beyers, *The Pattern of Photosynthesis and Respiration in Laboratory Microsystems* (Mem. 1st. Ital. Idrobiol., 1965).

Solution (a) CO_2 is being taken out of the water during the day and returned at night. The pond must therefore contain some plants. (The data is in fact from pond water containing both plants and animals.)

(b) Suppose t is the number of hours past dawn. The graph in Figure 3.48 shows that the CO_2 content changes at a greater rate for the first 6 hours of daylight, $0 < t < 6$, than it does for the final 6 hours of daylight, $6 < t < 12$. It turns out that plants photosynthesize more vigorously in the morning than in the afternoon. Similarly, CO_2 content changes more rapidly in the first half of the night, $12 < t < 18$, than in the 6 hours just before dawn, $18 < t < 24$. The reason seems to be that at night plants quickly use up most of the sugar that they synthesized during the day, and then their respiration rate is inhibited.

(c) We are now being asked about the total quantity of CO_2 in the pond, rather than the rate at which it is changing. We will let $f(t)$ denote the CO_2 content of the pond water (in mmol/l) at t hours past dawn. Then Figure 3.48 is a graph of the derivative $f'(t)$. There are 2.600 mmol/l of CO_2 in the water at dawn, so $f(0) = 2.600$.

The CO_2 content $f(t)$ decreases during the 12 hours of daylight, $0 < t < 12$, when $f'(t) < 0$, and then $f(t)$ increases for the next 12 hours. Thus, $f(t)$ is at a minimum when $t = 12$, at dusk. The total quantity of CO_2 at time $t = 12$ is equal to the amount present at time $t = 0$ plus the total change between $t = 0$ and $t = 12$. We have

$$f(12) = f(0) + \int_0^{12} f'(t)\,dt = 2.600 + \int_0^{12} f'(t)\,dt.$$

We must approximate the definite integral by a Riemann sum. From the graph in Figure 3.48, we estimate the values of the function $f'(t)$ in Table 3.8.

TABLE 3.8 *Rate, $f'(t)$, at which CO_2 is entering or leaving water*

t	$f'(t)$	t	$f'(t)$	t	$f'(t)$	t	$f'(t)$	t	$f'(t)$	t	$f'(t)$
0	0.000	4	−0.039	8	−0.026	12	0.000	16	0.035	20	0.020
1	−0.042	5	−0.038	9	−0.023	13	0.054	17	0.030	21	0.015
2	−0.044	6	−0.035	10	−0.020	14	0.045	18	0.027	22	0.012
3	−0.041	7	−0.030	11	−0.008	15	0.040	19	0.023	23	0.005

The left Riemann sum with $n = 12$ terms, corresponding to $\Delta t = 1$, gives

$$\int_0^{12} f'(t)\,dt \approx (0.000)(1) + (-0.042)(1) + (-0.044)(1) + \cdots + (-0.008)(1) = -0.346.$$

At 12 hours past dawn, the CO_2 content of the pond water reaches its lowest level, which is approximately

$$2.600 - 0.346 = 2.254 \text{ mmol/l}.$$

(d) The increase in CO_2 during the 12 hours of darkness equals

$$f(24) - f(12) = \int_{12}^{24} f'(t)\,dt.$$

Using Riemann sums to estimate this integral, we find that about 0.306 mmol/l of CO_2 was released into the pond during the night. In part (c) we calculated that about 0.346 mmol/l

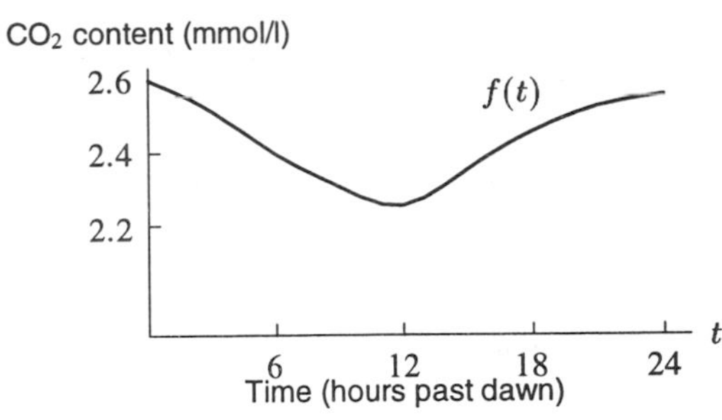

Figure 3.49: CO_2 content in pond water throughout the day

of CO_2 was absorbed from the pond during the day. If the pond is in equilibrium, we would expect the daytime absorption to equal the nighttime release. These quantities are sufficiently close (0.346 and 0.306) that the difference could be due to measurement error.
If the pond is in equilibrium, the area between the rate curve in Figure 3.48 and the axis for $0 \le t \le 12$ will equal the area between the rate curve and the axis for $12 \le t \le 24$. (The axis is the horizontal line at 0.00.) In this experiment the areas do look approximately equal.

(e) We must evaluate

$$f(b) = f(0) + \int_0^b f'(t)\,dt = 2.600 + \int_0^b f'(t)\,dt$$

for the values $b = 0, 3, 6, 9, 12, 15, 18, 21, 24$. Left Riemann sums with $\Delta t = 1$ give the values for the CO_2 content in Table 3.9. The graph is shown in Figure 3.49.

TABLE 3.9 CO_2 *content throughout the day*

b (hours after dawn)	0	3	6	9	12	15	18	21	24
$f(b)$ (CO_2 content)	2.600	2.514	2.396	2.305	2.254	2.353	2.458	2.528	2.560

Problems for Section 3.5

1. If $f(t)$ is measured in miles per hour and t is measured in hours, what are the units of measurement for

$$\int_a^b f(t)\,dt?$$

2. If the marginal cost function $C'(q)$ is measured in dollars per ton, and q gives the quantity in tons, what are the units of measurement for the following definite integral and what does it represent?

$$\int_a^b C'(q)\,dq$$

3. If $f(t)$ is measured in meters/second^2 and t is measured in seconds, what are the units of measurement for $\int_a^b f(t)\,dt$?
4. If $f(t)$ is measured in dollars per year and t is measured in years, what are the units of measurement for $\int_a^b f(t)\,dt$?
5. Oil is leaking out of a ruptured tanker at a rate of $r = f(t)$ gallons per minute. Write a definite integral expressing the total quantity of oil which leaks out of the tanker in the first hour.
6. The rate of growth of the net worth of a company is given by $r(t) = 5000te^{1-0.5t}$, where t is measured in years since 1980. How does the net worth of the company change between 1980 and 1990? If the company is worth $40,000 in 1980, what is it worth in 1990?
7. A news broadcast in early 1993 said the average American's annual income is changing at a rate given in dollars per month by $r(t) = 40(1.002)^t$ where t is in months from January 1, 1993. If this trend continues, what change in income can the average American expect during 1993?
8. A cup of coffee at 90°C is put into a 20°C room when $t = 0$. If the coffee's temperature is changing at a rate given in °C per minute by

$$r(t) = -7e^{-0.1t}, \quad t \text{ in minutes},$$

estimate, to one decimal place, the coffee's temperature when $t = 10$.

9. Water is leaking out of a tank at a rate of $R(t)$ gallons/hour, where t is measured in hours.
   (a) Write a definite integral that expresses the total amount of water that leaks out in the first two hours.
   (b) Figure 3.50 is a graph of $R(t)$. On a sketch, shade in the region whose area represents the total amount of water that leaks out in the first two hours.
   (c) Give an upper and lower estimate of the total amount of water that leaks out in the first two hours.

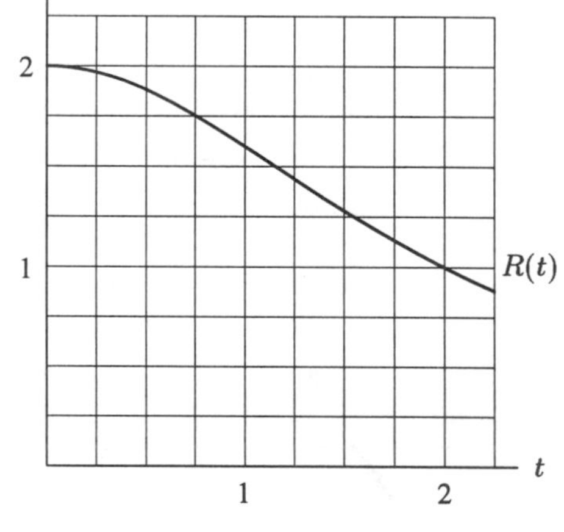

Figure 3.50

10. The rate at which the world's oil is being consumed is continuously increasing. Suppose the rate (in billions of barrels per year) is given by the function $r = f(t)$, where t is measured in years and $t = 0$ is the start of 1990.
    (a) Write a definite integral which represents the total quantity of oil used between the start of 1990 and the start of 1995.
    (b) Suppose $r = 32e^{0.05t}$. Using a left-hand sum with five subdivisions, find an approximate value for the total quantity of oil used between the start of 1990 and the start of 1995.
    (c) Interpret each of the five terms in the sum from part (b) in terms of oil consumption.

11. Two new salespeople have been hired by a company, and the number of sales per month is recorded for each in Figure 3.51, with the two curves labeled Salesperson A and Salesperson B. Which person has the most total sales after 6 months? After the first year? At approximately what times (if any) have they sold roughly equal total amounts? Approximately how many total sales has each made at the end of the first year?

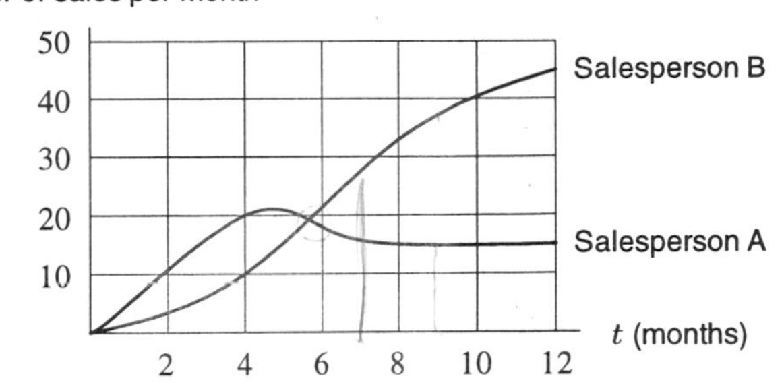

Figure 3.51: Which person sold the most items?

12. A marginal cost function $C'(q)$ is given in Figure 3.52. Assume fixed costs are $10,000.
    (a) Estimate the total cost to produce 30 units.
    (b) Approximate the additional cost if the company increases productions from 30 units to 40 units.
    (c) Find the value of $C'(25)$. Interpret you answer in terms of costs of production.

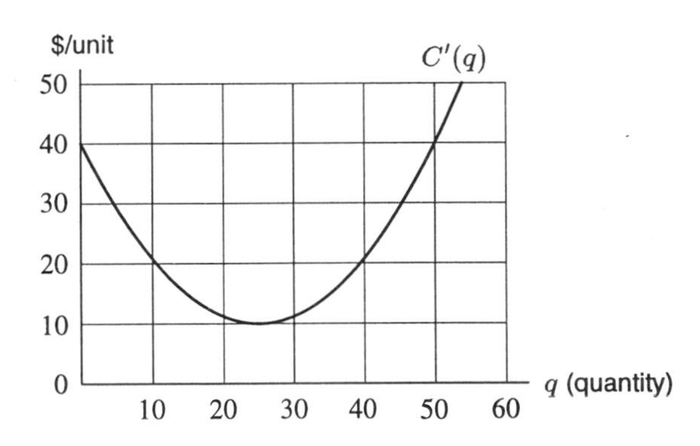

Figure 3.52: A marginal cost curve

13. The marginal cost function for a certain company is given by

$$C'(q) = \frac{10e^{0.08q}}{q+1}$$

where q is the quantity produced. It is known that $C(0) = 500$. Estimate the total cost in producing 20 units. Explain how you arrived at your answer. What are the fixed costs and what are the variable costs associated with this production size?

14. The marginal cost $C'(q)$ in producing q units is given in the table below.
    (a) If fixed costs are \$10,000, estimate the total cost in producing 400 units.
    (b) How much would the total cost increase if production were increased one unit, to 401 units?

TABLE 3.10 *Marginal cost in producing q units*

q	0	100	200	300	400	500	600
$C'(q)$	25	20	18	22	28	35	45

15. A car moves along a straight line with velocity, in feet/second, given by

$$v(t) = 6 - 2t \quad \text{for } t \geq 0.$$

    (a) Describe the car's motion in words. (When is it moving forward, backward, and so on?)
    (b) Suppose the car's position is measured from its starting point. When is it farthest forward? Backward?
    (c) Find s, the car's position measured from its starting point, as a function of time.

16. Suppose that the graph in Figure 3.53 represents your velocity, in miles per hour, on a long bicycle trip. Suppose that you start out 10 miles from home, and positive velocities take you towards home and negative velocities take you away from home. Write a paragraph describing your trip: Do you start out going towards or away from home? How long do you continue in that direction and how far are you from home when you turn around? How many times do you change direction? Do you ever get home? Where are you at the end of the four-hour bike ride? Make up a story to go with this bicycle trip.

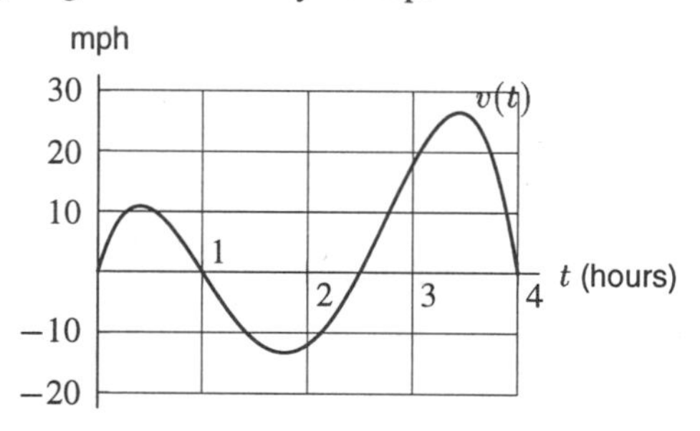

Figure 3.53: The velocity curve for a bicycle trip

17. A bicyclist is pedaling along a straight road with velocity, v, given in Figure 3.54. Suppose the cyclist starts 5 miles from a lake, and that positive velocities take her away from the lake and negative velocities towards the lake. When is the cyclist farthest from the lake, and how far away is she then?

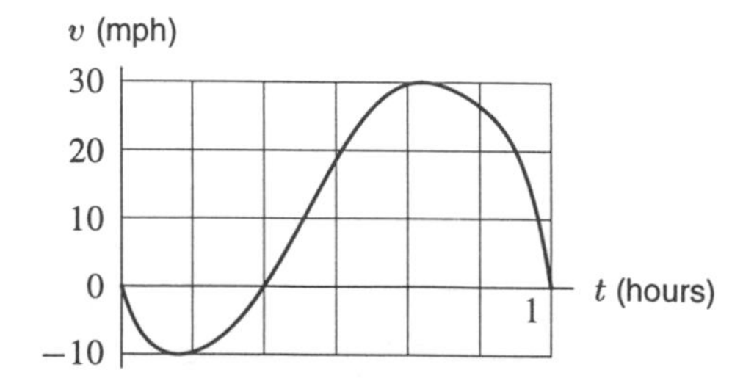

Figure 3.54

18. Suppose the rate of change of the price of stock in a certain company as a function of time, t, in weeks is in Figure 3.55.
    (a) At what time during this five-week period was the stock at its highest value?
    (b) At what time during this five-week period was the stock at its lowest value?
    (c) If the price of the stock as a function of time is given by $P(t)$, put the following quantities in increasing order:

$$P(0),\ P(1),\ P(2),\ P(3),\ P(4),\ P(5).$$

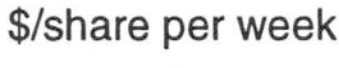

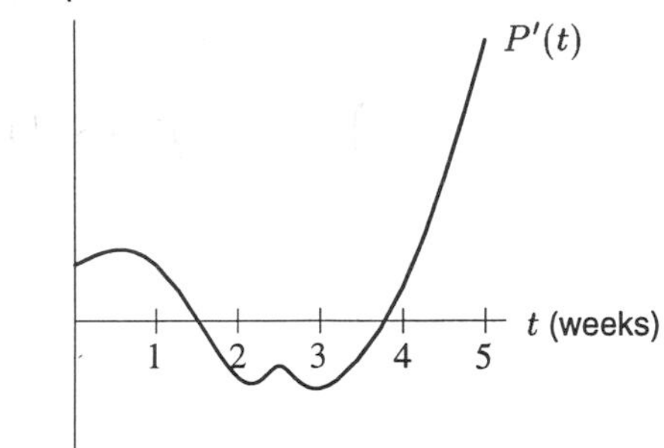

Figure 3.55: Rate of change of the price of a company's stock

3.6 THE FUNDAMENTAL THEOREM OF CALCULUS

Suppose a function $F(t)$ gives the total change of a quantity. In Chapter 2, we learned that the rate of change of the quantity is given by the derivative $F'(t)$. In Section 3.5, we learned that the definite integral of the rate of change $F'(t)$ is the total change $F(t)$. We have the following result:

$$F(b) - F(a) = \begin{matrix}\text{Total change in } F(t) \\ \text{between } t = a \text{ and } t = b\end{matrix} = \int_a^b F'(t)\,dt.$$

This result is one of the most important in calculus because it makes the connection between the derivative and the definite integral. It is called the Fundamental Theorem of Calculus and is often stated as follows:

The Fundamental Theorem of Calculus

If f is continuous and $f(t) = \dfrac{dF(t)}{dt}$, then

$$\int_a^b f(t)\,dt = F(b) - F(a).$$

In words:

The definite integral of a rate gives total change.

The Fundamental Theorem provides a precise way of computing certain definite integrals.

Example 1 Compute $\int_1^3 2x\,dx$ by two different methods.

Solution Using left- and right-hand sums, we can approximate this integral as accurately as we want. With $n = 100$, for example, the left-sum is 7.96 and the right sum is 8.04. Using $n = 500$ we learn

$$7.992 < \int_1^3 2x\,dx < 8.008.$$

The Fundamental Theorem, on the other hand, allows us to compute the integral exactly. We take $f(x) = 2x$. By Example 4, on page 153, we know that if $F(x) = x^2$, then $F'(x) = 2x$. So we take $f(x) = 2x$ and $F(x) = x^2$ and obtain

$$\int_1^3 2x\,dx = F(3) - F(1) = 3^2 - 1^2 = 8.$$

The Fundamental Theorem can also be used when the rate, $F'(t)$, is known and we want to find the total change $F(b) - F(a)$. If we know $F(a)$, the theorem enables us to reconstruct the function F from knowledge about its derivative $F' = f$.

Example 2 Suppose you are given that $F'(t) = \ln(t + 1)$ and $F(0) = 2$. Find the values of $F(b)$ at the points $b = 0, 0.1, 0.2, \ldots, 1.0$.

Solution We apply the Fundamental Theorem with $f(t) = \ln(t + 1)$ and $a = 0$ to get values for $F(b)$. Since

$$F(b) - F(0) = \int_0^b F'(t)\,dt = \int_0^b \ln(t + 1)\,dt$$

and $F([table : c = 2$, we have

$$F(b) = 2 + \int_0^b \ln(t + 1)\,dt.$$

Using Riemann sums to estimate the definite integral $\int_0^b \ln(t + 1)\,dt$ for each of the values $b = 0$, $0.1, 0.2, \ldots, 1.0$ gives the approximate values for F in Table 3.11:

TABLE 3.11 *Approximate values for F*

b	0	0.1	0.2	0.3	0.4	0.5	0.6	0.7	0.8	0.9	1.0
$F(b)$	2	2.005	2.018	2.040	2.069	2.105	2.148	2.196	2.251	2.310	2.375

Notice from the table that the function $F(b)$ is increasing between $b = 0$ and $b = 1$. This could have been predicted without the use of the Fundamental Theorem from the fact that $\ln(t + 1)$, the derivative of $F(t)$, is positive for t between 0 and 1.

Example 3 The graph of the derivative F' of some function F is given in Figure 3.56. If you are told that $F(20) = 150$, estimate the maximum value attained by F.

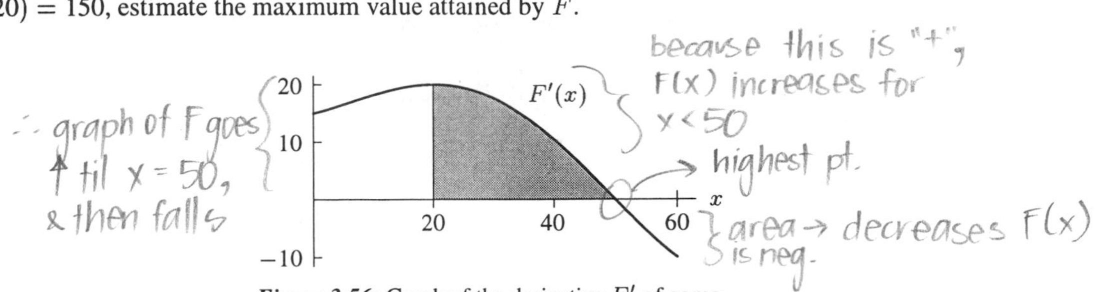

Figure 3.56: Graph of the derivative F' of some function F

Solution Let's begin by getting a rough idea of how F behaves. We know that $F(x)$ increases for $x < 50$ because the derivative of F is positive for $x < 50$. Similarly, $F(x)$ decreases for $x > 50$ because $F'(x)$ is negative for $x > 50$. Therefore, the graph of F rises until the point at which $x = 50$, and then it begins to fall. Evidently, the highest point on the graph of F is at $x = 50$, and so the maximum value attained by F is $F(50)$. To evaluate $F(50)$, we use the Fundamental Theorem:

$$F(50) - F(20) = \int_{20}^{50} F'(x)\, dx,$$

which gives

$$F(50) = F(20) + \int_{20}^{50} F'(x)\, dx = 150 + \int_{20}^{50} F'(x)\, dx.$$

The definite integral equals the area of the shaded region under the graph of F', which we can estimate is roughly 300. Therefore, the greatest value attained by F is $F(50) \approx 150 + 300 = 450$.

Example 4 The graph of the derivative F' of some function F is given in Figure 3.57. Assume $F(0) = 0$. Of the four numbers $F(1)$, $F(2)$, $F(3)$, and $F(4)$, which is largest? Which is smallest? How many of these four numbers are negative?

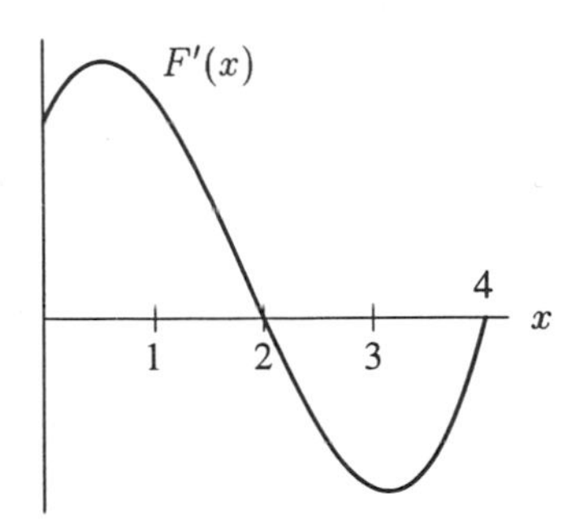

Figure 3.57: A graph of the derivative F'

Solution For every number b, the Fundamental Theorem tells us that

$$\int_0^b F'(x)\,dx = F(b) - F(0) = F(b) - 0 = F(b).$$

Therefore, the values of $F(1)$, $F(2)$, $F(3)$, and $F(4)$ are values of definite integrals. Recall that the value of a definite integral is equal to the area of the regions above the x-axis minus the area of the regions below the x-axis. Let A_1 represent the area of the region between $x = 0$ and $x = 1$, let A_2 represent the area of the region between $x = 1$ and $x = 2$, let A_3 represent the area of the region between $x = 2$ and $x = 3$, and let A_4 represent the area of the region between $x = 3$ and $x = 4$, as shown in Figure 3.58.

We see that the region between $x = 0$ and $x = 1$ lies above the x-axis, and so $F(1)$ is positive, and we have

$$F(1) = \int_0^1 F'(x)\,dx = A_1.$$

We see that the region between $x = 0$ and $x = 2$ also lies entirely above the x-axis, and so $F(2)$ is positive. We have

$$F(2) = \int_0^2 F'(x)\,dx = A_1 + A_2.$$

We see that $F(2) > F(1)$. The region between $x = 0$ and $x = 3$ includes parts above and below the x-axis. We have

$$F(3) = \int_0^3 F'(x)\,dx = (A_1 + A_2) - A_3.$$

Since the area A_3 appears to be approximately the same as the area A_2, we have $F(3) \approx F(1)$. Finally, we see that

$$F(4) = \int_0^4 F'(x)\,dx = (A_1 + A_2) - (A_3 + A_4).$$

Since the area $A_1 + A_2$ appears to be larger than the area $A_3 + A_4$, we see that $F(4)$ is still positive, but it is smaller than all the others.

We see that the largest value is $F(2)$ and the smallest value is $F(4)$. None of the numbers are negative.

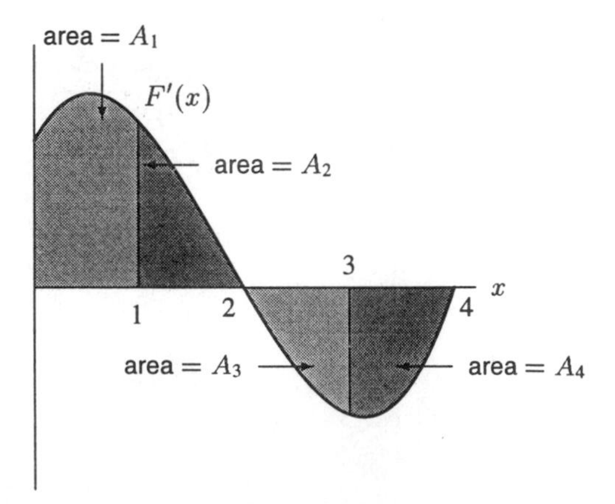

Figure 3.58: Using areas to understand definite integrals

Problems for Section 3.6

1. Figure 3.59 shows the graph of f. If $F' = f$ and $F(0) = 0$, find $F(b)$ for $b = 1, 2, 3, 4, 5, 6$.

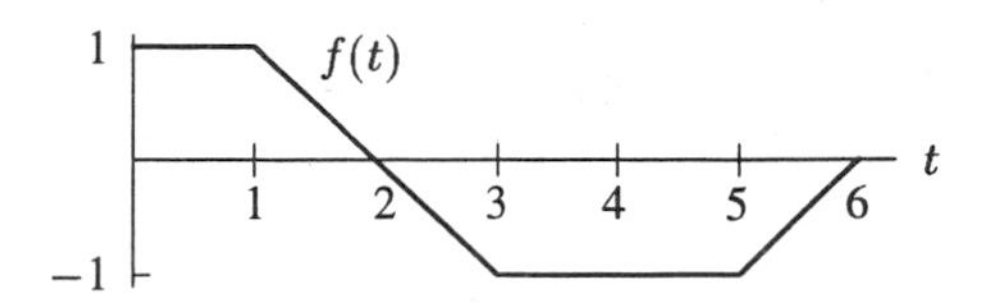

Figure 3.59

2. The graph of $y = f(x)$ is given in Figure 3.60.
   (a) What is $\int_{-3}^{0} f(x)\,dx$?
   (b) If the area of the shaded region is A, what is $\int_{-3}^{4} f(x)\,dx$?

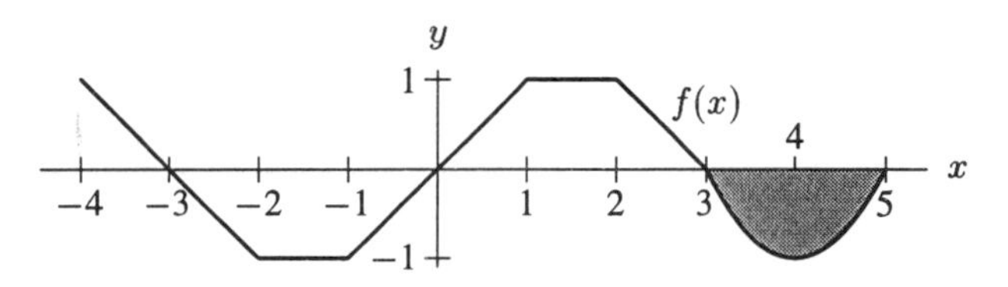

Figure 3.60

For Problems 3–5, suppose $F(0) = 0$ and $F'(x) = f(x) = 4 - x^2$ for $0 \le x \le 2.5$.

3. Approximate $F(b)$ for $b = 0,\ 0.5,\ 1,\ 1.5,\ 2,\ 2.5$.
4. Using a graph of F', decide where F is increasing and where F is decreasing.
5. Does F have a maximum value for $0 \le x \le 2.5$? If so, at what value of x does it occur, and approximately what is that maximum value?
6. If $F(t) = t(\ln t) - t$, it can be shown that $f(t) = F'(t) = \ln t$. Find $\int_{10}^{12} \ln t\,dt$ two ways:
   (a) Using left and right sums. (Approximate to one decimal place.)
   (b) Using the Fundamental Theorem of Calculus.
7. (a) Use left- and right-sums (and a computer or graphing calculator) to approximate the definite integral

$$\int_{1}^{4} 3x^2\,dx.$$

   (b) Use the fact that if $F(x) = x^3$ then $F'(x) = 3x^2$, and the Fundamental Theorem of Calculus to find the value of the integral in part (a) exactly.

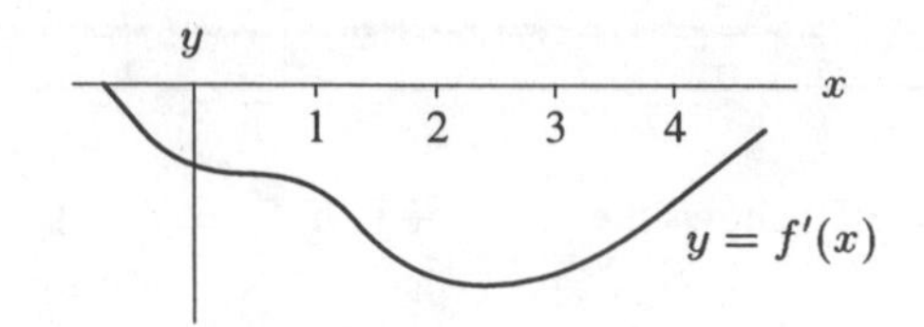

Figure 3.61: Note: This is the graph of f', not f.

Problems 8–9 concern the graph of f' in Figure 3.61.

8. Which is greater, $f(0)$ or $f(1)$?
9. List the following quantities in increasing order:

$$\frac{f(4)-f(2)}{2}, \quad f(3)-f(2), \quad f(4)-f(3).$$

For Problems 10–13, mark the following quantities on a copy of the graph of f in Figure 3.62.

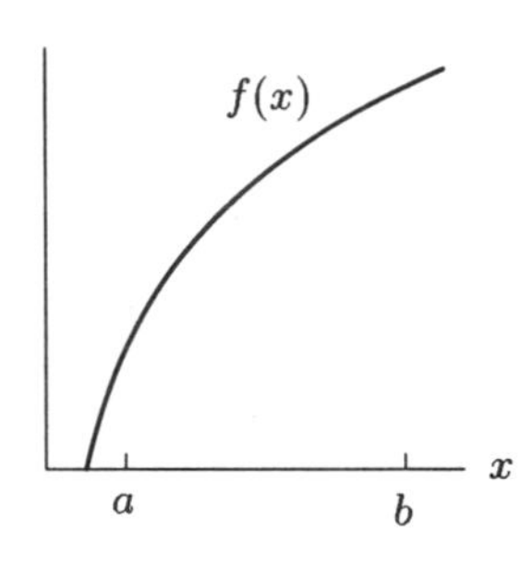

Figure 3.62

10. A length representing $f(b)-f(a)$.
11. A slope representing $\dfrac{f(b)-f(a)}{b-a}$.
12. An area representing $F(b)-F(a)$, where $F'=f$.
13. A length roughly approximating

$$\frac{F(b)-F(a)}{b-a}, \text{ where } F'=f.$$

14. In Section 2.1, page 122, we defined the average velocity of a particle over the interval $a \le t \le b$ as the change in position during this interval divided by $(b-a)$. If the velocity of the particle is $v=f(t)$, the average velocity is also given by $\frac{1}{b-a}\int_a^b f(t)dt$. Show why these two ways of calculating the average velocity always lead to the same answer.

REVIEW PROBLEMS FOR CHAPTER THREE

For Problems 1–6, find the integrals to one decimal place. In each case say how many subdivisions you used.

1. $\int_0^{10} 2^{-x}\, dx$

2. $\int_1^5 \ln(x+1)\, dx$

3. $\int_0^1 \sqrt{1+t^2}\, dt$

4. $\int_{-1}^1 \frac{x^2+1}{x^2-4}\, dx$

5. $\int_2^3 \frac{-1}{(r+1)^2}\, dr$

6. $\int_1^3 \frac{z^2+1}{z}\, dz$

7. Statisticians sometimes use values of the function

$$F(b) = \int_0^b e^{-x^2}\, dx.$$

(a) What is $F(0)$?
(b) Does the value of F increase or decrease as b increases? (Assume $b \geq 0$.)
(c) Estimate $F(1)$, $F(2)$, and $F(3)$.

8. If $F(x) = e^{x^2}$, it can be shown that $f(x) = F'(x) = 2xe^{x^2}$. Find $\int_0^1 2xe^{x^2}\, dx$ two ways:

(a) Using left and right sums. (Approximate to one decimal place.)
(b) Using the Fundamental Theorem of Calculus.

9. If $F(t) = t^2 + 3t$, it can be shown that $f(t) = F'(t) = 2t + 3$. Find $\int_1^4 (2t+3)\, dt$ two ways:

(a) Using left- and right-sums. (Approximate to one decimal place.)
(b) Using the Fundamental Theorem of Calculus.

10. A car accelerates smoothly from 0 to 60 mph in 10 seconds. Suppose the car's velocity as a function of time is given in Figure 3.63. Estimate how far the car travels during the 10-second period.

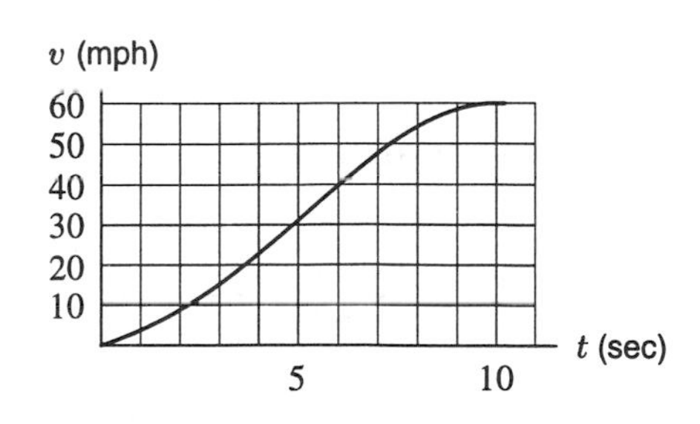

Figure 3.63

11. The Environmental Protection Agency was recently asked to investigate a spill of radioactive iodine. Measurements showed the ambient radiation levels at the site to be four times the maximum acceptable limit, so the EPA ordered an evacuation of the surrounding neighborhood.

It is known that the level of radiation from an iodine source decreases according to the formula

$$R(t) = R_0 e^{-0.004t}$$

where R is the radiation level (in millirems/hour) at time t, R_0 is the initial radiation level (at $t = 0$), and t is the time measured in hours.

(a) How long will it take for the site to reach an acceptable level of radiation?
(b) How much total radiation (in millirems) will have been emitted by that time, assuming the maximum acceptable limit is 0.6 millirems/hour?

12. Each of the graphs in Figure 3.64 represents the velocity, v, of a particle moving along the x-axis for time $0 \le t \le 5$. The vertical scales of all graphs are the same. Identify the graph(s) showing which particle(s)

(a) has a constant acceleration.
(b) ends up farthest to the left of where it started.
(c) ends up the farthest from its starting point.
(d) experiences the greatest initial acceleration.
(e) has the greatest average velocity.
(f) has the greatest average acceleration.

(I) 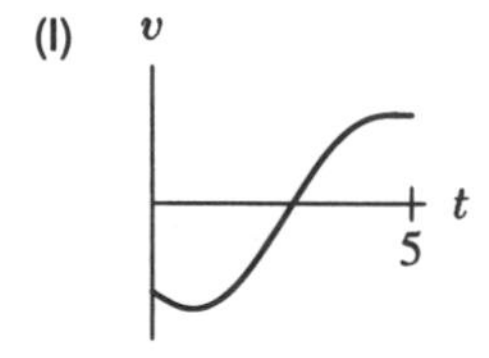

(II) 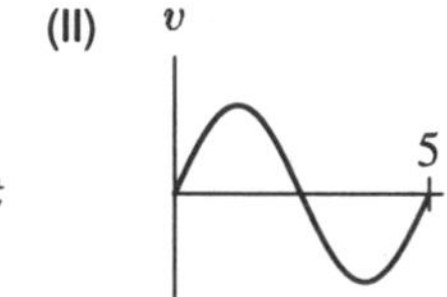

(III) 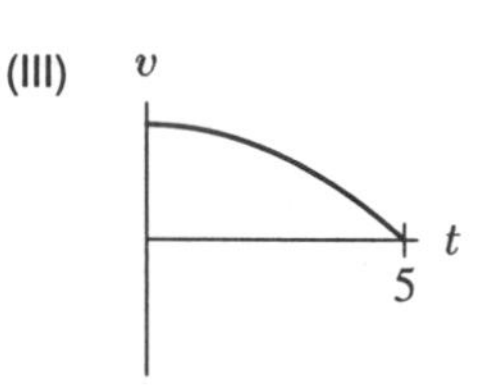

(IV) 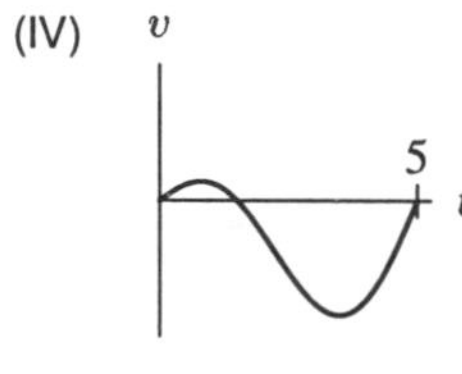

(V) 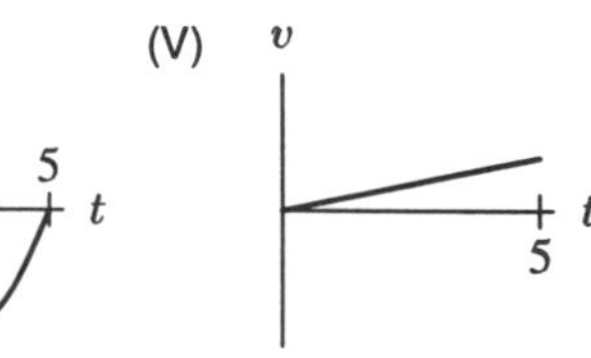

Figure 3.64

13. Assume the population, P, of Mexico (in millions), is given by

$$P = 67.38(1.026)^t,$$

where t is the number of years since 1980.

(a) What was the average population of Mexico between 1980 and 1990?
(b) What is the average of the population of Mexico in 1980 and the population in 1990?
(c) Explain, in terms of the concavity of the graph of P (see Figure 1.4 on page 38), why your answer to part (b) is larger or smaller than your answer to part (a).

14. Suppose the cost function $C(q)$ represents the total cost to produce a quantity q units of a certain product. The fixed costs for the production are $20,000. The marginal cost function is given by

$$C'(q) = 0.005q^2 - q + 56.$$

(a) On a graph of $C'(q)$, illustrate graphically the total variable cost of producing 150 units.
(b) Estimate $C(150)$, the total cost to produce 150 units.
(c) Find the value of $C'(150)$ and interpret your answer in terms of costs of production.
(d) Use your answers to parts (b) and (c) to estimate $C(151)$.

15. As this country's relatively rich coal deposits are depleted, a larger fraction of the country's coal will come from strip-mining, and it will become necessary to strip-mine larger and larger areas for each ton of coal. The graph in Figure 3.65 shows an estimate of the number of acres/million tons that will be defaced during strip-mining as a function of the number of million tons removed, starting from the present day.

(a) Estimate the total number of acres defaced in extracting the next 4 million tons of coal (measured from the present day). Draw four rectangles under the curve, and compute their area.

(b) Reestimate the number of acres defaced using rectangles above the curve.

(c) Use your answers to parts (a) and (b) to get a better estimate of the actual number of acres defaced.

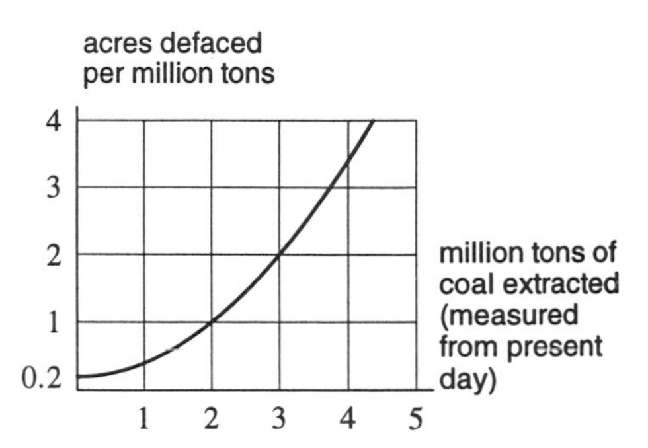

Figure 3.65

16. The Montgolfier brothers (Joseph and Etienne) were eighteenth-century pioneers in the field of hot-air ballooning. Had they had the appropriate instruments, they might have left us a record of one of their early experiments, like that shown in Figure 3.66. The graph shows their vertical velocity, v, with upward as positive.

(a) Over what intervals was the acceleration positive? Negative? Zero?
(b) What was the greatest altitude achieved, and at what time?
(c) At what time was the upward acceleration greatest?
(d) At what time was the deceleration greatest?
(e) What might have happened during this flight to explain the answer to part (d)?
(f) This particular flight ended on top of a hill. How do you know that it did, and what was the height of the hill above the starting point?

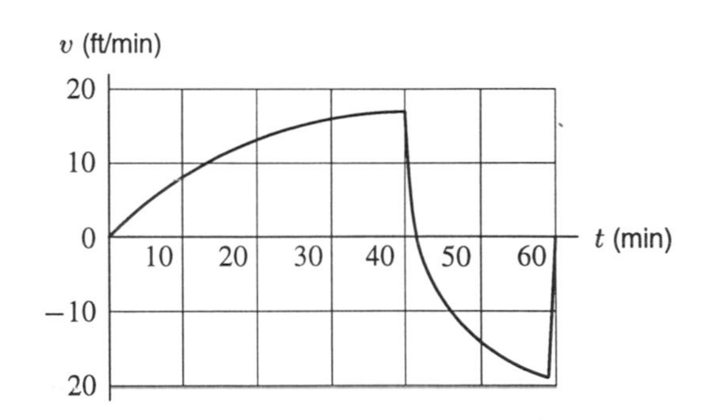

Figure 3.66

17. A mouse moves back and forth in a tunnel, attracted to bits of cheddar cheese alternately introduced to and removed from the ends (right and left) of the tunnel. The graph of the mouse's velocity, v, is given in Figure 3.67, with velocity being positive moving towards the right end of the tunnel, and negative towards the left end. Assuming that the mouse starts ($t = 0$) at the center of the tunnel, use the graph to estimate the time(s) at which:

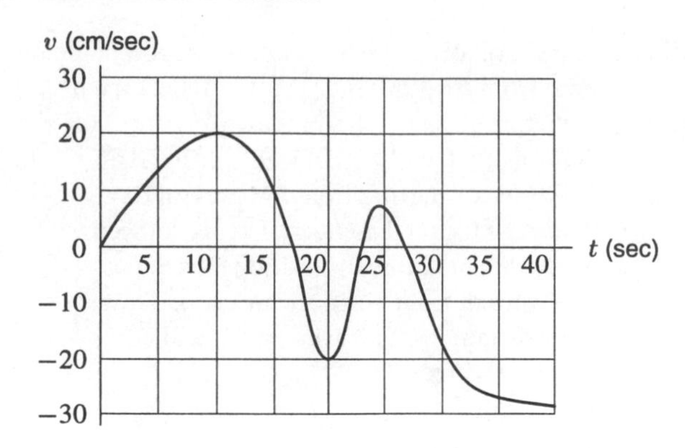

Figure 3.67

(a) The mouse changes direction.
(b) The mouse is moving most rapidly to the right; to the left.
(c) The mouse is farthest to the right of center; farthest to the left.
(d) The mouse's speed (i.e., the magnitude of its velocity) is decreasing.
(e) The mouse is at the center of the tunnel.

18. The graph of some function f is given in Figure 3.68. List, from *least* to *greatest*:

(a) $f'(1)$.
(b) The average value of $f(x)$, $0 \leq x \leq a$.
(c) The average value of the rate of change in $f(x)$, for $0 \leq x \leq a$.
(d) $\int_0^a f(x)\,dx$.

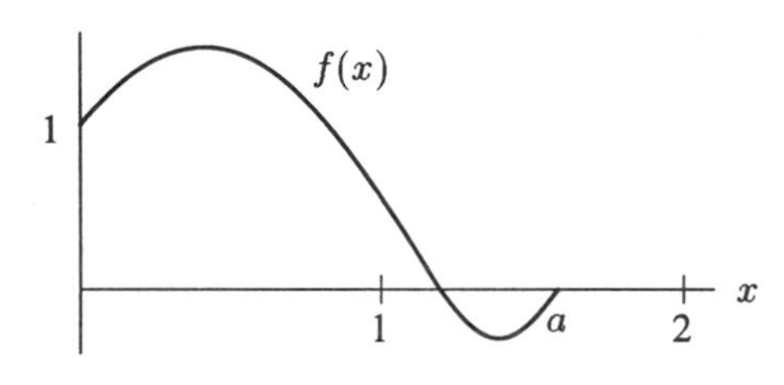

Figure 3.68

19. If you jump out of an airplane and your parachute fails to open, your downward velocity t seconds after the jump is approximated by

$$v(t) = \frac{g}{k}(1 - e^{-kt})$$

where $g = 9.8$ m/sec^2 and $k = 0.2$ sec^{-1}.

(a) Write an expression for the distance you fall in T seconds.
(b) If you jump from 5000 meters above the ground, estimate, using left- and right-hand sums, how many seconds you fall before hitting the ground.

CHAPTER FOUR

SHORT-CUTS TO DIFFERENTIATION

In this chapter we make a systematic study of the derivatives of functions given by formulas. These functions include powers, polynomials, exponential, logarithmic, and trigonometric functions. The chapter also contains general rules, such as the product, quotient, and chain rules, which allow us to differentiate combinations of functions.

4.1 DERIVATIVE FORMULAS FOR POWERS AND POLYNOMIALS

In Chapter 2, we defined the derivative function

$$f'(x) = \lim_{h \to 0} \frac{f(x+h) - f(x)}{h}$$

and saw how the derivative represented a slope and a rate of change. We also learned how to find the derivative of a function given by a graph (by estimating the slope of the tangent line at each point) and how to estimate the derivative of a function given by a table (by finding the average rate of change of the function between data values).

Useful Notation: We will write $\frac{d}{dx}(x^3)$, for example, to mean the derivative of x^3 with respect to x. Similarly, $\frac{d}{dt}\left(e^{2t}\right)$ denotes the derivative of e^{2t} with t as the variable.

Derivative of a Constant Function

The graph of a constant function $f(x) = c$ is a horizontal line, with a slope of 0 everywhere. Therefore, its derivative is 0 everywhere. (See Figure 4.1.)

> If $f(x) = c$, then $f'(x) = 0$.

For example, $\frac{d}{dx}(5) = 0$, $\frac{d}{dx}(\pi) = 0$.

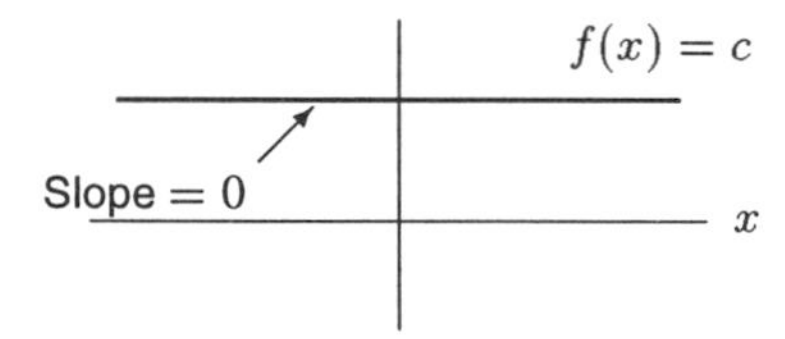

Figure 4.1: A constant function

Derivative of a Linear Function

We already know that the slope of a straight line is constant. This tells us that the derivative of a linear function is constant.

> If $f(x) = b + mx$, then $f'(x) = \text{Slope} = m$.

For example, $\frac{d}{dx}(5 - \frac{3}{2}x) = -\frac{3}{2}$.

To find the second derivative of a linear function, we must differentiate m, which is a constant. Thus, if f is linear,

$$f''(x) = 0.$$

So, as we would expect for a straight line, f is neither concave up nor concave down.

Derivative of a Constant Times a Function

Consider $y = f(x) = x$ and $y = g(x) = 5x$. We know that $f'(x) = 1$, the slope, and $g'(x) = 5$, the slope. In this case, g is 5 times f and the derivative of g is 5 times the derivative of f. Does this hold for any multiple of a function? In Figure 4.2, you see the graph of $y = f(x)$ and of three multiples: $y = 3f(x)$, $y = \frac{1}{2}f(x)$, and $y = -2f(x)$. What is the relationship between the derivatives of these functions? In other words, for a particular x value, how are the slopes of these graphs related?

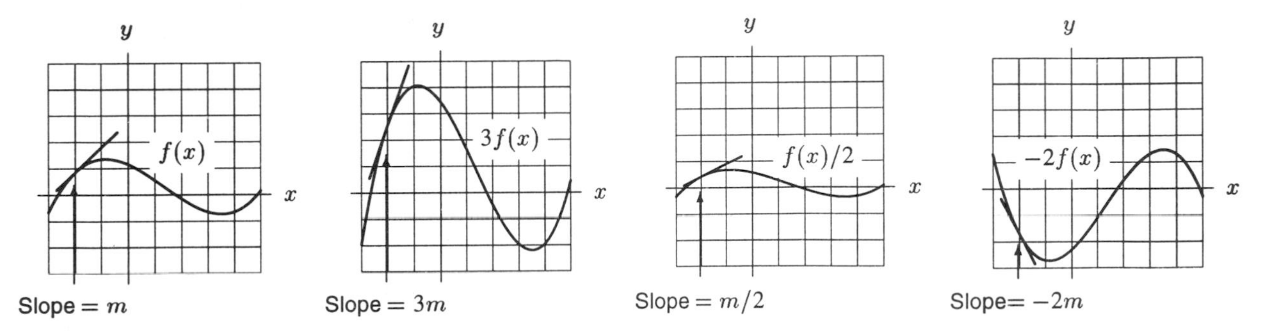

Figure 4.2: A function and its multiples: Derivative of multiple is multiple of derivative

Multiplying by a constant stretches or shrinks the graph (and flips it over the x-axis if the constant is negative). The zeros of the function remain the same and the peaks and valleys occur at the same x values. What does change is the slope of the curve at each point. If the graph has been stretched, the "rises" have all been increased by the same factor, whereas the "runs" remain the same. Thus, the slopes are all steeper by the same factor. If the graph has been shrunk, the slopes are all smaller by the same factor. If the graph has been flipped over the x-axis, the slopes will all have their signs reversed. In other words, if a function is multiplied by a constant, c, so is its derivative:

Derivative of a Constant Multiple: $\frac{d}{dx}\left[cf(x)\right] = cf'(x).$

Derivatives of Sums and Differences

Suppose we have two functions, $f(x)$ and $g(x)$, with the values listed in Table 4.1. Values of the sum $f(x) + g(x)$ are in the same table.

TABLE 4.1 *Sum of Functions*

x	$f(x)$	$g(x)$	$f(x)+g(x)$
0	100	0	100
1	110	0.2	110.2
2	130	0.4	130.4
3	160	0.6	160.6
4	200	0.8	200.8
5	250	1.0	251.0
6	310	1.2	311.2
7	380	1.4	381.4

You can see that when the increments of $f(x)$ and the increments of $g(x)$ are added together, they give the increments of $f(x) + g(x)$. For example, as x increases from 0 to 1, $f(x)$ increases by 10 and $g(x)$ increases by 0.2, while $f(x) + g(x)$ increases by $110.2 - 100 = 10.2$. Similarly, as x increases from 5 to 6, $f(x)$ increases by 60 and $g(x)$ by 0.2, while $f(x) + g(x)$ increases by $311.2 - 251.0 = 60.2$.

From this example, you can see that the rate at which $f(x) + g(x)$ is increasing is the sum of the rates at which $f(x)$ and $g(x)$ are increasing. Stating this with derivatives:

Derivative of Sum

$$\frac{d}{dx}[f(x) + g(x)] = f'(x) + g'(x).$$

Similarly for the difference:

Derivative of Difference

$$\frac{d}{dx}[f(x) - g(x)] = f'(x) - g'(x).$$

Derivatives of Positive Integral Powers of x

First, let's look at what graphs can tell us about the derivative of $f(x) = x^n$, with n a positive integer. We'll start with $n = 2$ and $n = 3$. The graphs of $f(x) = x^2$ and $g(x) = x^3$ are in Figure 4.3.

Derivative of x^2 and x^3

The shape of the graph of $f(x) = x^2$ in Figure 4.3 shows that its derivative is negative when $x < 0$, zero when $x = 0$, and positive for $x > 0$. The graph of $g(x) = x^3$, on the other hand, suggests that its derivative is positive when $x < 0$, zero when $x = 0$, and becomes positive again when $x > 0$.

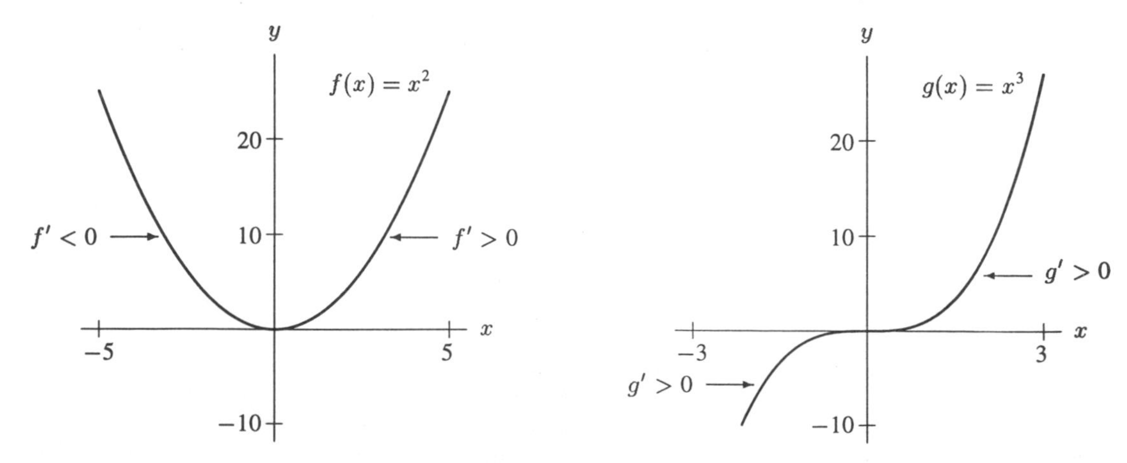

Figure 4.3: Graphs of $f(x) = x^2$ and $g(x) = x^3$

We'll calculate the derivative of $f(x) = x^2$ using the definition:

$$\begin{aligned} f'(x) &= \lim_{h\to 0} \frac{f(x+h)-f(x)}{h} = \lim_{h\to 0} \frac{(x+h)^2 - x^2}{h} \\ &= \lim_{h\to 0} \frac{x^2+2xh+h^2-x^2}{h} = \lim_{h\to 0} \frac{2xh+h^2}{h} \\ &= \lim_{h\to 0} \frac{h(2x+h)}{h}. \end{aligned}$$

To take the limit, we look at what happens when h is close to 0, but we do not let $h = 0$. Therefore, we can divide by h and say

$$f'(x) = \lim_{h\to 0} \frac{h(2x+h)}{h} = \lim_{h\to 0} (2x+h) = 2x,$$

because as h gets close to zero, $2x + h$ gets close to $2x$. So

$$f'(x) = \frac{d}{dx}(x^2) = 2x.$$

Now $f'(x) = 2x$ has the behavior we expected: it is negative for $x < 0$, zero when $x = 0$, and positive for $x > 0$. The graph of f' is in Figure 4.4, along with the graph of f.

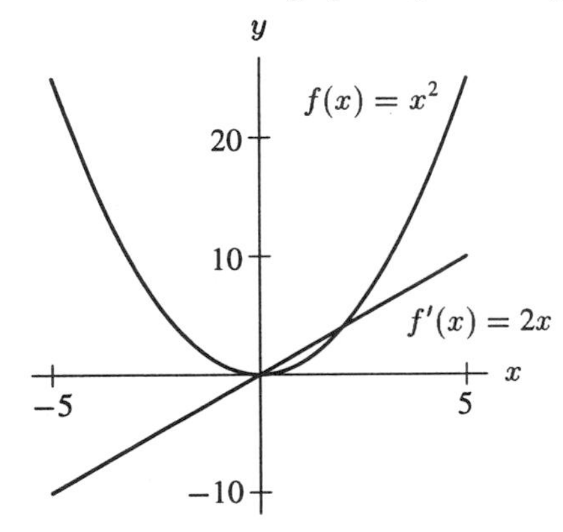

Figure 4.4: Graphs of $f(x) = x^2$ and its derivative

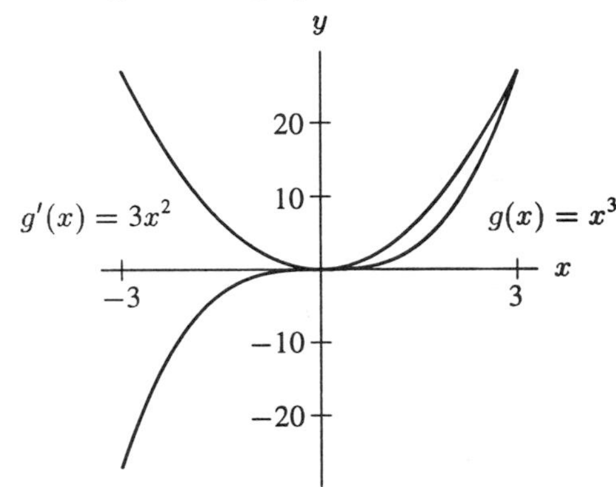

Figure 4.5: Graphs of $g(x) = x^3$ and its derivative

Example 1 Find the derivative of $g(x) = x^3$.

Solution

$$\begin{aligned} g'(x) &= \lim_{h\to 0} \frac{g(x+h)-g(x)}{h} = \lim_{h\to 0} \frac{(x+h)^3 - x^3}{h} \\ \text{Multiplying out} \longrightarrow &= \lim_{h\to 0} \frac{x^3+3x^2h+3xh^2+h^3-x^3}{h} \\ &= \lim_{h\to 0} \frac{3x^2h+3xh^2+h^3}{h} \\ \text{Dividing by } h \longrightarrow &= \lim_{h\to 0} (3x^2+3xh+h^2) = 3x^2, \end{aligned}$$

Looking at what happens as $h \to 0$

So $g'(x) = \frac{d}{dx}(x^3) = 3x^2$. Again, $g'(x) = 3x^2$ is zero when $x = 0$, but positive everywhere else. (See Figure 4.5.)

Further Examples

Similar calculations will show you that

$$\frac{d}{dx}(x^4) = 4x^3, \quad \frac{d}{dx}(x^5) = 5x^4,$$

and so on. These results might well lead you to conjecture that, for any positive integer n, the following general relation holds:

$$\boxed{\frac{d}{dx}(x^n) = nx^{n-1},}$$

and indeed this turns out to be true.

Example 2 Differentiate $5x^2 - 7x^3$.

Solution

$$\begin{aligned} \frac{d}{dx}(5x^2 - 7x^3) &= \frac{d}{dx}(5x^2) - \frac{d}{dx}(7x^3) && \text{Derivative of difference} \\ &= 5\frac{d}{dx}(x^2) - 7\frac{d}{dx}(x^3) && \text{Derivative of multiple} \\ &= 5(2x) - 7(3x^2) = 10x - 21x^2. \end{aligned}$$

Derivatives of Negative and Fractional Powers

We can examine the derivatives of functions such as $g(x) = x^{-1}$ and $k(x) = x^{-2}$ by looking at the graphs of these functions.

Example 3 The graph of $k(x) = x^{-2}$ is shown in Figure 4.6. Describe the graph of $k'(x)$.

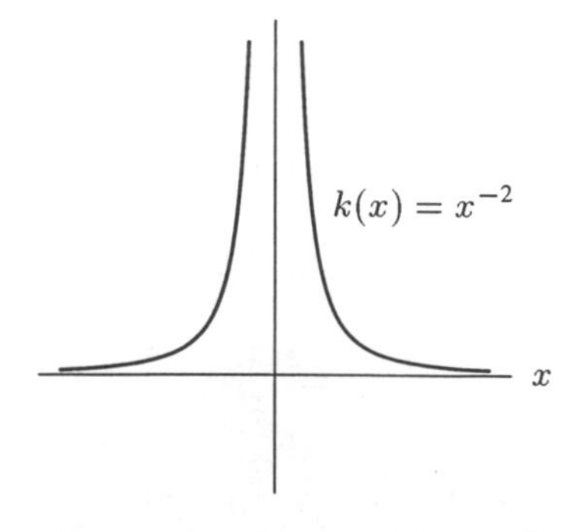

Figure 4.6: Graph of $k(x) = x^{-2}$

Solution What features do we expect to find in the graph of $f'(x)$? Since $k(x)$ is undefined at $x = 0$, we expect $k'(x)$ to be undefined at $x = 0$. $k(x)$ is increasing for $x < 0$ and decreasing for $x > 0$, so $k'(x)$ will be positive for negative x and negative for positive x. The graph of $k(x)$ gets steeper and steeper as x approaches zero, and so we expect the graph of $k'(x)$ to approach positive or negative infinity as x approaches zero. Finally, what about the derivative as x approaches positive or negative infinity? Since the function becomes more horizontal at its ends, the slope is approaching zero. We expect the derivative to approach zero as x approaches positive or negative infinity. The graph of $-2x^{-3}$ is shown in Figure 4.7, and this graph has all the features we expect in the graph of the derivative of x^{-2}. In fact, it can be shown that the derivative of x^{-2} is $-2x^{-3}$.

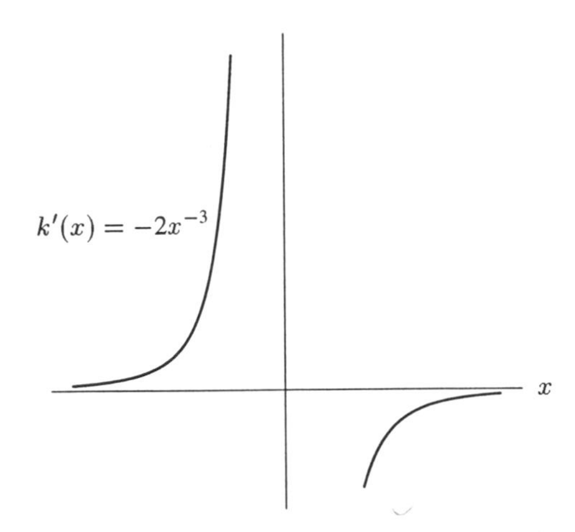

Figure 4.7: Graph of derivative of $k(x) = x^{-2}$

This example suggests that the rule

$$\frac{d}{dx}(x^n) = nx^{n-1}$$

may hold for n a negative integer also, and indeed it does. In fact, this rule holds for any constant real number n.

Example 4 Differentiate (a) $\frac{1}{x^3}$, (b) $x^{1/2}$, (c) $\frac{1}{\sqrt[3]{x}}$.

Solution (a) For $n = -3$: $\frac{d}{dx}\left(\frac{1}{x^3}\right) = \frac{d}{dx}(x^{-3}) = 3x^{-3-1} = -3x^{-4} = -\frac{3}{x^4}$, provided $x \neq 0$.

(b) For $n = 1/2$: $\frac{d}{dx}\left(x^{1/2}\right) = \frac{1}{2}x^{(1/2)-1} = \frac{1}{2}x^{-1/2} = \frac{1}{2\sqrt{x}}$.

(c) For $n = -1/3$: $\frac{d}{dx}\left(\frac{1}{\sqrt[3]{x}}\right) = \frac{d}{dx}\left(x^{-1/3}\right) = -\frac{1}{3}x^{(-1/3)-1} = -\frac{1}{3}x^{-4/3} = -\frac{1}{3x^{4/3}}$.

Derivatives of Polynomials

We already know how to differentiate powers, constant multiples, and sums. For example,

$$\frac{d}{dx}(3x^5) = 3\frac{d}{dx}(x^5) = 3 \cdot 5x^4 = 15x^4$$

and $$\frac{d}{dx}(x^5 + x^3) = \frac{d}{dx}(x^5) + \frac{d}{dx}(x^3) = 5x^4 + 3x^2.$$

Using these rules together, we can differentiate any polynomial.

Example 5 Find the derivatives of

(a) $5x^2 + 3x + 2$, (b) $3x^7 - \dfrac{x^5}{5} + 2x^2 - 13$. (c) $x^3 - 0.2x^2 + 1.3x - 2.51$

Solution (a)

$$\begin{aligned}\frac{d}{dx}(5x^2 + 3x + 2) &= 5\frac{d}{dx}(x^2) + 3\frac{d}{dx}(x) + \frac{d}{dx}(2) \\ &= 5 \cdot 2x + 3 \cdot 1 + 0 \qquad \text{Since the derivative of a constant, } d(2)/dx, \text{ is zero} \\ &= 10x + 3.\end{aligned}$$

(b)

$$\begin{aligned}\frac{d}{dx}\left(3x^7 - \frac{x^5}{5} + 2x^2 - 13\right) &= 3\frac{d}{dx}(x^7) - \frac{1}{5}\frac{d}{dx}(x^5) + 2\frac{d}{dx}(x^2) - \frac{d(13)}{dx} \\ &= 3 \cdot 7x^6 - \frac{1}{5} \cdot 5x^4 + 2 \cdot 2x - 0 \qquad \text{Since 13 is a constant, } d(13)/dx = 0 \\ &= 21x^6 - x^4 + 4x.\end{aligned}$$

(c)

$$\begin{aligned}\frac{d}{dx}\left(x^3 - 0.2x^2 + 1.3x - 2.51\right) &= \frac{d}{dx}(x^3) - 0.2\frac{d}{dx}(x^2) + 1.3\frac{d}{dx}(x) - \frac{d}{dx}(2.51) \\ &= 3x^2 - 0.2 \cdot 2x + 1.3 \cdot 1 - 0 \qquad \text{Since 2.51 is a constant, } d(2.51)/dx = 0 \\ &= 3x^2 - 0.4x + 1.3\end{aligned}$$

We can also use these rules to differentiate expressions which are not polynomials.

Example 6 Differentiate (a) $5\sqrt{x} - \dfrac{10}{x^2} + \dfrac{1}{2\sqrt{x}}$, (b) $0.1x^3 + 2x^{\sqrt{2}}$.

Solution (a)

$$\begin{aligned}\frac{d}{dx}\left(5\sqrt{x} - \frac{10}{x^2} + \frac{1}{2\sqrt{x}}\right) &= \frac{d}{dx}\left(5x^{1/2} - 10x^{-2} + \frac{1}{2}x^{-1/2}\right) \\ &= 5 \cdot \frac{1}{2}x^{-1/2} - 10(-2)x^{-3} + \frac{1}{2}\left(-\frac{1}{2}\right)x^{-3/2} \\ &= \frac{5}{2\sqrt{x}} + \frac{20}{x^3} - \frac{1}{4x^{3/2}}.\end{aligned}$$

(b)

$$\frac{d}{dx}(0.1x^3 + 2x^{\sqrt{2}}) = 0.1\frac{d}{dx}(x^3) + 2\frac{d}{dx}(x^{\sqrt{2}}) = 0.3x^2 + 2\sqrt{2}x^{\sqrt{2}-1}.$$

Problems for Section 4.1

For Problems 1–21, find the derivative of the given function.

1. $y = 5$
2. $y = 3x$
3. $y = 5x + 13$
4. $y = x^{12}$
5. $y = x^{-12}$
6. $y = x^{4/3}$
7. $y = x^{3/4}$
8. $y = x^{-3/4}$
9. $f(x) = \frac{1}{x^4}$
10. $f(x) = \sqrt[4]{x}$
11. $f(x) = Cx^2$, C a constant.
12. $y = 4x^{3/2} - 5x^{1/2}$
13. $y = 6x^3 + 4x^2 - 2x$
14. $y = -3x^4 - 4x^3 - 6x + 2$
15. $y = 3x^2 + 7x - 9$
16. $y = 8t^3 - 4t^2 + 12t - 3$
17. $y = 4.2q^2 - 0.5q + 11.27$
18. $y = 3t^5 - 5\sqrt{t} + \frac{7}{t}$
19. $y = 3t^2 + \frac{12}{\sqrt{t}} - \frac{1}{t^2}$
20. $y = z^2 + \frac{1}{2z}$
21. $y = ax^4 + bx^3 + cx^2 + dx + k$ (a, b, c, d, and k are constants.)
22. Find the functions in Problems 4–12 whose derivatives do not exist at $x = 0$.
23. Let $f(x) = x^3 - 4x^2 + 7x - 11$. Find $f'(0)$, $f'(2)$, and $f'(-1)$.
24. Let $f(x) = x^2 + 3x - 5$. Find $f'(0)$, $f'(3)$, and $f'(-2)$.
25. Let $f(x) = -3x + 2$ and $g(x) = 2x + 1$.
    (a) If $k(x) = f(x) + g(x)$, find a formula for $k(x)$ and verify the sum rule by comparing $k'(x)$ with $f'(x) + g'(x)$.
    (b) If $j(x) = f(x) - g(x)$, determine a formula for $j(x)$ and compare $j'(x)$ with $f'(x) - g'(x)$.
26. Let $r(t) = 2t - 4$. If $s(t) = 3r(t)$, verify the constant multiple rule by showing that $s'(t) = 3r'(t)$.
27. If $r(x) = f(x) + 2g(x) + 3$, and $f'(x) = g(x)$ and $g'(x) = r(x)$, express
    (a) $r'(x)$ in terms of $f(x)$ and $g(x)$. Your answer should not involve r, r', f', or g'.
    (b) $f'(x)$ in terms of $f(x)$ and $r(x)$. Your answer should not involve r', f', g, or g'.

For Problems 28–37, determine if the derivative rules from this section apply. If they do, find the derivative. If they don't apply, indicate why.

28. $y = \sqrt{x}$
29. $y = (x + 3)^{1/2}$
30. $y = 3x^2 + 4$
31. $y = \frac{1}{3z^2} + \frac{1}{4}$
32. $y = \frac{1}{3\sqrt{x}} + \frac{1}{4}$
33. $y = \frac{1}{3x^2 + 4}$
34. $y = 3^x$
35. $y = x^3$
36. $y = x^\pi$
37. $y = \pi^x$

4.2 USING THE DERIVATIVE FORMULAS FOR POLYNOMIALS

In Chapter 2, we learned that the derivative represents a slope and a rate of change, and we learned how to approximate a derivative numerically and graphically. In Section 4.1, we learned how to use short-cut formulas to find the derivative function analytically. In this section, we put these two pieces together.

Example 1 Let

$$f(x) = x^2 + 1$$

Compute the derivatives $f'(0)$, $f'(1)$, $f'(2)$, and $f'(-1)$. Check your answers graphically.

Solution The graph of $f(x) = x^2 + 1$ is given in Figure 4.8. The derivative $f'(x)$ gives the slope of the tangent line to this curve at the point x. We will compute the derivatives analytically, and then check the answers by looking at the slope of the tangent line to this function at the different values of x.

Since

$$f(x) = x^2 + 1$$

we see

$$f'(x) = 2x.$$

Thus, $f'(0) = 2(0) = 0$. At the point on the curve where $x = 0$, the slope of the tangent line is 0. We confirm this by seeing in Figure 4.8 that the tangent line at $x = 0$ is horizontal.

We have $f'(1) = 2(1) = 2$ and $f'(2) = 2(2) = 4$. The slope is 2 at $x = 1$ and the slope is 4 at $x = 2$. The slope is positive at both points and the curve is steeper at $x = 2$ than it is at $x = 1$. Again, this matches what we see in the graph in Figure 4.8.

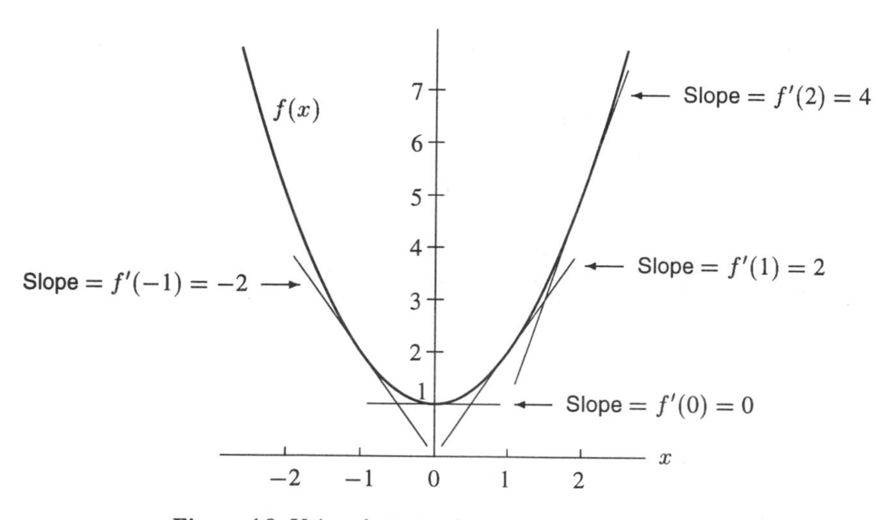

Figure 4.8: Using slopes to check values for derivatives

We have $f'(-1) = 2(-1) = -2$. This tells us that where $x = -1$, the slope is -2. Again, this agrees with the graph in Figure 4.8.

Example 2 Find the equation of the tangent line to the graph of

$$y = x^3 + 2x^2 - 5x + 7$$

at the point where $x = 1$. Sketch the graph of the curve and its tangent line on the same axes.

Solution The slope of a tangent line is given by the derivative, so we begin by finding dy/dx:

$$\frac{dy}{dx} = 3x^2 + 2(2x) - 5(1) + 0 = 3x^2 + 4x - 5.$$

Therefore, the slope of the tangent line at $x = 1$ is given by

$$\left.\frac{dy}{dx}\right|_{x=1} = 3(1)^2 + 4(1) - 5 = 2.$$

Thus, the equation of the tangent line is of the form

$$y = b + 2x.$$

When $x = 1$, we have

$$y = 1^3 + 2(1^2) - 5(1) + 7 = 5,$$

so the point $x = 1$, $y = 5$ lies on the tangent line. Substituting gives

$$5 = b + 2(1), \qquad \text{so} \qquad b = 3.$$

Thus, the equation of the tangent line is

$$y = 3 + 2x.$$

The graphs of the curve and the tangent line are given in Figure 4.9.

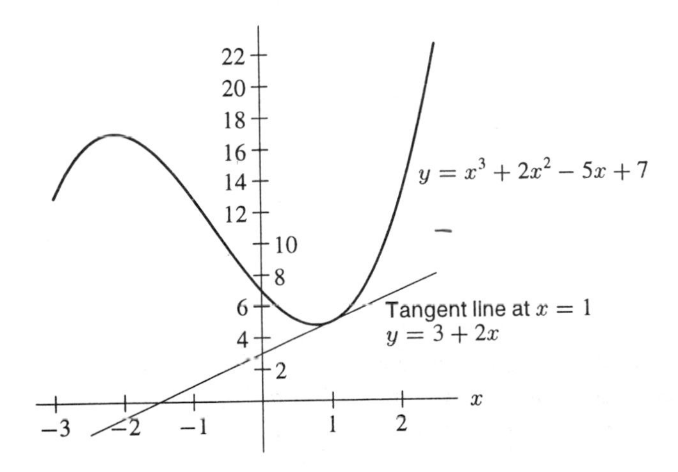

Figure 4.9: Finding the equation for this tangent line

Example 3 Find and interpret the second derivatives of

(a) $f(x) = x^2$, (b) $g(x) = x^3$, (c) $k(x) = x^{1/2}$.

Solution (a) For $f(x) = x^2$, $f'(x) = 2x$, so $f''(x) = \dfrac{d}{dx}(2x) = 2$. Since f'' is always positive, f is concave up, as expected for a parabola opening upwards. (See Figure 4.10.)

(b) For $g(x) = x^3$, $g'(x) = 3x^2$, so $g''(x) = \dfrac{d}{dx}(3x^2) = 3\dfrac{d}{dx}(x^2) = 3 \cdot 2x = 6x$. This is positive for $x > 0$ and negative for $x < 0$, which means x^3 is concave up for $x > 0$ and concave down for $x < 0$. (See Figure 4.11.)

(c) For $k(x) = x^{1/2}$,

$$k'(x) = \frac{1}{2}x^{(1/2)-1} = \frac{1}{2}x^{-1/2}$$

so

$$k''(x) = \frac{d}{dx}\left(\frac{1}{2}x^{-1/2}\right) = \frac{1}{2} \cdot (-\frac{1}{2})x^{-(1/2)-1} = -\frac{1}{4}x^{-3/2}.$$

Now k' and k'' only make sense on the domain of k, i.e. $x \geq 0$. In fact, k' and k'' are undefined when $x = 0$, since $x^{-1/2}$ and $x^{-3/2}$ are not defined for $x = 0$. We see that $k''(x)$ is negative for $x > 0$, so k is concave down. (See Figure 4.12.)

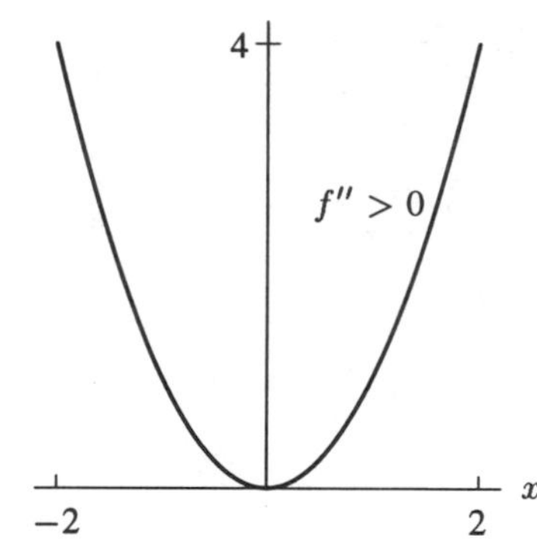

Figure 4.10: $f(x) = x^2$ and $f''(x) = 2$

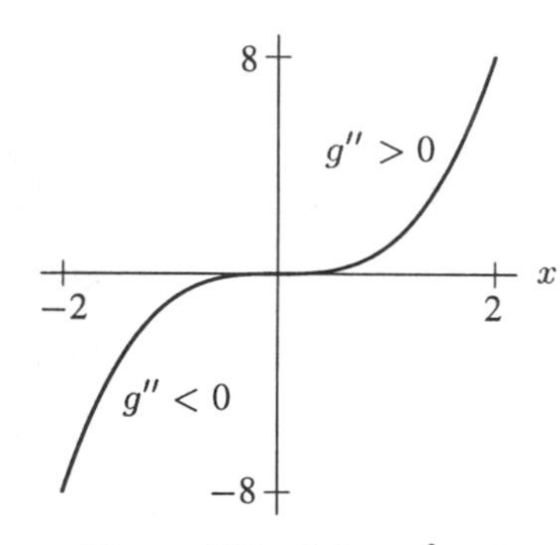

Figure 4.11: $g(x) = x^3$ and $g''(x) = 6x$

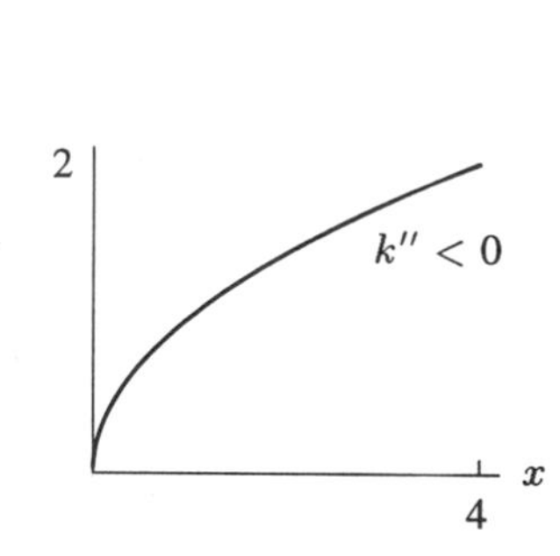

Figure 4.12: $k(x) = x^{1/2}$ and $k''(x) = -\frac{1}{4}x^{-3/2}$

Example 4 A real estate office in North Carolina rents beach houses by the week. They receive most of their inquiries for information between January and March. They know that, during this peak time, the average number of daily inquiries they receive, N, is a function of their advertising expenditures, a. Suppose N is approximated by

$$N = f(a) = 5 + 2a + 6\sqrt{a},$$

where a is in thousands of dollars. Find $f(4)$ and $f'(4)$, and interpret your answers in terms of inquiries and advertising expenditures.

Solution The fact that $f(4) = 5 + 2(4) + 6(\sqrt{4}) = 25$ tells us that when the real estate company spends \$4000 on advertising, they can expect about 25 inquiries per day.

Differentiating gives

$$f'(a) = 2(1) + 6\left(\frac{1}{2}a^{-1/2}\right) = 2 + \frac{3}{\sqrt{a}}$$

so

$$f'(4) = 2 + \frac{3}{\sqrt{4}} = 2 + \frac{3}{2} = 3.5.$$

This means that if a goes up by 1 (from 4 to 5), we can expect N to go up by about 3.5. In other words, if the company increased advertising expenditures from \$4000 to \$5000, they would expect inquiries to go up by about 3.5 inquiries per day.

Example 5 In Problem 2 of Section 2.7, you were given the graph of a cost function and asked to estimate the marginal cost at $q = 5$, at $q = 20$, and at $q = 40$. The graph of this cost function is given in Figure 4.13. The formula for this cost function is

$$C(q) = 0.0125q^3 - 0.75q^2 + 15q + 150.$$

Find the marginal cost at each of the quantities $q = 5$, $q = 20$ and $q = 40$, and interpret these marginal costs in terms of costs of production.

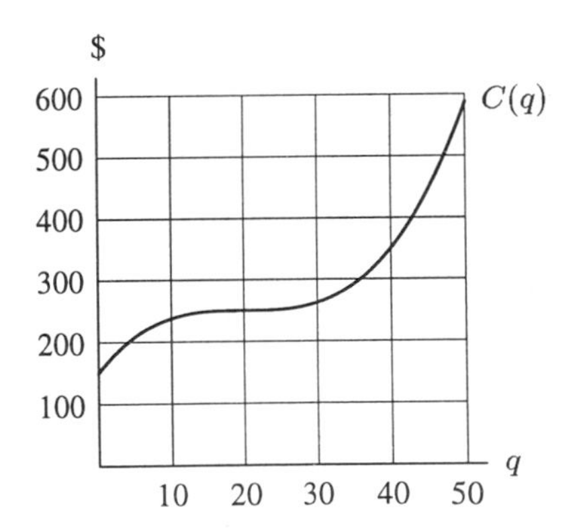

Figure 4.13: A cost function

Solution The marginal cost is the derivative of the cost function, and so we have

$$\text{Marginal cost} = C'(q) = 0.0125(3q^2) - 0.75(2q) + 15 = 0.0375q^2 - 1.5q + 15.$$

When $q = 5$, the marginal cost is given by

$$C'(5) = 0.0375(5^2) - 1.5(5) + 15 = 8.4375.$$

Thus, the additional cost in producing the 5th item is about \$8.44. When $q = 20$, the marginal cost is given by

$$C'(20) = 0.0375(20^2) - 1.5(20) + 15 = 0.$$

Thus, there is no additional cost in producing the 20th item. This corresponds to the fact that the cost curve in Figure 4.13 has a horizontal tangent line at the point where $q = 20$. Finally, when $q = 40$, the marginal cost is equal to

$$C'(40) = 0.0375(40^2) - 1.5(40) + 15 = 15.$$

Thus, the additional cost in producing the 40th item is about \$15. We see in Figure 4.13 that the cost curve is steeper at $q = 40$ than at $q = 5$, corresponding to the fact that $C'(40) > C'(5)$.

Example 6 If the position of a body, in meters, is given as a function of time t, in seconds, by

$$s = -4.9t^2 + 5t + 6,$$

find the velocity and acceleration of the body at time t.

Solution The velocity, v, is the derivative of the position:

$$v = \frac{ds}{dt} = \frac{d}{dt}(-4.9t^2 + 5t + 6) = -9.8t + 5,$$

and the acceleration, a, is the derivative of the velocity:

$$a = \frac{dv}{dt} = \frac{d}{dt}(-9.8t + 5) = -9.8.$$

Example 7 Figure 4.14 shows the graph of a cubic polynomial. Both graphically and algebraically, describe the behavior of the derivative of this cubic.

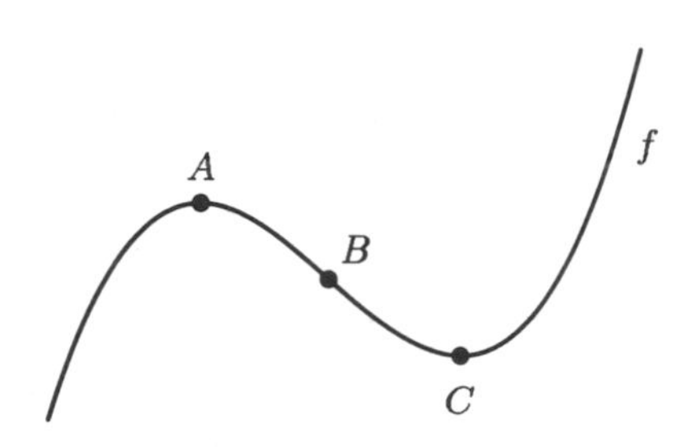

Figure 4.14: The cubic of Example 7

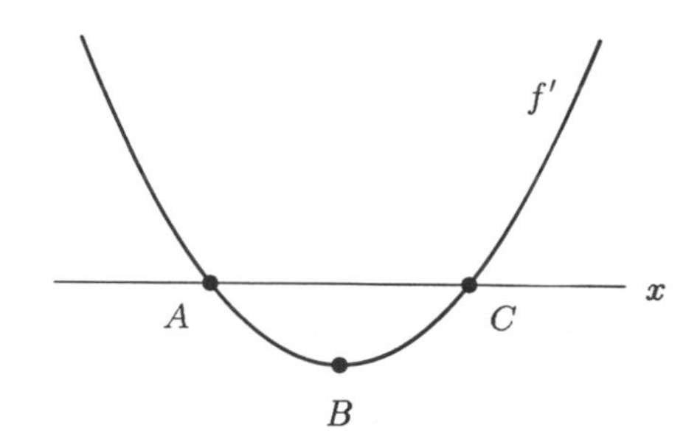

Figure 4.15: Derivative of the cubic of Example 7

Solution Graphical approach: Suppose you move along the curve from left to right. To the left of A, the slope is positive; it starts very positive and decreases until the curve reaches A, where the slope is 0. Between A and C the slope is negative. Between A and B the slope is decreasing (getting more negative); it is most negative at B. Between B and C the slope is negative but increasing; at C the slope is zero. From C to the right, the slope is positive and increasing. The graph of the derivative function is shown in Figure 4.15.

Algebraic approach: f is a cubic that goes to $+\infty$ as $x \to +\infty$ so

$$f(x) = ax^3 + bx^2 + cx + d$$

with $a > 0$. Hence,

$$f'(x) = 3ax^2 + 2bx + c,$$

whose graph is a parabola opening upward, as in Figure 4.15.

Problems for Section 4.2

1. Let $f(x) = x^3 - 2x^2 + 3x + 2$.
   (a) Find $f'(x)$.
   (b) Find $f'(-1)$, $f'(0)$, $f'(1)$, and $f'(2)$.
   (c) Sketch a graph of $f(x)$. Consider the slope of a tangent line to the graph of this function at each of the points $x = -1$, $x = 0$, $x = 1$, and $x = 2$. Do these slopes appear to match your answers to part (b)? Explain.
2. If $f(t) = 2t^3 - 4t^2 + 3t - 1$, find $f'(t)$ and $f''(t)$.
3. (a) Find the *eighth* derivative of $f(x) = x^7 + 5x^5 - 4x^3 + 6x - 7$. Think ahead! (The nth derivative is the result of differentiating n times.)
   (b) Find the seventh derivative of $f(x)$.
4. Find the equation of the line tangent to the graph of f at $(1, 1)$, where f is given by $f(x) = 2x^3 - 2x^2 + 1$.
5. Find the equation of the line tangent to the graph of $f(x) = 2x^3 - 5x^2 + 3x - 5$ at the point where $x = 1$.
6. A manufacturer has cost function $C(q) = 1000 + 2q^2$ where q is the quantity produced. Find the marginal cost of producing the 25th item, and interpret your answer in terms of costs of production.
7. The demand function for a certain product is given by $q = 300 - 3p$, where p is the price of the product and q is the quantity consumers will buy at that price. Recall that revenue equals price times quantity sold.
   (a) Write the revenue function as a function of price.
   (b) Find the marginal revenue when the price of the product is \$10, and interpret your answer in terms of revenues.
   (c) For what prices is the marginal revenue positive? For what prices is it negative?
8. Zebra mussels are freshwater shellfish that attach themselves to anything they can find. They first appeared in the St. Lawrence River in the early 1980s. They are moving upriver, and may spread throughout the Great Lakes. Suppose that in one small bay, the number of zebra mussels at time t is given by $Z(t) = 300t^2$, where t is measured in months since the zebra mussel first appeared at that location. How many zebra mussels are in the bay after four months, and at what rate is the population growing? Give units with your answers.

9. The yield, Y, of an apple orchard (measured in bushels of apples per acre) is a function of the amount x of fertilizer in pounds used per acre. Suppose

$$Y = f(x) = 320 + 140x - 10x^2$$

   (a) What is the yield if 5 pounds of fertilizer is used per acre?
   (b) Find $f'(5)$. Give units with your answer and interpret it in terms of apples and fertilizer.
   (c) Given your answer to part (b), should more or less fertilizer be used? Explain.

10. The graph of the equation $y = x^3 - 9x^2 - 16x + 1$ has a slope equal to 5 at exactly two points. Find the coordinates of the points.

11. Let $f(x) = x^3 - 12x$.
   (a) Use $f'(x)$ to determine the intervals on which $f(x)$ is decreasing.
   (b) Use $f''(x)$ to determine the intervals on which $f(x)$ is concave down.
   (c) On what intervals is $f(x)$ both decreasing and concave down?

12. On what intervals is the function $f(x) = x^4 - 4x^3$ both decreasing and concave up?

13. Given $p(x) = x^n - x$, find the intervals over which p is a decreasing function when: (a) $n = 2$ (b) $n = \frac{1}{2}$ (c) $n = -1$

14. Using a graph to help you, find the equations of all lines through the origin tangent to the parabola

$$y = x^2 - 2x + 4.$$

   Sketch your solutions.

15. If the demand equation is linear, we can write $p = b + mq$, where p is the price of the product, q is the quantity sold at that price, and b and m are constants.
   (a) Write the revenue function as a function of quantity sold.
   (b) Find the marginal revenue function.

16. A ball is dropped from the top of the Empire State building to the ground below. The height, y, of the ball above the ground (in feet) is given as a function of time, t, (in seconds) by

$$y = 1250 - 16t^2.$$

   (a) Find the velocity of the ball at time t. What is the sign of the velocity? Why is this to be expected?
   (b) Show that the acceleration of the ball is a constant. What are the value and sign of this constant?
   (c) When does the ball hit the ground, and how fast is it going at that time? Give your answer in feet per second and in miles per hour (1 ft/sec = 15/22 mph).

17. (a) Use the formula for the area of a circle of radius r, $A = \pi r^2$, to find $\dfrac{dA}{dr}$.
   (b) The result from part (a) should look familiar. What does $\dfrac{dA}{dr}$ represent geometrically? Draw a picture.
   (c) Use the difference quotient to explain the observation you made in part (b).

18. What is the formula for $V(r)$, the volume of a sphere of radius r? Find $\dfrac{dV}{dr}$. What is the geometrical meaning of $\dfrac{dV}{dr}$?

4.3 EXPONENTIAL, LOGARITHMIC, AND PERIODIC FUNCTIONS

The Exponential Function

What would you expect the graph of the derivative of the exponential function $f(x) = a^x$ to look like? The graph of the exponential function is shown in Figure 4.16. The function increases slowly when $x < 0$ and more rapidly for $x > 0$, so the values of f' are small for $x < 0$ and larger for $x > 0$. Since the function is increasing for all real values of x, the graph of the derivative must lie above the x-axis. After some reflection, you may decide that the graph of f' must resemble the graph of f itself. We will see how this observation holds for $f(x) = 2^x$ and $g(x) = 3^x$.

The Derivatives of 2^x and 3^x

In Chapter 2, we saw that the derivative of $f(x) = 2^x$ at $x = 0$ is given by

$$f'(0) = \lim_{h\to 0} \frac{2^h - 2^0}{h} = \lim_{h\to 0} \frac{2^h - 1}{h} \approx 0.6931,$$

where we found the limit by evaluating $(2^h - 1)/h$ for small values of h. We can use this fact to find the derivative of $f(x) = 2^x$. Using the definition of derivative, we have

$$f'(x) = \lim_{h\to 0} \left(\frac{2^{x+h} - 2^x}{h}\right) = \lim_{h\to 0} \left(\frac{2^x 2^h - 2^x}{h}\right) = \lim_{h\to 0} 2^x \left(\frac{2^h - 1}{h}\right)$$

$$= 2^x \lim_{h\to 0} \left(\frac{2^h - 1}{h}\right). \qquad \text{Since } x \text{ and } 2^x \text{ are fixed during this calculation}$$

We have already shown $\lim_{h\to 0} \frac{2^h - 1}{h} \approx 0.6931$, so we have shown that

$$\frac{d}{dx}(2^x) = f'(x) \approx (0.6931)2^x.$$

The graphs of $f(x) = 2^x$ and $f'(x) \approx (0.6931)2^x$ are shown in Figure 4.17. Notice that they do indeed resemble one another.

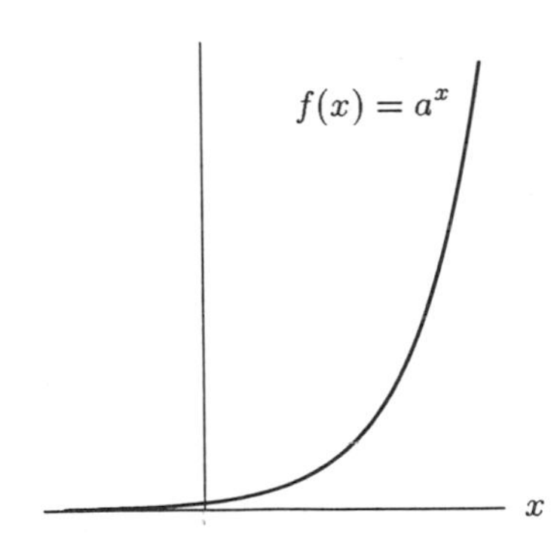

Figure 4.16: $f(x) = a^x$, with $a > 1$

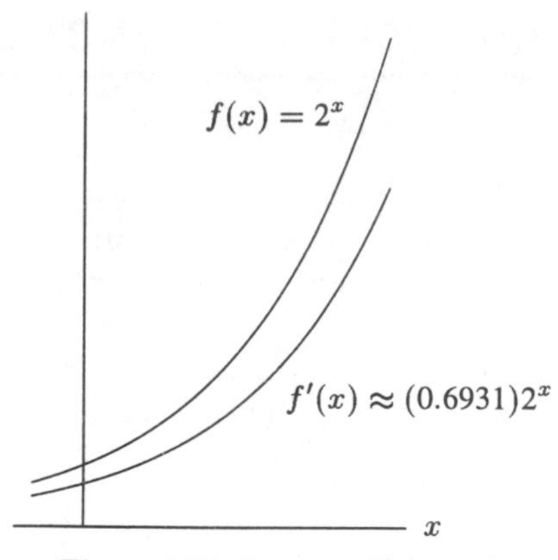

Figure 4.17: Graph of $f(x) = 2^x$ and its derivative

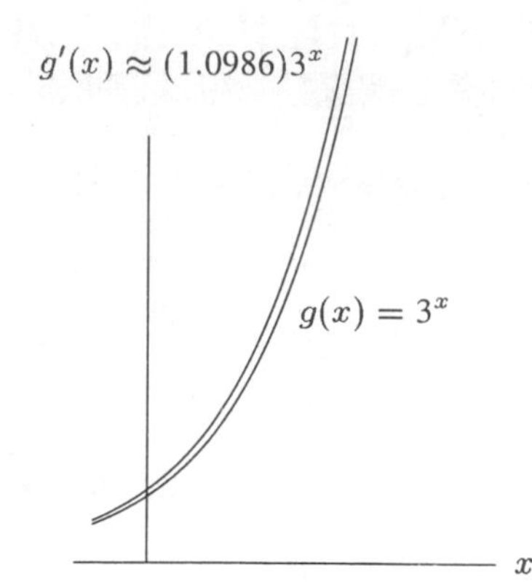

Figure 4.18: Graph of $g(x) = 3^x$ and its derivative

Example 1 Find the derivative of $g(x) = 3^x$ and plot g and g' on the same axes.

Solution As before

$$g'(x) = \lim_{h\to 0} \frac{3^{x+h} - 3^x}{h} = \lim_{h\to 0} \frac{3^x 3^h - 3^x}{h} = \lim_{h\to 0} 3^x \left(\frac{3^h - 1}{h}\right).$$

Using a calculator gives

$$\lim_{h\to 0} \left(\frac{3^h - 1}{h}\right) \approx 1.0986,$$

so

$$g'(x) \approx (1.0986)3^x.$$

The graphs are shown in Figure 4.18.

der - proportional*

Notice that for both $f(x) = 2^x$ and $g(x) = 3^x$, *the derivative is proportional to the original function.* For $f(x) = 2^x$, $f'(x) \approx (0.6931)2^x$, so the constant of proportionality is less than 1, and the graph of the derivative is below the graph of the original function. For $g(x) = 3^x$, $g'(x) \approx (1.0986)3^x$ and the constant is greater than 1, so the graph of the derivative is above that of the original function.

The Derivative of e^x

We have seen that the derivatives of 2^x and 3^x are proportional to the original function. Is there an exponential function which is its own derivative? In Figure 4.17, we see that the graph of the derivative of $f(x) = 2^x$ lies below the graph of $f(x)$. In Figure 4.18, we see that the graph of the derivative of $f(x) = 3^x$ lies above the graph of $f(x)$. Therefore, if there is an exponential function for which the derivative exactly matches the original function, we would expect the base to be between 2 and 3. If you experiment with different bases, you will find that a base, a, of approximately 2.7 gives an exponential function a^x whose derivative almost exactly matches the original function, a^x. In fact, the base $e = 2.718\ldots$ has the remarkable property that its exponential function is its own derivative.

$$\frac{d}{dx}(e^x) = e^x$$

The Derivative of a^x, where a is a constant

It turns out that the constants involved in the derivatives of 2^x and 3^x are natural logarithms. In fact, since $0.6931 \approx \ln 2$ and $1.0986 \approx \ln 3$, this suggests that:

$$\frac{d}{dx}(2^x) = (\ln 2)2^x \qquad \text{and} \qquad \frac{d}{dx}(3^x) = (\ln 3)3^x.$$

In Section 4.4, we will show that, in general,

$$\boxed{\frac{d}{dx}(a^x) = (\ln a)a^x.}$$

Figure 4.19 shows the graph of the derivative of 2^x below the graph of the function, and the graph of the derivative of 3^x above the graph of the function. With $e \approx 2.718$, the function e^x and its derivative are identical.

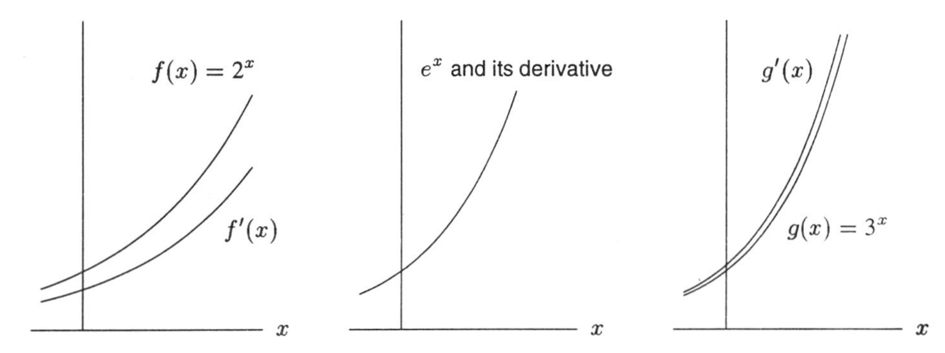

Figure 4.19: Graphs of the functions 2^x, e^x, and 3^x and their derivatives

---

Example 2 Differentiate $2 \cdot 3^x + 5e^x$.

Solution

$$\begin{aligned}
\frac{d}{dx}(2 \cdot 3^x + 5e^x) &= \frac{d}{dx}(2 \cdot 3^x) + \frac{d}{dx}(5e^x) \\
&= 2\frac{d}{dx}(3^x) + 5\frac{d}{dx}(e^x) \\
&\approx 2(1.0986)3^x + 5e^x \\
&= (2.1972)3^x + 5e^x.
\end{aligned}$$

The Derivative of $\ln x$

What would you expect the graph of the derivative of the logarithmic function $f(x) = \ln x$ to look like? A graph of the logarithmic function is shown in Figure 4.20. (Recall that the domain of the natural logarithm function is $x > 0$.) The function is increasing everywhere, so we expect the derivative to be positive for all $x > 0$. The logarithmic function is concave down everywhere, and so we expect the derivative to be decreasing for all $x > 0$. What else can we learn? The slope of the logarithmic function is very large near $x = 0$, and is very small for large x, so we expect the graph of the derivative to tend to $+\infty$ for x near 0 and to tend to 0 for very large x. A graph of the derivative of $f(x) = \ln x$ is shown in Figure 4.21.

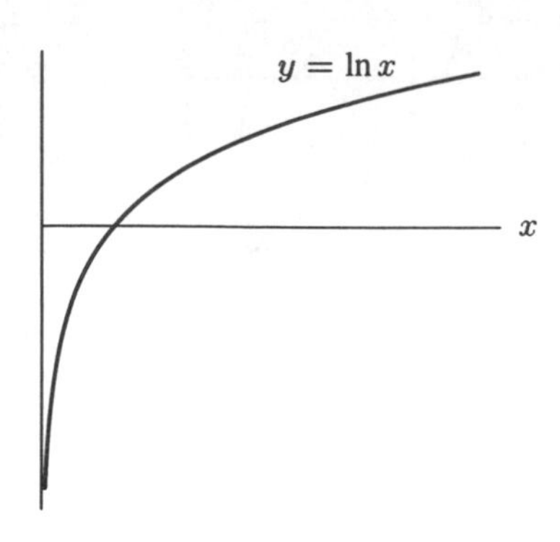

Figure 4.20: A graph of $y = \ln x$.

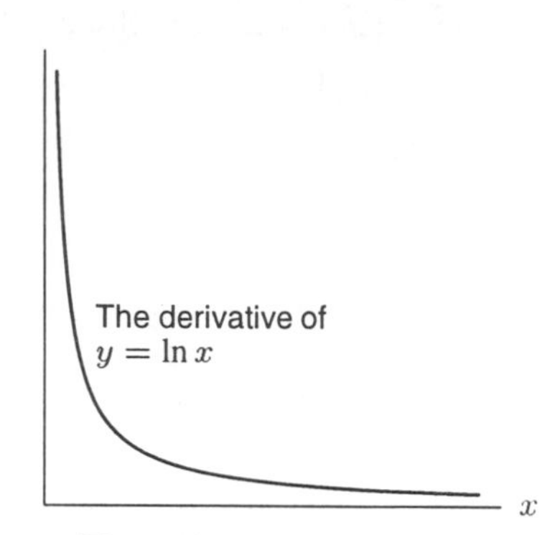

Figure 4.21: The derivative of $y = \ln x$.

The graph in Figure 4.21 should remind you of the function $y = 1/x$, since this is exactly what $f'(x)$ turns out to be. We will see an algebraic justification for this rule in Section 4.5.

$$\boxed{\frac{d}{dx}(\ln x) = \frac{1}{x}.}$$

Example 3 Differentiate $y = 5\ln t + 7e^t - 4t^2 + 12$.

Solution

$$\begin{aligned}\frac{d}{dt}(5\ln t + 7e^t - 4t^2 + 12) &= 5\frac{d}{dt}(\ln t) + 7\frac{d}{dt}(e^t) - 4\frac{d}{dt}(t^2) + \frac{d}{dt}(12)\\ &= 5\left(\frac{1}{t}\right) + 7(e^t) - 4(2t) + 0\\ &= \frac{5}{t} + 7e^t - 8t\end{aligned}$$

The Derivatives of Sine and Cosine

Since the sine and cosine functions are periodic, their derivatives must be periodic also. (Why?) Let's look at the graph of $f(x) = \sin x$ in Figure 4.22 and estimate the derivative function graphically.

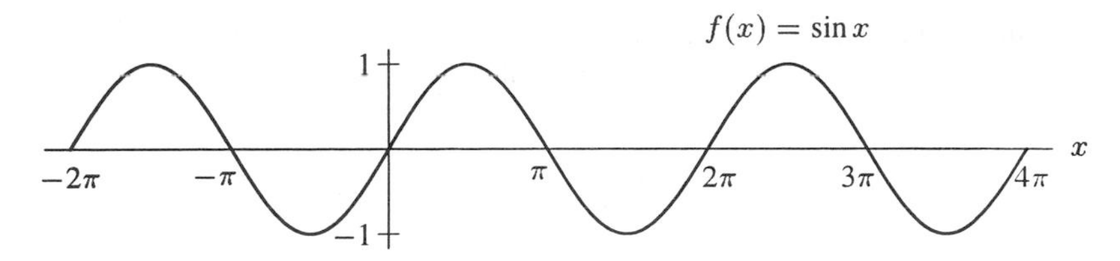

Figure 4.22: The sine function

First you might ask yourself where the derivative is zero. (At $x = \pm\pi/2$, $\pm 3\pi/2$, $\pm 5\pi/2$, etc.) Then ask yourself where the derivative is positive and where it is negative. (Positive for $-\pi/2 < x < \pi/2$; negative for $\pi/2 < x < 3\pi/2$, etc.) Since the largest positive slopes are at $x = 0, 2\pi$, and so on, and the largest negative slopes are at $x = \pi, 3\pi$, and so on, you get something like the graph in Figure 4.23.

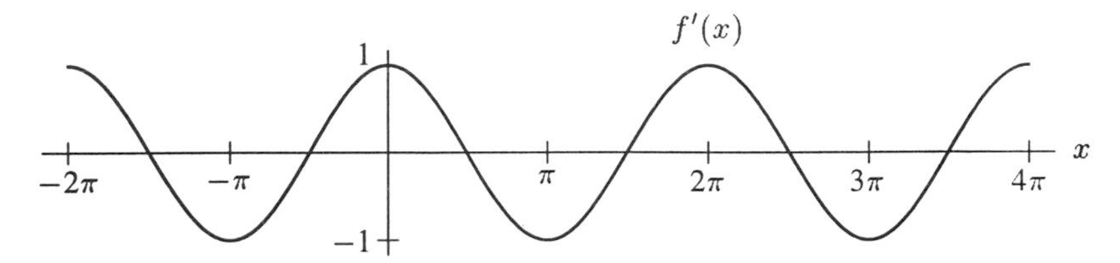

Figure 4.23: Derivative of $f(x) = \sin x$

The graph of the derivative looks suspiciously like the graph of the cosine function and this might lead you to conjecture, quite correctly, that the derivative of the sine is the cosine.

Of course, we cannot be sure, just from the graphs, that the derivative of the sine really is the cosine but it turns out that this is true. One thing we can do now is to check that the derivative function in Figure 4.23 has amplitude 1 (as it ought to if it is the cosine). That means we have to show that the derivative of $f(x) = \sin x$ is 1 when $x = 0$. The next example shows this is true when x is in radians.

Example 4 Using a calculator, estimate the derivative of $f(x) = \sin x$ at $x = 0$. Make sure your calculator is set in radians.

Solution Since $f(x) = \sin x$,

$$f'(0) = \lim_{h \to 0} \frac{\sin(0+h) - \sin 0}{h} = \lim_{h \to 0} \frac{\sin h}{h}.$$

Table 4.2 contains values of $(\sin h)/h$ which suggest that this limit is 1, so

$$f'(0) = \lim_{h \to 0} \frac{\sin h}{h} = 1.$$

TABLE 4.2

h (radians)	$\frac{\sin h}{h}$
± 0.1	0.99833
± 0.01	0.99998
± 0.001	1.00000
± 0.0001	1.00000

Warning: It is important to notice that in the previous example h was in *radians*; any conclusions we have drawn about the derivative of $\sin x$ are valid *only* when x is in radians.

Example 5 Starting with the graph of the cosine function, sketch a graph of its derivative.

Solution The graph of $g(x) = \cos x$ is in Figure 4.24(a). Its derivative is 0 at $x = 0, \pm\pi, \pm 2\pi$, and so on; it is positive for $-\pi < x < 0$, $\pi < x < 2\pi$, and so on, and it is negative for $0 < x < \pi$, $2\pi < x < 3\pi$, and so on. The derivative is in Figure 4.24(b).

(a)

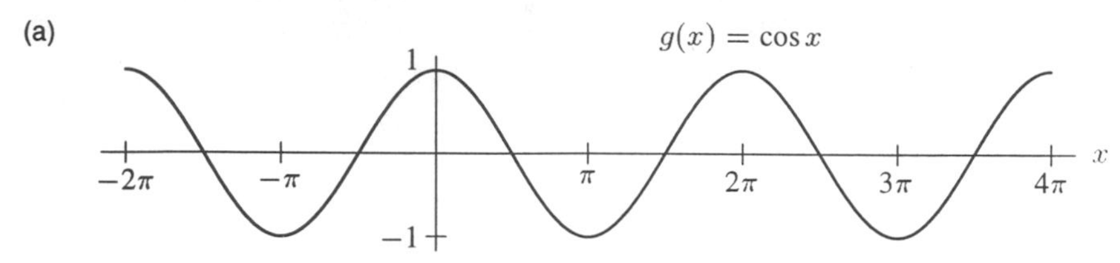

(b)

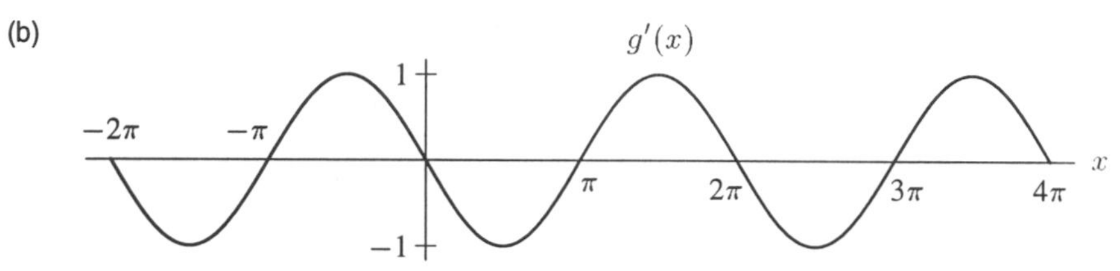

Figure 4.24: $g(x) = \cos x$ and its derivative, $g'(x)$

As we did with the sine, we'll use the graphs to make a conjecture. The derivative of the cosine in Figure 4.24(b) looks exactly like the graph of sine, except reflected about the x-axis. It turns out that the derivative of $\cos x$ is $-\sin x$.

> For x in radians,
>
> $$\frac{d}{dx}(\sin x) = \cos x \quad \text{and} \quad \frac{d}{dx}(\cos x) = -\sin x.$$

Example 6 Differentiate each of the following:

(a) $5\sin t - 8\cos t$ (b) $5 - 3\sin x + x^3$

Solution (a)

$$\begin{aligned}\frac{d}{dt}(5\sin t - 8\cos t) &= 5\frac{d}{dt}(\sin t) - 8\frac{d}{dt}(\cos t)\\ &= 5(\cos t) - 8(-\sin t)\\ &= 5\cos t + 8\sin t\end{aligned}$$

(b)

$$\begin{aligned}\frac{d}{dx}(5 - 3\sin x + x^3) &= \frac{d}{dx}(5) - 3\frac{d}{dx}(\sin x) + \frac{d}{dx}(x^3)\\ &= 0 - 3(\cos x) + 3x^2\\ &= -3\cos x + 3x^2.\end{aligned}$$

Problems for Section 4.3

Differentiate the functions in Problems 1–21. You may assume that A, B, and C, where they appear, are constants.

1. $y = 5t^2 + 4e^t$
2. $f(x) = 2e^x + x^2$
3. $f(x) = 2^x + 2 \cdot 3^x$
4. $y = 4 \cdot 10^x - x^3$
5. $y = 3x - 2 \cdot 4^x$
6. $y = \dfrac{3^x}{3} + \dfrac{33}{\sqrt{x}}$
7. $f(x) = x^3 + 3^x$
8. $y = 5 \cdot 5^t + 6 \cdot 6^t$
9. $P(t) = Ce^t$.
10. $D = 10 - \ln p$
11. $R = 3\ln q + \sin q$
12. $y = t^2 + 5\cos t$
13. $y = B + A\sin t$
14. $f(x) = Ae^x - Bx^2 + C$
15. $P = 3\sin t + 2\cos t$
16. $P(t) = 3000(1.02)^t$
17. $P(t) = 12.41(0.94)^t$
18. $y = 5\sin x - 5x + 4$
19. $R(q) = q^2 - 2\cos q$
20. $y = x^2 + 4x - 3\ln x$
21. $f(t) = A\sqrt{t} - B\sin t$
22. Let $f(t) = 4 - 2e^t$. Find $f'(-1)$, $f'(0)$, and $f'(1)$. Sketch a graph of $f(t)$, and draw tangent line segments at $t = -1, t = 0$, and $t = 1$. Do the slopes of the line segments appear to match the derivatives you found?
23. Let $y = \sin x$. Find the equation of the tangent line to this function at the point where $x = \pi$. Check your work by sketching a graph of the function and the tangent line on the same coordinate system.
24. Suppose the cost function for a certain product is given by $C = 1000 + 300\ln q$ where q is the quantity produced. Find the cost and the marginal cost at a production level of 500. Interpret your answers in economic terms.
25. (a) Find the slope of the graph of $f(x) = 1 - e^x$ at the point where it crosses the x-axis.
    (b) Find the equation of the tangent line to the curve at this point.
    (c) Find the equation of the line which is perpendicular to the tangent line at this point. (This line is known as the *normal* line.)

26. With an inflation rate of 5%, prices are described by

$$P = P_0(1.05)^t$$

where P_0 is the price in dollars when $t = 0$ and t is time in years. Suppose $P_0 = 1$. How fast (in cents/year) is the price of the good rising when $t = 10$?

27. Since January 1, 1960, the population of Slim Chance has been described by the formula

$$P = 35{,}000(0.98)^t$$

where P is the population of the city t years after 1960. At what rate was the population of the city changing on January 1, 1983?

28. Certain pieces of antique furniture increased very rapidly in price in the 1970s and 1980s. For example, the value of a particular rocking chair is well approximated by

$$V = 75(1.35)^t,$$

where V is in dollars and t is the number of years since 1975. Find the rate, in dollars per year, that the price is increasing as a function of time.

29. The *Global 2000 Report* gave the world's population, P, as 4.1 billion in 1975 and growing at 2% annually.
    (a) Give a formula for P in terms of time, t, measured in years since 1975.
    (b) Find $\frac{dP}{dt}$, $\left.\frac{dP}{dt}\right|_{t=0}$, and $\left.\frac{dP}{dt}\right|_{t=15}$. What do each of these represent in practical terms?

30. Hungary is one of the few countries in the world where the population is decreasing, currently at about 0.2% a year. Thus, if t is time in years since 1990, the population P, in millions, of Hungary can be approximated by

$$P = 10.8(0.998)^t.$$

    (a) What does this model predict the population of Hungary will be in the year 2000?
    (b) How fast (in people/year) does this model predict Hungary's population will be decreasing in the year 2000?

31. Using the tangent line to e^x at $x = 0$, show that

$$e^x \geq 1 + x$$

for all values of x. A sketch will be helpful.

32. (a) Find the equation of the tangent line to $y = \ln x$ at $x = 1$.
    (b) Use it to calculate approximate values for $\ln(1.1)$ and $\ln(2)$.
    (c) Using a graph, explain whether the approximate values you have calculated are smaller or larger than the true values. Would the same result have held if you had used the tangent line to estimate $\ln(0.9)$ and $\ln(0.5)$? Why?

33. For $f(x) = \sin x$, find the equations of the tangent lines at $x = 0$ and at $x = \pi/3$. Use each tangent line to approximate $\sin \pi/6$. Would you expect these results to be equally accurate, since they are taken equally far away from $x = \pi/6$ but on opposite sides? If the accuracy is different, can you account for the difference?

34. In this section, we stated that, for $a > 0$

$$\frac{d}{dx}(a^x) = (\ln a)a^x.$$

Use this expression for the derivative to explain for what values of a the function a^x is increasing and for what values it is decreasing.

35. Find all solutions of the equation

$$2^x = 2x.$$

How do you know that you found all solutions?

36. Find the value of c in Figure 4.25, where the line l tangent to the graph of $y = 2^x$ at $(0, 1)$ intersects the x-axis.

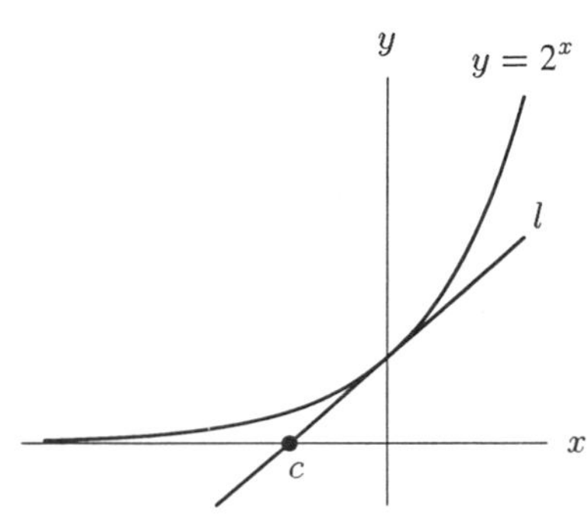

Figure 4.25

37. Find the quadratic function $g(x) = ax^2 + bx + c$ which best fits the function $f(x) = e^x$ at $x = 0$, in the sense that

$$g(0) = f(0), \quad \text{and} \quad g'(0) = f'(0), \quad \text{and} \quad g''(0) = f''(0).$$

Using a computer or calculator, sketch graphs of f and g on the same axes. What do you notice?

4.4 THE CHAIN RULE

Composite functions such as $\ln(3t)$ or e^{-x^2} occur frequently in practice. In this section we will see how to differentiate such functions.

The Derivative of a Composition of Functions

Suppose $f(g(x))$ is a composite function, with f being the outside function and g being the inside. Let us write

$$z = g(x) \quad \text{and} \quad y = f(z), \quad \text{so} \quad y = f(g(x)).$$

Then a small change in x, called Δx, generates a small change in z, called Δz. In turn, Δz generates a small change in y called Δy. Provided Δx and Δz are not zero, we can say:

$$\frac{\Delta y}{\Delta x} = \frac{\Delta y}{\Delta z} \cdot \frac{\Delta z}{\Delta x}.$$

Since $\frac{dy}{dx} = \lim_{\Delta x \to 0} \frac{\Delta y}{\Delta x}$, this suggests that in the limit as Δx, Δy, and Δz get smaller and smaller, we have:

The Chain Rule

$$\frac{dy}{dx} = \frac{dy}{dz} \cdot \frac{dz}{dx}.$$

Since $\dfrac{dy}{dz} = f'(z)$ and $\dfrac{dz}{dx} = g'(x)$, this suggests that

$$\frac{d}{dx} f(g(x)) = f'(z) \cdot g'(x)$$

or, substituting $z = g(x)$, we have the following version:

The Chain Rule

$$\frac{d}{dx} f(g(x)) = f'(g(x)) \cdot g'(x).$$

In words:

The derivative of a composite function is the product of the derivatives of the outside and inside functions. The derivative of the outside function must be evaluated at the inside function.

Example 1 Suppose the length, L cm, of a steel bar depends on the air temperature, H°C, which itself depends on time t, measured in hours. If the length increases by 2 cm for every degree increase in temperature and the temperature is increasing at 3°C per hour, how fast is the length of the bar increasing?

Solution We are told that

$$\text{Rate length increasing with respect to temperature} = \frac{dL}{dH} = 2 \text{ cm/°C}$$

$$\text{Rate temperature increasing with respect to time} = \frac{dH}{dt} = 3\text{°C/hr}.$$

We want to calculate the rate at which the length is increasing with respect to time, or dL/dt. We think of L as a function of H and H as a function of t. By the chain rule we know that

$$\frac{dL}{dt} = \frac{dL}{dH} \cdot \frac{dH}{dt} = \left(2\frac{\text{cm}}{\text{°C}}\right) \cdot \left(3\frac{\text{°C}}{\text{hr}}\right) = 6 \text{ cm/hr}.$$

Thus, the length is increasing at 6 cm/hr.

Example 1 above shows us how to interpret the chain rule in practical terms. The next examples show how it is used to compute derivatives.

Example 2 Find the derivative of the following functions: (a) $(4x^2+1)^7$, (b) e^{3x}.

Solution (a) Here $z = g(x) = 4x^2 + 1$ is the inside function; $f(z) = z^7$ is the outside function. Since $g'(x) = 8x$ and $f'(z) = 7z^6$, we have

$$\frac{d}{dx}\left[(4x^2+1)^7\right] = 7z^6 \cdot 8x = 7(4x^2+1)^6 \cdot 8x = 56x(4x^2+1)^6.$$

(b) Let $z = g(x) = 3x$ and $f(z) = e^z$. Then $g'(x) = 3$ and $f'(z) = e^z$, so

$$\frac{d}{dx}\left(e^{3x}\right) = e^z \cdot 3 = 3e^{3x}.$$

Example 3 Differentiate (a) $(x^2+1)^{100}$, (b) $\sqrt{3x^2+5x-2}$, (c) $\sqrt{e^x+1}$, (d) e^{x^2}.

Solution (a) Here $z = g(x) = x^2 + 1$ is the inside function; $f(z) = z^{100}$ is the outside function. Now $g'(x) = 2x$ and $f'(z) = 100z^{99}$, so

$$\frac{d[(x^2+1)^{100}]}{dx} = 100z^{99} \cdot 2x = 100(x^2+1)^{99} \cdot 2x = 200x(x^2+1)^{99}.$$

(b) Here $z = g(x) = 3x^2 + 5x - 2$ and $f(z) = \sqrt{z}$, so $g'(x) = 6x + 5$ and $f'(z) = \dfrac{1}{2\sqrt{z}}$. Hence

$$\frac{d(\sqrt{3x^2+5x-2})}{dx} = \frac{1}{2\sqrt{z}} \cdot (6x+5) = \frac{1}{2\sqrt{3x^2+5x-2}} \cdot (6x+5).$$

(c) Let $z = g(x) = e^x + 1$ and $f(z) = \sqrt{z}$. Therefore $g'(x) = e^x$ and $f'(z) = \dfrac{1}{2\sqrt{z}}$. We get

$$\frac{d(\sqrt{e^x+1})}{dx} = \frac{1}{2\sqrt{z}} e^x = \frac{e^x}{2\sqrt{e^x+1}}.$$

(d) To evaluate e^{x^2} you must first evaluate x^2 and then take e to that power. So $z = g(x) = x^2$ and $f(z) = e^z$. Therefore, $g'(x) = 2x$, and $f'(z) = e^z$, giving

$$\frac{d(e^{x^2})}{dx} = e^z \cdot 2x = e^{x^2} \cdot 2x = 2xe^{x^2}.$$

Using the Chain Rule to Establish Derivative Formulas

We will use the chain rule to verify the formulas for derivatives of logarithms and derivatives of exponentials with arbitrary base a.

Derivative of $\ln x$

We'll use the chain rule, and differentiate an identity which involves $\ln x$. Since $e^{\ln x} = x$, using the chain rule gives:

$$\frac{d}{dx}(e^{\ln x}) = \frac{d}{dx}(x),$$

$$e^{\ln x} \cdot \frac{d}{dx}(\ln x) = 1. \qquad \text{Since } e^x \text{ is outside function and } \ln x \text{ is inside function.}$$

Thus,

$$\frac{d}{dx}(\ln x) = \frac{1}{e^{\ln x}} = \frac{1}{x},$$

so

$$\boxed{\frac{d(\ln x)}{dx} = \frac{1}{x}.}$$

Derivative of a^x

Earlier we showed that the derivative of a^x is proportional to a^x. Now we show that the constant of proportionality is $\ln a$. We will use the identity

$$\ln(a^x) = x \ln a.$$

Differentiating, using $\frac{d}{dx}(\ln x) = \frac{1}{x}$ and the chain rule on the left and remembering that $\ln a$ is a constant, we obtain:

$$\frac{d}{dx}(\ln a^x) = \frac{1}{a^x} \cdot \frac{d}{dx}(a^x) = \ln a.$$

Solving gives the result we expected in Section 4.3:

$$\boxed{\frac{d(a^x)}{dx} = (\ln a)a^x.}$$

Problems for Section 4.4

Find the derivative of the functions in Problems 1–30.

1. $f(x) = (x+1)^{99}$
2. $f(x) = \sqrt{1-x^2}$
3. $w = (t^2+1)^{100}$
4. $w = (t^3+1)^{100}$
5. $w = (\sqrt{t}+1)^{100}$
6. $f(t) = e^{3t}$
7. $y = e^{3w/2}$
8. $y = e^{-4t}$
9. $y = \sqrt{s^3+1}$
10. $w = e^{\sqrt{s}}$

11. $P = e^{-0.2t}$

12. $w = e^{-3t^2}$

13. $y = \ln(5t + 1)$

14. $P = 50e^{-0.6t}$

15. $z = 2^{5t-3}$

16. $y = \sin(x^2)$

17. $y = 2\cos(5t)$

18. $f(x) = 6e^{5x} + e^{-x^2}$

19. $y = 6\sin(2t) + 3\cos(4t)$

20. $f(x) = \ln(1 - x)$

21. $f(t) = \ln(t^2 + 1)$

22. $f(x) = \ln(1 - e^{-x})$

23. $f(x) = \ln(e^x + 1)$

24. $f(t) = 5\ln(5t + 1)$

25. $f(x) = \sin(3x)$

26. $z = \cos(4\theta)$

27. $w = \sin(e^t)$

28. $f(x) = e^{\cos x}$

29. $y = (5 + e^x)^2$

30. $y = \sqrt{1 + \sin x}$

31. Find the equation of the tangent line to the graph of

$$y = e^{-2t}$$

at $t = 0$. Check your work by sketching the graphs of $y = e^{-2t}$ and the tangent line on the same coordinate system.

32. Find the equation of the tangent line at $x = 1$ to $y = f(x)$ where $f(x)$ is the function in Problem 18.

33. For what values of x is the graph of $y = e^{-x^2}$ concave down?

34. Assume that the demand function for a certain product is given by

$$q = f(p) = 10,000e^{-0.25p},$$

where q is the quantity sold and p is the price of the product, in dollars. Find $f(2)$ and $f'(2)$. Explain in economic terms what information each of these answers gives you.

35. Assume the cost function for a certain product is

$$C(q) = 1000 + 30e^{0.05q}$$

where q is the quantity produced. Find the cost and the marginal cost when $q = 50$. Explain in economic terms what information each of these answers gives you.

36. Given: $\left\{\begin{array}{ll} F(2) = 1 & G(4) = 2 \\ F(4) = 3 & G(3) = 4 \\ F'(2) = 5 & G'(4) = 6 \\ F'(4) - 7 & G'(3) - 8 \end{array}\right\}$ find: $\left\{\begin{array}{lll} \text{(a)} & H(4) & \text{if } H(x) = F(G(x)) \\ \text{(b)} & H'(4) & \text{if } H(x) = F(G(x)) \\ \text{(c)} & H(4) & \text{if } H(x) = G(F(x)) \\ \text{(d)} & H'(4) & \text{if } H(x) = G(F(x)) \\ \text{(e)} & H'(4) & \text{if } H(x) = F(x)/G(x) \end{array}\right.$

37. One gram of radioactive carbon-14 decays according to the formula

$$Q = e^{-0.000121t}$$

where Q is the number of grams of carbon-14 remaining after t years.

(a) Find the rate at which carbon-14 is decaying (in grams/year).
(b) Sketch the rate you found in part (a) against time.

38. The temperature, H, in degrees Fahrenheit (°F), of a can of soda that is put into a refrigerator to cool is given as a function of time, t, in hours, by

$$H = 40 + 30e^{-2t}.$$

(a) Find the rate at which the temperature of the soda is changing (in °F/hour).
(b) What is the sign of $\frac{dH}{dt}$? Why?
(c) When, for $t \geq 0$, is the magnitude of $\frac{dH}{dt}$ largest? In terms of the can of soda, why is this?

39. If you invest P dollars in a bank account at an annual interest rate of $r\%$, after t years you will have B dollars, where

$$B = P\left(1 + \frac{r}{100}\right)^t.$$

(a) Find $\frac{dB}{dt}$, assuming P and r are constant. In terms of money, what does $\frac{dB}{dt}$ represent?
(b) Find $\frac{dB}{dr}$, assuming P and t are constant. In terms of money, what does $\frac{dB}{dr}$ represent?

40. A boat at anchor is bobbing up and down in the sea. The vertical distance, y, in feet, between the sea floor and the boat is given as a function of time, t, in minutes, by

$$y = 15 + \sin 2\pi t.$$

(a) Find the vertical velocity, v, of the boat at time t.
(b) Make a rough sketch of y and v against t.

41. On page 101 the depth, y, of water in Boston harbor was given by

$$y = 5 + 4.9\cos\left(\frac{\pi}{6}t\right),$$

where t is the number of hours since midnight.

(a) Find $\frac{dy}{dt}$. What does $\frac{dy}{dt}$ represent, in terms of water level?
(b) For $0 \leq t \leq 24$, when is $\frac{dy}{dt}$ zero? (Figure 1.11 on page 102 may help.) Explain what it means (in terms of water level) for $\frac{dy}{dt}$ to be zero.

42. Imagine you are zooming in on the graph of each of the following functions near the origin:

$$y = x \qquad y = \sqrt{x} \qquad y = x^2 \qquad y = \sqrt{x/(x+1)}$$

$$y = x^3 \qquad y = \ln(x+1) \qquad y = \tfrac{1}{2}\ln(x^2+1) \qquad y = \sqrt{2x - x^2}$$

Which of them look the same? Group together those functions which become indistinguishable, and give the equations of the lines they look like.

4.5 THE PRODUCT RULE AND QUOTIENT RULES

You now know how to find derivatives of powers and exponentials, of sums and constant multiples of functions, and of compositions of functions. This section will show you how to find the derivatives of products and quotients.

The Product Rule

Suppose we know the derivatives of $f(x)$ and $g(x)$ and want to calculate the derivative of the product, $f(x)g(x)$. We start by looking at an example. Suppose $f(x) = 3x + 1$ and $g(x) = 5x - 4$. Then $f'(x) = 3$ and $g'(x) = 5$. We first find the product

$$\begin{aligned} f(x)g(x) &= (3x+1)(5x-4) \\ &= 15x^2 - 7x - 4. \end{aligned}$$

Thus, the derivative of the product is $30x - 7$. Notice that the derivative of the product is *not* equal to the product of the derivatives, since $f'(x)g'(x) = (3)(5) = 15$. In general, we have the following rule.

> **The Product Rule**
>
> $$(fg)' = f'g + fg'.$$
>
> In words:
>
> The derivative of a product is the derivative of the first factor multiplied by the second, plus the first factor multiplied by the derivative of the second.

In the alternative notation, if $u = f(x)$ and $v = g(x)$, we have the following:

> **The Product Rule**
>
> $$\frac{d(uv)}{dx} = \frac{du}{dx} \cdot v + u \cdot \frac{dv}{dx}$$

We verify this rule for the example above with $f(x) = 3x + 1$ and $g(x) = 5x - 4$. Using the product rule, we see that the derivative of $f(x)g(x)$ is

$$\begin{aligned} f'(x)g(x) + f(x)g'(x) &= (3)(5x-4) + (3x+1)(5) \\ &= (15x - 12) + (15x + 5) \\ &= 30x - 7. \end{aligned}$$

We see that the formula for the product rule gives us the same answer as first multiplying out the functions and then finding the derivative.

Example 1 Differentiate (a) x^2e^x, (b) $(3x^2 + 5x)e^x$, (c) $\dfrac{e^x}{x^2}$.

Solution (a)

$$\frac{d(x^2e^x)}{dx} = \left(\frac{d(x^2)}{dx}\right) e^x + x^2\frac{d(e^x)}{dx} = 2xe^x + x^2e^x = (2x + x^2)e^x.$$

(b)

$$\begin{aligned}\frac{d((3x^2+5x)e^x)}{dx} &= \left(\frac{d(3x^2+5x)}{dx}\right)e^x + (3x^2+5x)\frac{d(e^x)}{dx}\\ &= (6x+5)e^x + (3x^2+5x)e^x = (3x^2+11x+5)e^x.\end{aligned}$$

(c) First we must write $\frac{e^x}{x^2}$ as the product $x^{-2}e^x$:

$$\begin{aligned}\frac{d}{dx}\left(\frac{e^x}{x^2}\right) = \frac{d(x^{-2}e^x)}{dx} &= \left(\frac{d(x^{-2})}{dx}\right)e^x + x^{-2}\frac{d(e^x)}{dx}\\ &= -2x^{-3}e^x + x^{-2}e^x = (-2x^{-3}+x^{-2})e^x.\end{aligned}$$

Example 2 Differentiate

(a) $5xe^{x^2}$ (b) $x^2\sin(3x)$ (c) $x\ln x$

Solution (a)

$$\begin{aligned}\frac{d}{dx}(5xe^{x^2}) = \left(\frac{d}{dx}(5x)\right)e^{x^2} + 5x\frac{d(e^{x^2})}{dx} &= (5)e^{x^2} + 5x(e^{x^2}\cdot 2x)\\ &= 5e^{x^2} + 10x^2e^{x^2}.\end{aligned}$$

(b)

$$\begin{aligned}\frac{d}{dx}(x^2\sin 3x) = \left(\frac{d(x^2)}{dx}\right)\sin 3x + x^2\frac{d(\sin 3x)}{dx} &= (2x)\sin 3x + x^2(\cos 3x)\cdot(3)\\ &= 2x\sin 3x + 3x^2\cos 3x.\end{aligned}$$

(c)

$$\begin{aligned}\frac{d(x\ln x)}{dx} = \left(\frac{d(x)}{dx}\right)\ln x + x\frac{d(\ln x)}{dx} &= 1\cdot\ln x + x\cdot\frac{1}{x}\\ &= \ln x + 1.\end{aligned}$$

The Quotient Rule

Suppose we want to differentiate a function of the form $Q(x) = \frac{f(x)}{g(x)}$. We'll find a formula for Q' in terms of f' and g'.

Since $f(x) = Q(x)g(x)$, we can use the product rule:

$$\begin{aligned}f'(x) &= Q'(x)g(x) + Q(x)g'(x)\\ &= Q'(x)g(x) + \frac{f(x)}{g(x)}g'(x).\end{aligned}$$

Solving for $Q'(x)$:

$$Q'(x) = \frac{f'(x) - \dfrac{f(x)}{g(x)}g'(x)}{g(x)}.$$

Multiplying the top and bottom by $g(x)$ to simplify gives

$$\left[\frac{f(x)}{g(x)}\right]' = \frac{f'(x)g(x) - f(x)g'(x)}{(g(x))^2}$$

So we have the following rule:

The Quotient Rule

$$\left(\frac{f}{g}\right)' = \frac{f'g - fg'}{g^2}.$$

In words:

The derivative of a quotient is the derivative of the numerator times the denominator minus the numerator times the derivative of the denominator all over the denominator squared.

Alternatively, if $u = f(x)$ and $v = g(x)$, this can be written as follows:

The Quotient Rule

$$\frac{d}{dx}\left(\frac{u}{v}\right) = \frac{\dfrac{du}{dx} \cdot v - u \cdot \dfrac{dv}{dx}}{v^2}.$$

Example 3 Differentiate (a) $\dfrac{5x^2}{x^3+1}$, (b) $\dfrac{1}{1+e^x}$, (c) $\dfrac{e^x}{x^2}$.

Solution (a)

$$\begin{aligned}\frac{d}{dx}\left(\frac{5x^2}{x^3+1}\right) &= \frac{\left(\dfrac{d}{dx}(5x^2)\right)(x^3+1) - 5x^2\dfrac{d}{dx}(x^3+1)}{(x^3+1)^2} \\ &= \frac{10x(x^3+1) - 5x^2(3x^2)}{(x^3+1)^2} \\ &= \frac{-5x^4+10x}{(x^3+1)^2}.\end{aligned}$$

(b)

$$\begin{aligned}
\frac{d}{dx}\left(\frac{1}{1+e^x}\right) &= \frac{\left(\frac{d}{dx}(1)\right)(1+e^x) - 1\frac{d}{dx}(1+e^x)}{(1+e^x)^2} \\
&= \frac{0(1+e^x) - 1(0+e^x)}{(1+e^x)^2} \\
&= \frac{-e^x}{(1+e^x)^2}.
\end{aligned}$$

(c) This is the same as part (c) of Example 1, but this time we will do it by the quotient rule.

$$\begin{aligned}
\frac{d}{dx}\left(\frac{e^x}{x^2}\right) &= \frac{\left(\frac{d(e^x)}{dx}\right)x^2 - e^x\left(\frac{d(x^2)}{dx}\right)}{(x^2)^2} = \frac{e^x x^2 - e^x(2x)}{x^4} \\
&= e^x\left(\frac{x^2-2x}{x^4}\right) = e^x\left(\frac{x-2}{x^3}\right).
\end{aligned}$$

This is, in fact, the same answer as before, although it looks different. Can you show that it is the same?

Problems for Section 4.5

1. If $f(x) = x^2(x^3+5)$, find $f'(x)$ two ways: by using the product rule and by multiplying out. Do you get the same result? Should you?
2. If $f(x) = (2x+1)(3x-2)$, find $f'(x)$ two ways: by using the product rule and by multiplying out. Do you get the same result?

For Problems 3–20, find the derivative. In some cases, it may be to your advantage to simplify first.

3. $f(x) = xe^x$
4. $f(x) = \dfrac{x}{e^x}$
5. $y = x \cdot 2^x$
6. $w = (t^3+5t)(t^2-7t+2)$
7. $y = (t^2+3)e^t$
8. $z = \dfrac{3t+1}{5t+2}$
9. $y = (t^3-7t^2+1)e^t$
10. $z = \dfrac{t^2+5t+2}{t+3}$
11. $f(x) = \dfrac{x^2+3}{x}$
12. $f(z) = \dfrac{3z^2}{5z^2+7z}$
13. $y = te^{-t^2}$
14. $f(z) = \sqrt{z}e^{-z}$
15. $f(z) = \dfrac{\sqrt{z}}{e^z}$
16. $f(t) = te^{5-2t}$
17. $f(w) = (5w^2+3)e^{w^2}$
18. $f(x) = x^2\cos x$
19. $f(x) = 2x\sin(3x)$
20. $f(x) = e^{-2x}\cdot\sin x$
21. Find the equation of the tangent line to $f(x) = x^2e^{-x}$ at $x = 0$. Check your work by graphing this function and the tangent line on the same coordinate system.

22. Assume the demand function for a certain product is given by

$$q = 1000e^{-0.02p}$$

where p is the price of the product and q is the quantity sold at that price.
   (a) Write revenue, R, as a function of price.
   (b) Find the marginal revenue function.
   (c) Find the revenue and marginal revenue when the price is \$10, and interpret your answers in economic terms.

23. The quantity, q, of a certain skateboard sold depends on the selling price, p, so we write $q = f(p)$. You are given that $f(140) = 15{,}000$ and $f'(140) = -100$.
   (a) What does $f(140) = 15{,}000$ and $f'(140) = -100$ tell you about the sale of the skateboards?
   (b) The total revenue, R, earned by the sale of skateboards is given by $R = pq$. Find $\left.\dfrac{dR}{dp}\right|_{p=140}$.
   (c) What is the sign of $\left.\dfrac{dR}{dp}\right|_{p=140}$? If the skateboards are currently selling for \$140, how should the price be changed to increase revenues?

24. Find the equation of the tangent line at $x = 1$ to $y = f(x)$ where $f(x) = \dfrac{3x^2}{5x^2 + 7x}$.

25. Given: $\left\{\begin{array}{ll} H(3) = 1 & F(3) = 5 \\ H'(3) = 3 & F'(3) = 4 \end{array}\right\}$ find: $\left\{\begin{array}{ll} \text{(a) } G'(3) & \text{if } G(z) = F(z) \cdot H(z) \\ \text{(b) } G'(3) & \text{if } G(w) = F(w)/H(w) \end{array}\right.$

26. (a) Suppose

$$f(x) = \frac{x}{x^2 - 1}.$$

   For what values of x does this formula define f? Find a formula for $f'(x)$. Explain why $f(1.01)$ is so large, and why $f'(1.01)$ is even larger in magnitude.
   (b) Now suppose

$$g(x) = \frac{x^2 + 3x - 4}{x^2 - 1}.$$

   For what values of x does this formula define g? Find a formula for $g'(x)$. Explain why $g(1.01)$ and $g'(1.01)$ don't seem to be very large.

27. Let $f(v)$ be the gas consumption (in liters/km) of a car going at velocity v (in km/hr). In other words, $f(v)$ tells you how many liters of gas the car uses to go one kilometer, if it is going at velocity v. You are told that

$$f(80) = 0.05 \text{ and } f'(80) = 0.0005.$$

   (a) Let $g(v)$ be the distance the same car goes on one liter of gas at velocity v. What is the relationship between $f(v)$ and $g(v)$? Find $g(80)$ and $g'(80)$.
   (b) Let $h(v)$ be the gas consumption in liters per hour. In other words, $h(v)$ tells you how many liters of gas the car uses in one hour if it is going at velocity v. What is the relationship between $h(v)$ and $f(v)$? Find $h(80)$ and $h'(80)$.
   (c) How would you explain the practical meaning of the values of these functions and their derivatives to a driver who knows no calculus?

28. A museum has decided to sell one of its paintings and to invest the proceeds. The price the painting will fetch changes with time, and is denoted by $P(t)$, where t is the number of years since 1990. If the picture is sold between 1990 and 2010 (so $0 \leq t \leq 20$), and the money from the sale is invested in a bank account earning 5% annual interest compounded once a year, the balance, $B(t)$, in the account in the year 2010 is given by

$$B(t) = P(t)(1.05)^{20-t}.$$

(a) Explain why $B(t)$ is given by this formula.
(b) Show that the formula for $B(t)$ is equivalent to

$$B(t) = (1.05)^{20}\frac{P(t)}{(1.05)^t}.$$

(c) Find $B'(10)$, given that $P(10) = 150{,}000$ and $P'(10) = 5000$.

REVIEW PROBLEMS FOR CHAPTER FOUR

Find the derivatives for the functions in Problems 1–18.

1. $f(x) = x^3 - 3x^2 + 5x - 12$
2. $y = 5e^{-0.2t}$
3. $s(t) = (t^2 + 4)(5t - 1)$
4. $g(t) = e^{(1+3t)^2}$
5. $f(t) = 2te^t - \dfrac{1}{\sqrt{t}}$
6. $h(r) = \dfrac{r^2}{2r+1}$
7. $f(z) = \dfrac{z^2+1}{\sqrt{z}}$
8. $f(z) = \ln(z^2 + 1)$
9. $y = xe^{3x}$
10. $q = 100e^{-0.05p}$
11. $y = x^2 \ln x$
12. $s(t) = t^2 + 2\ln t$
13. $g(x) = \dfrac{x^2 + \sqrt{x} + 1}{x^{3/2}}$
14. $h(t) = \ln\left(e^{-t} - t\right)$
15. $f(x) = \sin(2x)$
16. $y = x^2 \cos x$
17. $h(r) = 5\sin^2 r$
18. $p(\theta) = \dfrac{\sin(5-\theta)}{\theta^2}$
19. Given: $r(2) = 4$, $s(2) = 1$, $s(4) = 2$, $r'(2) = -1$, $s'(2) = 3$, $s'(4) = 3$. Compute the following derivatives, or state what additional information you would need to be able to compute the derivative.
(a) $H'(2)$ if $H(x) = r(x) \cdot s(x)$
(b) $H'(2)$ if $H(x) = \sqrt{r(x)}$
(c) $H'(2)$ if $H(x) = r(s(x))$
(d) $H'(2)$ if $H(x) = s(r(x))$
20. In 1975, the population of Mexico was about 84 million and growing at 2.6% annually, while the population of the US was about 250 million and growing at 0.7% annually. Which population was growing faster, if we measure growth rates in people/year? Explain your answer.

21. Suppose the distance, s, of a moving body from a fixed point is given as a function of time by $s = 20e^{t/2}$.

(a) Find the velocity, v, of the body as a function of t.
(b) Find v as a function of s, and hence show that s satisfies the differential equation $s' = \frac{1}{2}s$.

22. Suppose that demand for a certain product is given by

$$q = 5000e^{-0.08p},$$

where p is the price of the product and q is the quantity sold at that price.

(a) What quantity is sold at a price of $10?
(b) Find the derivative when the price is $10 and interpret your answer in terms of demand for the product.

23. Suppose the demand equation is given in Problem 22. Find the revenue and marginal revenue at a price of $10. Interpret your answers in economic terms.

24. Given a number $a > 1$, the equation

$$a^x = 1 + x$$

has the solution $x = 0$ for all a. Are there any other solutions? How does your answer depend on the value of a? [Hint: Graph the functions on both sides of the equation.]

25. Suppose you put a yam in a hot oven, maintained at a constant temperature 200°C. Suppose that at time $t = 30$ minutes, the temperature T of the yam is 120° and is increasing at a (instantaneous) rate of 2°/min. Newton's law of cooling (or, in our case, warming) implies that the temperature at time t will be given by a formula of the form

$$T(t) = 200 - ae^{-bt}.$$

Find a and b.

26. Suppose the total number of people, N, who have contracted a disease by a time t days after its outbreak is given by

$$N = \frac{1{,}000{,}000}{1 + 5{,}000e^{-0.1t}}.$$

(a) In the long run, how many people will have had the disease?
(b) Is there any day on which more than a million people fall sick? Half a million? Quarter of a million? (Note: You do not have to try to find out on what days these things happen.)

27. Suppose the depth of the water, y, in meters, in the Bay of Fundy, Canada, is given as a function of time, t, in hours after midnight, by the function

$$y = 10 + 7.5\cos(0.507t).$$

Is the tide rising or falling, and how fast (in meters/hour), at each of the following times?
(a) 6:00 am (b) 9:00 am (c) Noon (d) 6:00 pm

CHAPTER FIVE

USING THE DERIVATIVE

In Chapter 2 we introduced the derivative and some of its interpretations. In Chapter 4 we saw how to differentiate all of the standard functions. Now we will use first and second derivatives to find optimum values, to analyze the behavior of families of functions, and to consider applications in economics.

As we saw in Chapter 2, the connection between the derivative and the original function is given by the following:

- If $f' > 0$ on an interval, then f is increasing on that interval.
- If $f' < 0$ on an interval, then f is decreasing on that interval.
- If $f'' > 0$ on an interval, then the graph of f is concave up on that interval.
- If $f'' < 0$ on an interval, then the graph of f is concave down on that interval.

We can do more with these principles now than we could in Chapter 2 because we now have formulas for the derivatives of the standard functions.

5.1 USING THE FIRST DERIVATIVE

Why Is it Useful to Know Where a Function is Increasing and Decreasing?

The following example shows how to use the derivative of a function to understand its graph. When we graph a function on a computer or graphing calculator, we see only part of the picture. The derivative can often direct our attention to important features on the graph.

Example 1 Sketch a helpful graph of the function $f(x) = x^3 - 9x^2 - 48x + 52$.

Solution Since f is a cubic polynomial, we expect a graph that is roughly S-shaped. Graphing this function with $-10 \leq x \leq 10$, $-10 \leq y \leq 10$, gives the two nearly vertical lines in Figure 5.1. We know that there is more going on than this, but how do we know where to look?

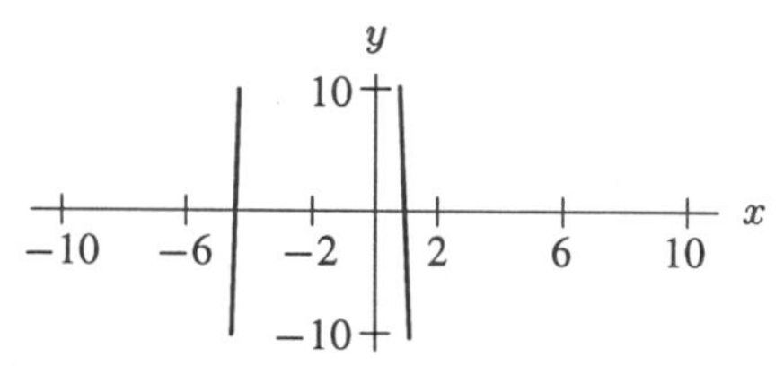

Figure 5.1: Unhelpful graph of $f(x) = x^3 - 9x^2 - 48x + 52$

We'll use the derivative to determine where the function is increasing and where it is decreasing. The derivative of f is

$$f'(x) = 3x^2 - 18x - 48.$$

To find where $f' > 0$ or $f' < 0$, we first find where $f' = 0$, that is, where $3x^2 - 18x - 48 = 0$. Solving this quadratic, we get $x = -2$ and $x = 8$. Since $f' = 0$ *only* at $x = -2$ and $x = 8$, and since f' is continuous, f' cannot change sign on any of the intervals $x < -2$, $-2 < x < 8$, or $x > 8$. How can we tell the sign of f' on each of these intervals? The easiest way is to pick a point and substitute into f'. For example, since $f'(-3) = 33 > 0$, we know f' is positive for $x < -2$, so f is increasing for $x < -2$. Similarly, since $f'(0) = -48$ and $f'(10) = 72$, we know that f decreases between $x = -2$ and $x = 8$ and increases for $x > 8$. Summarizing the behavior of f on each interval:

		$x = -2$		$x = 8$		
f	Increasing ↗		Decreasing ↘		Increasing ↗	x
f'	+	0	−	0	+	

We find that $f(-2) = 104$ and $f(8) = -396$. Hence on the interval $-2 < x < 8$ the function decreases from a high of 104 to a low of -396. (Now we see why not much showed up in our first calculator graph.) One more point on the graph is easy to get: the y intercept, $f(0) = 52$. With just these three points we can get a much more helpful graph. By setting the domain and range in our calculator to $-10 \leq x \leq 20$ and $-400 \leq y \leq 400$, we get Figure 5.2. Where does the graph in Figure 5.1 fit into this one?

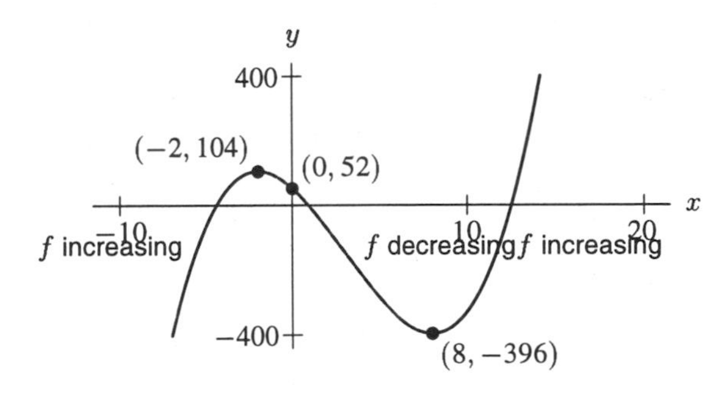

Figure 5.2: Helpful graph of $f(x) = x^3 - 9x^2 - 48x + 52$

Critical Points

In the preceding example, the points $x = -2$ and $x = 8$, where $f'(x) = 0$, played a key role. Now we will give a name to such points.

For any function f, a point p in the domain of f where $f'(p) = 0$ or $f'(p)$ is undefined is called a **critical point** of the function. In addition, the point $(p, f(p))$ on the graph of f is also called a critical point. A **critical value** of f is the value, $f(p)$, of the function at a critical point, p.

Notice that "critical point of f" can refer either to special points in the domain of f or to special points on the graph of f. You will know which meaning is intended from the context.

What Do the Critical Points Tell Us?

Geometrically, at a critical point where $f'(p) = 0$, the line tangent to the graph of f at p is horizontal. At a critical point where $f'(p)$ is undefined, there is no horizontal tangent to the graph—there's either a vertical tangent or no tangent at all. (For example, $x = 0$ is a critical point for the absolute value function $f(x) = |x|$.) However, most of the functions we will work with will be differentiable everywhere, and therefore most of our critical points will be of the $f'(p) = 0$ variety.

The points where $f'(p) = 0$ (or $f'(p)$ is undefined) divide the domain of f into intervals on which the sign of the derivative stays the same, either positive or negative. Therefore, *between two successive critical points the graph of a function cannot change direction; it either goes up or down.*

A function may have any number of critical points or none at all. (See Figures 5.3–5.5.)

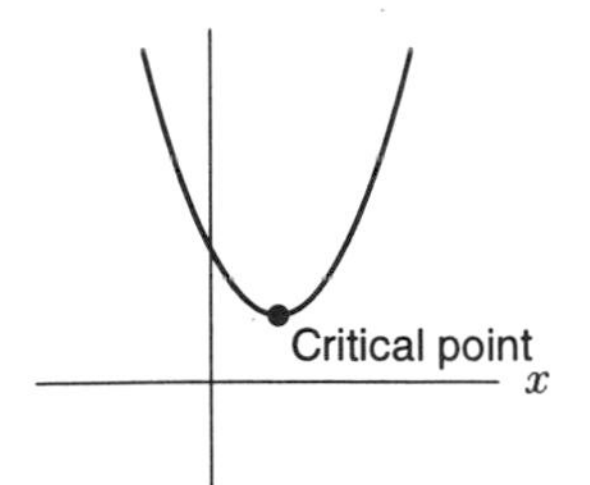

Figure 5.3: A quadratic: One critical point

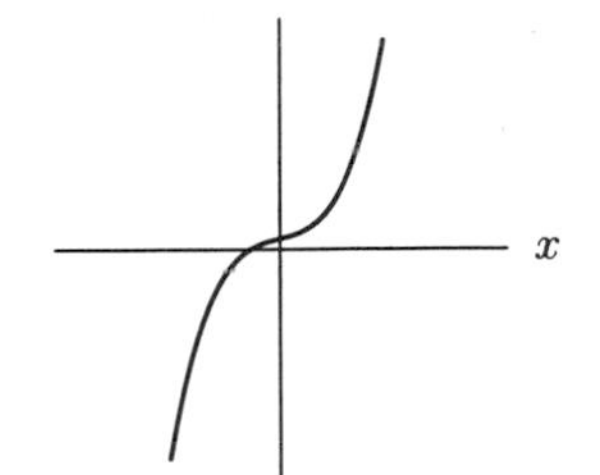

Figure 5.4: $f(x) = x^3 + x + 1$: No critical points

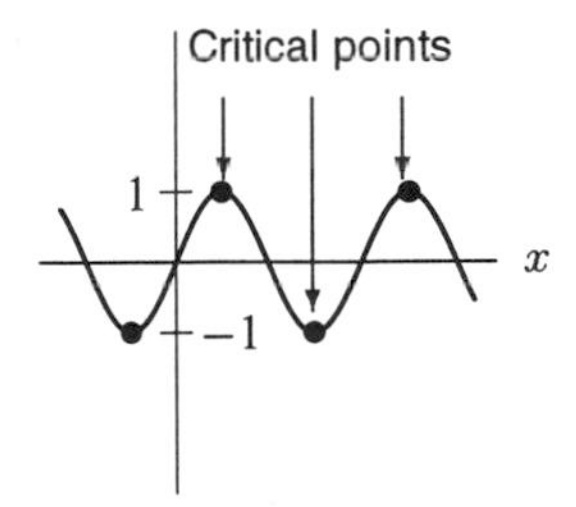

Figure 5.5: Infinitely many critical points

Local Maxima and Minima

What happens to a function at a critical point? Suppose that $f'(p) = 0$. We know that *at* p the graph has a horizontal tangent, but what happens *near* p? If f' has different signs on either side of p, then the graph changes direction at p, so the graph must look like one of those in Figure 5.6.

In the left-hand graph in Figure 5.6 we say that f has a local minimum at p, and in the right-hand graph we say that f has a local maximum at p. We use the adjective "local" because we are describing only what happens near p.

Suppose p is a critical point of a function f. Then

- f has a **local minimum** at p if, near p, the values of f get no smaller than $f(p)$.
- f has a **local maximum** at p if, near p, the values of f get no larger than $f(p)$.

How Do We Decide Which Critical Points Are Local Maxima and Which Are Local Minima?

As Figure 5.6 illustrates, we have the following criterion.

The First-Derivative Test for Local Maxima and Minima

Suppose p is a critical point in the domain of f and f is continuous at p. If f' changes sign at p, then f has either a local minimum or a local maximum at p.

- If f' is negative to the left of p and positive to the right of p, then f has a local minimum at p.
- If f' is positive to the left of p and negative to the right of p, then f has a local maximum at p.

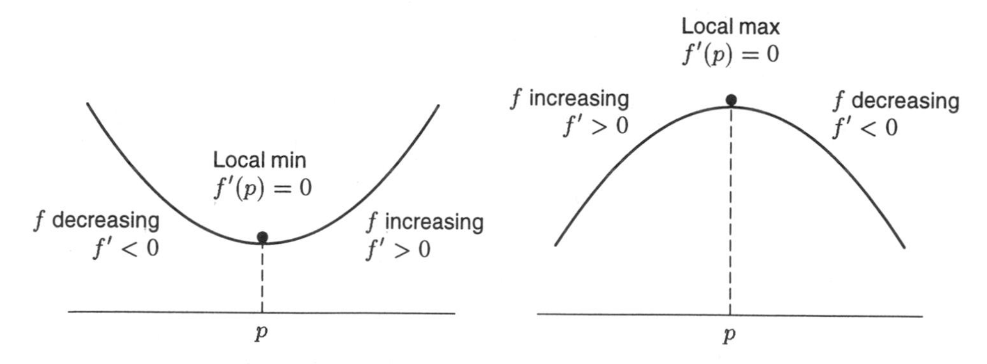

Figure 5.6: Changes in direction: Local maxima and minima

Example 2 Identify the local maxima and minima of $f(x) = x^3 - 9x^2 - 48x + 52$. (This is a continuation of Example 1.)

Solution The function has derivative $f'(x) = 3x^2 - 18x - 48 = 3(x^2 - 6x - 16) = 3(x - 8)(x + 2)$. To the left of $x = -2$, f' is positive. Between $x = -2$ and $x = 8$, f' is negative. To the right of $x = 8$, f' is positive. By the first-derivative test, f has a local maximum at $x = -2$ and a local minimum at $x = 8$. (See Figure 5.7.)

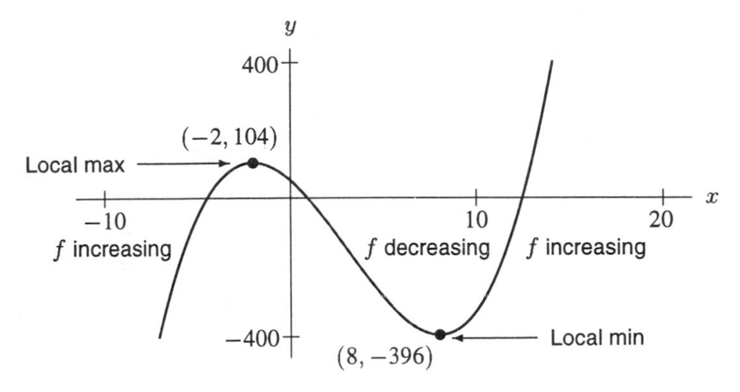

Figure 5.7: Local maxima and minima of $f(x) = x^3 - 9x^2 - 48x + 52$

Warning!

The sign of f' doesn't *have* to change at a critical point. Consider $f(x) = x^3$, whose graph is in Figure 5.8. The derivative , $f'(x) = 3x^2$, is positive on both sides of $x = 0$, so f increases on both sides of $x = 0$, and there is neither a local maximum nor a local minimum at $x = 0$. In other words, *a function doesn't have to have a local maximum or local minimum at every critical point.*

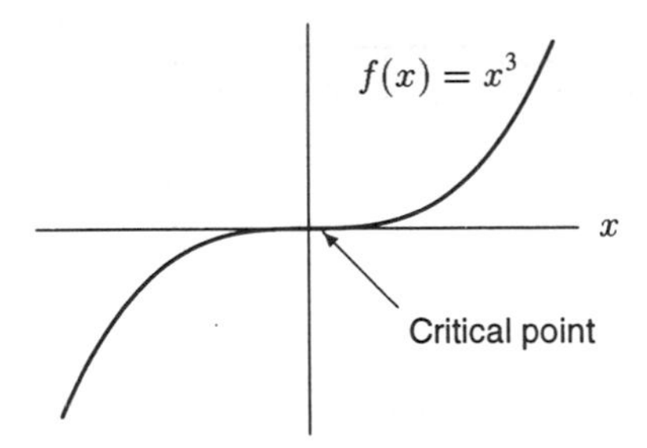

Figure 5.8: Critical point which is not a local maximum or minimum.

Global Maxima and Minima

The local maxima and minima tell us where a function is locally largest or smallest. However we are often more interested in where the function is absolutely largest or smallest in a given domain. We say

- f has a **global minimum** at p if all values of f are greater than or equal to $f(p)$.
- f has a **global maximum** at p if all values of f are less than or equal to $f(p)$.

How Do We Find Global Maxima and Minima?

If f is a continuous function defined on a closed interval $a \leq x \leq b$ (i.e., an interval containing its endpoints), Figure 5.9 shows that the global maximum or minimum of f occurs either at a local maximum or a local minimum respectively, or at one of the endpoints, $x = a$ or $x = b$.

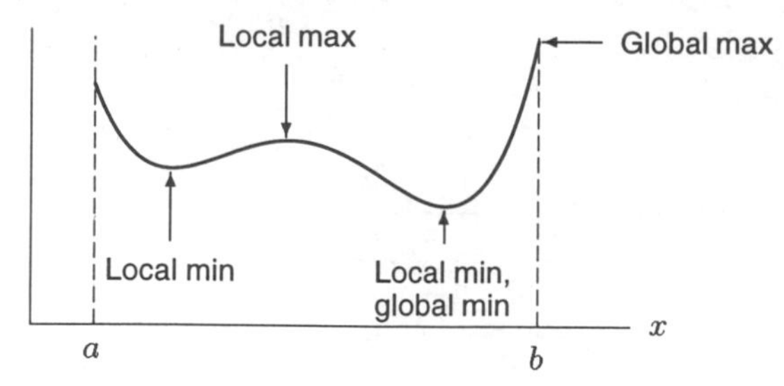

Figure 5.9: Global maximum and minimum on a closed interval $a \leq x \leq b$

> **To find the global maximum and minimum of a continuous function on a closed interval:** Compare values of the function at all the critical points in the interval and at the endpoints.

What if the function is defined on an open interval $a < x < b$ (i.e., an interval not including its endpoints), or on the entire real line? We say there is no global maximum in Figure 5.10 because the function has no actual largest value. The global minimum in Figure 5.10 coincides with the local minimum and is marked. Figures 5.10 and 5.11 show there may not be a global maximum on an open interval, and it is also possible to not have a global minimum on an open interval.

> **To find the global maximum and minimum of a continuous function on an open interval or on the entire real line:** Find the value of the function at all the critical points and sketch a graph.

Warning!

Notice that local maxima or minima of a function f occur at critical points, where $f'(p) = 0$ (or $f'(p)$ is undefined). Since global maxima or minima can occur at endpoints (where f' is not necessarily 0 or undefined), *not every global maximum or minimum is a local maximum or minimum.*

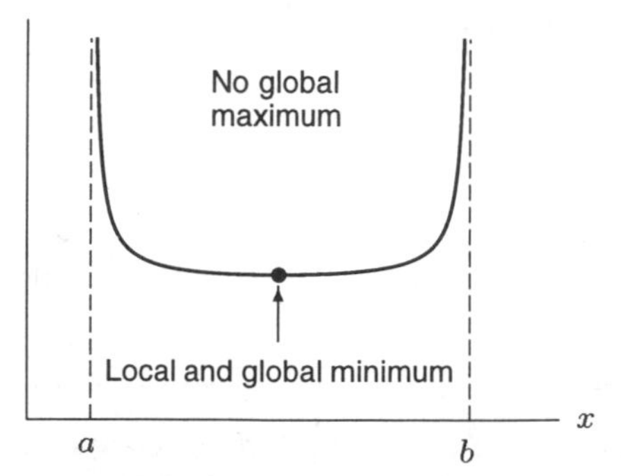

Figure 5.10: Global maximum and minimum on $a < x < b$

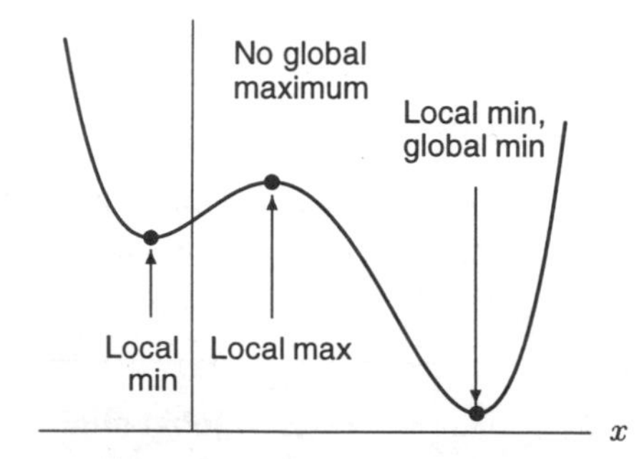

Figure 5.11: Global maximum and minimum when the domain is the real line.

Example 3 Find the global maxima and minima of $f(x) = x^3 - 9x^2 - 48x + 52$ on the following intervals:

(a) $-5 \leq x \leq 12$ (b) $-5 \leq x \leq 14$ (c) $-5 \leq x < \infty$.

Solution (a) We have calculated the critical points of this function previously using

$$f'(x) = 3x^2 - 18x - 48 = 3(x+2)(x-8)$$

so $x = -2$ and $x = 8$ are critical points. Evaluating f at the critical points and the endpoints, we discover

$$\begin{aligned} f(-5) &= (-5)^3 - 9(-5)^2 - 48(-5) + 52 = -58 \\ f(-2) &= (-2)^3 - 9(-2)^2 - 48(-2) + 52 = 104 \\ f(8) &= (8)^3 - 9(8)^2 - 48(8) + 52 = -396 \\ f(12) &= (12)^3 - 9(12)^2 - 48(12) + 52 = -92. \end{aligned}$$

So the global maximum on $[-5, 12]$ is 104 and occurs at $x = -2$, and the global minimum on $[-5, 12]$ is -396 and occurs at $x = 8$.

(b) For the interval $[-5, 14]$, we compare

$$f(-5) = -58, \quad f(-2) = 104, \quad f(8) = -396$$

with the value of the function at the new endpoint:

$$f(14) = 360.$$

The global maximum is now 360 and occurs at $x = 14$, and the global minimum is still -396 and occurs at $x = 8$. Notice that since the function is increasing for $x > 8$, changing the right-hand end of the interval from $x = 12$ to $x = 14$ alters the global maximum but not the global minimum. See Figure 5.12.

(c) Figure 5.12 shows that for $-5 \leq x < \infty$ there is no global maximum, because we can make $f(x)$ as large as we please by choosing x large enough. The global minimum remains -396 at $x = 8$.

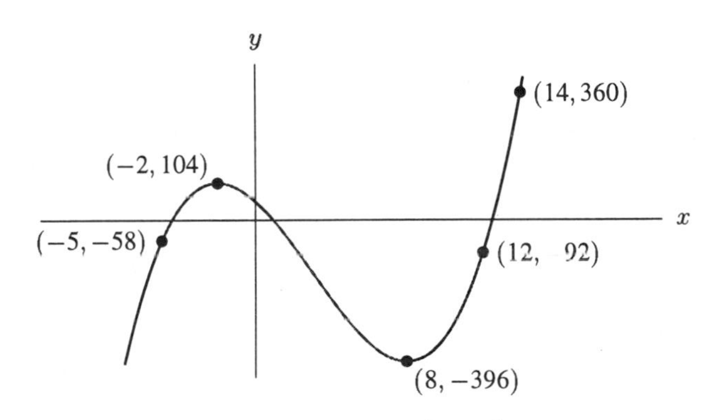

Figure 5.12: Graph of $f(x) = x^3 - 9x^2 - 48x + 52$

Example 4 Use a graph of the function $f(x) = xe^{-2x}$ to find its local maxima and minima, and confirm your result analytically. What are its global maximum and minimum on the interval $0 \le x \le 2$? On the interval $-1 \le x \le 4$? On the interval $-1 \le x < \infty$?

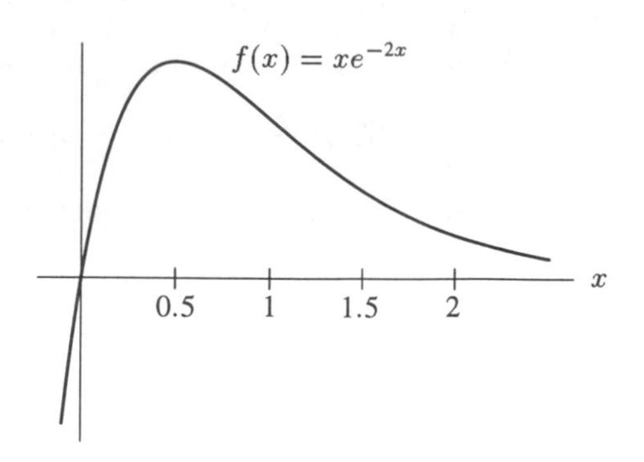

Figure 5.13: Find maxima and minima

Solution The graph in Figure 5.13 shows that this function has no local minima but that there is a local maximum near $x = 1/2$. Confirming this result analytically means showing that this is what we find when we use the formula for the derivative to find the critical points. The product rule gives us:

$$\begin{aligned} f'(x) &= x(e^{-2x}(-2)) + (1)(e^{-2x}) \\ &= e^{-2x}(-2x+1). \end{aligned}$$

Since e^{-2x} is never equal to zero, we have $f'(x) = 0$ only where $-2x + 1 = 0$. Thus, $x = 1/2$ is the only critical point of the function. The formula for f' shows that $f' > 0$ for $x < 1/2$ and $f' < 0$ for $x > 1/2$. Thus, by the first derivative test, there is a local maximum at $x = 1/2$. The graph shows that it is also a global maximum.

The global maximum and minimum of f on the interval $0 \le x \le 2$ must occur at either a critical point of f or an endpoint of the interval. Checking the values of the function at the critical point and at both endpoints gives:

$$f(0) = 0, \quad f(1/2) = 0.18394, \quad f(2) = 0.03663.$$

The global minimum for $0 \le x \le 2$ is 0 and occurs at $x = 0$, and the global maximum is 0.18394 and occurs at $x = 1/2$.

If we look at a graph of $f(x)$ on the interval $-1 \le x \le 4$, we see that the local minimum occurs at the left endpoint ($x = -1$) and the global maximum remains at $x = 1/2$. To check this analytically, we check the endpoints and the critical point:

$$f(-1) = -7.389, \quad f(1/2) = 0.18394 \qquad f(4) = 0.00134.$$

The global minimum for $-1 \le x \le 4$ is -7.389 and occurs at $x = -1$, and the global maximum is 0.18394 and occurs at $x = 1/2$. This confirms what we see graphically.

On the interval $-1 \le x < \infty$, we see on a graph of $f(x)$ that as x gets large, $f(x)$ stays positive and gets closer and closer to zero. Thus, the global maximum and minimum will remain at $x = 1/2$ and $x = -1$, respectively.

Problems for Section 5.1

1. Indicate all critical points on the graph of f in Figure 5.14 and determine which correspond to local maxima of f, which to local minima, and which to neither.

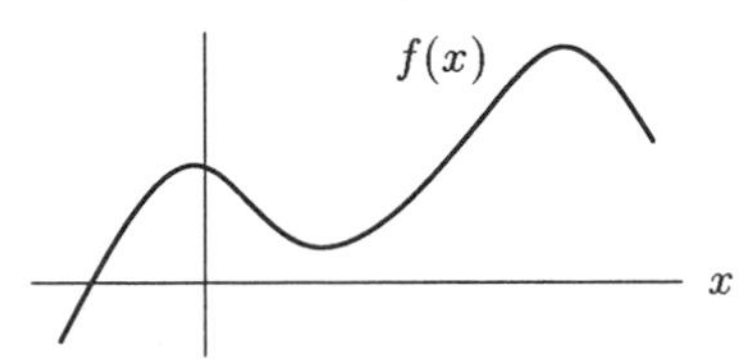

Figure 5.14

2. Sketch graphs of two continuous functions f and g, each of which has exactly five critical points, the points A–E in Figure 5.15, and which satisfy the following conditions:

(a) $\lim_{x \to -\infty} f(x) = \infty$ and $\lim_{x \to \infty} f(x) = \infty$

(b) $\lim_{x \to -\infty} g(x) = -\infty$ and $\lim_{x \to \infty} g(x) = 0$

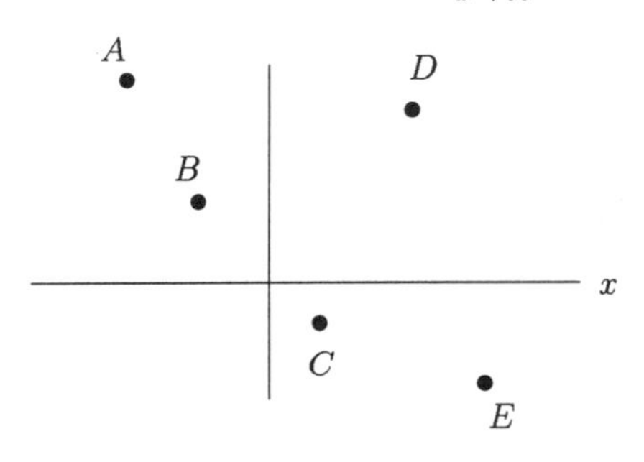

Figure 5.15

3. Indicate on the graph of the derivative function f' in Figure 5.16 the x-values that are critical points of the function f itself. At which critical points does f have local maxima, local minima, or neither?

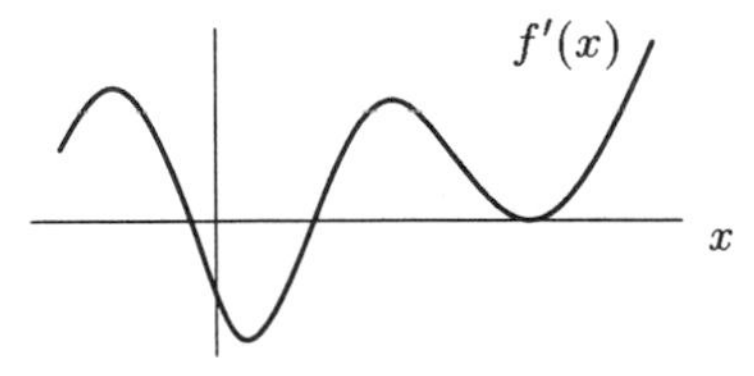

Figure 5.16

In each of Problems 4–9, find all critical points and then use the first-derivative test to determine local maxima and minima. When you are finished, use a calculator or computer to sketch a graph of each function to check your work.

4. $f(x) = 2x^3 + 3x^2 - 36x + 5$
5. $f(x) = 3x^4 - 4x^3 + 6$
6. $f(x) = (x^2 - 4)^7$
7. $f(x) = (x^3 - 8)^4$
8. $f(x) = x^4 - 8x^2 + 5$
9. $f(x) = 2x^2e^{5x} + 1$

10. How many real roots does the equation $x^5 + x + 7 = 0$ have? How do you know? [Hint: How many critical points does this function have?]

11. Plot the graph of $f(x) = x^3 - e^x$ using a graphing calculator or computer to find all local and global maxima and minima for: (a) $-1 \le x \le 4$ (b) $-3 \le x \le 2$

12. For $y = f(x) = x^{10} - 10x$, and $0 \le x \le 2$, find the value(s) of x for which:
    (a) $f(x)$ has a local maximum or local minimum. Indicate which ones are maxima and which are minima.
    (b) $f(x)$ has a global maximum or global minimum. Indicate which ones are maxima and which are minima.

13. For $f(x) = x - \ln x$, and $0.1 \le x \le 2$, find the value(s) of x for which:
    (a) $f(x)$ has a local maximum or local minimum. Indicate which ones are maxima and which are minima.
    (b) $f(x)$ has a global maximum or global minimum. Indicate which ones are maxima and which are minima.

14. For $f(x) = \sin^2 x - \cos x$, and $0 \le x \le \pi$, find, to two decimal places, the value(s) of x for which:
    (a) $f(x)$ has a local maximum or local minimum. Indicate which ones are maxima and which are minima.
    (b) $f(x)$ has a global maximum or global minimum. Indicate which ones are maxima and which are minima.

15. (a) Water is flowing at a constant rate into a cylindrical container standing vertically. Sketch a graph showing the depth of water against time.
    (b) Water is flowing at a constant rate into a cone-shaped container standing on its point. Sketch a graph showing the depth of the water against time.

16. Choose the constants a and b in the function $f(x) = x^2 + ax + b$ so that the global minimum for this parabola is at the point $(-2, -3)$.

17. Choose the constants a and b in the function

$$f(x) = axe^{bx}$$

such that $f(\frac{1}{3}) = 1$ and the function has a maximum at $x = \frac{1}{3}$.

18. Sketch the graph of $f(x) = 2x^3 - 9x^2 + 12x + 1$. Use the graph to decide how many solutions the following equations have: (a) $f(x) = 10$ (b) $f(x) = 5$ (c) $f(x) = 0$ (d) $f(x) = 2e$

    You need not find these solutions.

19. Suppose that consumer demand for a certain product is changing over time, and the rate of change of this demand, $r(t)$, is given in Table 5.1.

TABLE 5.1 *Rate of change of demand for a certain product*

Time, t (weeks)	0	1	2	3	4	5	6	7	8	9	10
Rate, $r(t)$ (units/week)	12	10	4	−2	−3	−1	3	7	11	15	10

    (a) When is the demand for this product increasing? When is it decreasing?
    (b) Estimate the times at which demand is at a local maximum, and the times at which demand is at a local minimum.

20. Suppose f has a continuous derivative everywhere. From the values of $f'(\theta)$ in the table below, estimate the θ values with $1 < \theta < 2.1$ at which $f(\theta)$ has a local maximum or minimum, and identify which is which.

θ	1.0	1.1	1.2	1.3	1.4	1.5	1.6	1.7	1.8	1.9	2.0	2.1
$f'(\theta)$	2.37	0.31	−2.00	−3.45	−3.34	−1.70	0.76	2.80	3.61	2.76	0.69	−1.62

21. (a) On a computer or calculator, graph $f(\theta) = \theta - \sin\theta$. Can you tell whether the function has any zeros in the interval $0 \le \theta \le 1$?
(b) Find f'. What does the sign of f' tell you about the zeros of f in the interval $0 \le \theta \le 1$?

22. Assume the function f is differentiable everywhere and has just one critical point, at $x = 3$. In parts (a) through (d), you are given additional conditions. In each case decide whether $x = 3$ is a local maximum, a local minimum, or neither. Explain your reasoning. Also sketch possible graphs for all four cases.
(a) $f'(1) = 3$ and $f'(5) = -1$
(b) $\lim_{x\to\infty} f(x) = \infty$ and $\lim_{x\to-\infty} f(x) = \infty$
(c) $f(1) = 1, f(2) = 2, f(4) = 4, f(5) = 5$
(d) $f'(2) = -1, f(3) = 1, \lim_{x\to\infty} f(x) = 3$

23. (a) Show that $x > 2\ln x$ for all $x > 0$.
[Hint: Find the minimum of $f(x) = x - 2\ln x$.]
(b) Use the above result to show that $e^x > x^2$ for all positive x.
(c) Is $x > 3\ln x$ for all positive x?

24. Show that for all values of x

$$x^4 - 4x > -4.$$

25. The rabbit population on a small Pacific island is approximated by

$$P(t) = \frac{2000}{1 + e^{(5.3-0.4t)}}$$

with t measured in years since 1774, when Captain James Cook left 10 rabbits on the island. Using a calculator or computer:
(a) Graph P. Does the population level off?
(b) Estimate when the rabbit population grew most rapidly. How large was the population at that time?
(c) What natural causes could lead to the shape of the graph of P?

26. For what values of a and b will the function $f(x) = a(x - b\ln x)$ have a global minimum at the point $(2, 5)$? See Figure 5.17 for a graph of $f(x)$ when $a = 1$ and $b = 1$.

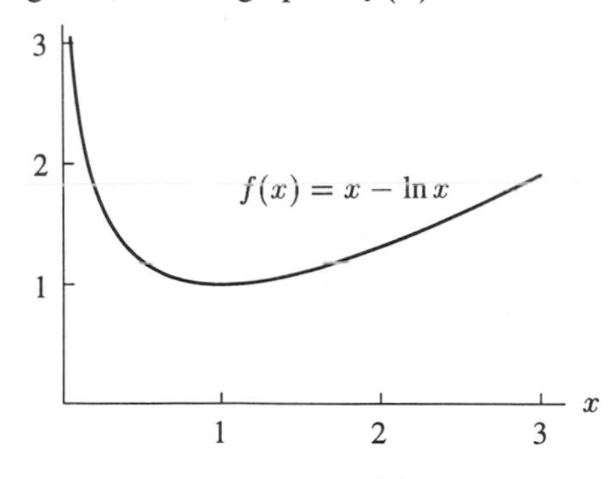

Figure 5.17: Graph of $f(x) = x - \ln x$

5.2 USING THE SECOND DERIVATIVE

Recall that the sign of the second derivative, f'', tells us whether the graph of f is concave up or concave down. If $f'' > 0$ on an interval, then the graph of f is concave up on the interval; and if $f'' < 0$ then the graph is concave down on the interval.

Why Is It Useful to Know the Concavity of f?

Geometrically, if a curve is concave up near some point then it lies above its tangent line at that point and if it is concave down it lies below its tangent line at that point. See Figure 5.18.

The next example shows how we can use this to get estimates for functions.

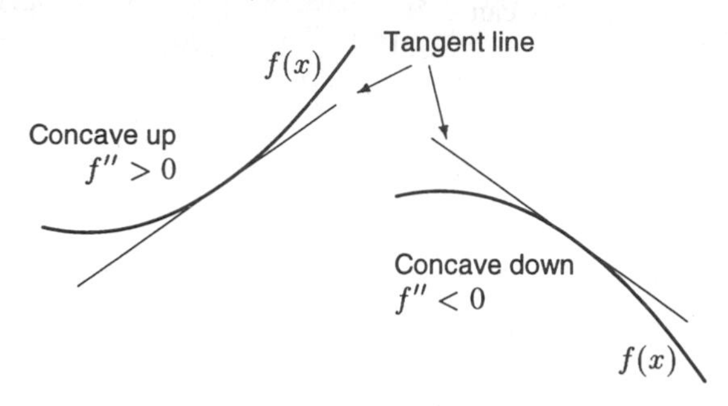

Figure 5.18: Concavity and the tangent line

Example 1 Explain why we know that $e^x \geq 1 + x$ for all values of x.

Solution Figure 5.19 shows that the graph of the function $f(x) = e^x$ is concave up everywhere, and the equation of its tangent line at the point $(0, 1)$ is $y = x + 1$. Since the graph always lies above its tangent, we have the inequality

$$e^x \geq 1 + x.$$

Problem 28 on page 305 shows how this inequality can be used to compute e.

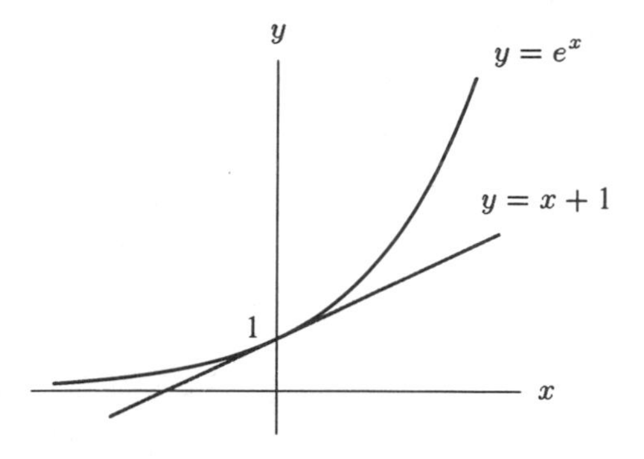

Figure 5.19: Graph showing that $e^x \geq 1 + x$

In the next example, we will see that knowing the concavity of a function f can be enough to determine the end behavior of f.

Example 2 If f is increasing and its graph is concave up at all points, what can you conclude about $\lim_{x\to\infty} f(x)$? What can you say about $\lim_{x\to\infty} g(x)$ if g is decreasing and its graph is concave down at all points?

Solution Since the graph of f must stay above a tangent line with positive slope, we must have $\lim_{x\to\infty} f(x) = \infty$. Since the graph of g must stay below a tangent line with negative slope, we must have $\lim_{x\to\infty} g(x) = -\infty$. Can you draw a sketch to show why this must be true?

The Second-Derivative Test for Local Maxima and Minima

Knowing the concavity can also be useful in testing if a critical point is a local maximum or a local minimum. Suppose p is a critical point of f, so that $f'(p) = 0$ and the graph of f has a horizontal tangent line at p. If the graph is concave up at p, then f has a local minimum at p. Likewise, if the graph is concave down, f has a local maximum. (See Figure 5.20.) Thus we have the following result:

The Second-Derivative Test for Local Maxima and Minima

- If $f'(p) = 0$ and $f''(p) > 0$ then f has a local minimum at p.
- If $f'(p) = 0$ and $f''(p) < 0$ then f has a local maximum at p.

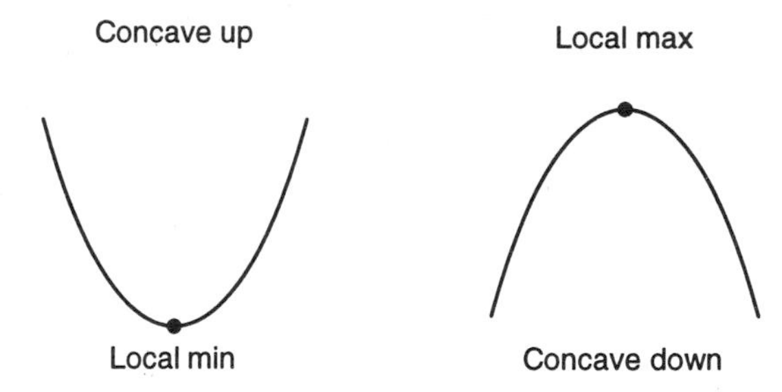

Figure 5.20: Local maxima and minima and concavity

Here are some examples of the use of the second-derivative test.

Example 3 Classify as local maxima or local minima the critical points of

(a) $f(x) = x^3 - 9x^2 - 48x + 52$ (b) $g(x) = xe^{-x}$ (c) $h(x) = x + \dfrac{1}{x}$

Solution (a) As we saw in Example 2 on page 288,

$$f'(x) = 3x^2 - 18x - 48$$

and the critical points of f are $x = -2$ and $x = 8$. We have

$$f''(x) = 6x - 18.$$

Thus $f''(8) = 6(8) - 18 = 30 > 0$. So f has a local minimum at $x = 8$. Since $f''(-2) = 6(-2) - 18 = -30 < 0$, f has a local maximum at $x = -2$. (A graph of this function is in Figure 5.22.)

(b) We have

$$g'(x) = e^{-x} - xe^{-x} = (1 - x)e^{-x}.$$

Hence $x = 1$ is the only critical point. We see that g' changes from positive to negative at $x = 1$ since e^{-x} is always positive, so by the first-derivative test g has a local maximum at $x = 1$. If we wish to use the second-derivative test, we compute

$$g''(x) = (x - 2)e^{-x}$$

and thus $g''(1) = (-1)e^{-1} < 0$, so again $x = 1$ gives a local maximum. (See Figure 5.23.)

(c) For $h(x) = x + \frac{1}{x}$ we calculate

$$h'(x) = 1 - \frac{1}{x^2}$$

and so the critical points of h are at $x = \pm 1$. Now

$$h''(x) = \frac{2}{x^3}$$

so $h''(1) = 2 > 0$ and $x = 1$ gives a local minimum. On the other hand $h''(-1) = -2 < 0$ so $x = -1$ gives a local maximum.

Warning!

The second-derivative test does not tell us anything if both $f'(p) = 0$ and $f''(p) = 0$. For example, if $f(x) = x^3$ and $g(x) = x^4$, both $f'(0) = g'(0) = 0$ and $f''(0) = g''(0) = 0$. However, the point $x = 0$ is a minimum for g but is neither a maximum nor a minimum for f. When the second derivative test fails to give information about a critical point p because $f''(p) = 0$, the first-derivative test can still be useful. For example, g' changes sign from negative to positive at 0, so we know g has a local minimum there.

Inflection Points

In Section 5.1 we studied points where the slope changes sign, which led us to critical points. Now we will look at points where the concavity changes.

> A point at which the graph of a function changes concavity is called an **inflection point** of f.

The words "inflection point of f" can refer either to a point in the domain of f or to a point on the graph of f. The context of the problem will tell you which is meant.

How Do You Locate an Inflection Point?

Since the concavity changes at an inflection point, the sign of f'' changes there. It is positive on one side of the inflection point, and negative on the other; so at the inflection point, f'' is zero or undefined. (See Figure 5.21.)

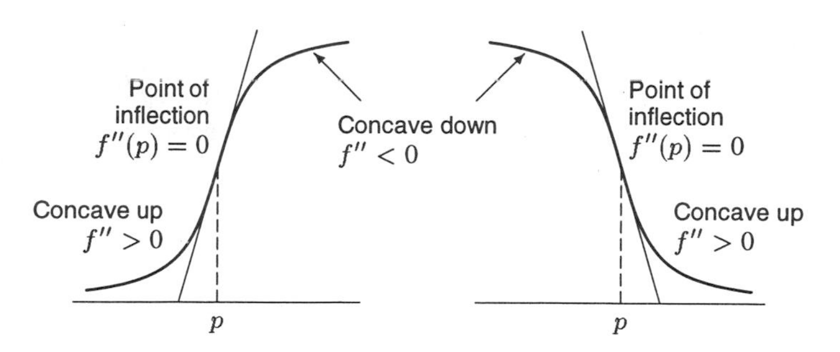

Figure 5.21: Change in concavity

Example 4 Find the inflection points of $f(x) = x^3 - 9x^2 - 48x + 52$.

Solution From the graph of $f(x)$ in Figure 5.22 we see that part of the graph of f is concave up and part is concave down, and so the function must have an inflection point. We calculate $f'(x) = 3x^2 - 18x - 48$, and $f''(x) = 6x - 18$, so $f''(x) = 0$ when $x = 3$. Further, $f''(x) < 0$ when $x < 3$ and $f''(x) > 0$ when $x > 3$, so the graph changes concavity at $x = 3$. Hence $x = 3$ is an inflection point.

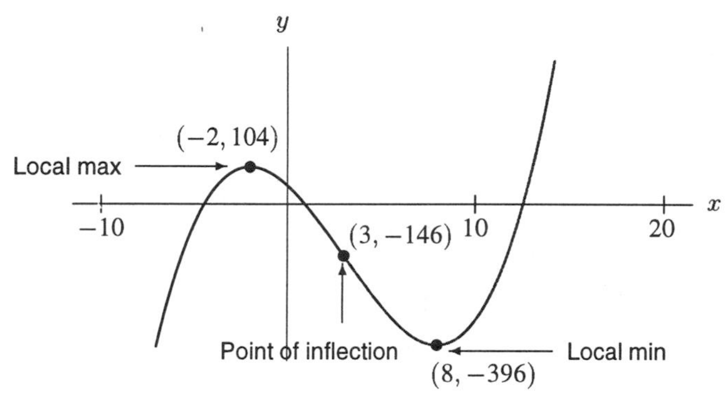

Figure 5.22: Graph of $f(x) = x^3 - 9x^2 - 48x + 52$, again

Example 5 Find the inflection points for $g(x) = xe^{-x}$ and sketch the graph.

Solution We have $g'(x) = (1 - x)e^{-x}$ and $g''(x) = (x - 2)e^{-x}$. Hence $g'' < 0$ when $x < 2$ and $g'' > 0$ when $x > 2$, so $x = 2$ is an inflection point. The graph is sketched in Figure 5.23.

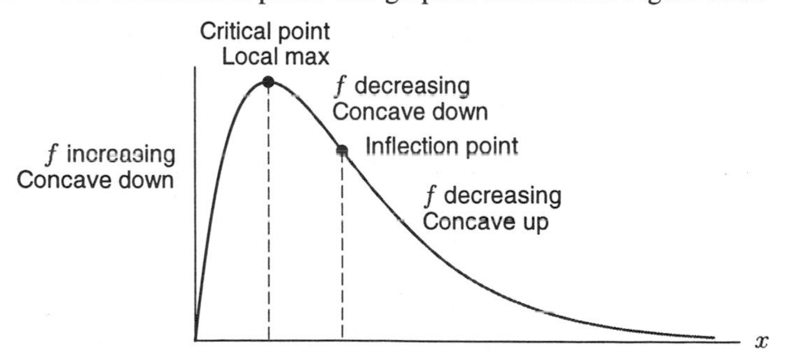

Figure 5.23: Graph of $g(x) = xe^{-x}$

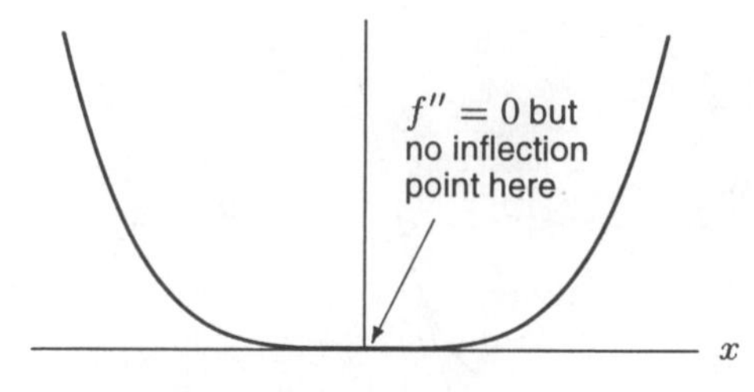

Figure 5.24: Graph of $f(x) = x^4$

Warning!

Not every point x where $f''(x) = 0$ (or f'' is undefined) is an inflection point (just as not every point where $f' = 0$ is a local maximum or minimum). For instance $f(x) = x^4$ has $f''(x) = 12x^2$ so $f''(0) = 0$, but $f'' > 0$ when $x > 0$ and when $x < 0$, so there is *no* change in concavity at $x = 0$. See Figure 5.24.

The Relation Between Inflection Points and Local Maxima and Minima of the Derivative

Inflection points can also be interpreted in terms of first derivatives. Recall that the graph of f is concave up where f' is increasing and concave down where f' is decreasing. So the concavity changes where f' changes from increasing to decreasing or from decreasing to increasing; that is, where f' has a local maximum or a local minimum. Another way to see this is to notice that if p is an inflection point of f, then $f''(p) = 0$ (or $f''(p)$ is undefined) and hence p is a critical point of the derivative function f'. Further, since f'' changes sign at p, the first derivative test shows that this critical point is a local maximum or minimum of f'.

A function f has an inflection point at p
- If f' has a local minimum or a local maximum at p
- If f'' changes sign at p

Example 6 Sketch the graph of $f(x) = (x^2 + 1)e^x$ on the interval $-5 \leq x \leq -0.5$, and find the points where it is increasing most rapidly and where it is increasing least rapidly.

Solution The graph of $f(x)$ is shown in Figure 5.25. To determine where f is increasing most rapidly and where it is increasing least rapidly, we find where the maximum and minimum values of its derivative occur because the derivative gives the rate of increase of f. Using the product rule, we have

$$f'(x) = (x^2 + 1)e^x + (2x)e^x = e^x(x^2 + 2x + 1) = e^x(x + 1)^2.$$

The graph of f' is shown in Figure 5.26. We see that the global maximum of f' on the interval $-5 \leq x \leq -0.5$ occurs at $x = -3$ and the global minimum of f' on the interval occurs at $x = -1$. Thus f is increasing most rapidly at $x = -3$ and least rapidly at $x = -1$. Notice in Figure 5.25 that $x = -3$ and $x = -1$ are both inflection points of f. This is because they are local maxima or minima of f', so f' changes from increasing to decreasing or from decreasing to increasing there. Hence f changes concavity at these points which makes them points of inflection.

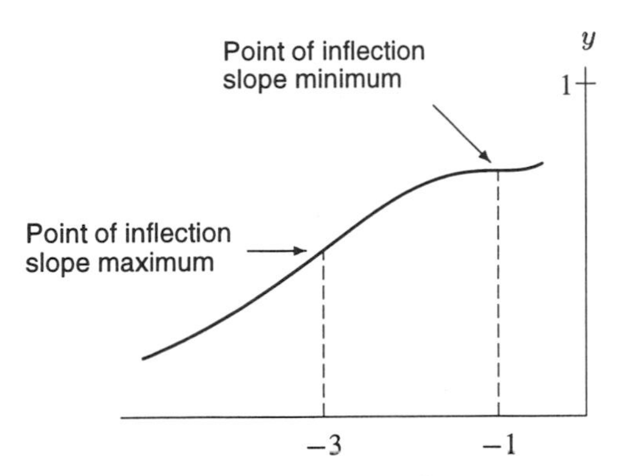

Figure 5.25: Graph of $f(x) = (x^2 + 1)e^x$ on the window $-5 \le x \le -0.5, 0 \le y \le 1$

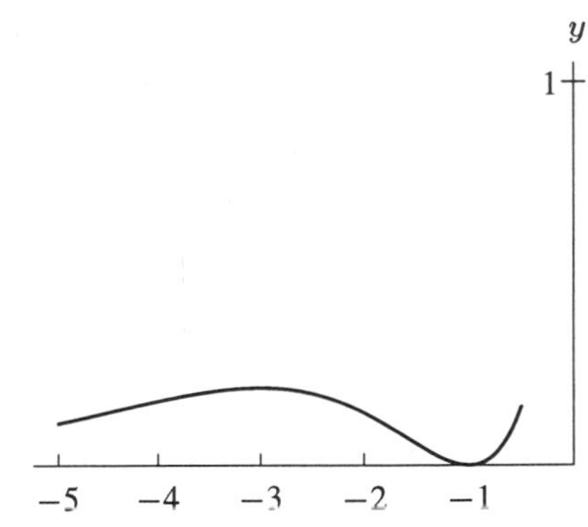

Figure 5.26: Graph of $f'(x) = e^x(x + 1)^2$ on the window $-5 \le x \le -0.5, 0 \le y \le 1$

Example 7 Suppose that water is being poured into the vase in Figure 5.27 at a constant rate measured in volume per unit time. Graph $y = f(t)$, the depth of the water against time, t. Explain the concavity, and indicate the inflection points.

Solution At first the water level, y, rises quite slowly because the base of the vase is wide, and so it takes a lot of water to make the depth increase. However, as the vase narrows, the rate at which the water is rising increases. This means that initially y is increasing at an increasing rate, and the graph is concave up. The rate of increase in the water level is at a maximum when the water reaches the middle of the vase, where the diameter is smallest; this is an inflection point. After that, the rate at which y increases starts to decrease again, and so the graph is concave down. (See Figure 5.28.)

Figure 5.27: A vase

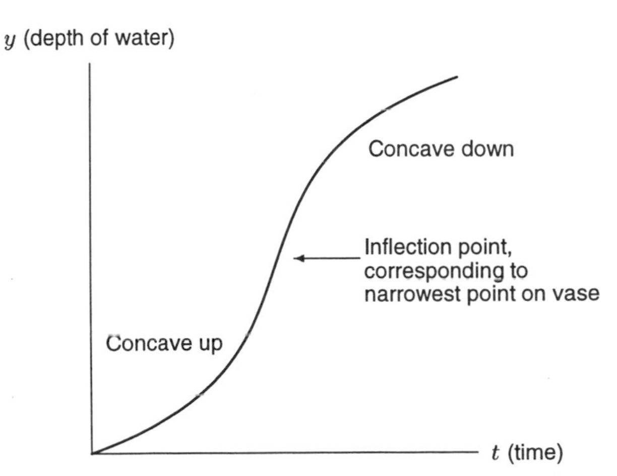

Figure 5.28: Graph of depth of water in the vase, y, against time, t

Problems for Section 5.2

1. How many inflection points does the function f shown in Figure 5.29 have? Indicate approximately where the inflection points are.

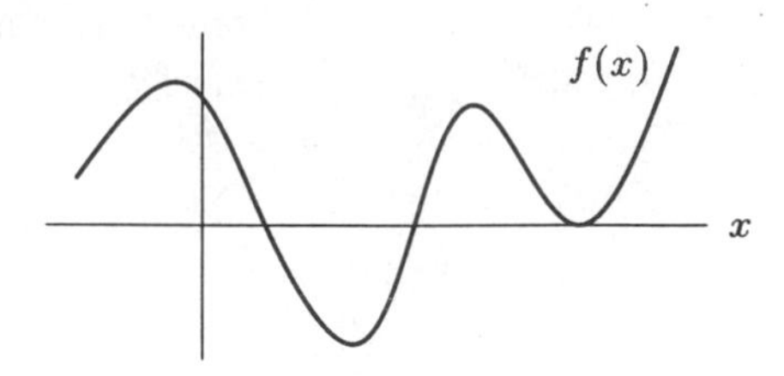

Figure 5.29

2. Indicate on the graph of the derivative f' in Figure 5.30 the x values that are inflection points of the function f itself.

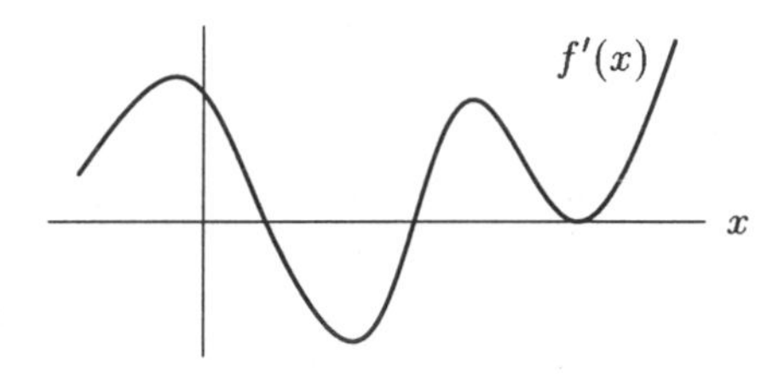

Figure 5.30

3. Indicate on the graph of the second derivative f'' in Figure 5.31 the x values that are inflection points of the function f itself.

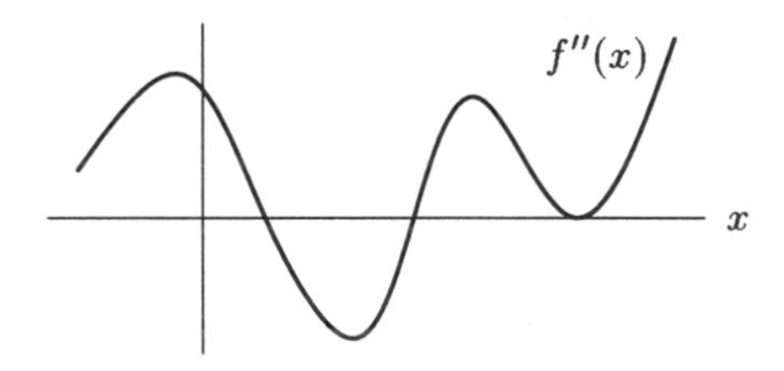

Figure 5.31

4. A function $f(x)$ has derivative $f'(x) = e^{x^2} + 2x - 1$. Show that $x = 0$ is a zero of this derivative and therefore that $x = 0$ is a critical point of f. Use the second derivative test to determine whether $x = 0$ is a local maximum, a local minimum, or neither.

Determine analytically the coordinates of all maxima and minima and points of inflection for the functions in Problems 5–8. Sketch the functions.

5. $y = x^4 - 4x^3 + 10$
6. $y = 2 + 3\cos x, \quad$ for $\quad 0 \leq x \leq 6\pi$
7. $y = \dfrac{1}{2}xe^{-10x}$
8. $f(x) = 3x^5 - 5x^3$, for $\quad -1.5 \leq x \leq 1.5$

For Problems 9–12, sketch graphs of f, f' and f''. Use these graphs to help you answer the questions.

9. For $f(x) = e^{x/2} - \ln(x^2+1)$, find the coordinates of all intercepts, maxima, minima, and inflection points to two decimal places.

10. For $f(x) = x^{10} - 10x, \quad 0 \leq x \leq 2$, find the value(s) of x for which $f(x)$ is increasing most rapidly; decreasing most rapidly.

11. For $f(x) = x - \ln x, \quad 0.1 \leq x \leq 2$, find the value(s) of x for which $f(x)$ is increasing most rapidly; decreasing most rapidly.

12. For $f(x) = x - \cos x, \quad 0 \leq x \leq \pi$, find the value(s) of x for which
    (a) $f(x)$ is greatest; $f(x)$ is least.
    (b) $f(x)$ is increasing most rapidly; decreasing most rapidly.
    (c) The slopes of the lines tangent to the graph of f are increasing most rapidly.

13. For the function, f, given in the graph in Figure 5.32:
    (a) Sketch $f'(x)$.
    (b) Where does $f'(x)$ change its sign?
    (c) Where does $f'(x)$ have local maxima or minima?

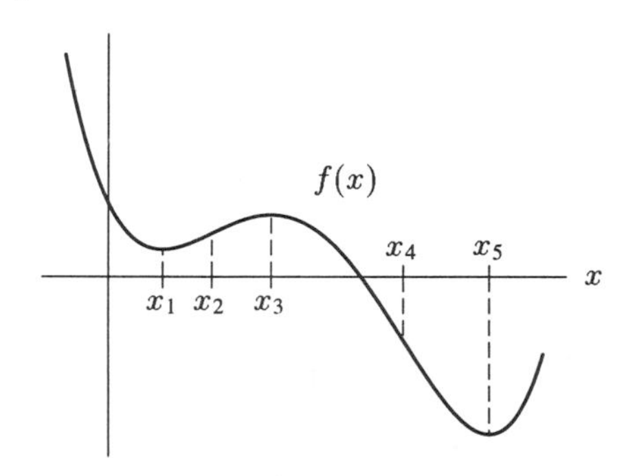

Figure 5.32

14. Using your answer to Problem 13 as a guide, write a short paragraph (using complete sentences) which describes the relationship between the following features of a function f:
    (a) The local maxima and minima of f.
    (b) The points at which f changes concavity.
    (c) The sign changes of f'.
    (d) The local maxima and minima of f'.

15. Suppose you know that a certain function f has $f(0) = 0$ and derivative given by

$$f'(x) = (\ln x)^2 - 2(\sin x)^4 \qquad \text{on the interval} \qquad 0 < x \leq 7.5$$

Use a calculator or computer to graph f' and its derivative. Clearly state where $f(x)$ is increasing and where it is decreasing, and where $f(x)$ is concave up and where it is concave down. Use this information to sketch a graph of $f(x)$.

16. Use concavity to show that $\ln x \leq x - 1$.
[Hint: What is the equation of the tangent line to $y = \ln x$ at $x = 1$?]

17. Sketch a graph of $f(x) = e^{1/x}$, with $x \neq 0$, on the interval $-5 \leq x \leq 5$.
    (a) For what values of x is the function increasing? Decreasing? Neither?
    (b) For what values of x is the graph of the function concave up? Concave down? Neither?

18. Let $f(x) = \ln(x^3 + 1)$, on the interval $-1 \le x \le 5$. Use graphs of f, f' and f'' to answer the following questions:
    (a) For what values of x is the function increasing? Decreasing? Neither?
    (b) For what values of x is the the graph of the function concave up? Concave down? Neither?

19. Assume the polynomial f has exactly two local maxima and one local minimum.
    (a) Sketch a possible graph of f.
    (b) What is the largest number of zeros f could have?
    (c) What is the least number of zeros f could have?
    (d) What is the least number of inflection points f could have?
    (e) Is the degree of f even or odd? How can you tell?
    (f) What is the smallest degree f could have?
    (g) Find a possible formula for $f(x)$.

For Problems 20–25, sketch a possible graph of $y = f(x)$, using the given information about the derivatives $y' = f'(x)$ and $y'' = f''(x)$. Assume that the function is defined and continuous for all real x.

20.

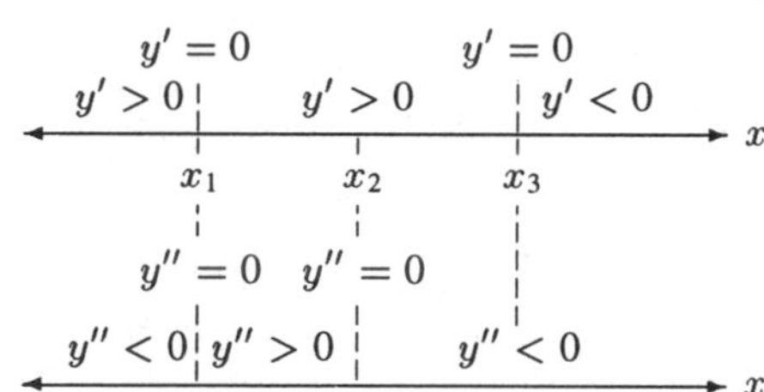

21.

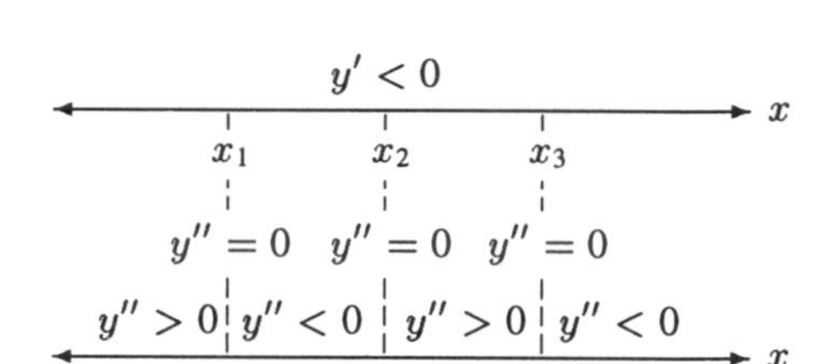

22.

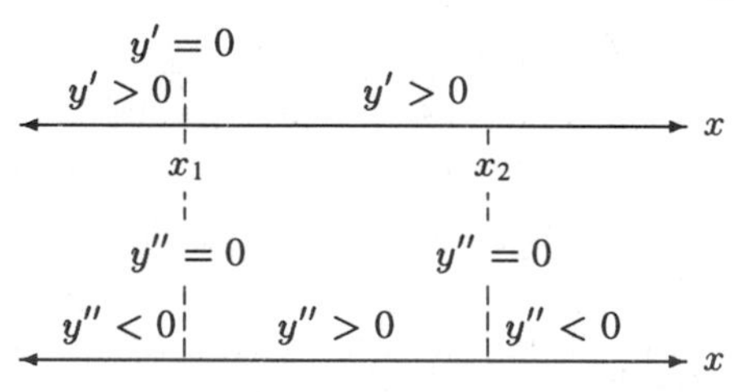

23.

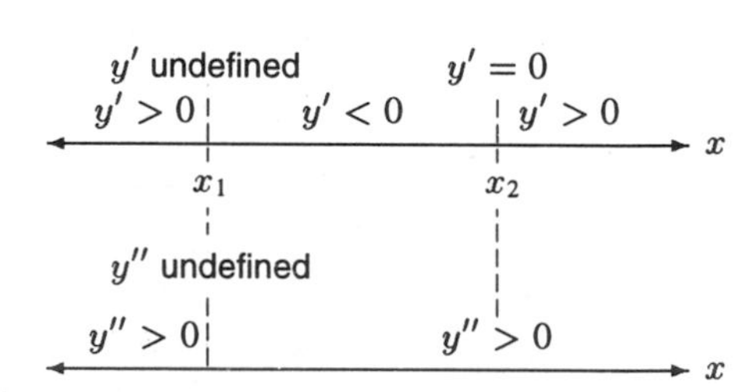

24.

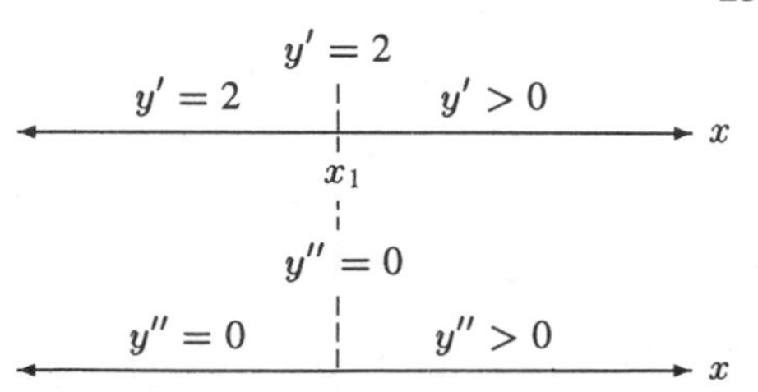

25.

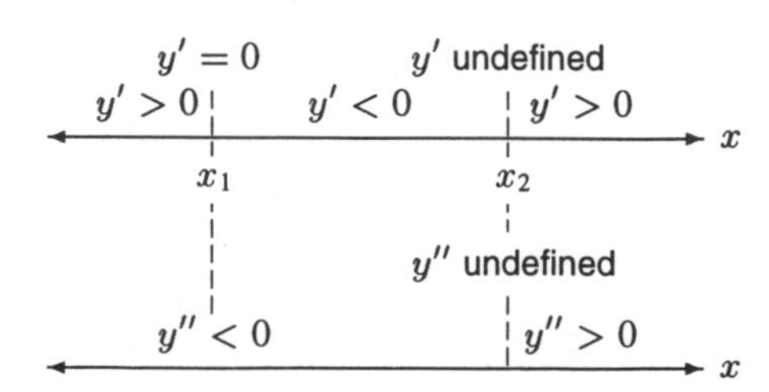

26. If water is flowing at a constant rate (i.e., constant volume per unit time) into the Grecian urn in Figure 5.33, sketch a graph of the depth of the water against time. Mark on the graph the time at which the water reaches the widest point of the urn.

Figure 5.33

27. If water is flowing at a constant rate (i.e., constant volume per unit time) into the vase in Figure 5.34, sketch a graph of the depth of the water against time. Mark on the graph the time at which the water reaches the corner of the vase.

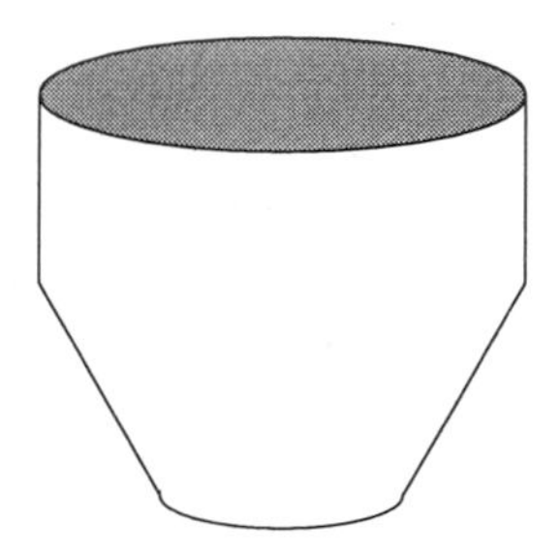

Figure 5.34

28. Given that $e^x \geq 1 + x$ for all x, let $x = 1/n$ and show that for every positive integer n,

$$e > \left(1 + \frac{1}{n}\right)^n.$$

Consider $x = -1/(n+1)$, and show that

$$e < \left(1 + \frac{1}{n}\right)^{n+1}$$

for all positive integers n.

Use your calculator with some specific choices of n to prove that $2.7 < e < 2.8$. Be clear about what choice of n you are making. Suppose you want to calculate e to 10 decimal places this way. Figure out a specific value of n that will enable you to pinch e into an interval of length less than 10^{-10}. (Try this on your calculator. Did you get a reasonable answer? If not, why not?)

5.3 FAMILIES OF CURVES: A QUALITATIVE STUDY

We saw in Chapter 1 that knowledge of one function can provide knowledge of the graphs of many others. The shape of the graph of $y = x^2$ also tells us, indirectly, about the shape of the graphs of $y = x^2 + 2$, $y = (x + 2)^2$, $y = 2x^2$, and countless other functions. We say that all functions of the form $y = a(x + b)^2 + c$ form a *family of functions*; their graphs are identical to that of $y = x^2$ except for shifts and stretches determined by the values of a, b, and c. The constants a, b, c are called *parameters*. Different values of the parameters give different members of the family.

A reason for studying families of functions is their use in mathematical modeling. Confronted with the problem of modeling some phenomenon, a crucial first step involves recognizing families of functions which might fit the available data.

Curves of the Form $y = ax^3 + bx^2 + cx + d$

This is the family of cubic polynomials. From the specific examples we encountered in earlier chapters, we expect the graphs in this family to be S-shaped.

The effect of the parameter d is easy to see. It is the y-intercept of the curve, and changing d shifts the entire curve up or down without changing its shape. Since this is easy enough to account for, we set $d = 0$ and study the family $y = ax^3 + bx^2 + cx$; each of these curves passes through the origin.

One effect of the parameter a, called the leading coefficient, is also easy to observe. No matter what the size of a, provided $a \neq 0$, for large positive and negative values of x the term ax^3 dominates $bx^2 + cx$, and so for such values the curve looks like $y = ax^3$. If $a = 0$, we have a quadratic polynomial whose graph will be a parabola. For simplicity, we will assume that $a > 0$.

For the family $y = ax^3 + bx^2 + cx$, we have

$$\frac{dy}{dx} = 3ax^2 + 2bx + c, \qquad \frac{d^2y}{dx^2} = 6ax + 2b.$$

The formula for dy/dx is a quadratic polynomial, so it may equal 0 at exactly zero, one, or two points. Hence a given cubic may have *no* critical points (e.g., $y = x^3 + x$), *one* critical point (e.g., $y = x^3$), or *two* critical points (e.g., $y = x^3 - x$).

However, every cubic has exactly one inflection point. It occurs where $d^2y/dx^2 = 0$, which is at

$$x = \frac{-b}{3a}.$$

The second derivative changes from negative to positive at this point, so the graph changes from concave down to concave up there.

Since the graph of a cubic is S-shaped, it must look like one of the three possible cases in Figure 5.35. (Remember, we're assuming that $a > 0$.) The reasoning behind these sketches is as follows: in case (a), there is no critical point and one inflection point where the concavity changes. The curve must be always increasing. This is so because dy/dx can never change sign and hence must be always positive, or always negative. Since $a > 0$ and therefore $y \to +\infty$ as $x \to +\infty$, dy/dx must be always positive. A similar argument goes for case (b) where there is one critical point. The curve has a horizontal tangent at one point. It is not hard to show that in this case the critical point and the inflection point coincide. For case (c) where there are two critical points, the curve must get from $y = -\infty$ to $y = +\infty$ as x goes from $-\infty$ to $+\infty$, changing concavity once and having two horizontal tangents. If we disregard the size of the critical values (how high the peak, how low the valley) the shape must be something like case (c).

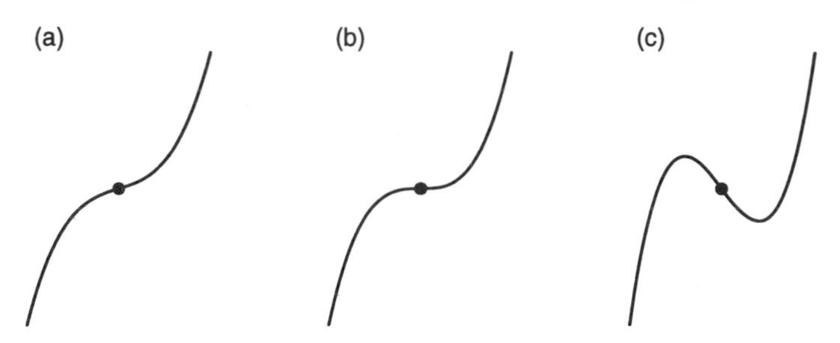

Figure 5.35: A cubic with (a) no critical points, (b) one critical point, (c) two critical points (The dots mark the inflection points.)

Curves of the Form $y = xe^{-bx}$

We will consider only $x \geq 0$. The graph of one member of the family, with $b = 1$, is shown in Figure 5.36.

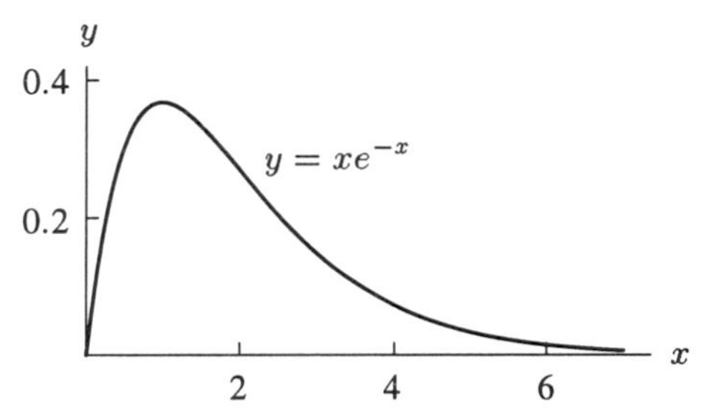

Figure 5.36: One member of the family $y = xe^{-bx}$

Such a graph represents a quantity that increases rapidly at first, and then decreases toward zero. For example, the number of bacteria in a person during the course of an infection, from onset to cure, might be described in this way. The question is, how exactly does the shape of the curve change when b varies? For simplicity we will consider only curves with $b > 0$. Note that all curves of the family pass through the origin. The behavior of the curves as $x \to \infty$ is also pretty clear. Since $b > 0$ (by assumption) and e^{bx} grows much faster than x as $x \to \infty$, we must have

$$xe^{-bx} = \frac{x}{e^{bx}} \to 0 \quad \text{as} \quad x \to \infty.$$

Geometrically, this means that the x-axis is a horizontal asymptote.

What effect does the parameter b have on the graph? Try sketching the graphs for different positive values of b. What do you notice? You should see that the general shape of the graphs is the same, but for small b, the maximum is higher and more to the right than for larger values of b. See Figure 5.37. We use analytical methods to determine precisely how b affects the maximum.

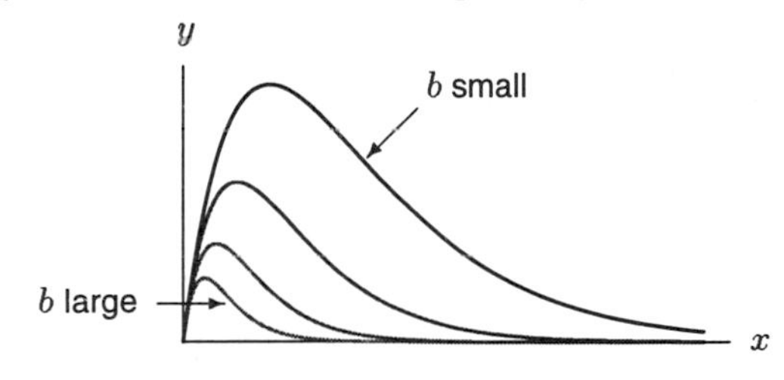

Figure 5.37: Graph of $y = xc^{-bx}$, with b varying.

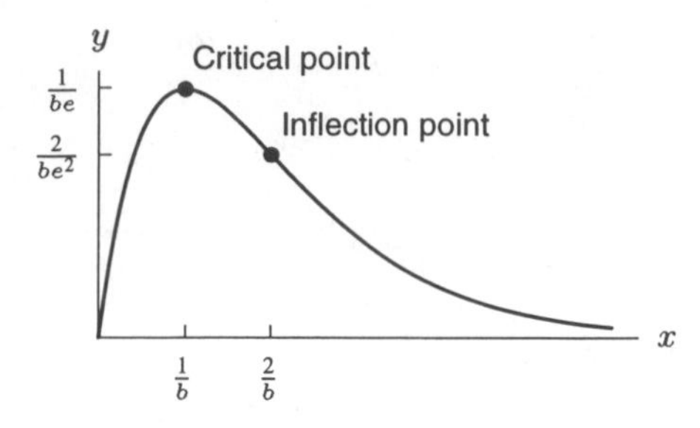

Figure 5.38: Graph of $y = xe^{-bx}$

First we find and classify critical points. We compute

$$\frac{dy}{dx} = x\left(-be^{-bx}\right) + e^{-bx} = (1 - bx)e^{-bx},$$

There is a critical point where $dy/dx = 0$, that is , where $(1 - bx)e^{-bx} = 0$. Since e^{-bx} is never zero, the only critical point is $x = 1/b$. Looking at the expression for dy/dx, we can see that if $x < 1/b$ then $dy/dx > 0$, and if $x > 1/b$, then $dy/dx < 0$. Hence there is a local maximum at $x = 1/b$. The maximum value is

$$y = \frac{1}{b}e^{-b(1/b)} = \frac{1}{be}.$$

Second we check concavity. We compute

$$\frac{d^2y}{dx^2} = (1 - bx)\left(-be^{-bx}\right) + (-b)e^{-bx} = -b(2 - bx)e^{-bx}.$$

The second derivative shows that $d^2y/dx^2 = 0$ only where $2 - bx = 0$. Solving for x gives $x = 2/b$. It is easy to see from the expression for d^2y/dx^2 that the second derivative changes from negative to positive at $x = 2/b$, so the graph changes from concave down to concave up at $x = 2/b$, which is therefore an inflection point. (See Figure 5.38.) At that point

$$y = \frac{2}{b}e^{-b(2/b)} = \frac{2}{be^2}.$$

The most obvious effect of varying b is to move the critical point (local maximum) of the curve. If b is large, then $1/b$ is small and so xe^{-bx} reaches a low maximum very quickly. If b is small, then $1/b$ is very large, so the curve rises very high before the decay towards zero begins. The inflection point moves as b varies, too. It will always be twice as far from the y-axis as the critical point. See Figure 5.39.

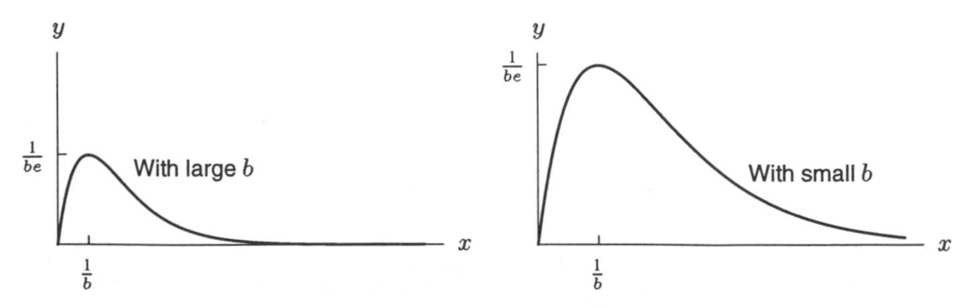

Figure 5.39: Graph of $y = xe^{-bx}$ with large b and small b

Figure 5.37 shows several curves in the family on the same set of axes. Note also that all the curves have the same tangent line at the origin. This is because the derivative dy/dx has the same value at $x = 0$ for every b, namely $dy/dx = 1$. (Note that the scales on the x and y axes are different.)

Curves of the Form $y = a(1 - e^{-bx})$

The graph of one member of the family, with $a = 2$ and $b = 1$, is shown in Figure 5.40. Such a graph represents a quantity which is increasing but leveling off. For example, a body falling against air resistance speeds up and its velocity levels off as it approaches the terminal velocity. Similarly, if a pollutant pouring into a lake builds up toward a saturation level, its concentration may be described in this way. The graph might also represent the temperature of a cooking yam.

How does the shape change if a and b vary? For simplicity, we will assume that a and b are positive. We are interested in how the shape of the graph changes when one parameter is fixed and the other is allowed to vary.

First, we examine the effect of a on the graph. Fix b at some positive number, say $b = 1$. Substitute different values for a and look at the graphs as in Figure 5.41. How do the graphs change? We see that as x gets larger and larger, y approaches a. Thus $y = a$ is a horizontal asymptote. To see this analytically, notice that as $x \to \infty$, $e^{-bx} \to 0$ and so $y = a(1 - e^{-bx})$ approaches $y = a$ from below. Physically, a may represent the terminal velocity of a falling object or the saturation level of a pollutant in a pond.

Now we examine the effect of b on the graph. Fix a at some positive number, say $a = 2$. Substitute different values for b and determine how the graphs change. Notice that the parameter b determines how sharply the curve rises and how soon it gets close to the line $y = a$. The larger b is, the more rapid is the rise. See Figure 5.42.

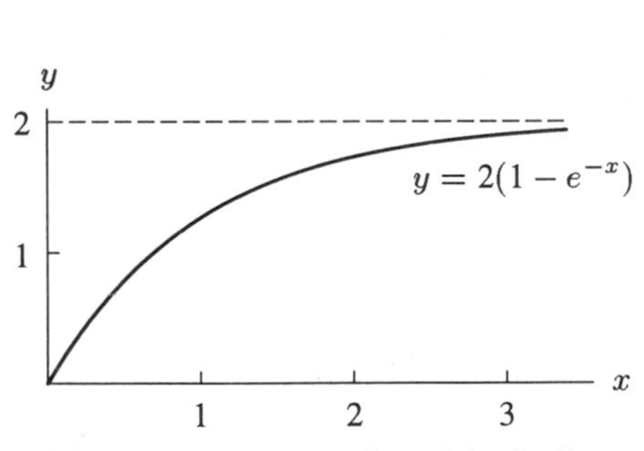

Figure 5.40: One member of the family $y = a(1 - e^{-bx})$

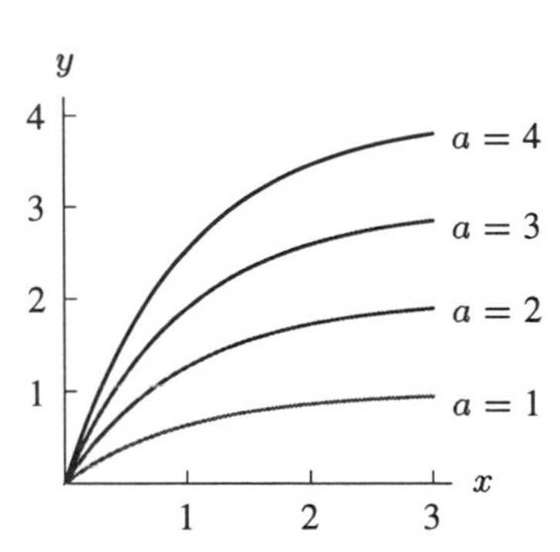

Figure 5.41: $y = a(1 - e^{-x})$ for various a

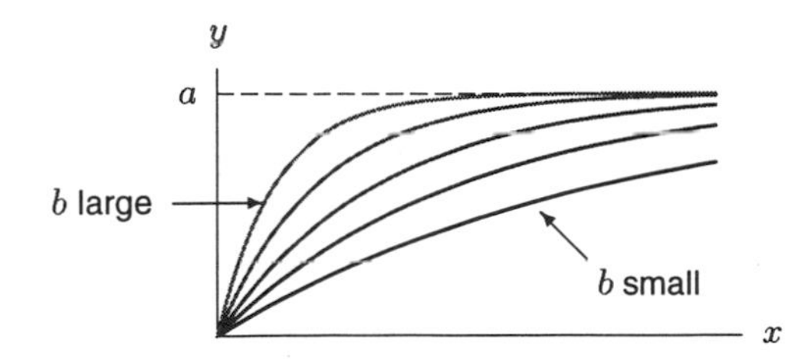

Figure 5.42: $y = a(1 - e^{-bx})$, with a fixed, b varying

Curves of the Form $y = e^{-(x-a)^2/b}$

This is a multiple of the *normal density* function, used in probability and statistics. In applications, it is common to scale the function by multiplying it by a positive constant, but we will just consider the two parameter family depending on a and b, and we will assume that $b > 0$. The graph of any member of the family is a bell-shaped curve as shown in Figure 5.43 (where $a = 1$ and $b = 2$).

As we saw in Chapter 1, the role of a is to move the curve to the right or left. Notice that for any curve in the family, y is always positive. Since $b > 0$, $y \to 0$ as $x \to \pm\infty$ for any choice of a. Thus, the x-axis is always a horizontal asymptote. The graph is a bell-shaped curve with peak at $x = a$, and with two inflection points. See Figure 5.44.

What effect does the parameter b have on this curve? By keeping a fixed and experimenting with different values for b, we see that b influences how narrow or wide the bell is. When b is small, the bell is narrow, and when b is large, the bell is wider. See Figure 5.45

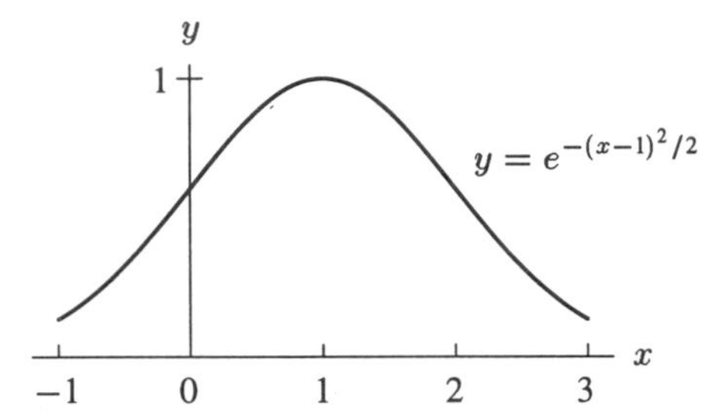

Figure 5.43: Graph of one member of the family $y = e^{-(x-a)^2/b}$

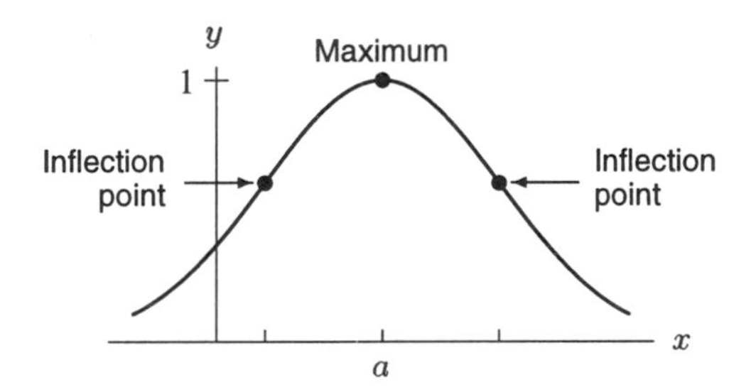

Figure 5.44: Graph of $y = e^{-(x-a)^2/b}$: bell-shaped curve with peak at $x = a$

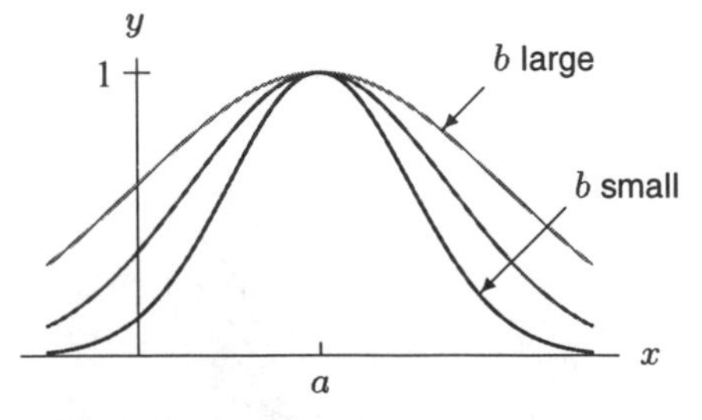

Figure 5.45: Graph of $y = e^{-(x-a)^2/b}$ with various values of b (a fixed)

Curves of the Form $y = \dfrac{a}{1+100e^{-bt}}$

We will consider only $t \geq 0$, $a > 0$, and $b > 0$. Functions in this family are known as *logistic curves.*

The graph of one member of this family of curves, with $a = 1$ and $b = 1$ is shown in Figure 5.46. A logistic curve is used to model many situations, from sales of a new product to the spread of a virus. Notice that the graph is everywhere increasing. There is one inflection point, and the graph is concave up to the left of the inflection point and concave down to the right of the inflection point. The graph has a horizontal asymptote. We will study logistic curves in more detail in Chapter 7.

We first determine the effect of the parameter a on the graph. Fix b at some positive number, and substitute different values for a as in Figure 5.47. What do you observe in the graph? As t gets larger and larger, y approaches a and so there is a horizontal asymptote at $y = a$. (You are asked in Problem 11 to explain analytically why this is so.)

What is the effect of the parameter b on the graph? We fix a and experiment with different values for b. We notice that as b increases, the curve levels off at the asymptote much more rapidly, and the slope at the inflection point is larger. See Figure 5.48.

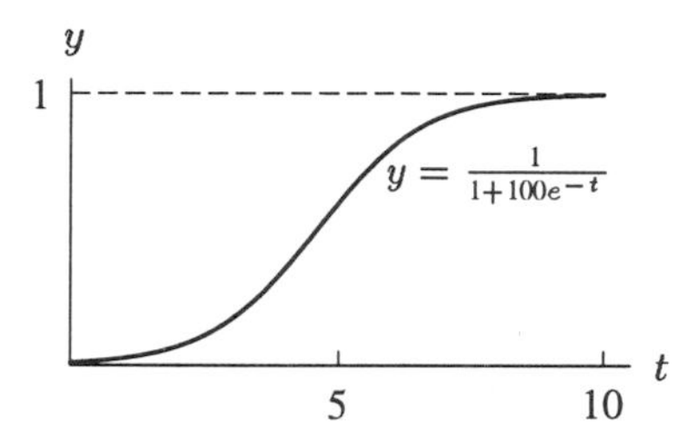

Figure 5.46: Graph of one member of the family $y = \dfrac{a}{1+100e^{-bt}}$

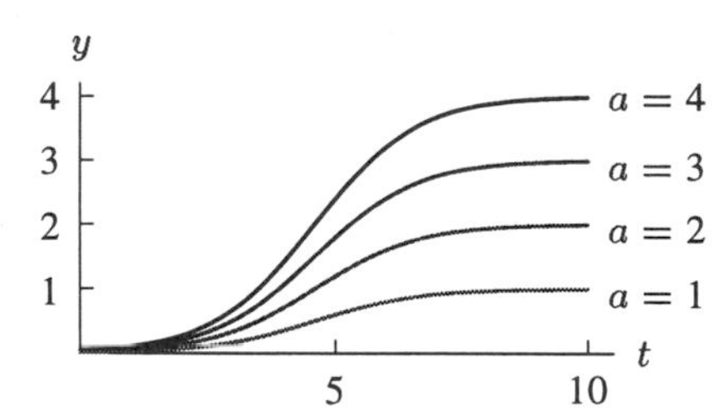

Figure 5.47: Graph of $y = \dfrac{a}{1+100e^{-t}}$ for various values of a

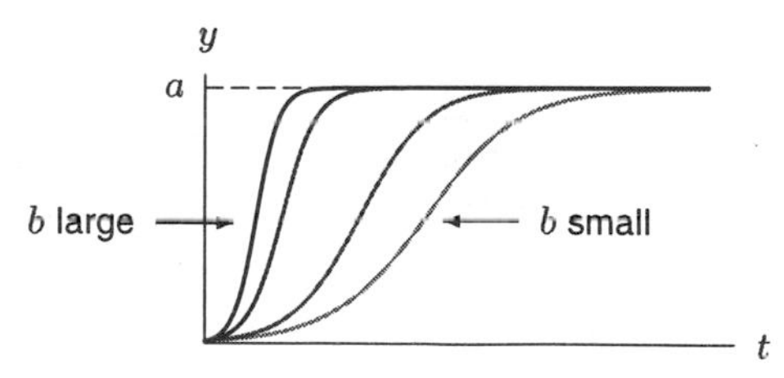

Figure 5.48: Graph of $y = \dfrac{1}{1+100e^{-bt}}$ with various values of b (a fixed)

Problems for Section 5.3

1. What effect do the parameters a and b have on the parabola $y = x^2 + ax + b$?

2. Consider the family of curves
$$y = A(x + B)^2$$
   (a) Assuming $B = 0$, what is the effect of varying A on the graph? Consider:
      (i) $A > 0$ and $A < 0$.
      (ii) $A > 1$ and $0 < A < 1$.
   (b) Assuming $A = 1$, what is the effect of varying B on the graph?
   (c) Write a couple of sentences describing the role played by A and B in determining the shape of the graph. Illustrate your answer with sketches.

3. Consider the family
$$y = \frac{A}{x + B}.$$
   (a) If $B = 0$, what is the effect of varying A on the graph?
   (b) If $A = 1$, what is the effect of varying B?
   (c) On one set of axes, graph the function for several values of A and B.

4. Consider the family of curves
$$y = ae^{-bx^2}, \qquad \text{for positive } a,\ b.$$
   Analyze the effect of varying a and b on the shape of the curve. Illustrate your answer with sketches.

5. Consider the family of curves
$$y = A\sin(Bx + C).$$
   What is the effect of varying A, B, and C on the shape of the graph? For each parameter, sketch some graphs for various values (holding the others constant).

6. Consider the *surge function*
$$y = axe^{-bx}, \qquad \text{for positive } a,\ b.$$
   (a) Find the maxima, minima, and points of inflection.
   (b) How does varying a and b affect the shape of the graph?
   (c) On one set of axes, graph this function for several values of a and b.

7. Consider the family of curves
$$y = e^{-ax}\sin(bx)$$
   with $x \geq 0$ and a and b both positive.
   (a) Sketch a graph of the function with $a = 1$ and $b = 2$. Describe the general shape of this graph and explain in terms of the formula why this shape makes sense.
   (b) What effect does the parameter a have on the graph?
   (c) What effect does the parameter b have on the graph?

8. Determine analytically under what conditions on a, b, and c, the cubic
$$f(x) = x^3 + ax^2 + bx + c$$
   is increasing everywhere.

9. Consider the family of curves

$$y = a(x - b \ln x)$$

with $x \geq 0$ and a and b both positive. What is the effect of varying a and b on the shape of the graph? For each parameter, sketch some graphs for various values (holding the other parameter constant). Label the x-coordinate of the minimum on your graphs.

10. In this section, we studied the logistic curve $y = \dfrac{a}{1 + 100e^{-bt}}$. Now we introduce a third parameter, C. Consider the family of functions

$$y = \frac{a}{1 + Ce^{-bt}}.$$

If a and b are fixed, what is the effect of the paramter C on the curve? Explain your answer and illustrate your answer with a sketch.

11. Explain analytically (by considering what happens to y as $x \to \infty$) why the logistic curve $y = \dfrac{a}{1 + 100e^{-bt}}$ with $b > 0$ has a horizontal asymptote at $y = a$.

12. Consider the function $p(x) = x^3 - ax$, where a is constant.
   (a) If $a < 0$, show analytically that $p(x)$ is always increasing.
   (b) If $a > 0$, show analytically that $p(x)$ has a local maximum and a local minimum.
   (c) Sketch and label typical graphs for the cases when $a < 0$ and when $a > 0$.

13. Consider the function $p(x) = x^3 - ax$, where a is constant and $a > 0$.
   (a) Find the local maxima and minima of p.
   (b) What effect does increasing the value of a have on the positions of the maxima and minima?
   (c) On the same axes, sketch and label the graphs of p for three positive values of a.

14. (a) Find all critical points of $f(x) = x^4 + ax^2 + b$.
   (b) Under what conditions on a and b will this function have exactly one critical point? What is the one critical point, and is it a local maximum, a local minimum, or neither?
   (c) Under what conditions on a and b will this function have exactly three critical points? What are they and which are local maxima and which are local minima?
   (d) Is it ever possible for this function to have two critical points? No critical points? More than three critical points? Give an explanation in each case.

15. The number, N, of people who have heard a rumor spread by mass media is modeled by the following function of time, t:

$$N = a(1 - e^{-kt}).$$

Suppose that there are 200,000 people in the population who hear the rumor eventually. If 10% of them heard in the first day, find a and k, assuming t is measured in days.

16. Suppose the temperature, T, of a yam put into a hot oven maintained at 200°C is given as a function of time by

$$T = a(1 - e^{-kt}) + b$$

where T is in degrees Celsius and t is in minutes.
   (a) If the yam starts at 20°C, find a and b.
   (b) If the temperature of the yam is initially increasing at 2°C per minute, find k.

5.4 APPLICATIONS TO ECONOMICS: OPTIMIZATION

The fundamental issue for a producer of goods is how to maximize profit. Recall that the revenue function, $R(q)$, gives total revenue from selling a quantity of goods q, and that the cost function, $C(q)$, gives the total cost of producing a quantity q. The profit $\pi(q)$ is the difference: $\pi(q) = R(q) - C(q)$.

Maximizing Profit

We begin by revisiting Example 4 in Section 2.7, in which we were asked to find the maximum profit if the total revenue and total cost functions are given by the curves R and C, respectively, in Figure 5.49. Since profit = revenue − cost = $R(q) - C(q)$, we see that profit is maximized when $R(q) > C(q)$ and the vertical distance between the two curves is at a maximum. This occurs at about $q = 140$.

In Section 2.7, we determined that this is the point where the slopes of the two curves $R(q)$ and $C(q)$ are equal. Does this always have to be so? The slopes of the curves are given, respectively, by the marginal revenue, $MR = R'(q)$, and the marginal cost $MC = C'(q)$. If $R' > C'$, we increase our profit by producing more items – the revenue from the additional items exceeds the cost of producing them. If $R' < C'$, we increase our profit by producing fewer items – the cost of producing the items outweighs the revenues and so we are better off not making the items. The profit, $R(q) - C(q)$, is largest when the slopes of the curves are equal, which is at the point where marginal revenue equals marginal cost.

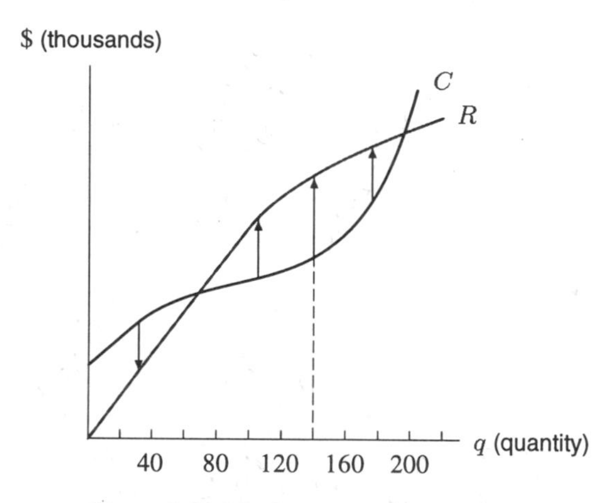

Figure 5.49: Maximum profit at $q = 140$

Let's look at the general situation: To maximize or minimize profit over an interval, we optimize the profit function π where

$$\pi(q) = R(q) - C(q).$$

According to Section 5.1, global maxima and minima of a function can only occur at critical points of the function or at the endpoints, if any, of the interval. To find critical points of π, look for zeros of the derivative:

$$\pi'(q) = R'(q) - C'(q) = 0.$$

So

$$R'(q) = C'(q),$$

that is, the slopes are equal. In economic language:

> The maximum (or minimum) profit can only occur when
>
> $$\text{Marginal cost} = \text{Marginal revenue}.$$

Of course, maximal or minimal profit does not *have* to occur where $MR = MC$; there are always the endpoints to consider. However, this relationship is quite powerful, because it is the condition that helps determine the maximum (or minimum) profit in general.

Example 1 Find the quantity q which will maximize profit if the total revenue and total cost (in dollars) are given by

$$R(q) = 5q - 0.003q^2$$
$$C(q) = 300 + 1.1q$$

where $0 \leq q \leq 1000$ units. What production level will give the minimum profit?

Solution We can begin by looking for production levels that give Marginal revenue = Marginal cost:

$$MR = R'(q) = 5 - 0.006q$$
$$MC = C'(q) = 1.1.$$

So

$$5 - 0.006q = 1.1$$
$$q = \frac{3.9}{0.006} = 650 \text{ units.}$$

Does this represent a local maximum or minimum of π? We can tell by looking at what is going on at production levels of 649 units and 651 units. When $q = 649$ we have $MR = \$1.106$, which is greater than the (constant) marginal cost of \$1.10. This means that producing one more unit will bring in more revenue than its cost, so profit will increase. When $q = 651$, $MR = \$1.094$, which is *less* than MC, so it is not profitable to produce the 651st unit. We conclude that $q = 650$ is a local maximum for the profit function π. The profit earned by producing and selling this quantity is $\pi(650) = R(650) - C(650) = \$1982.50 - \$1015 = \$967.50$.

To check for global maxima we need to look at the endpoints. If $q = 0$, the only cost is \$300 (the fixed costs) and there is no revenue, so $\pi(0) = -300$. At the upper limit of $q = 1000$, $R(1000) = \$2000$, $C(1000) = \$1400$ and so $\pi(1000) = \$600$. Therefore, the maximum profit is where $MR = MC$, at $q = 650$, and the minimum profit occurs when $q = 0$.

Example 2 The total revenue and total cost curves for a product are given in Figure 5.50.

(a) Sketch the curves for marginal revenue and marginal cost on the same axes, and indicate on this graph for what quantity profit is maximized.

(b) Sketch the general shape of the profit function $\pi(q)$. Assume that the fixed costs of production are positive.

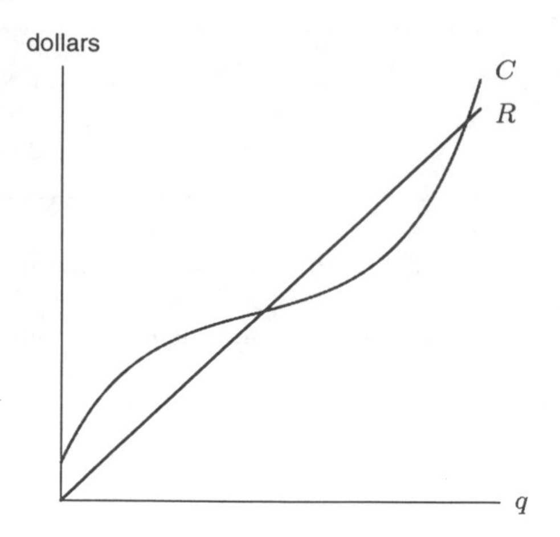

Figure 5.50: Total revenue and total cost

Solution (a) The graphs of total revenue $R(q)$ and total cost $C(q)$ are given in Figure 5.50. To graph marginal revenue and marginal cost, we sketch graphs of the derivatives of $R(q)$ and $C(q)$, respectively. Since $R(q)$ is a straight line with positive slope, the graph of marginal revenue, MR, is a horizontal line. (See Figure 5.51.) We see that $C(q)$ is always increasing, so marginal cost, MC, is always positive. As q increases, the cost function goes from concave down to concave up, so the slope of the cost function goes from decreasing to increasing. Therefore, the graph of marginal cost, MC, is decreasing and then increasing, as shown in Figure 5.51. The local minimum on the marginal cost curve corresponds to the inflection point of $C(q)$.

Where is profit maximized? We know that the maximum profit can occur when marginal revenue equals marginal cost. We see in Figure 5.51 that the marginal cost curve crosses the marginal revenue curve at two points, labeled q_1 and q_2, and so there are two points for which $MR = MC$. Which of these gives maximum profit?

We first consider a production size of q_1. To the left of the point q_1, we have $MC > MR$. What does this tell us about our profit function π? Since $\pi = R - C$, we have $\pi' = R' - C' = MR - MC$. Since $MR < MC$ to the left of q_1, we see that π' is negative there and the profit

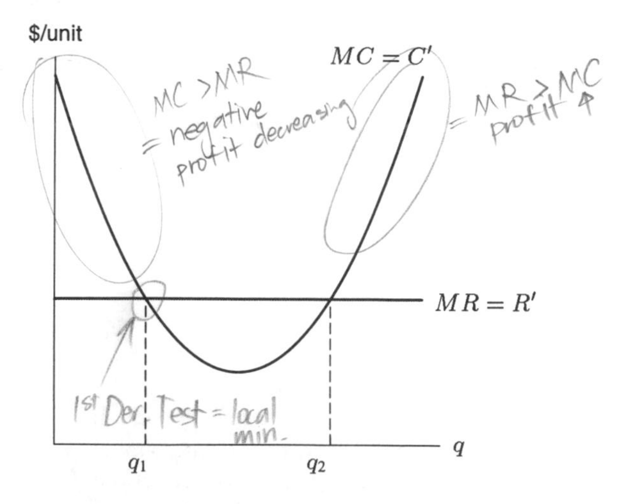

Figure 5.51: Marginal revenue and marginal cost

function is decreasing. To the right of q_1, we have $MR > MC$, so π' is positive and the profit function is increasing. By the first derivative test, the profit function has a local minimum at q_1. This is certainly not the production level we want.
What happens at q_2? To the left of q_2, we have $MR > MC$, so π' is positive and the profit function is increasing. To the right of q_2, we have $MC > MR$, so π' is negative and the profit function is decreasing. By the first derivative test, the profit function has a local maximum at q_2. The global maximum we seek will occur either at an endpoint (the largest and smallest possible production levels) or at the production level q_2.

(b) We saw that the profit function is decreasing to the left of q_1, increasing between q_1 and q_2, and decreasing to the right of q_2. Thus, as production increases, the profit function decreases, then increases, and then decreases again. What is the vertical intercept for the profit function? Since $\pi(q) = R(q) - C(q)$, the vertical intercept is $\pi(0) = R(0) - C(0)$. We know that $R(0) = 0$ and that $C(0)$ is equal to the fixed costs of production. Therefore the vertical intercept of the profit function is a negative number, equal in magnitude to the size of the fixed cost. Thus, the graph of the profit function has the general shape shown in Figure 5.52.

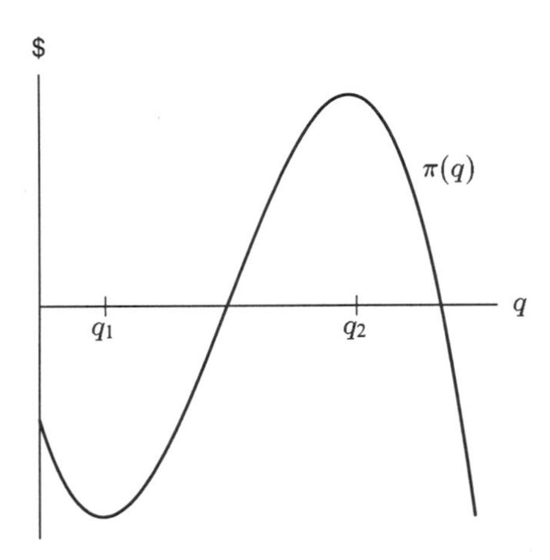

Figure 5.52: Profit function

Minimizing Average Costs

An individual company in a large industry always wants to maximize its profit. However it alone cannot affect demand for its product because other companies beyond its control also produce and sell the same product. The mechanism of the market will tend to force the supply of the product for the industry as a whole to the level at which price equals the minimum average cost of each firm, even though this does not maximize the profit for the industry as a whole. (This is why companies have an incentive to collude - by doing so they can increase their profit.) In this section we study the quantity of production that yields the minimum average cost.

As we saw in Chapter 2, the average cost of producing a good is equal to the total cost of production divided by the quantity produced. If $a(q)$ represents the average cost per item in producing q items, we have $a(q) = C(q)/q$. Figure 5.53 shows that $C(q)/q$ is the slope of a line from the origin $(0, 0)$ to the point $(q, C(q))$ on the graph of the cost function. This enables us to visualize average cost as a slope.

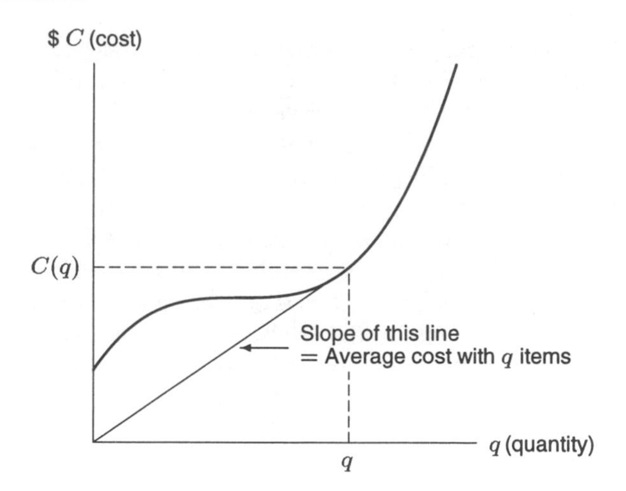

Figure 5.53: Visualizing average cost as a slope

How does average cost change as the quantity produced increases? Figure 5.54 shows the graph of a cost function, with several lines drawn from the origin to the cost curve. The lines with larger slopes correspond to production levels with higher average costs, and those with smaller slopes correspond to production levels with lower average costs. The slopes start out steep, at low production levels, and get less steep as production increases up to a certain level. Beyond this level, the slopes get steeper again. This tells us that average cost decreases and then increases as production increases. In Figure 5.54, we see that the slope of the line through the origin, and hence the average cost, is minimized at production level q_0.

Let's examine the production level q_0 in Figure 5.54 more closely. At this point, a line from the origin to the cost curve has minimum slope. We see in Figure 5.54 that this minimum occurs at the point where the line from the origin is tangent to the cost curve. At this point, the slope of the line from the origin to the cost curve is equal to the slope of the tangent line to the cost curve. Since the slope of the line from the origin to the cost curve is equal to average cost, and the slope of the tangent line to the cost curve is equal to marginal cost, we see that this is the point where average cost equals marginal cost. We see graphically that the minimum average cost occurs at the production level for which average cost equals marginal cost.

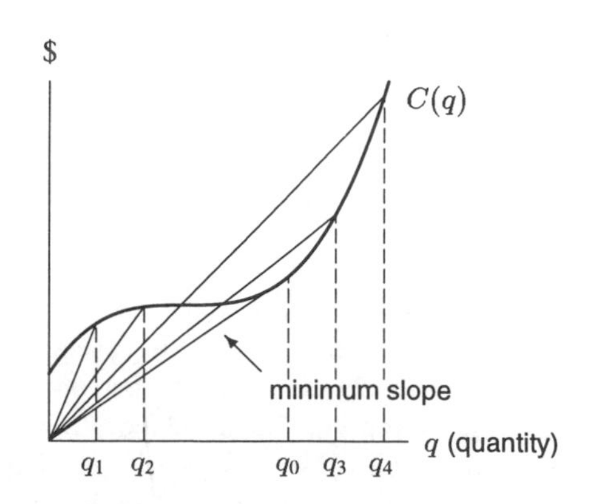

Figure 5.54: Visualizing average cost

In the next example we consider what happens when marginal cost and average cost are not equal.

Example 3 Suppose 100 items are produced at an average cost of \$2 per item. Find the average cost of producing 101 items:

(a) if the marginal cost to produce the 101st item is \$1.

(b) if the marginal cost to produce the 101st item is \$3.

Solution (a) If 100 items are produced at an average cost of \$2 per item, the total cost of producing the items is \$200. Since the marginal cost to produce the 101st item is \$1, it costs \$1 more to produce the additional item. Total costs for producing 101 items will therefore be \$201. The average cost to produce these items is 201/101, or \$1.99 per item. The average cost has gone down. This makes sense: if it costs less than the average to produce additional items, producing them will decrease the average cost.

(b) In this case, the marginal cost to produce the 101st item is \$3. The total cost to produce 101 items will be \$203 and the average cost will be 203/101, or \$2.01 per item. Our average cost has gone up. Again, this makes sense: if it costs more than the average to produce additional items, average costs will increase when these items are produced.

Our graphical analysis of Figure 5.54 and the results of Example 3 reveal the relationship between marginal cost and average cost.

> Average cost and marginal cost are related as follows:
> (1) Minimum average cost occurs when marginal cost equals average cost.
> (2) If marginal cost is less than average cost, average cost is reduced by increasing production.
> (3) If marginal cost is greater than average cost, average cost is increased by increasing production.

We can demonstrate (1) analytically. Recall that average cost is given by $a(q) = C(q)/q$. To minimize average cost, we find the critical points of $a(q)$.

Use the quotient rule to find $a'(q)$:

$$a'(q) = \frac{q \cdot C'(q) - C(q) \cdot 1}{q^2}.$$

Set $a'(q) = 0$ to find the critical points:

$$\begin{aligned}
\frac{q \cdot C'(q) - C(q)}{q^2} &= 0 \\
q \cdot C'(q) - C(q) &= 0 \\
q \cdot C'(q) &= C(q) \\
C'(q) &= \frac{C(q)}{q} \\
\text{Marginal cost} &= \text{Average cost.}
\end{aligned}$$

Example 4 Assume that the total cost function is given by $C(q) = q^3 - 6q^2 + 12q$, in thousands of dollars, where q is measured in thousands and $0 \leq q \leq 5$.

(a) Sketch a graph of $C(q)$ and estimate visually the quantity at which average cost is minimized.

(b) Determine analytically the exact value of q at which average cost is minimized.

(c) Sketch a graph of the average cost function, and use the graph to find the minimum average cost.

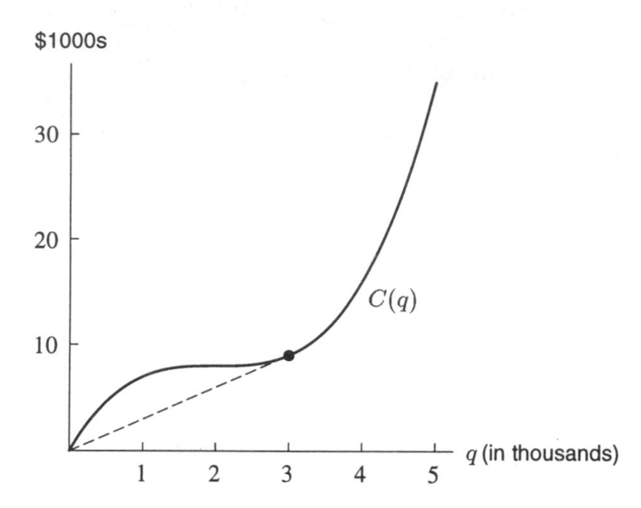

Figure 5.55: Minimizing average cost

Solution (a) A graph of $C(q)$ is given in Figure 5.55. Average cost is minimized at the point where a line between the origin and a point on the curve has minimum slope. This minimum slope occurs where the line through the origin is tangent to the curve. In Figure 5.55, this point is at approximately $q = 3$, corresponding to a production of 3000 units.

(b) We must find the quantity for which marginal cost equals average cost. Marginal cost equals the derivative $C'(q) = 3q^2 - 12q + 12$, so we have:

$$\begin{aligned}
\text{Marginal cost} &= \text{Average cost.} \\
3q^2 - 12q + 12 &= \frac{q^3 - 6q^2 + 12q}{q} \\
3q^2 - 12q + 12 &= q^2 - 6q + 12 \\
2q^2 - 6q &= 0 \\
2q(q-3) &= 0
\end{aligned}$$

There are two roots, at $q = 0$ and at $q = 3$. Since $q = 0$ is not a meaningful answer, we see that the average cost is minimized at $q = 3$. To minimize average costs, production should be set at 3000 units.

(c) Since average cost is equal to total cost divided by quantity, we have

$$a(q) = \frac{C(q)}{q} = \frac{q^3 - 6q^2 + 12q}{q} = q^2 - 6q + 12.$$

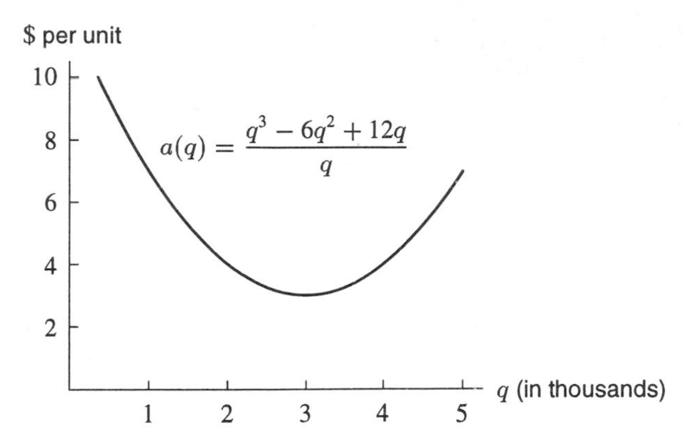

Figure 5.56: Graph of the average cost function

A graph of this function is shown in Figure 5.56. We see graphically that the minimum average cost occurs at $q = 3$, which is a production of 3000 units. At this production level, the average cost is 3 dollars per unit, and the total cost is $9000.

Maximizing Revenue

Suppose you own a city bus company with a fixed schedule. Your costs will be the same no matter how many people ride the buses, so you will maximize profit by maximizing your revenue. What price should you charge for a ticket to maximize revenue?

In practice, the quantity sold of a good will often depend on the price charged. We know that revenue is equal to price times quantity. Recall that the demand equation introduced in Chapter 1 tells us how consumer demand depends on the price of the product. Using the demand equation, we can write the quantity explicitly as a function of price, $q = f(p)$. If we substitute $f(p)$ for q in our revenue function, we obtain revenue as a function of price. Similarly, if we write the demand equation with price as an explicit function of quantity and substitute this for price in our revenue function, we will have revenue as a function of quantity. In either case, it makes sense to talk about the price (or quantity, respectively) that will maximize revenue.

Example 5 Suppose the demand equation for all-day passes to an amusement park is given by $p = 70 - 0.02q$, where p is the price of a pass in dollars and q is the number of people attending at that price.

(a) What price corresponds to an attendance of 3000 people? What is the total revenue at that price? What is the total revenue if the price is $20?
(b) Write the revenue function as a function of attendance at the amusement park, q.
(c) What attendance will maximize revenue?
(d) What price should be charged to maximize revenue?
(e) What is the maximum revenue?

Solution (a) If $q = 3000$, we see from the demand equation that $p = 70 - 0.02(3000) = 10$. A price of $10 corresponds to an attendance of 3000 people. At this price, total revenue is $(3000)(10)$, or $30,000. In order to find total revenue at a price of $20, we first find the attendance corresponding to this price. We have

$$\begin{aligned} p &= 70 - 0.02q \\ 20 &= 70 - 0.02q \\ -50 &= -0.02q \\ 2500 &= q. \end{aligned}$$

At a price of \$20, attendance will be 2500, and so revenue will be $(2500)(20)$, or \$50,000. Notice that, although demand is reduced, the revenue is still higher at a price of \$20 than at a price of \$10.

(b) Since Revenue = price × quantity, we have

$$\begin{aligned} R(q) &= p \cdot q \\ &= (70 - 0.02q)q \\ &= 70q - 0.02q^2. \end{aligned}$$

(c) To maximize revenue, we find the critical points of the revenue function $R(q) = 70q - 0.02q^2$:

$$\begin{aligned} R'(q) &= 70 - 0.02(2q) \\ 0 &= 70 - 0.04q \\ 70 &= 0.04q \\ 1750 &= q. \end{aligned}$$

Maximum revenue is achieved when attendance at the amusement park is 1750 people.

(d) We find the price corresponding to an attendance of 1750, using the demand equation:

$$p = 70 - 0.02(1750) = 70 - 35 = 35.$$

The optimal price for an all-day pass at the amusement park is \$35.

(e) When the optimal price of \$35 is charged, the attendance at the park is 1750 people. Thus, the maximum revenue is $R = pq = (35)(1750) = 61250$. The maximum revenue is \$61,250. The corresponding profit cannot be determined without knowledge of the costs.

Example 6 A white-water rafting company knows that at a price of \$80 for a half-day trip, they will attract 300 customers. For every \$5 decrease in price, they attract about an additional 30 customers. What price should the company charge in order to maximize revenue?

Solution We first find the demand equation relating price to demand. Let's make a table of values first to be sure that we understand what is going on. If price, p, in dollars, is 80, the number of trips sold, q, is 300. If p is 75, then q is 330, and so on as in Table 5.2.

TABLE 5.2 *Demand for rafting trips*

Price, p (dollars)	80	75	70	65	...
Trips sold, q	300	330	360	390	...

You should notice that this table of values corresponds to a linear function, so the demand q is a linear function of the price p. The slope is $30/(-5) = -6$, so the demand function is $q = -6p + b$, where b is the vertical intercept. Since $(80, 300)$ is a point on the line, we have

$$\begin{aligned} q &= -6p + b \\ 300 &= -6(80) + b \\ 300 &= -480 + b \\ 780 &= b \end{aligned}$$

The demand function is $q = -6p + 780$.

Since revenue $R = p \cdot q$, the revenue function as a function of price is $R = p(-6p + 780) = -6p^2 + 780p$. The graph of this revenue function is given in Figure 5.57, and we see the maximum there. To find the maximum analytically, we differentiate the revenue function and find the critical points:

$$\begin{aligned} R' &= -12p + 780 = 0 \\ p &= \frac{780}{12} = 65. \end{aligned}$$

Maximum revenue is achieved when the price is \$65.

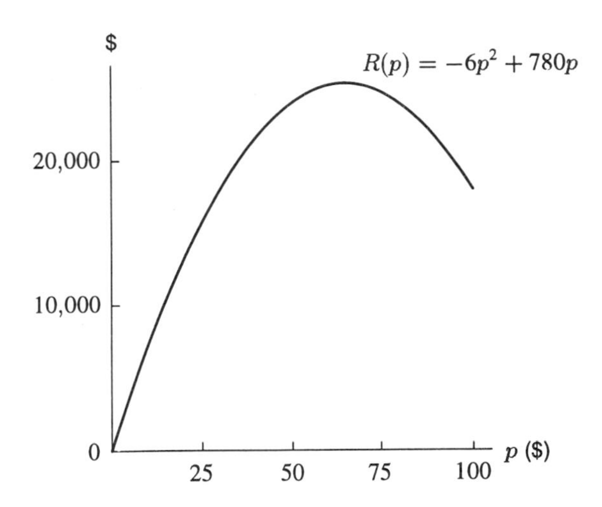

Figure 5.57: Revenue for a rafting company

Problems for Section 5.4

1. If $R(q) = 450q$ and $C(q) = 10000 + 3q^2$, at what quantity is profit maximized? What is the total profit at this production level?

2. Assume the cost function is given by $C(q) = q^3 - 60q^2 + 1200q + 1000$ for $0 \leq q \leq 50$. If the product sells for \$588 each, what production level will maximize profit? Find the total cost, total revenue, and total profit at this production level. Sketch a graph of the cost and revenue functions on the same axes, and label the production level and corresponding cost, revenue, and profit on your graph. (As you sketch the graph, it may help to know that the costs can go as high as \$35,000.)

3. Cost and revenue functions for a certain product are shown in Figure 5.58, with quantity q in thousands of units and cost and revenue in thousands of dollars. For what production levels is the profit function positive? Negative? Estimate the point at which profit would be maximized.

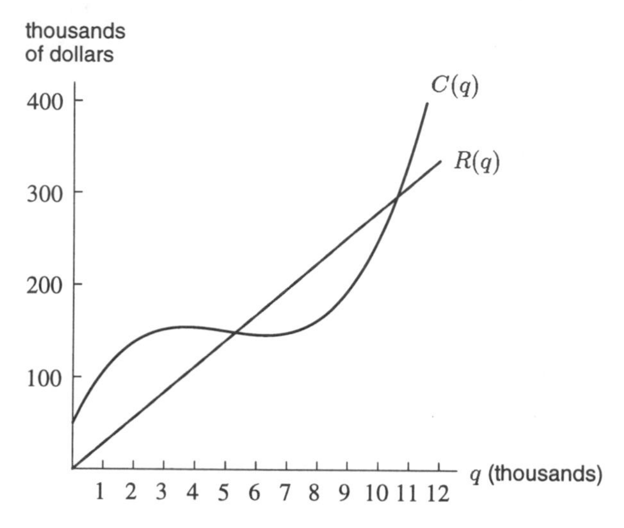

Figure 5.58: At what quantity is profit maximized?

4. Suppose that at a production level of 2000 for a given product, marginal revenue is \$4 per unit and marginal cost is \$3.25 per unit. Do you expect maximum profit to occur at a production level above or below 2000? Explain.

5. In Figure 5.51, the points where marginal revenue equals marginal cost are labeled q_1 and q_2. On the graph of the corresponding total cost and total revenue functions given in Figure 5.59, label the corresponding points q_1 and q_2. Explain in terms of slopes the significance of these points, and clearly explain why one is a local minimum and one is a local maximum.

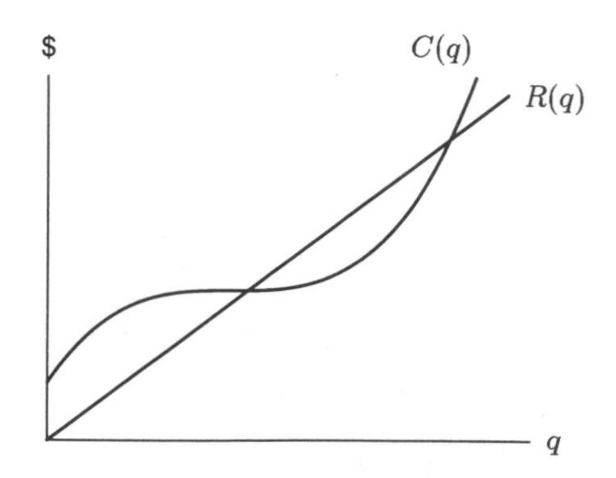

Figure 5.59: Total cost and total revenue

6. Graphs of the marginal revenue and marginal cost curves for a certain product are shown in Figure 5.60. At approximately what quantity is profit maximized? Explain.

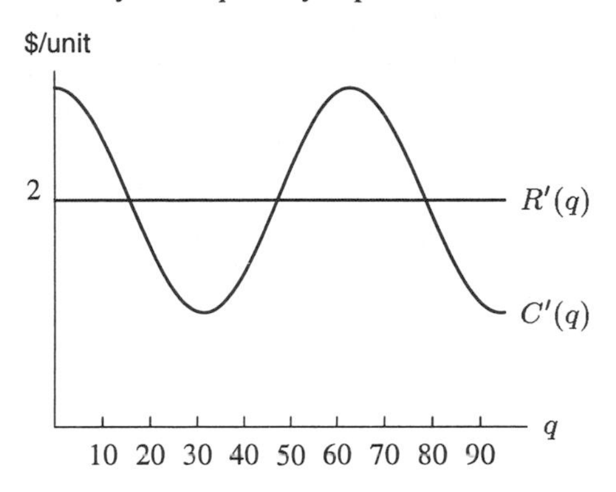

Figure 5.60: Marginal cost and marginal revenue

7. Suppose the total cost $C(q)$ of producing q goods is given by:

$$C(q) = 0.01q^3 - 0.6q^2 + 13q.$$

   (a) What is the fixed cost?
   (b) What is the maximum profit if each item is sold at a price of $7? (Assume you can sell everything you produce.)
   (c) At a fixed production level of 34 goods, for each $1 increase in price, 2 fewer goods are sold. Should you raise the price, and if so by how much?

8. Assume that the total cost function for a product is given by $C(q) = q^3 - 12q^2 + 48q$, in thousands of dollars, where q is measured in thousands, and $0 \leq q \leq 12$.

   (a) Sketch a graph of $C(q)$ and visually estimate on your graph the quantity where average cost is minimized.
   (b) Determine analytically the exact value of q at which average cost is minimized.

9. A reasonably realistic model of a firm's costs is given by the *short-run Cobb-Douglas cost curve*

$$C(q) = Kq^{1/a} + F,$$

where a is a positive constant, F is the fixed costs, and K measures the technology available to the firm.

   (a) Show that C is concave down if $a > 1$.
   (b) Assuming that average cost is minimized when average cost equals marginal cost, find what value of q minimizes the average cost.

10. Suppose a firm produces a quantity q of some good and that the average cost per item is given by:

$$a(q) = 0.01q^2 - 0.6q + 13, \quad \text{for} \quad q > 0.$$

   (a) What is the total cost, $C(q)$, of producing q goods?
   (b) What is the minimum marginal cost? What is the practical interpretation of this result?
   (c) At what production level is the average cost a minimum? What is the lowest average cost?
   (d) Compute the marginal cost at $q = 30$. How does this relate to your answer to part (c)? Explain this relationship both analytically and qualitatively.

11. Assume that the demand equation for a certain product is given by $p = 45 - 0.01q$. Write the revenue function as a function of q, and find the quantity that will maximize revenue. What price corresponds to this quantity, and what is the total revenue at this point?

12. At a price of \$8 per ticket, a musical theater group can fill every seat in the theater, which has a maximum capacity of 1500. For every additional dollar charged, the number of people buying tickets goes down by 75. What ticket price will maximize revenue?

13. Suppose you run a small independent furniture business. Your assistant signs a deal with a customer to deliver however many chairs the customer orders. The price will be \$90 per chair up to 300 chairs, and above 300, the price will be reduced by \$0.25 per chair (on the whole order) for every additional chair over 300 ordered, up to 100. What are the largest and smallest revenues your company can make under this deal?

14. Assume that the demand equation for a product is $p = b_1 - a_1 q$, and that the cost function is $C(q) = b_2 + a_2 q$, where p is the price of the product and q is the quantity sold. (Assume b_1, a_1, b_2, a_2 are all positive.) Find the value of q, in terms of other variables, that will maximize profit.

15. Suppose you are given the graph of the average cost $a(q)$ in Figure 5.61.
    (a) Sketch a graph of the marginal cost $C'(q)$.
    (b) Show that if
$$a(q) = b + mq$$
    then
$$C'(q) = b + 2mq.$$

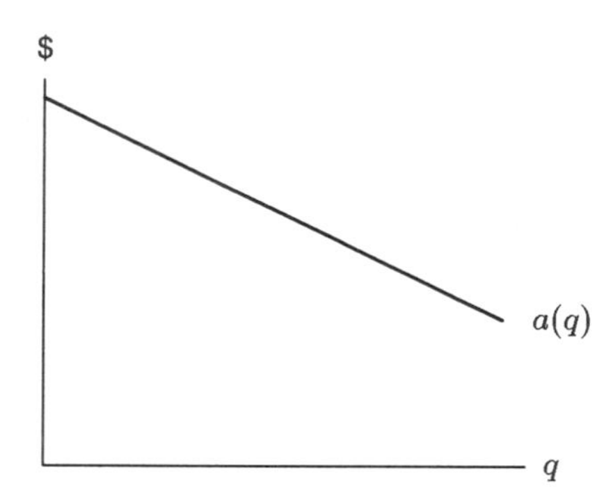

Figure 5.61: Average cost of a good

16. Suppose a company manufactures only one product. The quantity produced, q, of this product depends on the amount of capital, K, invested (i.e., the number of machines the company owns, the size of its building, and so on) and the amount of labor, L, available. It is often assumed that q can be expressed as a function of K and L by a *Cobb-Douglas production function*:
$$q = cK^{\alpha}L^{\beta}$$
where c, α, β are positive constants, with $0 < \alpha < 1$ and $0 < \beta < 1$.

In this problem we will see how the Russian government could use a Cobb-Douglas function to estimate how many people a newly privatized industry might employ. A company in such an industry will have only a small amount of capital available to it, and will need to use all of it; K is therefore fixed. Suppose L is measured in man-hours, and that each man-hour costs the company w rubles (a ruble is the unit of Russian currency). Suppose that the company has no other costs besides labor, and that each unit of the good can be sold for a fixed price of p rubles. How many man-hours of labor should the company use in order to maximize its profit?

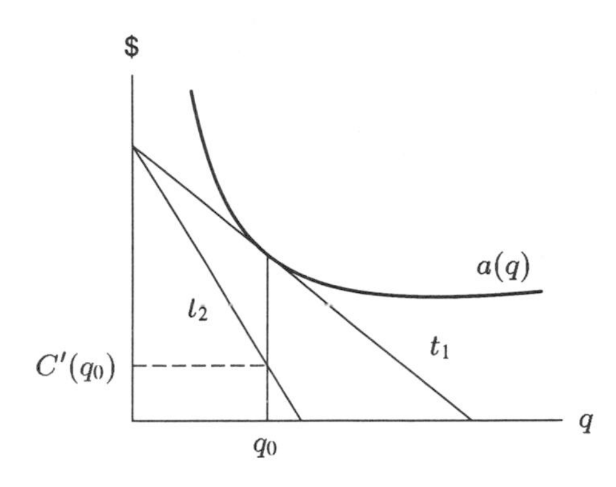

Figure 5.62

17. $C(q)$ is the total cost of producing a quantity q. The average cost $a(q)$ is given in Figure 5.62. The following rule is used by economists to determine the marginal cost $C'(q_0)$, for any q_0:
    - Construct the tangent t_1 to $a(q)$ at q_0.
    - Let t_2 be the line with the same y-intercept as t_1 but with twice the slope of t_1.

    Then $C'(q_0)$ is as shown in Figure 5.62. Explain why this rule works.

5.5 MORE APPLICATIONS TO ECONOMICS: ELASTICITY

If a company changes the price of an item it sells, you would expect the number of sales to change. In general, a higher price will result in fewer sales and a lower price will result in more sales. But how much will the sales be affected? In other words, how much will a change in the price of a product affect the demand for that product?

Elasticity of Demand

The sensitivity of demand to changes in price varies with different products, and even varies at different prices for the same product. For example, a change in the price of light bulbs will probably not affect the demand for light bulbs very much, whereas a change in the price of a certain brand of car may have a significant effect on the demand for that car.

We wish to find a way to measure this sensitivity of demand to price changes. Our measure should work for products as diverse as light bulbs and cars. The prices of these two items are so different that it makes little sense to talk about absolute changes in price: changing the price of light bulbs by $1 is a substantial change, whereas changing the price of a car by $1 is unlikely to have any effect. We use, instead, the percent change in price: how does, for example, a 1% increase in price (of either light bulbs or cars) affect the demand for that product? Similarly, as we look at the effect of price changes on demand, it is the percent change in demand rather than the absolute change that we use.

Recall that Δp denotes the change in the price p of a product, and Δq denotes the corresponding change in quantity q demanded. The percent change in the price equals $\Delta p/p$, and the percent change in demand is $\Delta q/q$. The measure we are looking for is the

$$\text{Ratio of percent change in demand to percent change in price} = \frac{\Delta q/q}{\Delta p/p}$$

Notice that this ratio will be negative since an increase in price will usually cause a decrease in the quantity demanded. If this ratio is large (in absolute value), it tells us that a change in price will cause a relatively larger change in the demand for the product. We rewrite this ratio as follows:

$$\text{Ratio of percent change in demand to percent change in price} = \frac{\Delta q/q}{\Delta p/p} = \frac{\Delta q}{q} \cdot \frac{p}{\Delta p} = \frac{p}{q} \cdot \frac{\Delta q}{\Delta p}$$

For small changes in p, we can approximate $\Delta q/\Delta p$ with the derivative dq/dp. This leads to the following definition:

> The **elasticity of demand** for a product, denoted E, is the ratio of percent change in demand to percent change in price.[1] We have:
>
> $$E = \frac{p}{q} \cdot \frac{dq}{dp}.$$

When $E = -1$, a one percent change in price corresponds approximately to a one percent change in demand. If $E < -1$, changes in price result in relatively larger changes in demand, and we say the demand is *elastic*. If $-1 < E \leq 0$, changes in price result in relatively smaller changes in demand, and we say the demand is *inelastic*.

Example 1 A hotel owner suspects that raising the price of his rooms from \$75 to \$80 per night will reduce his weekly sales from 100 rooms to 90 rooms.

a) What is the elasticity of demand for rooms at a price of \$75?
b) Should he raise the price?

Solution a) The percent change in the price is

$$\Delta p/p = 5/75 = 0.067 = 6.7\%$$

and the percent change in demand is

$$\Delta q/q = -10/100 = -0.1 = -10\%$$

The elasticity of demand is the ratio

$$E = -0.10/0.067 = -1.5$$

Note that the elasticity has absolute value greater than 1 because the percent change in the demand is greater than the percent change in the price.

[1] Some economists want elasticity to be a positive quantity and hence use the definition $E = -\frac{p}{q} \cdot \frac{dq}{dp}$

b) At a price of \$75 per room the weekly revenue will be

$$(100 \text{ rooms})(\$75 \text{ per room}) = \$7500$$

At a price of \$80 per room the weekly revenue will be

$$(90 \text{ rooms})(\$80 \text{ per room}) = \$7200$$

A price rise results in loss of revenue, and so the price should not be raised. A \$5 price rise just affects the sales too strongly.

Example 2 If the demand equation for a product is given by $q = 1000 - 2p^2$, find the elasticity at $p = 10$ and at $p = 15$, and interpret your answers.

Solution We first find the derivative $dq/dp = -4p$. At a price of 10, the quantity demanded is $q = 1000 - 2(10^2) = 800$, and $dq/dp = -4(10) = -40$. At this price, the elasticity is

$$E = \frac{p}{q} \cdot \frac{dq}{dp} = \frac{10}{800}(-40) = -0.5.$$

The demand is inelastic: at a price of \$10, a 1% increase in price will result in approximately a 0.5% decrease in demand.

At a price of \$15, we have $q = 550$ and $dq/dp = -60$. The elasticity is

$$E = \frac{p}{q} \cdot \frac{dq}{dp} = \frac{15}{550}(-60) = -1.64.$$

The demand is elastic: a 1% increase in price will result in approximately a 1.64% decrease in demand.

Elasticity of Demand and Maximum Revenue

We saw that for the product in Example 2 the elasticity at $p = 10$ is -0.5. This means that if prices go up from \$10 by 1%, demand will decrease by about 0.5%. This tells us that we can raise the price without hurting demand very much, so revenue will probably increase if we raise the price. On the other hand, we also saw that the elasticity at $p = 15$ is -1.64. A 1% increase in price from \$15 will cause demand to fall by about 1.64% (or, alternatively, a 1% decrease in price will cause demand to increase by about 1.64%.) Since a change in the price will have a relatively larger effect on demand, we can increase revenue by lowering the price. What is the optimal price for the product?

The optimal price is the price that brings in the greatest profit. We will assume that cost is independent of quantity so that maximum profit is achieved at the price that brings in the greatest revenue. In Table 5.3 we examine the demand q, revenue R, and elasticity E for the product in Example 2 at several prices. As expected, we see that as price increases, the demand decreases. The revenue $R = pq$ is the product of two factors, the price and the demand, and as one goes up, the other goes down. Elasticity measures the relative significance of these two competing factors. From the table it appears that maximum revenue is achieved by setting the price at about \$13, and that elasticity at that price is about -1. At prices below \$13, we have $E > -1$, indicating that the reduction in demand caused by a price increase will be relatively small, and so we might as well raise the price. On the other hand, at prices above \$13, we have $E < -1$, indicating that the increase in demand caused by a price decrease will be relatively large, and so we ought to lower the price. We have discovered the relationship between elasticity and revenue.

TABLE 5.3 *Revenue and elasticity at different points*

price p	10	11	12	13	14	15
demand q	800	758	712	662	608	550
revenue R	8000	8338	8544	8606	8512	8250
elasticity E	-0.5	-0.64	-0.81	-1.02	-1.29	-1.64
	inelastic	inelastic	inelastic	elastic	elastic	elastic

> Elasticity and revenue are related as follows:
> Maximum revenue occurs when $E = -1$.
> If $E > -1$, demand is inelastic and revenue is increased by raising the price.
> If $E < -1$, demand is elastic and revenue is increased by lowering the price.

We can demonstrate part of this result analytically. Revenue is maximized as a function of price at a critical point of the revenue function. Using the product rule to differentiate $R = pq$, we have

$$\frac{dR}{dp} = p\frac{dq}{dp} + \frac{dp}{dp}q = p\frac{dq}{dp} + q$$

At a critical point the derivative equals zero, and so we have

$$\begin{aligned} p\frac{dq}{dp} + q &= 0 \\ p\frac{dq}{dp} &= -q \\ \frac{p}{q}\frac{dq}{dp} &= -1 \\ E &= -1 \end{aligned}$$

At a critical point the elasticity is -1, and so revenue is maximized at a price where the elasticity of demand is -1.

Elasticity of Demand for Different Products

We have seen that elasticity of demand for a certain product can vary with the price of the product. We can also compare elasticity of demand for different products. If there are close substitutes for a product, or if the product is a luxury rather than a necessity, we would expect that changes in price would have a large effect on demand, and that the demand for the product would be elastic. On the other hand, if there are no close substitutes or if the product is a necessity, we expect that changes in price will have a relatively small effect on demand, and that the demand will be inelastic. We expect, for example, that demand for such things as salt, penicillin, eyeglasses, and lightbulbs will be inelastic over the usual range of prices for these products, and this turns out to be so. Table 5.4 lists the elasticity of demand with repect to price for selected products.[2]

TABLE 5.4 *Elasticity of demand for selected products*

Air Transportation	-1.10
Automobiles	-1.50
Automobile parts	-0.50
Lightbulbs	-0.33
Flour	-0.79
Furniture	-3.04
Furs	-2.30
Jewelry	-2.60
Milk	-0.31
Oranges	-0.97
Poultry	-0.27
Radios	-1.50
Sporting goods	-1.2
Sugar	-0.44

Problems for Section 5.5

1. Assume that the demand equation for yams is given by $q = 5000 - 10p^2$, where q is in pounds of yams and p is the price of a pound of yams.
   (a) If the current price of yams is \$2 per pound, how many pounds will be sold?
   (b) Is the demand at \$2 elastic or inelastic? Is it more accurate to say "People must have their yams and will buy them no matter what the price" or "Yams are a luxury item and people will stop buying them if the price gets too high"?

2. Assume that the demand for yams is as given in Problem 1.
   (a) At a price of \$2 per pound, what is the total revenue for the yam farmer?
   (b) Write revenue as a function of price, and then find the price that will maximize revenue.
   (c) What quantity is sold at the price you found in part (b), and what is the total revenue?
   (d) Show that $E = -1$ at the price you found in part (b).

3. What is the elasticity for jewelry in Table 5.4? Explain what this number tells you about the effect of price increases on the demand for jewelry. Is the demand for jewelry elastic or inelastic? Is this what you expect? Explain.

[2] David R. Kamerschen & Lloyd M. Valentine, *Intermediate Microeconomic Theory*, 1977, South-Western Publishing Co.

4. What is the elasticity for milk in Table 5.4? Explain what this number tells you about the effect of price increases on the demand for milk. Is the demand for milk elastic or inelastic? Is this what you expect? Explain.

5. Would you expect the demand for high definition television sets to be elastic or inelastic? Explain.

6. The demand function for a certain product is $q = 2000 - 5p$ where q is units sold at a price of p dollars. Find the elasticity if the price is \$20, and interpret your answer in terms of demand for the product.

7. School organizations will sometimes raise money by selling candy door to door. For a certain brand of candy, the following data have been collected, where p is the price of the candy and q is the quantity sold at that price.

TABLE 5.5

p	\$1.00	\$1.25	\$1.50	\$1.75	\$2.00	\$2.25	\$2.50
q	2765	2440	1980	1660	1175	950	480

(a) Estimate the elasticity of demand at a price of \$1.00. Is the demand for this candy elastic or inelastic?
(b) Estimate the elasticity at each of the prices shown. What do you notice? Give an explanation for why this might be so.
(c) At approximately what price is elasticity equal to -1?
(d) Find the total revenue at each of the prices shown, and confirm that the total revenue appears to be maximized at approximately the price where $E = -1$, found in part (c).

8. Find the exact price that maximizes revenues for sales of the product in Example 2.

9. Show analytically that if elasticity of demand $E < -1$, then the derivative of revenue with respect to price $dR/dp < 0$.

10. Show analytically that if elasticity of demand $E > -1$, then the derivative of revenue with respect to price $dR/dp > 0$.

REVIEW PROBLEMS FOR CHAPTER FIVE

For each of the functions in Problems 1–3, do the following:
(a) Find f' and f''.
(b) Find the critical points of f.
(c) Find any inflection points of f.
(d) Evaluate f at its critical points and at the endpoints of the given interval. Identify the local and global maxima and minima of f in the interval.
(e) Sketch a graph of f. Indicate clearly where f is increasing or decreasing, and its concavity.

1. $f(x) = x^3 - 3x^2 \quad (-1 \leq x \leq 3)$

2. $f(x) = 4x^3 - x^4 \quad (-4 \leq x \leq 4)$

3. $f(x) = x^2e^{-x} \quad (-1 \leq x \leq 4)$

For each of the functions in Problems 4–6, do the following:
(a) Find f' and f''.
(b) Find the critical points of f.
(c) Find any inflection points of f.
(d) Evaluate f at its critical points.
(e) Find the limit of $f(x)$ as x tends to $+\infty$ and $-\infty$. Identify the local and global maxima and minima of f for $-\infty < x < \infty$.
(f) Sketch a graph of f over the interval. Indicate clearly where f is increasing or decreasing, and its concavity.

4. $f(x) = 2x^3 - 9x^2 + 12x + 1$

5. $f(x) = \ln(x^2 + 1)$

6. $f(x) = xe^{-x}$

7. Sketch a graph of $e^{-x^2/2}$, marking local maxima and minima and points of inflection.

For each of the functions in Problems 8–13, use derivatives to find and identify local maxima, minima, and points of inflection. Sketch the graphs. Confirm your answers using a calculator or computer.

8. $f(x) = x^3 + 3x^2 - 9x - 15$

9. $f(x) = x^5 - 15x^3 + 10$

10. $f(x) = x - 2\ln x$ for $x > 0$

11. $f(x) = x^2e^{5x}$

12. $f(x) = e^{-x^2}$

13. $f(x) = \dfrac{x^2}{x^2 + 1}$

From the graphs of f' in Problems 14–17 determine:
(a) Over what intervals is f increasing? Decreasing?
(b) Whether f has maxima or minima? If so, which, and where?

14.

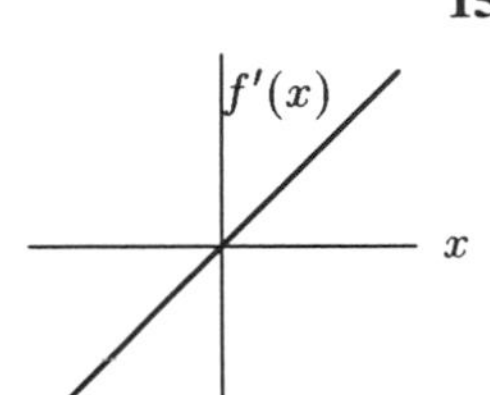

15.

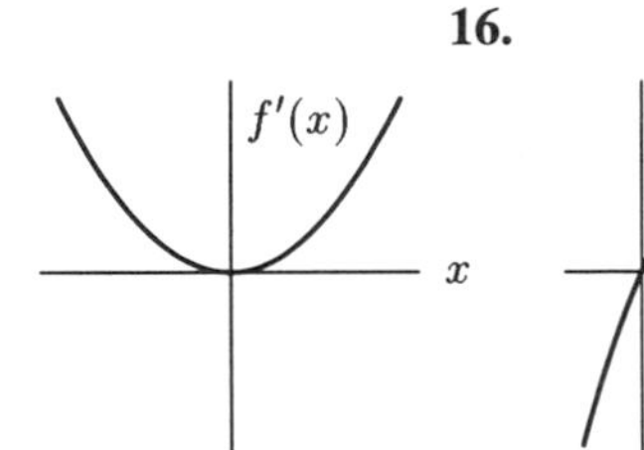

16.

$f'(x)$ x 2 4

17.

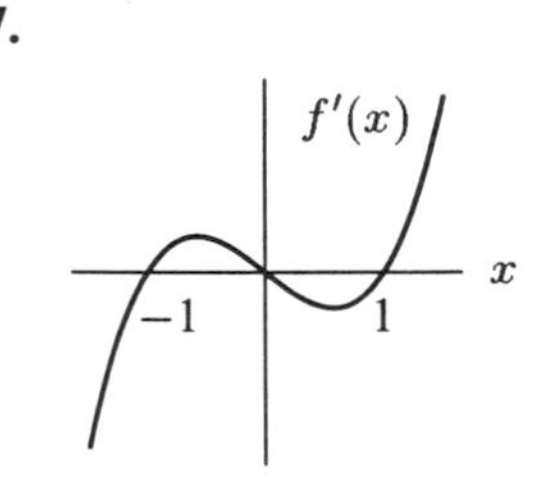

18. Sketch several members of the family $y = x^3 - ax^2$ on the same coordinate plane. Discuss the effect of the parameter a on the graph. Find all critical points for this function.

19. Suppose $g(t) = (\ln t)/t$ for $t > 0$.

(a) Does g have either a global maximum or a global minimum on $0 < t < \infty$? If so, where, and what are their values?

(b) What does your answer to part (a) tell you about the number of solutions to the equation

$$\frac{\ln x}{x} = \frac{\ln 5}{5}?$$

(Note: There are many ways to investigate the number of solutions to this equation. We are asking you to draw a conclusion from your answer to part (a).)

(c) Find the solution(s).

20. Consider the vase in Figure 5.63. Assume the vase is filled with water at a constant rate (i.e., constant volume per unit time).

(a) Graph $y = f(t)$, the depth of the water, against time, t. Show on your graph the points at which the concavity changes.

(b) Where does $y = f(t)$ grow fastest? Slowest? Estimate the ratio between these two growth rates.

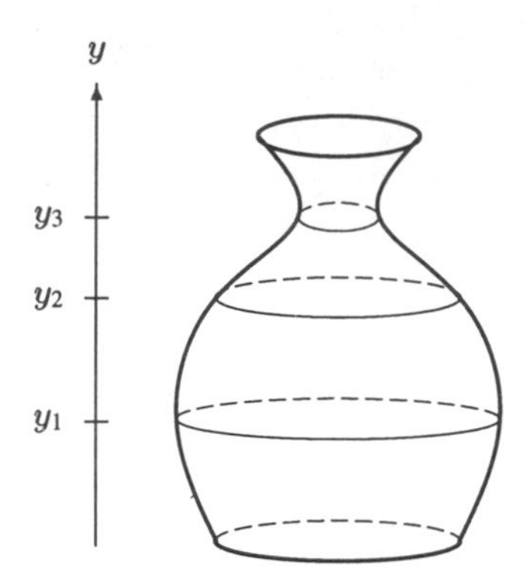

Figure 5.63

21. When birds lay eggs, they do so in clutches of several at a time. When the eggs hatch, each clutch gives rise to a brood of baby birds. We want to determine the clutch size which maximizes the number of birds surviving to adulthood per brood. If the clutch is small, there are few baby birds in the brood; if the clutch is large, there are so many baby birds to feed that most die of starvation. The number of surviving birds per brood as a function of clutch size is shown by the benefit curve in Figure 5.64.[3]

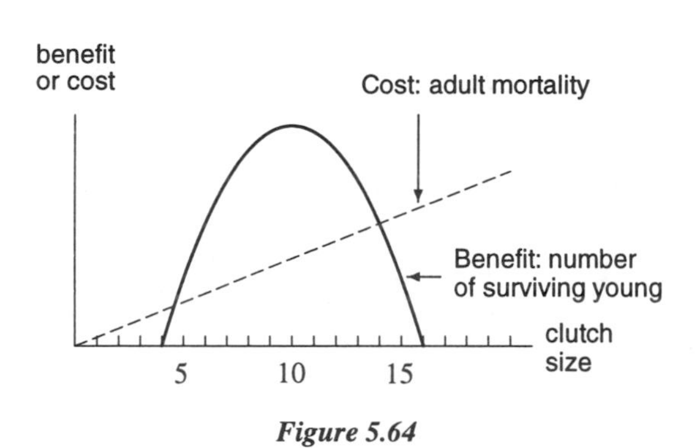

Figure 5.64

(a) Estimate the clutch size which maximizes the number of survivors per brood.

(b) Suppose also that there is a biological cost to having a larger clutch: the female survival rate is reduced by large clutches. This cost is represented by the dotted line in Figure 5.64. If we take cost into account by assuming that the optimal clutch size in fact maximizes the vertical distance between the curves, what is the new optimal clutch size?

[3]Data from C. M. Perrins and D. Lack reported by J. R. Krebs and N. B. Davies in *An Introduction to Behavioural Ecology* (Oxford: Blackwell, 1987).

22. Let $f(v)$ be the amount of energy consumed by a flying bird, measured in joules per second (a joule is a unit of energy), as a function of its speed v (in meters/sec). See Figure 5.65.

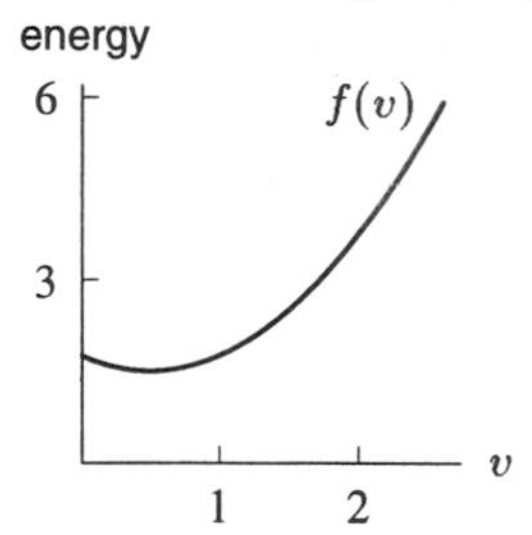

Figure 5.65: Amount of energy consumed by a flying bird

(a) Suggest a reason for the shape of this graph (in terms of the way birds fly).

Now let $a(v)$ be the amount of energy consumed by the same bird, measured in joules *per meter*.

(b) What is the relationship between $f(v)$ and $a(v)$?
(c) Where is $a(v)$ a minimum?
(d) Should the bird try to minimize $f(v)$ or $a(v)$ when it is flying? Why?

23. A bird such as a starling feeds worms to its young. To collect worms, the bird flies to a site where worms are to be found, picks up several in its beak, and flies back to its nest. The *loading curve* in Figure 5.66 shows how the number of worms (the load) a starling collects depends on the time it has been searching for them.[4] The curve is concave down because the bird can pick up worms more efficiently when its beak is empty; when its beak is partly full, the bird becomes much less efficient. The traveling time (from nest to site and back) is represented by the distance PO in Figure 5.66. Suppose the bird wants to maximize the rate at which it brings worms to the nest, where

$$\text{Rate worms arrive at nest} = \frac{\text{Load}}{\text{Traveling time} + \text{Searching time}}$$

(a) Draw a line in Figure 5.66 whose slope is this rate.
(b) Using the graph, estimate the load which maximizes this rate.
(c) If the traveling time is increased, does the optimal load increase or decrease? Why?

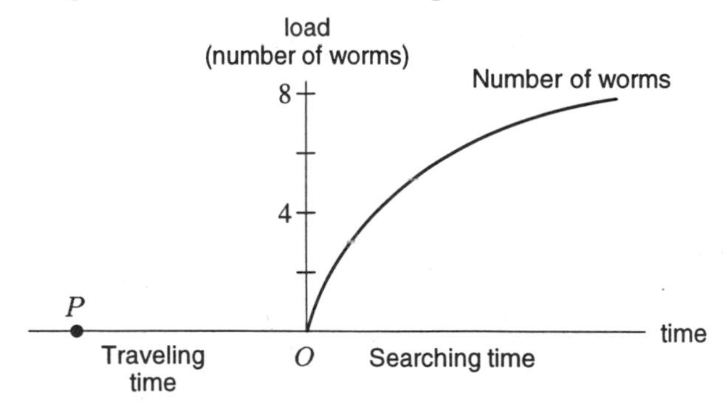

Figure 5.66: Bird's loading curve

[4]Alex Kacelnick(1984). Reported by J. R. Krebs and N. B. Davis, *An Introduction to Behavioural Ecology* (Oxford: Blackwell, 1987).

24. Figure 5.67 shows cost and revenue functions for a product.

(a) Estimate the production level that will maximize profit.

(b) Sketch graphs of marginal revenue and marginal cost for this product on the same coordinate system. Label on this graph the production level that will maximize profits.

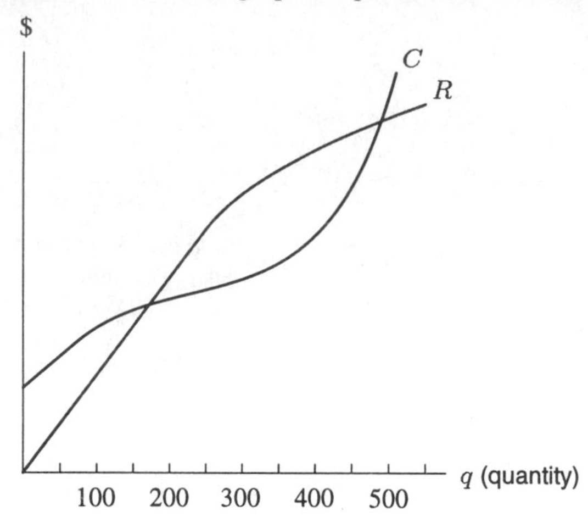

Figure 5.67: Where is profit maximized?

25. A premium ice cream company finds that at a price of \$4.00, demand for the product is 4000. For every \$0.25 decrease in price, demand increases by 200. Find the price and quantity sold that will maximize revenues.

26. Suppose that the demand function for a certain product is given by $q = 5000 - 3p^2$.

(a) Find the elasticity at a price of \$30. Is the demand at this price elastic or inelastic?

(b) Based on the elasticity found in (a), do you expect the price to maximize revenue to be above or below \$30? Explain.

(c) Find the price which maximizes revenue.

27. Let $C(q)$ be the total cost of producing a quantity q of a certain good. The *average cost* is $a(q) = C(q)/q$.

(a) Interpret $a(q)$ graphically, as the slope of a line in Figure 5.68.

(b) Find on the graph the quantity q_0 where $a(q)$ is minimal.

(c) What is the relationship between $a(q_0)$ and the marginal cost, $C'(q_0)$? Explain your result graphically. What does this result mean, in terms of economics?

(d) Graph $C'(q)$ and $a(q)$ on the same axes. Mark q_0 on the q-axis.

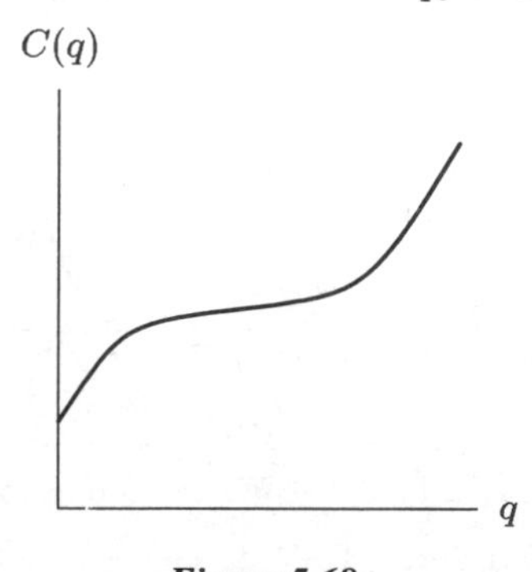

Figure 5.68

28. For $a > 0$, the line

$$a(a^2 + 1)y = a - x$$

forms a triangle in the first quadrant with the x- and y-axes.

(a) Find, in terms of a, the x- and y-intercepts of the line.
(b) Find the area of the triangle, as a function of a.
(c) Find the value of a making the area a maximum.
(d) What is this greatest area?
(e) If you want the triangle to have area $1/5$, what choices do you have for a?

29. Consider a large tank of water, with temperature $W(t)$. The ambient temperature $A(t)$ (i.e., the temperature of the surrounding air), is given by the graph in Figure 5.69.

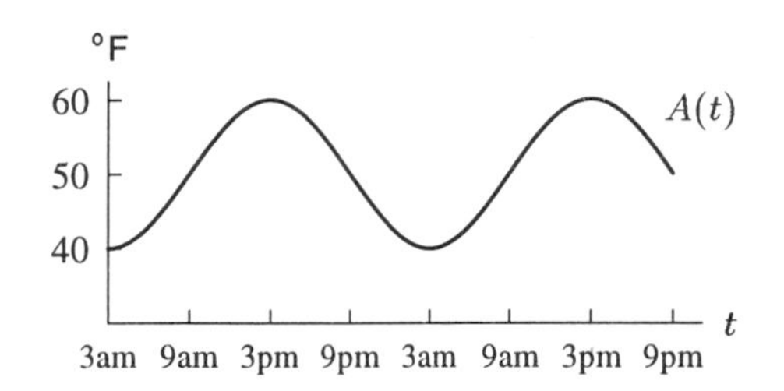

Figure 5.69

The temperature of the water is affected by the temperature of the surrounding air.

(a) How does the temperature of the water change if the water is colder than the surrounding air? What if the water is warmer?
(b) Using your answer in part (a), sketch a possible graph for $W(t)$ on the same axes as $A(t)$.
(c) Explain the relationship between the maxima and minima of $W(t)$ and the points where the two graphs intersect.
(d) What is the relationship between the rate at which the temperature of the water changes and the difference $A(t) - W(t)$?
(e) What is the relationship between the inflection points of $W(t)$ and the points where $A(t) - W(t)$ has a maxima or minima?
(f) Assume the tank is refilled with cold water (35°F) at 3 am. Sketch a possible graph for $W(t)$. Pay attention to the concavity.

CHAPTER SIX

USING THE DEFINITE INTEGRAL

We have seen that the derivative has many useful interpretations, and the same is true of the definite integral. In Chapter 3, we saw how a definite integral can represent an area, an average value, or a total change. In this chapter, we will see how the Fundamental Theorem of Calculus can be used to extract information about a function F from its derivative F'. We will use definite integrals to solve problems in life sciences, economics, and probability.

6.1 THE DEFINITE INTEGRAL REVISITED

In Chapter 3 we defined the definite integral to allow us to reconstruct from the velocity the distance traveled by an object. If f is bounded and continuous (except perhaps at a few points) on the interval $a \leq x \leq b$, we divide the interval into n equal subdivisions of length $\Delta x = (b-a)/n$ using the points $a = x_0, x_1, \ldots, x_n = b$, and then we form the Riemann sums:

$$\text{Left-hand sum} = f(x_0)\Delta x + f(x_1)\Delta x + \cdots + f(x_{n-1})\Delta x = \sum_{i=0}^{n-1} f(x_i)\Delta x.$$

$$\text{Right-hand sum} = f(x_1)\Delta x + f(x_2)\Delta x + \cdots + f(x_n)\Delta x = \sum_{i=1}^{n} f(x_i)\Delta x.$$

We expect that as n goes to infinity and the subdivisions become finer, both sums will approach the same number. This common limit is defined to be the *definite integral*:

$$\int_a^b f(x)\,dx = \lim_{n\to\infty}(\text{Left-hand sum}) = \lim_{n\to\infty}\sum_{i=0}^{n-1} f(x_i)\Delta x$$
$$= \lim_{n\to\infty}(\text{Right-hand sum}) = \lim_{n\to\infty}\sum_{i=1}^{n} f(x_i)\Delta x.$$

The function f is called the *integrand*, and the numbers a and b are the *limits of integration*.

In Chapter 3 we only considered the case $a < b$, but we now allow $a \geq b$. We still set $x_0 = a$, $x_n = b$, and $\Delta x = (b-a)/n$. If $a > b$, the quantity Δx is negative and the names left- and right-sum no longer apply. (The points $a = x_0, x_1, \ldots, x_n = b$ still divide the interval into equal subdivisions, but they go from right to left, not left to right.) However, the definition in the preceding box still applies.

Interpretations of the Definite Integral

Two interpretations of the integral $\int_a^b f(x)\,dx$ were discussed in Chapter 3:

- The integral represents the **area** between the graph of f and the x-axis from a to b, assuming $f \geq 0$ and $a < b$. (See Figure 6.1.)
- The **average value** of f over the interval $[a,b]$ is defined in terms of the integral:

$$\begin{matrix}\text{Average value of } f \\ \text{from } a \text{ to } b\end{matrix} = \frac{1}{b-a}\int_a^b f(x)\,dx.$$

The average value of f is the height of the rectangle with base $(b-a)$ and whose area equals the area under the graph of $f(x)$ between $x = a$ and $x = b$. (See Figure 6.2.)

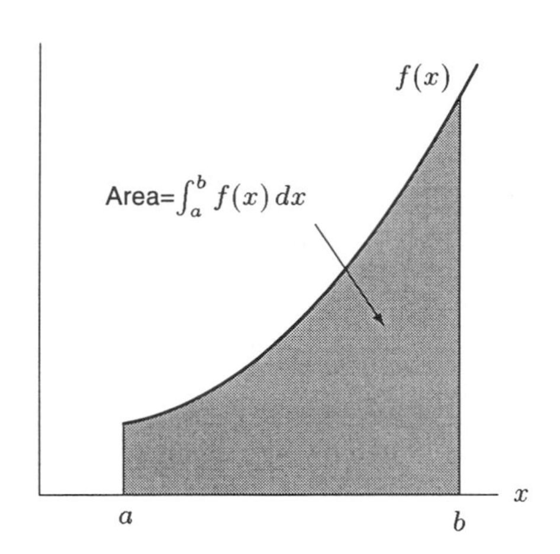

Figure 6.1: Area as an integral

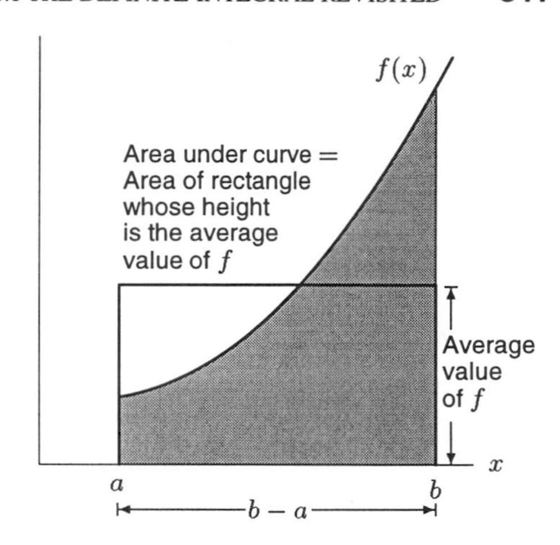

Figure 6.2: Average value

The Definite Integral of a Rate of Change: When $f = F'$

If f is the derivative of a function F, then we can think of $f(x)$ as the instantaneous rate of change of the quantity $F(x)$. The integral $\int_a^b f(x)\,dx$ then represents the total change in $F(x)$ as x changes from a to b:

$$\text{Total change in } F(x) \text{ between } a \text{ and } b = \int_a^b F'(x)\,dx = \int_a^b f(x)\,dx.$$

Since the change in $F(x)$ can also be expressed as $F(b) - F(a)$, we have the result presented at the end of Chapter 3:

The Fundamental Theorem of Calculus

If f is the derivative of F, that is, $f = F'$, and if f is continuous, then

$$\int_a^b f(x)\,dx = \int_a^b F'(x)\,dx = F(b) - F(a).$$

In words:

The definite integral of a rate of change gives total change.

If $f = F'$, we call F an **antiderivative** of f. This theorem has tremendous practical importance: it says that *if* we have a formula for F, we don't need to calculate the left and right sums to find a definite integral, but instead we can simply subtract $F(a)$ from $F(b)$. Equally important is the fact that if we estimate the integral $\int_a^b f(x)\,dx$ by a Riemann sum, this theorem gives us an estimate for $F(b) - F(a)$ which can be used to extract information about F from its derivative f.

Example 1 The graph of a function $f(x)$ is given in Figure 6.3.

a) Evaluate $\int_0^5 f(x)\,dx$.
b) Find the average value of $f(x)$ on the interval $x = 0$ to $x = 5$. Check your answer graphically.

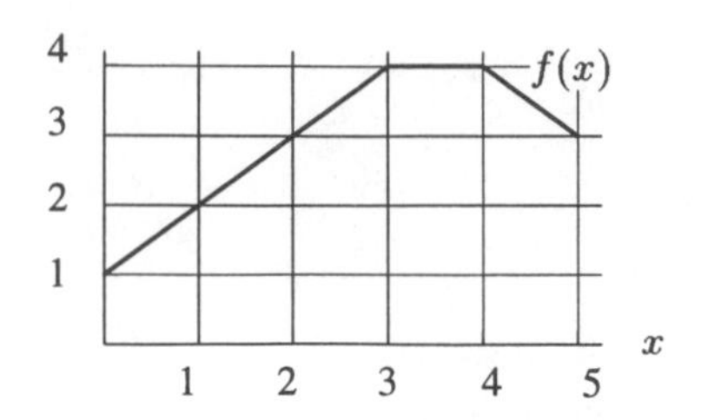

Figure 6.3: Estimate $\int_0^5 f(x)dx$

Solution a) Since $f(x) \geq 0$ on the interval from 0 to 5, the integral is equal to the area of the region under the graph of $f(x)$ between $x = 0$ and $x = 5$. From Figure 6.3 we see that this region consists of 13 full boxes and 4 half boxes, each of area 1, for a total area of 15, and so we have

$$\int_0^5 f(x)\, dx = 15.$$

b) The average value of $f(x)$ on the interval from 0 to 5 is given by

$$\text{Average value } = \frac{1}{5-0}\int_0^5 f(x)\, dx = \frac{1}{5}(15) = 3.$$

We can check our answer graphically by drawing a horizontal line at $y = 3$ on a graph of $f(x)$. (See Figure 6.4.) We see that, between $x = 0$ and $x = 5$, the area under the graph of $f(x)$ is equal to the area of the rectangle with height 3.

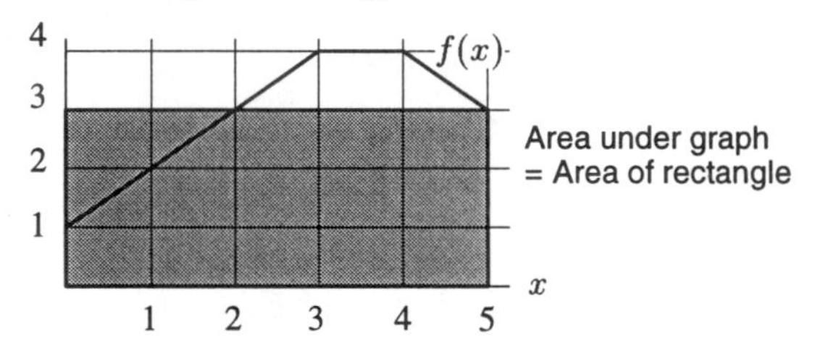

Figure 6.4: Average value of $f(x)$ is 3

Example 2 The graph of the derivative $F'(x)$ of a function $F(x)$ is given in Figure 6.5. If $F(0) = 100$, find $F(10)$.

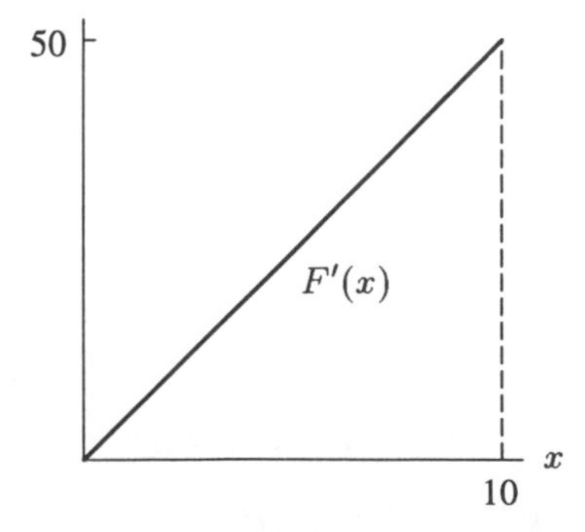

Figure 6.5: The graph of a derivative $F'(x)$

Solution The Fundamental Theorem of Calculus tells us that

$$F(10) - F(0) = \int_0^{10} F'(x)\,dx,$$

and so

$$F(10) = F(0) + \int_0^{10} F'(x)\,dx.$$

In other words, the total change in the function F between 0 and 10 is given by $\int_0^{10} F'(x)\,dx$. This integral is equal to the area under the graph of $F'(x)$ between $x = 0$ and $x = 10$. We see in Figure 6.5 that this area is equal to 250. Therefore,

$$\int_0^{10} F'(x)\,dx = 250.$$

Since $F(0) = 100$, we have $F(10) = 100 + 250 = 350$.

Graphing a Function Given a Graph of its Derivative

Suppose we have the graph of f' and we want to sketch the graph of f. We know that when f' is positive, f is increasing, and when f' is negative, f is decreasing. In other words, when the graph of f' lies above the x-axis, f is increasing, and when the graph of f' lies below the x-axis, f is decreasing. If we want to know exactly how much f increases or decreases, we compute the area between the x-axis and the graph of f'.

Example 3 The graph of the derivative $f'(x)$ of a function $f(x)$ is shown in Figure 6.6, and the values of some areas are given. If $f(0) = 10$, sketch a graph of the function $f(x)$ and give the coordinates of the local maxima and minima.

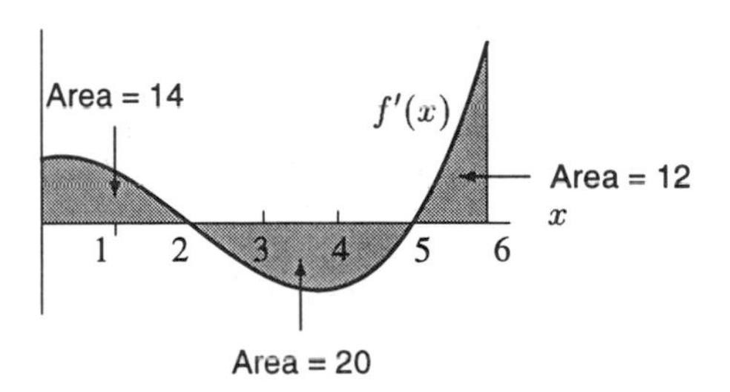

Figure 6.6: The graph of a derivative f'

Solution We see in Figure 6.6 that the derivative f' is positive between 0 and 2, negative between 2 and 5, and positive between 5 and 6. Therefore, the function f is increasing between 0 and 2, decreasing between 2 and 5, and increasing between 5 and 6. It must have the general shape shown in Figure 6.7. We see that there is a local maximum at $x = 2$ and a local minimum at $x = 5$. Notice that we can sketch the general shape of the graph of f without knowing any areas. We use the areas to sketch a more precise graph of this function.

We are told that $f(0) = 10$, so we begin by plotting the point $(0, 10)$. We see in Figure 6.6 that

$$\int_0^2 f'(x)\,dx = 14.$$

Therefore, the total change in f between 0 and 2 is 14, and the value of f increases 14 units between $x = 0$ and $x = 2$. Since $f(0) = 10$, we have $f(2) = 10 + 14 = 24$. The point $(2, 24)$ is a point on the graph of $f(x)$.

The total change in f between 2 and 5 is $\int_2^5 f'(x)\,dx$. We see in Figure 6.6 that the area between $x = 2$ and $x = 5$ is 20. Since this area lies entirely below the x-axis, we have

$$\int_2^5 f'(x)\,dx = -20.$$

The total change in f is -20. In other words, $f(x)$ decreases by 20 units between $x = 2$ and $x = 5$. Since $f(2) = 24$, we have $f(5) = 24 - 20 = 4$. Thus, the point $(5, 4)$ lies on the graph. Finally, between $x = 5$ and $x = 6$, the total change in $f(x)$ is an increase of 12 units, so $f(6) = 4 + 12 = 16$. We plot the point $(6, 16)$ on our graph. See Figure 6.8.

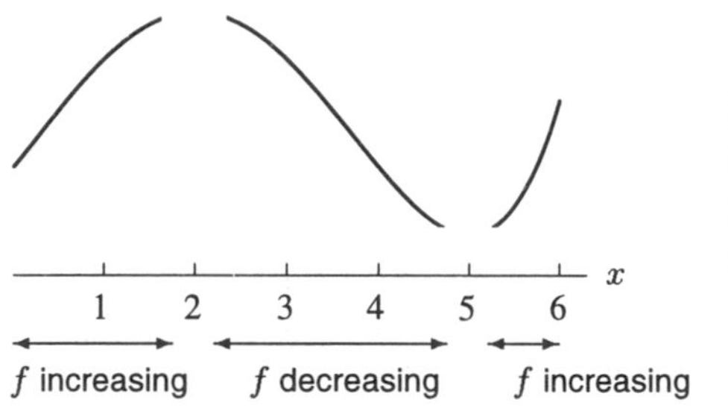

Figure 6.7: The shape of f

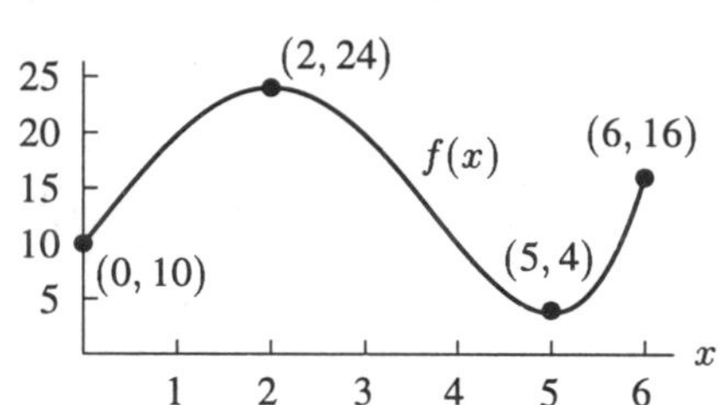

Figure 6.8: The graph of f

Applications of the Fundamental Theorem

Example 4 After a forest fire, vegetation grows back quickly. For a certain species of tree, the number of new trees per square mile taking root each year is approximated for 30 years after the fire by

$$r(t) = 70 - 9\sqrt{t},$$

where t is the number of years since the forest fire. How many trees will grow during the first 10 years after a fire?

Solution The function $r(t)$ gives the rate in trees per year at which new trees are appearing. The total number of new trees during the first 10 years is equal to the definite integral:

$$\text{Total number of new trees} = \int_0^{10} r(t)\,dt = \int_0^{10} (70 - 9\sqrt{t})\,dt.$$

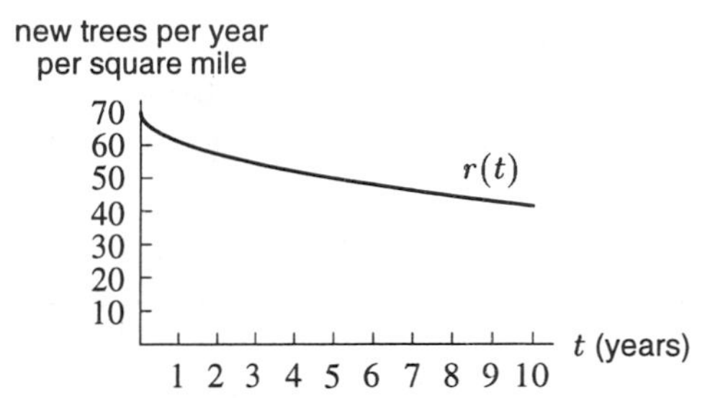

Figure 6.9: Rate of population growth of trees

We compute this integral using a calculator or by estimating the area under the graph of $r(t)$. (See Figure 6.9.) We have

$$\int_0^{10} (70 - 9\sqrt{t})\, dt = 510.3$$

About 510 new trees per square mile will appear during the first 10 years after the forest fire.

Example 5 The marginal cost function to produce q units of a certain product is approximated by

$$C'(q) = q^2 - 40q + 500 \text{ dollars per unit.}$$

If the fixed costs of production are \$2000, estimate the total cost to produce 30 units.

Solution The total cost to produce 30 units is equal to $C(30)$. Using the Fundamental Theorem of Calculus, we have

$$C(30) = C(0) + \int_0^{30} C'(q)\, dq.$$

Since the fixed costs of production are \$2000, we know $C(0) = 2000$. We can estimate $\int_0^{30} C'(q)\, dq$ using a calculator or by estimating the area under the graph of $C'(q)$. (See Figure 6.10.) We have

$$\int_0^{30} C'(q)\, dq = \int_0^{30} (q^2 - 40q + 500)\, dq = 6000.$$

Therefore, $C(30) = 2000 + 6000 = 8000$. The total cost to produce 30 items is approximately \$8000.

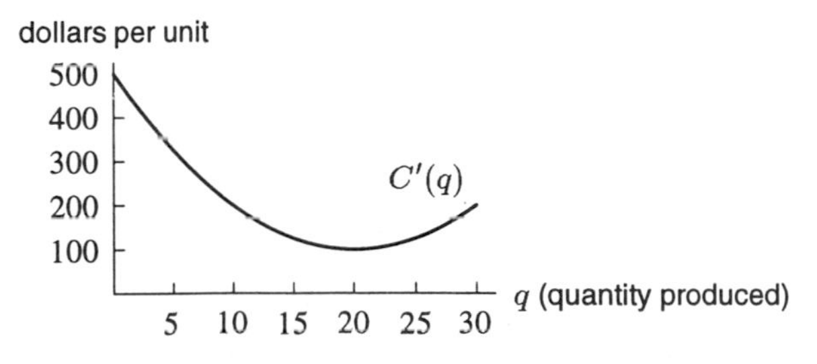

Figure 6.10: A marginal cost function

Example 6 Table 6.1[1] gives the emissions of nitrogen oxides in millions of metric tons per year in the United States from 1940 to 1990. Estimate the total emissions of nitrogen oxide during this 50 year period.

TABLE 6.1 *Emissions of nitrogen oxides*

Year	1940	1950	1960	1970	1980	1990
NO_x (millions of metric tons)	6.9	9.4	13.0	18.5	20.9	19.6

Solution Table 6.1 gives the rate of emissions of nitrogen oxide in millions of metric tons per year. To find the total emissions, we use left-hand and right-hand Riemann sums. We have

$$\begin{aligned}\text{Left-hand sum} &= (6.9)(10) + (9.4)(10) + (13.0)(10) + (18.5)(10) + (20.9)(10) = 687. \\ \text{Right-hand sum} &= (9.4)(10) + (13.0)(10) + (18.5)(10) + (20.9)(10) + (19.6)(10) = 814. \\ \text{Average of left- and right-hand sums} &= \frac{687 + 814}{2} = 750.5.\end{aligned}$$

The total emissions of nitrogen oxide between 1940 and 1990 is about 750 million metric tons.

Problems for Section 6.1

1. If the graph of f is in Figure 6.11:
   (a) What is $\int_0^6 f(x)\,dx$?
   (b) What is the average value of f on the interval $x = 0$ to $x = 6$?

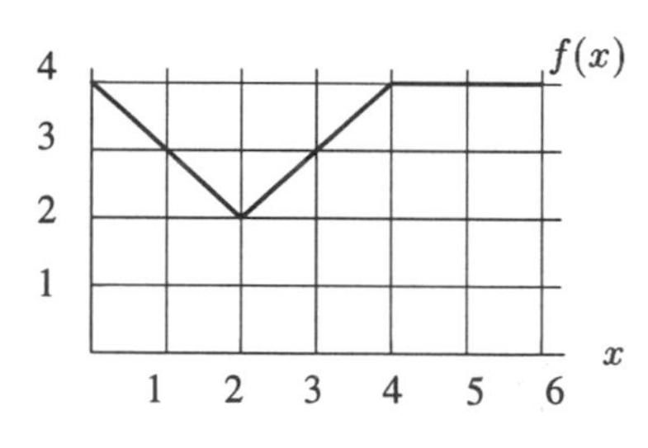

Figure 6.11

2. The graph of the derivative f' of a function f is given in Figure 6.12.
   (a) Where is f increasing and where is it decreasing? What are the x-coordinates of the local maxima and minima of f?

[1]Statistical Abstract of the U.S., 1992

(b) Give a possible rough sketch of f. (You don't need a scale on the vertical axis.)

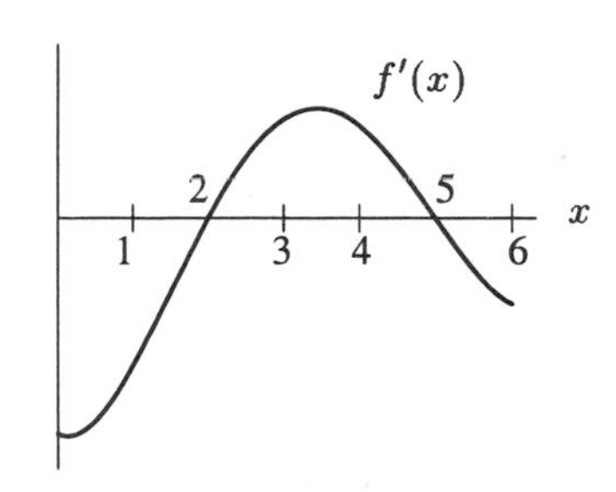

Figure 6.12

3. The graph of the derivative g' of g is given in Figure 6.13. If $g(0) = 0$, sketch a graph of g. Give (x, y)-coordinates for all local maxima and minima.

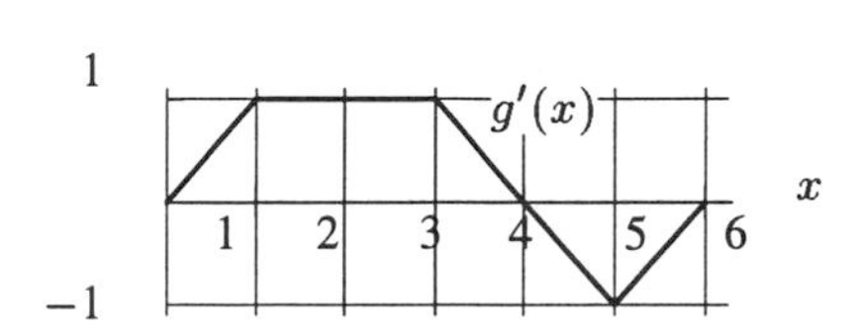

Figure 6.13

4. The graph of the derivative F' of F is given in Figure 6.14, with some areas labeled. If $F(0) = 14$, sketch a graph of F. Give (x, y)-coordinates for all local maxima and minima.

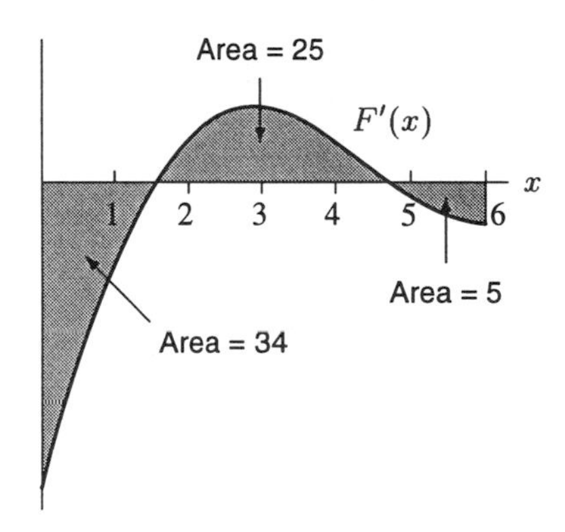

Figure 6.14

5. Consider the costs of drilling an oil well. The marginal costs depend on the depth at which you are drilling; drilling becomes more expensive, per meter, as you dig deeper into the earth. Suppose the fixed costs are 1,000,000 riyals (the riyal is the unit of currency of Saudi Arabia), and the marginal costs are

$$C'(x) = 4000 + 10x$$

in riyals/meter, where x is the depth in meters. Find the total cost of drilling a well 500 meters deep.

6. Ice is forming on a pond at a rate given by

$$\frac{dy}{dt} = \frac{\sqrt{t}}{2} \text{ inches per hour,}$$

where y is the thickness of the ice in inches at time t measured in hours since the ice started forming.

(a) Estimate the thickness of the ice after 8 hours.
(b) At what rate is the thickness of the ice increasing after 8 hours?

7. In 1987 the average per capita income in the US was \$26,000. Suppose that average per capita income is increasing at a rate in dollars per year given by

$$r(t) = 480(1.024)^t,$$

where t is the number of years since 1987. Estimate the average per capita income in 1995.

8. The marginal revenue function for selling a product is given by $R'(q) = 200 - 12\sqrt{q}$ dollars per unit, where q indicates the number of units sold.

(a) Sketch the graph of $R'(q)$.
(b) Estimate the total revenue in selling 100 units.
(c) What is the marginal revenue at 100 units? Use this value and your answer to part (b) to estimate the total revenue in selling 101 units.

9. A manufacturer of mountain bikes has marginal cost function

$$C'(q) = \frac{600}{0.3q + 5},$$

where q is the quantity of bicycles produced.

(a) If the fixed cost in producing the bicycles is \$2000, find the total cost to produce 30 bicycles.
(b) If the bikes are sold for \$200 each, what is the profit (or loss) on the first 30 bicycles?
(c) Find the marginal profit on the 31st bicycle.

10. The rate of sales (measured in sales per month) of a company during the year is given by

$$r(t) = t^4 - 20t^3 + 118t^2 - 180t + 200,$$

where t is measured in months since January 1.

(a) Sketch a graph of the rate of sales per month during the first year ($t = 0$ to $t = 12$). Does it appear that more sales were made during the first half of the year, or during the second half?
(b) Estimate the total sales of the company during the first 6 months of the year.
(c) Estimate the total sales of the company during the last 6 months of the year.
(d) What are the total sales for the entire year?
(e) What are the average sales per month for the company during the year?

11. The rate at which the world's oil is being consumed is continuously increasing. The rate of consumption (in billions of barrels per year) can be approximated by $r(t) = 32e^{0.05t}$, where t is measured in years since the start of 1990.

(a) Sketch the graph of $r(t)$.
(b) Represent the total quantity of oil used between the start of 1990 and the start of 1995 as a definite integral.
(c) Estimate the total quantity of oil used during this five year period.

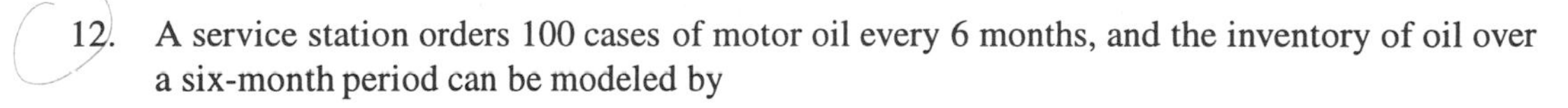

12. A service station orders 100 cases of motor oil every 6 months, and the inventory of oil over a six-month period can be modeled by

$$f(t) = 100e^{-0.5t} \text{ cases of oil at time } t,$$

where t is measured in months since the order arrives.

(a) How many cases are there at the start of the six-month period? How many cases are left at the end of the six-month period?
(b) Find the average number of cases in inventory over the six-month period.

13. Annual coal production in the United States (in quadrillion BTU per year) is given in Table 6.2. Estimate the total amount of coal produced in the United States between 1960 and 1990. [2]

TABLE 6.2

Year	1960	1965	1970	1975	1980	1985	1990
Coal production (quadrillion BTU)	10.82	13.06	14.61	14.99	18.60	19.33	22.46

14. One of the earliest pollution problems brought to the attention of the Environmental Protection Agency (EPA) was the case of the Sioux Lake in eastern South Dakota. For years a small paper plant located nearby had been discharging waste containing carbon tetrachloride (CCl_4) into the waters of the lake. At the time the EPA learned of the situation, the chemical was entering at a rate of 16 cubic yards/year.

The agency immediately ordered the installation of filters designed to slow (and eventually stop) the flow of CCl_4 from the mill. Implementation of this program took exactly three years, during which the flow of pollutant was steady at 16 cubic yards/year. Once the filters were installed, the flow declined. From the time the filters were installed until the time the flow stopped, the rate of flow was well approximated by

$$\text{Rate (in cubic yards/year)} = t^2 - 14t + 49$$

where t is time measured in years since the EPA learned of the situation (Thus $t \geq 3$).

(a) Draw a graph showing the rate of CCl_4 flow into the lake as a function of time, beginning at the time the EPA first learned of the situation.
(b) How many years elapsed between the time the EPA learned of the situation and the time the pollution flow stopped entirely?
(c) How much CCl_4 entered the waters during the time shown in the graph in part (a)?

6.2 APPLICATIONS TO LIFE SCIENCES

Population Growth Rates

Models describing exponential growth of populations have been analyzed in previous chapters. In this section, we discuss rate of growth of populations in general.

[2] *World Almanac*, 1995

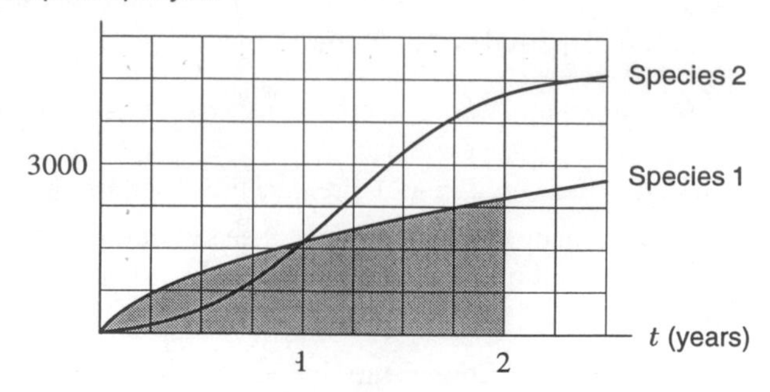

Figure 6.15: Population growth rates for two species

Example 1 The rates of growth of the populations of two species of plants (measured in new plants per year) are shown in Figure 6.15. Assume that the populations of the two species are equal at time $t = 0$.

a) Which population is larger after one year?
b) Which population is larger after two years?
c) How much does the population of species 1 increase during the first two years?

Solution a) The rate of growth of the population of species 1 is higher than that of species 2 throughout the first year, and so the population of species 1 is larger after one year.

b) After two years, the situation is less clear, since the population of species 1 was increasing faster for the first year and that of species 2 for the second. However, we can use the fact that, if $r(t)$ is the rate of growth of the population

$$\begin{array}{c}\text{Total change in population}\\ \text{during first two years}\end{array} = \int_0^2 r(t)dt.$$

This definite integral is the area under the graph of the function $r(t)$ between 0 and 2. Since the area in Figure 6.15 representing the total population change for species 2 is clearly larger than the area representing the total population change for species 1, the population of species 2 is larger after two years.

c) The population change for species 1 equals the area of the region under the graph of $r(t)$ between $t = 0$ and $t = 2$ in Figure 6.15. The region consists of approximately 16.5 boxes, each of area (750 plants/year)(0.25 year) = 187.5 plants, giving a total of (16.5)(187.5) = 3093.75. The population of species 1 goes up by about 3100 plants during the two years.

Relative Rates of Population Growth

When dealing with human populations, it is customary to give the *relative* rate of growth of a population rather than the rate of growth of the population. The difference is that the rate of growth gives the number of persons per year by which the population is increasing, but the relative rate gives the percentage per year by which the population is increasing. In Example 1, we were given the rates of growth of the populations, not the relative rate, and that is why population change could be represented by the integral of the rate.

How can you find total population change if you have the relative rate of growth of the population? The answer is that you cannot without more information. But you can find the percentage change in the population. Let's see how.

If $P(t)$ is the population, where t is in years, then the rate of growth of P is the derivative P', and the relative rate of growth is the quotient P'/P. Since

$$\frac{d}{dt}(\ln P) = \frac{P'}{P} = \text{relative growth rate}$$

we have:

> The relative growth rate of P is the rate of change of $\ln(P)$.
> The integral of the relative growth rate gives the total change in $\ln(P)$.

Example 2 The relative rate of growth P'/P of a population $P(t)$ over a 50 year period is given in Figure 6.16. What can you say about the population change during the period?

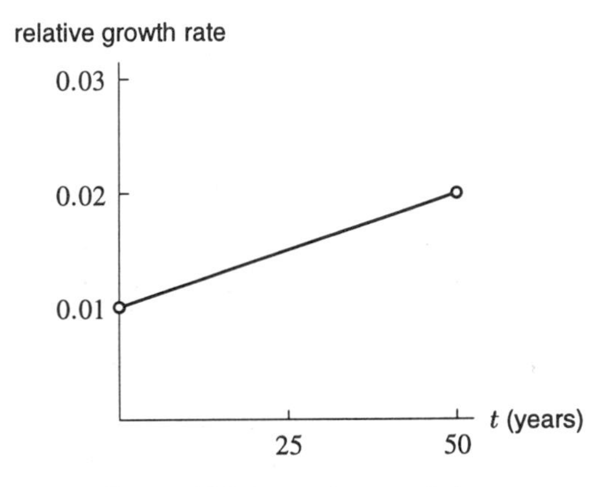

Figure 6.16: A growing population.

Solution The change in $\ln P$ equals the integral of the relative rate of growth of P, and so we have

$$\ln P(50) - \ln P(0) = \int_0^{50} P'/P\,dt$$

This integral equals the area under the graph of P'/P between $t = 0$ and $t = 50$, which is 0.750. Thus

$$\begin{aligned} \ln P(50) - \ln P(0) &= 0.75 \\ \ln(P(50)/P(0)) &= 0.750 \\ P(50)/P(0) &= e^{0.75} = 2.1 \end{aligned}$$

The population more than doubled during the 50 years, going up by about 210%. We can not determine the amount by which the population increased unless we know how large the population was initially.

What might cause the relative growth rate of a population to increase? If all other things remained constant, the relative growth rate would increase if either the relative birth rate increased or the relative death rate decreased. Even if the relative birth rate is decreasing, however, we will still see an increase in the relative population growth rate if the relative death rate is decreasing even faster. This is the case with the population of the world today. The important variable is the difference between the birth rate and the death rate.

Example 3 In Figure 6.17[3], we see relative birth rates and relative death rates for developed countries and developing countries.

a) Which is larger, the birth rate or the death rate? What does this tell you? Are the birth rates increasing or decreasing? Are the death rates increasing or decreasing? Is the birth rate or the death rate changing faster? What does this tell you about how the populations are changing?
b) By what percentage did the population of developing countries increase during the century 1800-1900?
c) By what percentage did the population of developing countries increase during the 27 years 1950-1977?

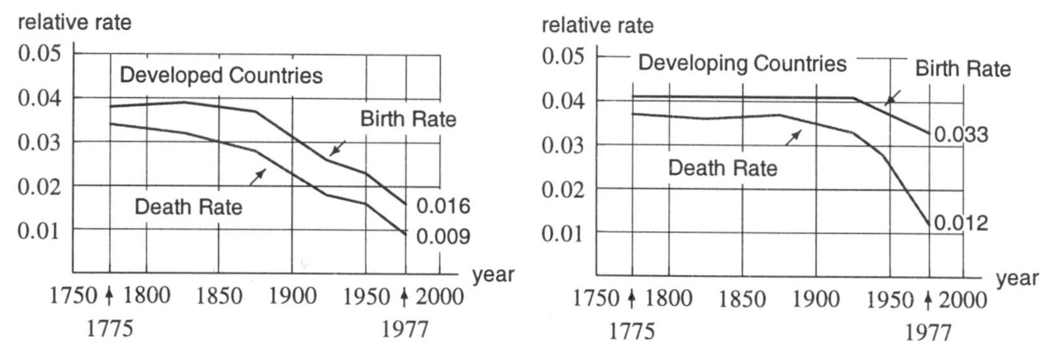

Figure 6.17: Birth and death rates in developed and developing countries, 1775-1977

Solution a) We first look at the situation in developed countries. The birth rate is higher and so the population is increasing. We see that the birth rate and death rate are both decreasing. Since the birth rate and death rate appear to be decreasing at approximately the same rate, the population of developed countries is increasing at a constant relative rate.

In developing countries, again we see that the birth rate and death rate are both decreasing (since about 1925) and that the birth rate is higher. In developing countries in recent years, the death rate is decreasing faster than the birth rate, and so the relative rate of population growth is increasing. Notice that this is happening in spite of the fact that the birth rate has decreased significantly since 1925. The decline in the death rate is the primary reason for the recent population explosion in developing countries.

b) The relative rate of growth of the population is the difference between the relative birth rate and the relative death rate, and so it is represented on Figure 6.17 by the vertical distance between the birth rate curve and the death rate curve. Therefore the area of the region between the two curves from 1800 to 1900, shown in Figure 6.18, gives the change in $\ln P$. The region is approximately a rectangle of height 0.005 and width 100, and thus of area 0.5. We have

$$\ln P(1900) - \ln P(1800) = 0.5$$
$$\ln(P(1900)/P(1800)) = 0.5$$
$$P(1900)/P(1800) = e^{0.5} = 1.65$$

[3] "Food and Population: A Global Concern", Elaine Murphy, Washington, DC: Population Reference Bureau, Inc., 1984, p.2.

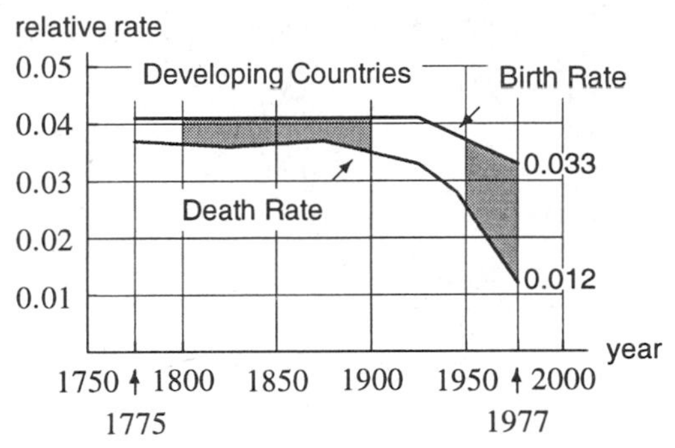

Figure 6.18: Relative rate of population increase = relative Birth rate − relative Death rate

During the nineteenth century the population of developing countries increased about 165%.

c) The region between the birth rate and death rate curves from 1950 to 1977 in Figure 6.18 consists of approximately 1.5 rectangles of area (0.010)(27) = 0.27, so the area of this region is approximately (1.5) (0.27) = 0.405. We have

$$\ln P(1977) - \ln P(1950) = 0.405$$

and so

$$P(1977)/P(1950) = e^{0.405} = 1.50$$

The population of developing countries increased by about 150% during the 27 years from 1950 to 1977.

Bioavailability in Drug Treatments

An important application of the definite integral in pharmacology and drug toxicity is to measure the overall presence of a drug in the bloodstream during the course of a treatment, known as the *bioavailability* of the drug. Unit bioavailability represents 1 unit concentration of the drug in the bloodstream for 1 hour. For example, a concentration of $3\mu\text{g/cm}^3$ in the blood for 2 hours would have bioavailability equal to $(3)(2) = 6(\mu\text{g/cm}^3)$-hours.

Ordinarily the concentration of a drug in the blood is not constant. If the drug is given orally or injected into the muscle, the concentration in the blood at first increases as the drug is absorbed into the bloodstream, and then the concentration gradually decreases back to zero as the drug is broken down and excreted. See Figure 6.19 for a typical graph of concentration as a function to time.

Suppose that we want to calculate the bioavailability of a drug that is in the blood with concentration $C(t)\mu\text{g/cm}^3$ at time t for the time period $0 \le t \le T$. In order to use what we know about bioavailability for constant concentrations, we first divide up the time interval into many short intervals during each of which the concentration changes very little and so is approximately constant. During the interval from t to $t + \Delta t$

$$\begin{aligned}\text{bioavailability} &\approx \text{Concentration} \times \text{Time}\\ &= C(t)\Delta t\end{aligned}$$

Summing over all subintervals gives

$$\text{Total bioavailability} \approx \sum C(t)\Delta t.$$

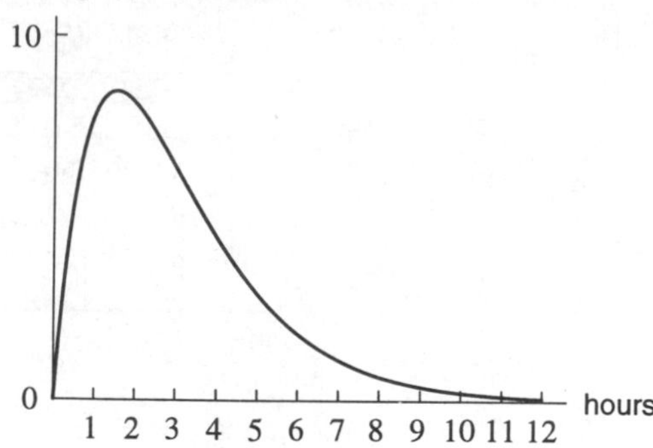

Figure 6.19: Drug concentration curve following oral dose[a]

[a] *Drug Treatment*, Graeme S. Avery (Ed.), Adis Press, 1976.

In the limit as $\Delta t \to 0$, we get the following integral:

$$\text{Bioavailability} = \int_0^T C(t)dt$$

It follows that the total bioavailability of a drug is equal to the area under the graph of the drug concentration function $C(t)$.

Example 4 Blood concentration profiles of two drugs given at time $t = 0$ are shown in Figure 6.20. Discuss the differences and similarities between the two drugs in terms of peak concentration, speed of absorption into the blood stream, and total bioavailability.

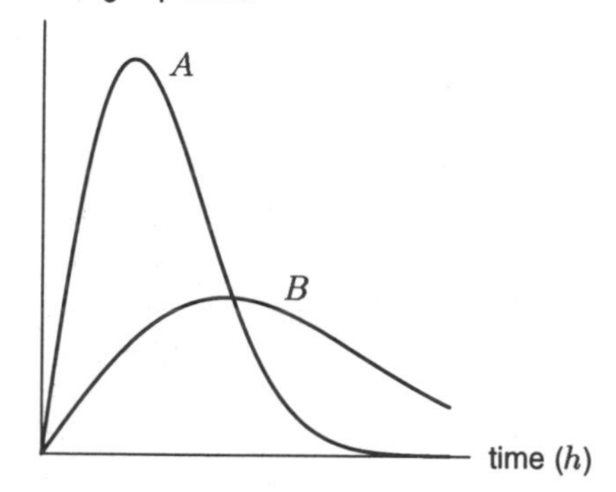

Figure 6.20: Comparing two drug products[a]

[a] *Drug Treatment*, Graeme S. Avery (Ed.), Adis Press, 1976.

Solution We see that Product A has a peak concentration more than twice as high as that of product B. Product A is absorbed more rapidly into the blood stream than Product B. However, the two products appear to have about the same total bioavailability, since we see that the areas under the graphs of the concentration functions are about the same for both products.

Problems for Section 6.2

1. Two species of plants have been introduced into a region. Past history has shown that the rates of growth of the populations of the two plants follow the patterns shown in the graph in Figure 6.21. The populations of the two species are equal at time $t = 0$, and time t is measured in years.
   (a) Which species has a larger population at the end of 5 years? At the end of 10 years?
   (b) Which species do you think will have the larger population after 20 years? Explain.

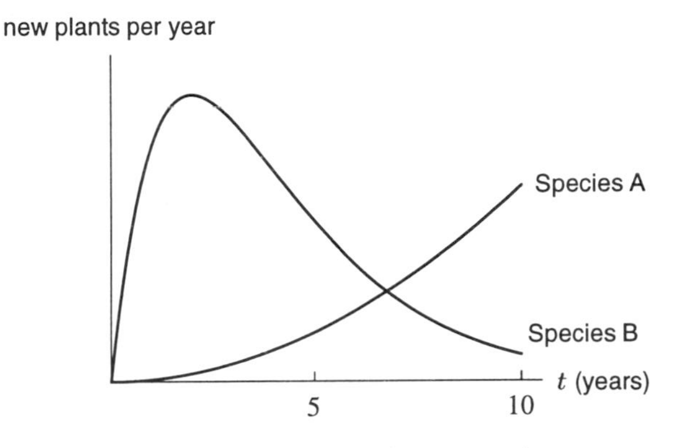

Figure 6.21: Growth rates for two species

2. Consider a bacteria population whose birth rate, B, is given by the curve in Figure 6.22 as a function of time in hours. The birth rate is in births per hour. The curve marked D gives the death rate (in deaths per hour) of the same population.
   (a) Explain what the shape of each of these graphs tells you about the population.
   (b) Use the graphs to find the time at which the net rate of increase of the population is at a maximum.
   (c) Suppose at time $t = 0$ the population has size N. Sketch the graph of the total number born by time t. Also sketch the graph of the number alive at time t. Use the given graphs to find the time at which the population size is a maximum.

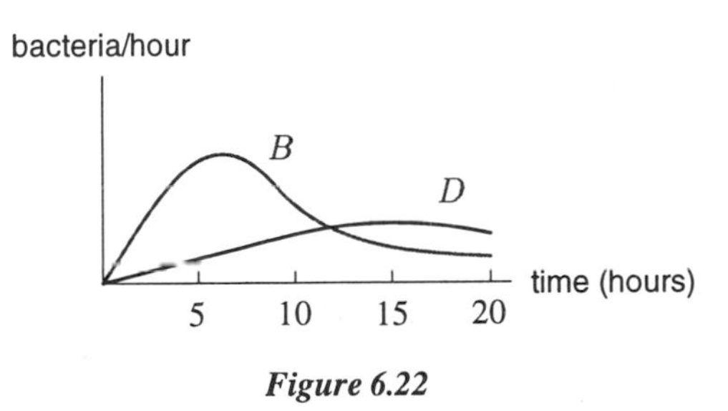

Figure 6.22

3. Birth and death rates are often reported as births or deaths per thousand members of the population. What is the relative rate of growth of a population with a birth rate of 30 births per 1000 and a death rate of 20 deaths per 1000?

4. Figure 6.23 shows plasma concentration curves for two pain relievers. Compare the two products in terms of level of peak concentration, time until peak concentration, and overall bioavailability.

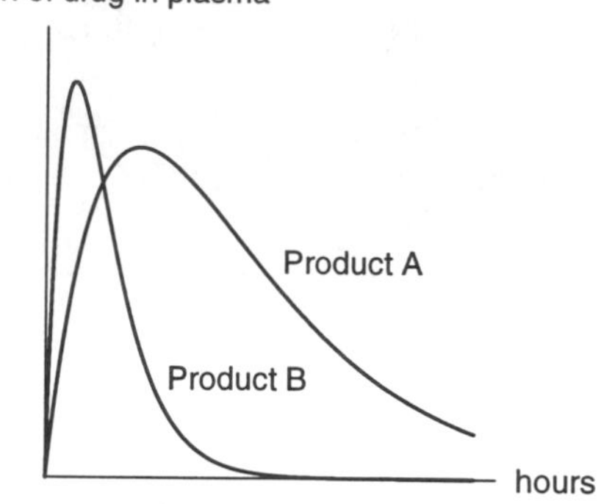

Figure 6.23: Plasma concentration curves for two pain relievers

5. Figure 6.24 shows plasma concentration curves for two drugs used to slow a rapid heart rate. Compare the two products in terms of level of peak concentration, time until peak concentration, and overall bioavailability.

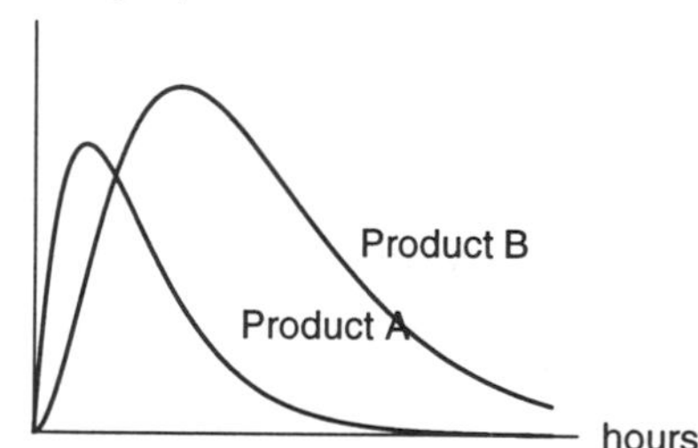

Figure 6.24: Plasma concentration curves for two drug products

6. Assume that, during the first four minutes after a foreign substance is introduced into the blood, the rate at which antibodies are made (in thousands of antibodies per minute) is given by

$$r(t) = \frac{t}{t^2 + 1},$$

where t is measured in minutes and $0 \leq t \leq 4$. Find the total quantity of antibodies in the blood at the end of the first four minutes.

7. The rate, r, at which people get sick during an epidemic of the flu can be approximated by $r = 1000te^{-0.5t}$ where r is measured in people/day and t is measured in days since the start of the epidemic.

(a) Sketch a graph of r as a function of t. (b) When are people getting sick fastest?
(c) How many people get sick altogether?

8. The Quabbin Reservoir in the western part of Massachusetts provides most of Boston's water. The graph in Figure 6.25 represents the flow of water in and out of the Quabbin Reservoir throughout 1993.

   (a) Sketch a possible graph for the quantity of water in the reservoir, as a function of time.

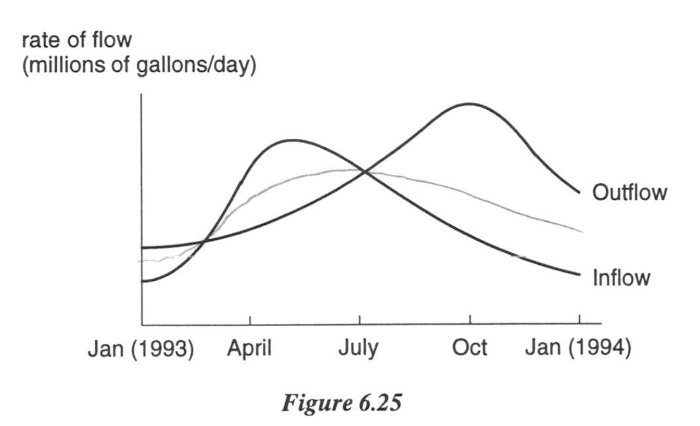

Figure 6.25

   (b) When, in the course of 1993, was the quantity of water in the reservoir largest? Smallest? Mark and label these points on the graph you drew in part (a).
   (c) When was the quantity of water decreasing most rapidly? Again, mark and label this time on both graphs.
   (d) By July 1994 the quantity of water in the reservoir was about the same as in January 1993. Draw plausible graphs for the flow into and the flow out of the reservoir for the first half of 1994. Explain your graph.

6.3 ECONOMIC APPLICATION - PRESENT AND FUTURE VALUE

Present and Future Value

Many business deals involve future payments. If you buy a car or furniture, for example, you may buy it on credit and pay over a period of time. If you are going to accept payment in the future under such a deal, you obviously need to know how much you should be paid. Being paid \$100 in the future is clearly worse than being paid \$100 today for many reasons. One is inflation: \$100 in the future may well buy less than \$100 today because prices are likely to rise. More important, if you are given the money today, you can do something else with it—for example, put it in the bank, invest it somewhere, or spend it. Thus, even without inflation, if you are to accept payment in the future, you would expect to be paid more to compensate for this loss of potential earnings. The question we will consider now is, how much more?

To simplify matters, we will only consider what you would lose by not earning interest; we will not consider the effect of inflation. Let's look at some specific numbers. Suppose you deposit \$100 in an account which earns 7% interest compounded annually, so that in a year's time you will have \$107. Thus, \$100 today will be worth the same as \$107 a year from now. We say that the \$107 is the *future value* of the \$100, and that the \$100 is the *present value* of the \$107. In general, we say the following:

- The **future value**, $\$B$, of a payment, $\$P$, is the amount to which the $\$P$ would grow by the future date if deposited in an interest bearing bank account.
- The **present value**, $\$P$, of a future payment, $\$B$, is the amount which would have to be deposited in a bank account today to produce exactly $\$B$ in the account at the relevant time in the future.

As you might expect, the present value will always be smaller than the future value. In our work on compound interest in Chapter 1, we saw that with an interest rate of r, compounded annually, and a time period of t years, a deposit of $\$P$ grows to a future balance of $\$B$, where

$$B = P(1+r)^t, \quad \text{or equivalently,} \quad P = \frac{B}{(1+r)^t}.$$

If the interest is compounded n times a year at rate r, and if $\$B$ is the *future value* of $\$P$ after t years and $\$P$ is the *present value* of $\$B$, then

$$B = P\left(1+\frac{r}{n}\right)^{nt}, \quad \text{or equivalently,} \quad P = \frac{B}{(1+r/n)^{nt}}.$$

Note that for a 7% interest rate, $r = 0.07$. In addition, as we saw in Chapter 1, when n tends towards infinity, we say that the compounding becomes continuous, and we get the following result:

If the interest is compounded continuously at rate r, and if $\$B$ is the future value of $\$P$ after t years and $\$P$ is the present value of $\$B$, then

$$B = Pe^{rt}, \quad \text{or equivalently,} \quad P = \frac{B}{e^{rt}} = Be^{-rt}.$$

In the rest of this section, we will assume that interest is compounded continuously.

Example 1 You win the lottery and are offered the choice between \$1 million in four yearly installments of \$250,000 each, starting now, and a lump-sum payment of \$920,000 now. Assuming a 6% interest rate, compounded continuously, and ignoring taxes, which should you choose?

Solution We will do the problem in two ways. First, we assume that you pick the option with the largest present value. The first of the four \$250,000 payments is made now, so

$$\text{Present value of first payment} = \$250{,}000.$$

The second payment is made one year from now and so

$$\text{Present value of second payment} = \$250{,}000e^{-0.06(1)}.$$

Calculating the present value of the third and fourth payments similarly, we find:

$$\begin{aligned}\text{Total present value} &= \$250{,}000 + \$250{,}000e^{-0.06(1)} + \$250{,}000e^{-0.06(2)} + \$250{,}000e^{-0.06(3)}\\ &\approx \$250{,}000 + \$235{,}441 + \$221{,}730 + \$208{,}818\\ &= \$915{,}989.\end{aligned}$$

Since the present value of the four payments is less than \$920,000, you are better off taking the \$920,000 right now.

Alternatively, we can do the problem by comparing the future values of the two pay schemes. The scheme with the highest future value is the best from a purely financial point of view. We calculate the future value of both schemes three years from now, on the date of the last \$250,000 payment. At that time,

$$\text{Future value of the lump sum payment} = \$920{,}000e^{0.06(3)} \approx \$1{,}101{,}440.$$

Now we calculate the future value of the first \$250,000 payment:

$$\text{Future value of the first payment} = \$250{,}000e^{0.06(3)}.$$

Calculating the future value of the other payments similarly, we find:

$$\begin{aligned}\text{Total future value} &= \$250{,}000e^{0.06(3)} + \$250{,}000e^{0.06(2)} + \$250{,}000e^{0.06(1)} + \$250{,}000\\ &\approx \$299{,}304 + \$281{,}874 + \$265{,}459 + \$250{,}000\\ &= \$1{,}096{,}637.\end{aligned}$$

The future value of the \$920,000 payment is greater, so you are better off taking the \$920,000 right now. Of course, since the present value of the \$920,000 payment is greater than the present value of the four separate payments, you would expect the future value of the \$920,000 payment to be greater than the future value of the four separate payments.

(**Note:** If you read the fine print, you will find that many lotteries do not make their payments right away, but often spread them out, sometimes far into the future. This is to reduce the present value of the payments made, so that the value of the prizes is much less than it might first appear!)

Income Stream

When we consider payments made to or by an individual, we usually think of *discrete* payments, that is, payments made at specific moments in time. However, we may think of payments made by or to a company as being *continuous*. The revenues earned by a huge corporation, your local phone company, for example, come in essentially all the time, and therefore they can be represented by a continuous *income stream*. Since the rate at which revenue is received may vary from time to time, the income stream is described by

$$S(t) \text{ dollars/year.}$$

Notice that $S(t)$ is a *rate* (its units are dollars per year, for example) and that the rate depends on the time, t, usually measured in years from the present.

Present and Future Values of an Income Stream

Just as we can find the present and future values of a single payment, so we can find the present and future values of a stream of payments. As before, the future value represents the total amount of money obtained by depositing the income stream into a bank account as the money is received and letting it gather interest. The present value represents the amount of money you would have to deposit today (in an interest-bearing bank account) in order to match what you would get from the income stream.

When we are working with a continuous income stream, we will assume that interest is compounded continuously. The reason for this is that the approximations we are going to make (of sums by integrals) are much simpler if both payments and interest are continuous.

Suppose that we want to calculate the present value of the income stream described by a rate of $S(t)$ dollars per year, and that we are interested in the period from now until T years in the future. In order to use what we know about single deposits to calculate the present values of an income stream, we must first divide the stream into many small deposits, each of which is made at approximately one instant. We divide the interval $0 \leq t \leq T$ into subintervals, each of length Δt:

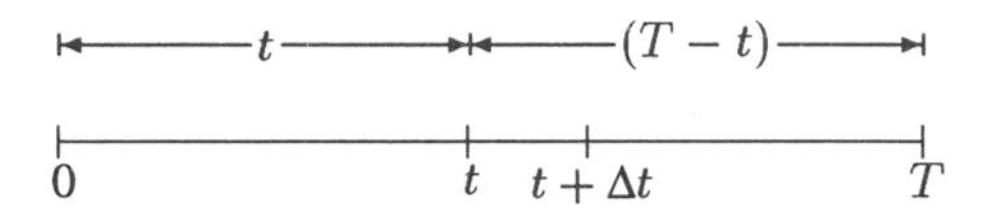

Assuming Δt is small, the rate, $S(t)$, at which deposits are being made will not vary much within one subinterval. Thus, between t and $t + \Delta t$:

$$\begin{aligned} \text{Amount paid} &\approx \text{Rate of deposits} \times \text{Time} \\ &\approx (S(t) \text{ dollars/year})(\Delta t \text{ years}) \\ &= S(t)\Delta t \text{ dollars.} \end{aligned}$$

Measured from the present, the deposit of $S(t)\Delta t$ is made t years in the future. Thus,

$$\begin{array}{c}\text{Present value of money} \\ \text{deposited in interval } t \text{ to } t + \Delta t\end{array} \approx S(t)\Delta t e^{-rt}.$$

Summing over all subintervals gives

$$\text{Total present value} \approx \sum S(t)e^{-rt}\Delta t \text{ dollars.}$$

In the limit as $\Delta t \to 0$, we get the following integral:

$$\boxed{\text{Present value of income stream } S(t) = \int_0^T S(t)e^{-rt}\,dt.}$$

In computing future value at time T, the deposit of $S(t)\Delta t$ at time t earns interest until the final time T. It therefore earns interest for a time period of $(T - t)$ years, and so

$$\text{Future value of money deposited in interval } t \text{ to } t + \Delta t \approx [S(t)\Delta t]\, e^{r(T-t)}.$$

Summing over all subintervals, we get:

$$\text{Total future value } \approx \sum S(t)\Delta t e^{r(T-t)} \text{ dollars.}$$

As the length of the subdivisions tends toward zero, the sum becomes an integral:

$$\text{Future value at time } T \text{ of income stream } S(t) = \int_0^T S(t)e^{r(T-t)}\,dt.$$

Example 2 Find the present and future values of a constant income stream of $100 per year over a period of 20 years, assuming an interest rate of 10% compounded continuously.

Solution Computing the integral using a calculator, we have

$$\text{Present value } = \int_0^{20} 100e^{-0.1t}dt \approx \$864.66.$$

Once we know the present value, it is easy to find the future value (or vice versa) since we have the following relationship, where B represents the future value and P represents the present value:

$$B = Pe^{rt}.$$

Thus,

$$\text{Future value} = 864.66e^{0.10(20)} = \$6389.02.$$

Alternatively, we could have found the future value first, using the integral formula given, and then calculated the present value from the future value.

We found that the future value is $6389.02. Deposits totaled $100 per year over a 20 year period, so the total amount deposited was $2000. The money has more than tripled because of the interest.

Example 3 Suppose that your account pays 6% interest, and you will need $50,000 in the account in 8 years.

a) If you deposit money into the account at a constant rate throughout the 8 year period, what rate (in dollars per year) must you use in order to meet your goal?

b) If you decide to make one lump sum deposit now instead of depositing money continually, what amount must you deposit now?

Solution a) We let S stand for the rate in dollars per year at which you will deposit money, so S is our (constant) income stream. We know that

$$\text{Future value} = \int_0^T Se^{r(T-t)}dt.$$

In this case, $r = 0.06, T = 8$, and the future value required is $B = \$50{,}000$. We have

$$50{,}000 = \int_0^8 Se^{0.06(8-t)}dt$$

Since S is a constant, we can pull it out in front of the integral sign:

$$50{,}000 = S \int_0^8 e^{0.06(8-t)}dt$$

Using a calculator to evaluate the integral, we have

$$50{,}000 \approx S(10.2679).$$

Solving for S, we have

$$S \approx 4869.54.$$

To meet your goal, you will have to deposit money at a continuous rate of \$4869.54 per year (or about \$400 a month.)

b) If, instead, you decide to make one lump sum deposit now the amount you must deposit now is the present value P of the \$50,000, so we solve

$$\begin{aligned} B &= Pe^{rt} \\ 50{,}000 &= Pe^{0.06(8)} \\ 30{,}939.17 &\approx P. \end{aligned}$$

You would have to deposit \$30,939.17 into the account now if you want it to total \$50,000 in eight years.

Problems for Section 6.3

1. Draw a graph, with time in years on the horizontal axis, of what an income stream might look like for a company that sells sunscreen in the Northeast United States.
2. A company that owes your company money offers to begin repaying the debt either by making four annual payments of \$5000, spread out over the next three years (one now, one in a year, one in two years, and one in three years), or by waiting and making a lump-sum payment of \$25,000 at the end of the three years. If you can assume an 8% annual return on the money, compounded continuously, and you use only financial considerations, which option for repayment should you choose? Justify your answer. What other considerations might you want to consider in making your decision?
3. Find the present and future values of a constant income stream of \$3000 per year over a period of 15 years, assuming a 6% annual interest rate compounded continuously.
4. A certain bond is guaranteed to pay $(100 + 10t)$ dollars per year for 10 years, where t is the number of years from the present. Find the present value of this income stream, given an interest rate of 5%, compounded continuously.
5. (a) A bank account earns 10% interest compounded continuously. At what (constant, continuous) rate must a parent deposit money into such an account in order to save \$100,000 in 10 years for a child's college expenses?
   (b) If the parent decides instead to deposit a lump sum now in order to attain the goal of \$100,000 in 10 years, how much must be deposited now?

6. The value of good wine increases with age. Thus, if you are a wine dealer, you have the problem of deciding whether to sell your wine now, at a price of P a bottle, or to sell it later at a higher price. Suppose you know that the price of your wine t years from now is well approximated by $P(1+20\sqrt{t})$. Assuming continuous compounding and a prevailing interest rate of 5% per year, when is the best time to sell your wine?

7. (a) If you deposit money continuously at a constant rate of \$1000 per year into a bank account that earns 5% interest, how many years will it take for the balance to reach \$10,000?
   (b) How many years would it take if the account had \$2000 in it initially?

8. (a) During the 1970s, ACME Widgets sold at a yearly rate, in widgets per year, given by $R = R_0 e^{0.15t}$, where t is time in years since January 1, 1970. Suppose they were selling widgets at a rate of 1000 per year on the first day of the decade. How many widgets did they sell during the decade? How many did they sell if the rate on January 1, 1970 was 150,000,000 widgets per year?
   (b) In the first case above (1000 widgets per year on January 1, 1970), how long did it take for half the widgets in the 1970s to be sold? In the second case (150,000,000 widgets per year on January 1, 1970), when had half the widgets in the 1970s been sold?
   (c) In 1980 ACME began an advertising campaign claiming that half the widgets it had sold in the previous ten years were still in use. In light of your answer to part (b), what must the average lifetime of a widget be in order to justify this claim?

9. In 1980 West Germany made a loan of 20 billion deutsche marks to the Soviet Union, to be used for the construction of a natural gas pipeline connecting Siberia to Western Russia, and continuing to West Germany (Urengoi – Uschgorod – Berlin). The deal was as follows: In 1985, upon completion of the pipeline, the Soviet Union would deliver natural gas to West Germany, at a constant rate, for all future times. Assuming a constant price of natural gas of 10 pfennig (= 0.10 deutsche mark) per cubic meter, and assuming West Germany expects 10% annual interest on its investment (compounded continuously), at what rate does the Soviet Union have to deliver the gas, in billions of cubic meters per year? Keep in mind that delivery of gas could not begin until the pipeline was completed. Thus, West Germany received no return on its investment until after five years had passed. (Note: A deal of this type was actually made between the two countries; obviously, the terms of the deal were much more complex than presented here.)

6.4 ECONOMIC APPLICATION - CONSUMER AND PRODUCER SURPLUS

Supply and Demand Curves

In a free market, the quantity of a certain item produced and sold can be described by the supply and demand curves of the item. As we saw in Chapter 1, the *supply curve* $S(q)$ shows what quantity q of the item the producers will supply at different price levels, and the *demand curve* $D(q)$ shows what quantity of goods are bought at various prices. Typical supply and demand curves are shown in Figure 6.26. As we would expect, the supply curve is increasing (as price increases, the quantity supplied goes up) and the demand curve is decreasing (as price increases, the demand for the product goes down.) Recall that it is traditional to place price along the vertical axis and quantity along the horizontal axis, as we see in Figure 6.26.

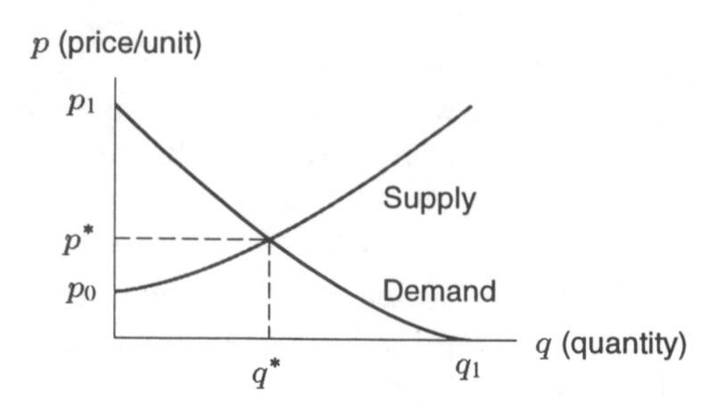

Figure 6.26: Supply and demand curves

It is assumed that the market will settle to the *equilibrium price* and *quantity*, p^* and q^*, where the graphs cross. This means that at the equilibrium point, a quantity q^* of an item will be produced and sold for a price of p^* each.

Consumer and Producer Surplus

Notice that at equilibrium, a number of consumers have bought the item at a lower price than they would have been willing to pay. (For example, there are some consumers who would have been willing to pay prices up to p_1.) Similarly, there are some suppliers who would have been willing to produce the item at a lower price (down to p_0, in fact). We define the following terms:

> The **consumer surplus** represents the buyers' gain from trade. It equals the total amount gained by consumers by buying the good at the current price, rather than at the price they would have been willing to pay.
> The **producer surplus** represents the suppliers' gain from trade. It equals the total amount gained by producers by selling at the current price, rather than at the price they would have been willing to accept.
> In the absence of artificial price controls, the current price will equal the equilibrium price.

The consumers and producers both individually and collectively are richer merely for having traded. The consumer and producer surplus measure how much richer they are.

Suppose that all consumers who would buy at the equilibrium price actually buy the good at the maximum price they are willing to pay. To see how much they would pay, divide the interval from 0 to q^* into subintervals of length Δq. Figure 6.27 shows that a quantity Δq of items are sold at a price of about p_1, another Δq are sold for a slightly lower price of about p_2, the next Δq for a price of about p_3, and so on. Thus, the consumers' total expenditure is about

$$p_1\Delta q + p_2\Delta q + p_3\Delta q + \cdots = \sum p_i \Delta q.$$

If D is the demand function given by $p = D(q)$, and if all consumers who were willing to pay more than p^* paid as much as they were willing, then as $\Delta q \to 0$, we would have

$$\begin{array}{c}\text{Consumer}\\\text{expenditure}\end{array} = \int_0^{q^*} D(q)\,dq = \begin{array}{c}\text{Area under demand}\\\text{curve from 0 to } q^*.\end{array}$$

Now if all goods are sold at the equilibrium price, the consumers' actual expenditure is only p^*q^*, the area of the rectangle between the axes and the lines $q = q^*$ and $p = p^*$. Thus the consumer surplus may be calculated as follows:

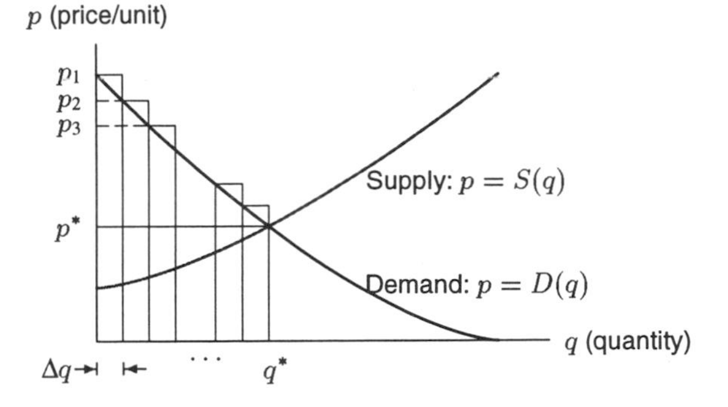

Figure 6.27: Calculation of consumer surplus

$$\text{Consumer surplus} = \left(\int_0^{q^*} D(q)dq\right) - p^*q^* = \text{Area between demand curve and line } p = p^*.$$

See Figure 6.28(a). Similarly, if the supply curve is given by the function $p = S(q)$, the producer surplus is represented in Figure 6.28(b), and defined as follows:

$$\text{Producer Surplus} = p^*q^* - \left(\int_0^{q^*} S(q)dq\right) = \text{Area between supply curve and line } p = p^*.$$

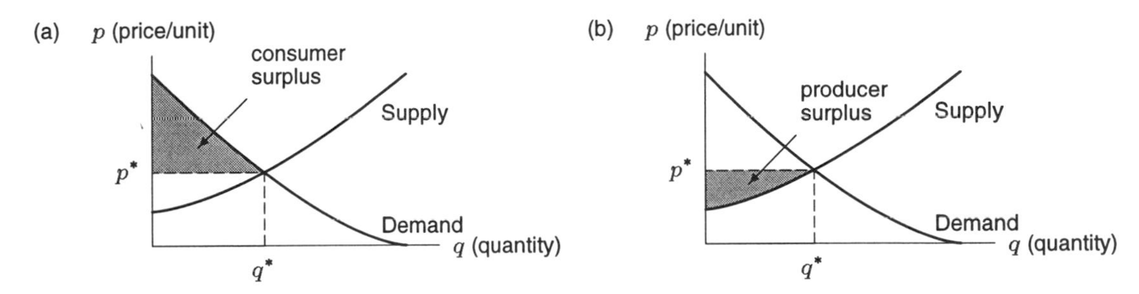

Figure 6.28: Consumer and producer surplus

Example 1 Suppose that the demand for a product is given by $D(q) = 100e^{-0.008q}$ and the supply curve is given by $S(q) = 4\sqrt{q} + 10$ for $0 \leq q \leq 500$, as in Figure 6.29.

a) At a price of \$50, what quantity will consumers be willing to buy and what quantity will producers be willing to supply? Do you think that the market will push prices up or down?
b) Find the equilibrium price and equilibrium quantity. Since market forces tend to push prices closer to the equilibrium price, was your answer to the last question in part (a) right?
c) If the price is equal to the equilibrium price, calculate the consumer surplus and interpret it.
d) If the price is equal to the equilibrium price, calculate the producer surplus and interpret it.

Solution

a) We compute the quantity demanded at a price of \$50 using the demand function $p = 100e^{-0.008q}$. Since $p = 50$, we set $50 = 100e^{-0.008q}$ and solve for q. Equivalently, we can work with the demand curve, locating the quantity q on the horizontal axis that corresponds to a price of \$50 on the vertical axis. We see that $q \approx 86.6$. In other words, at a price of \$50, consumer demand would be about 87 items.
We compute the quantity supplied at a price of \$50 using the supply equation $p = 4\sqrt{q} + 10$. Set $50 = 4\sqrt{q} + 10$, and solve for q, or use your graphing calculator or computer to find the corresponding point. We see that $q = 100$, so at a price of \$50, producers supply about 100 items.
At a price of \$50, the supply is larger than the demand, so some items will remain unsold. We can expect prices to be pushed down.

b) Graphs of $S(q)$ and $D(q)$ are shown in Figure 6.29. We see that the equilibrium price is about $p^* = 48$, and the equilibrium quantity is about $q^* = 91$ items. The market will indeed push prices downward from \$50, closer to the equilibrium price of \$48.

c) The consumer surplus is the area under the demand curve and above the line $p = 48$. (See Figure 6.30.) We have

$$\begin{aligned}\text{Consumer surplus} &= \int_0^{q^*} D(q)dq - p^*q^* \\ &= \int_0^{91} (100e^{-0.008q})dq - (48)(91) \\ &= 6464 - 4368 \\ &= 2096.\end{aligned}$$

The consumer surplus is 2096. Consumers gained \$2096 in buying goods at the equilibrium price instead of what they would have been willing to pay.

d) The producer surplus is the area above the supply curve and below the line $p = 48$. (See

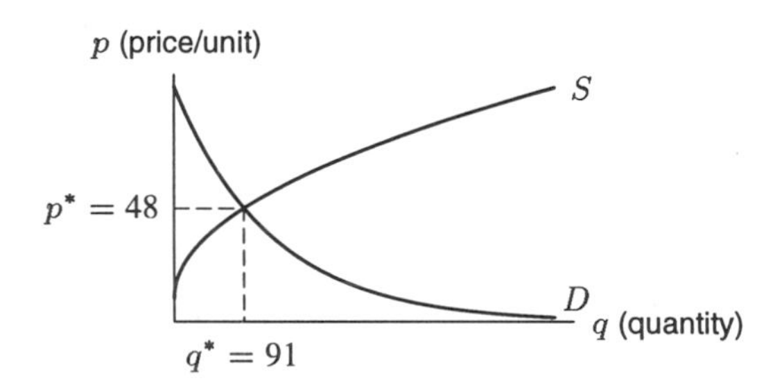

Figure 6.29: Demand and supply curves for a product

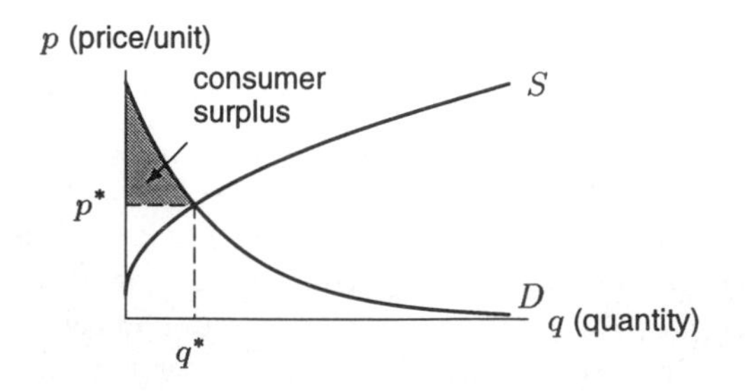

Figure 6.30: Consumer surplus

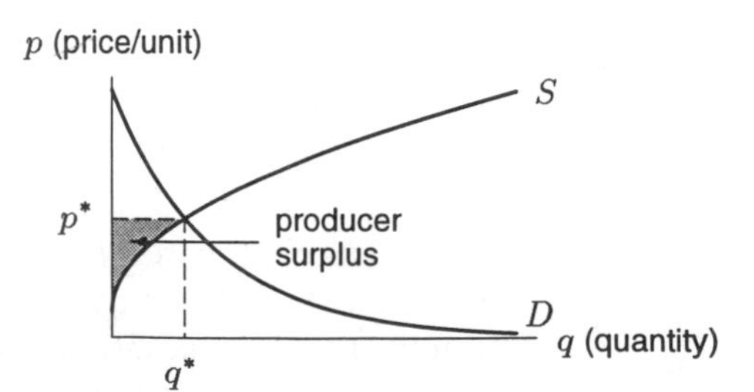

Figure 6.31: Producer surplus

Figure 6.31.) We have

$$\begin{aligned}
\text{Producer surplus} &= p^* q^* - \int_0^{q^*} S(q)\,dq \\
&= (48)(91) - \int_0^{91} (4\sqrt{q} + 10)\,dq \\
&\approx 4368 - 3225 \\
&= 1143.
\end{aligned}$$

The producer surplus is 1143. Producers gained \$1143 in supplying goods at the equilibrium price instead of the price at which they would have been willing to provide the goods. Notice that the consumer gains from trade are larger than the producer gains from trade for this product.

Wage and Price Controls

In a free market, the price of a product will generally move to the equilibrium price. In some circumstances, however, outside forces will keep the price artificially high or artificially low. Rent control, for example, will keep prices below what the market would bear, whereas cartel pricing or the minimum wage will raise prices above the market price. In these situations, the definitions of consumer and producer surplus are still valid and it makes sense to see what happens to these quantities in a situation where the price is artificially high or low.

Example 2 The dairy industry is a good example of cartel pricing: the government has set milk prices artificially high. What effect does forcing the price up from the equilibrium price have:

a) on the consumer surplus?
b) on the producer surplus?
c) on the total gains from trade (consumer surplus + producer surplus)?

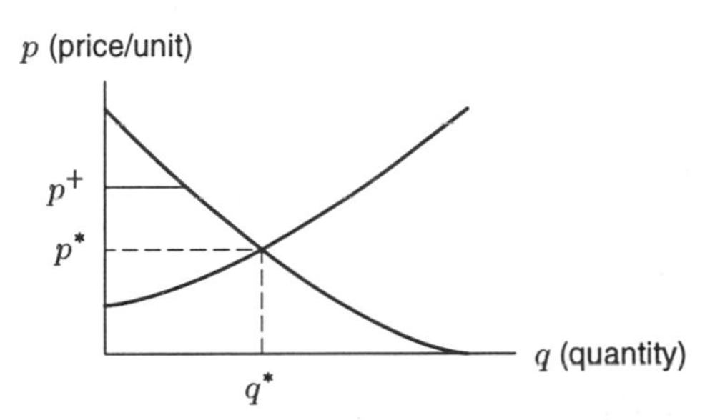

Figure 6.32: What effect does the artificially high price, p^+, have?

Solution a) A graph of possible demand and supply curves for the milk industry is given in Figure 6.32, with the equilibrium price and quantity labeled p^* and q^* respectively. Suppose that the price is fixed at the artificially high price labeled p^+ in Figure 6.32. Recall that the consumer surplus is the difference between the amount the consumers did pay (p^+) and the amount they would have been willing to pay (given on the demand curve). This is the area shaded in Figure 6.33(a).

Notice that this consumer surplus is clearly less than the consumer surplus at the equilibrium price, shown in Figure 6.33(b)

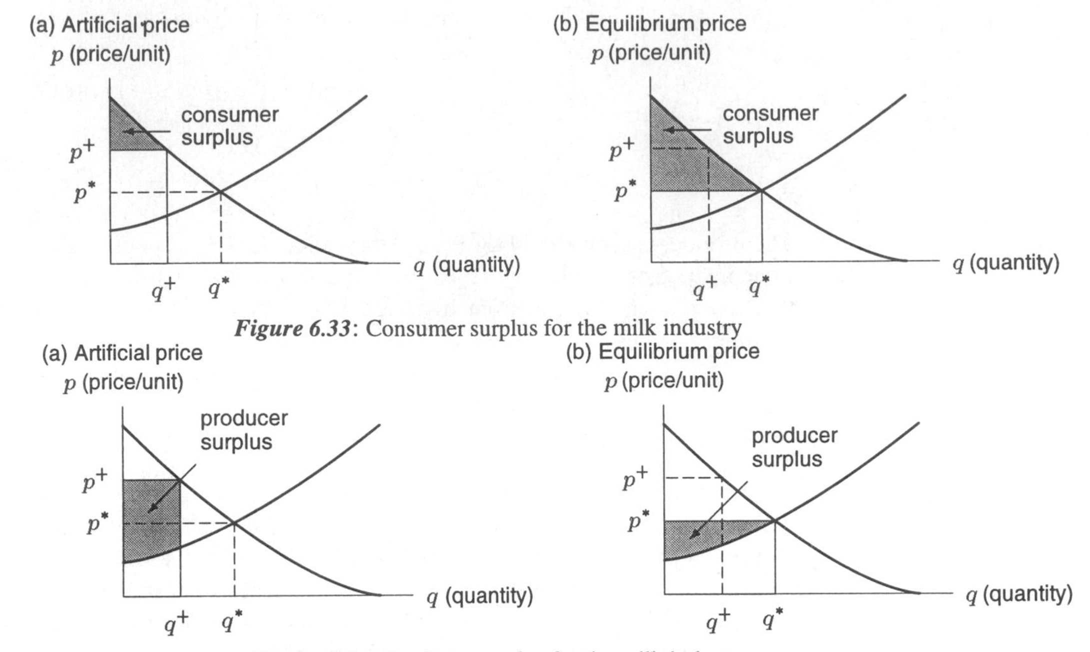

Figure 6.33: Consumer surplus for the milk industry

Figure 6.34: Producer surplus for the milk industry

b) At a price of p^+, the quantity sold, q^+, is less than it would have been at the equilibrium price. The producer surplus is the area between p^+ and the supply curve *at this reduced demand.* This area is shaded in Figure 6.34(a). Compare this producer surplus (at the artificially high price) to the producer surplus in Figure 6.34(b) (at the equilibrium price). It appears that in this case, producer surplus is greater at the artificial price than at the equilibrium price. (Different supply and demand curves might have led to a different answer.)

c) The total gains from trade (consumer surplus + producer surplus) at the artificially high price of p^+ is the area shaded in Figure 6.35(a). The total gains from trade at the equilibrium price of p^* is the area shaded in Figure 6.35(b). It is clear that, under artificial price conditions, total gains from trade go down. The total financial effect on all producers and consumers combined is a negative one.

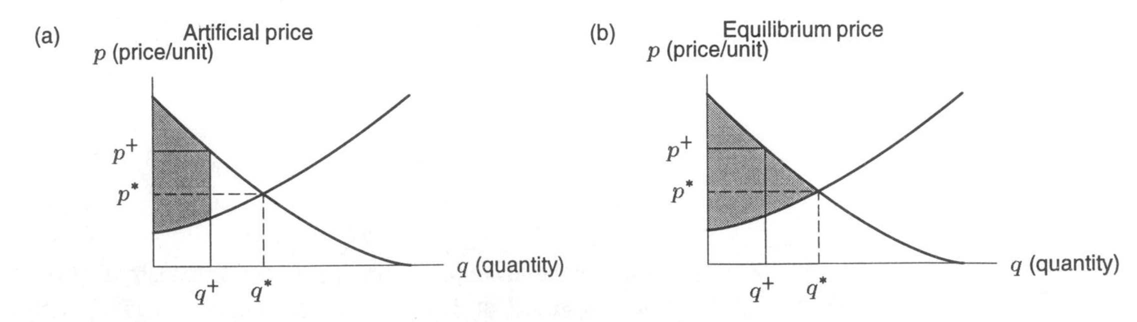

Figure 6.35: Total gains from trade

8. The graph of a cumulative distribution function is given in Figure 6.63.
   (a) Use the cumulative distribution function to find the median.
   (b) Describe the density function: For what values is it positive? Where is it increasing and where is it decreasing? Identify all local maximum and minimum values.

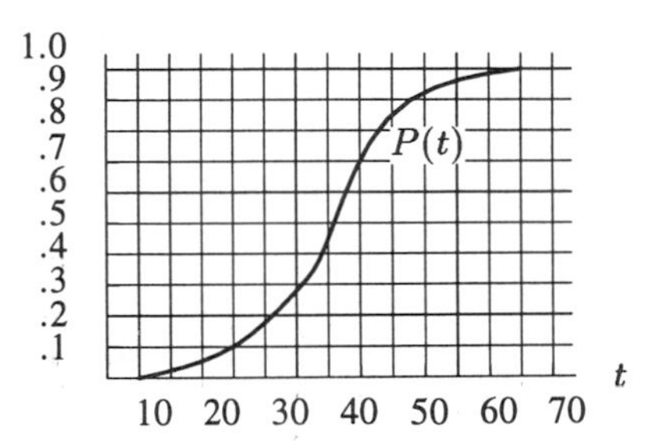

Figure 6.63: A cumulative distribution function

9. In 1950 an experiment was done observing the time gaps between successive cars on the Arroyo Seco Freeway. The data[5] show that the density function of these time gaps is approximately

$$p(x) = 0.122e^{-0.122x},$$

where x is the time in seconds and $0 \leq x \leq 40$. Find the median and mean time gap, and interpret them in terms of cars on the freeway.

10. Consider a group of people who have received treatment for a disease such as cancer. Let t be the *survival time*, the number of years a person lives after receiving treatment. The density function giving the distribution of t is $p(t) = Ce^{-Ct}$ for some positive constant C. What is the practical meaning for the cumulative distribution function $P(t) = \int_0^t p(x)\,dx$?

11. IQ scores are believed to be normally distributed with mean 100 and standard deviation 15.
    (a) Write a formula for the density distribution of IQ scores.
    (b) Estimate the fraction of the population with IQ between 115 and 120.

12. Show with a calculator that the area under the graph of the density function of the normal distribution

$$p(x) = \frac{1}{\sqrt{2\pi}}e^{-(x-15)^2/2}$$

is close to 1. Make it clear in your solution what limits of integration you used.

13. (a) Using a calculator or computer, sketch graphs of the density function of the normal distribution

$$p(x) = \frac{1}{\sigma\sqrt{2\pi}}e^{-(x-\mu)^2/(2\sigma^2)}$$

   (i) For fixed μ (say, $\mu = 5$) and varying σ (say, $\sigma = 1, 2, 3$).
   (ii) For varying μ (say, $\mu = 4, 5, 6$) and fixed σ (say, $\sigma = 1$).

   (b) Explain how the graphs confirm that μ is the mean of the distribution and that σ shows how closely the data is clustered around the mean.

[5]Reported by Daniel Furlough and Frank Barnes

REVIEW PROBLEMS FOR CHAPTER SIX

1. Find the area under the graph of $f(x) = x^2 + 2$ between $x = 0$ and $x = 6$.
2. Find the average value of the function $f(x) = 5 + 4x - x^2$ between $x = 0$ and $x = 3$.
3. Assume the marginal cost function in producing a certain product is $C'(q) = q^2 - 50q + 700$, for $0 \le q \le 50$. If fixed costs are \$500, find the total cost to produce 50 items.
4. The derivative of a function $f(t)$ is given by $f'(t) = t^3 - 6t^2 + 8t$ for $0 \le t \le 5$. Sketch a graph of this derivative function, and describe how the function $f(t)$ changes over the period $t = 0$ to $t = 5$. When is it increasing and when is it decreasing? When is it at its maximum and when is it at its minimum?

For Problems 5–6 the graph of $f'(x)$ is given. Sketch a possible graph for $f(x)$. Mark the points $x_1 \ldots x_4$ on your graph and label local maxima, local minima and points of inflection.

5.

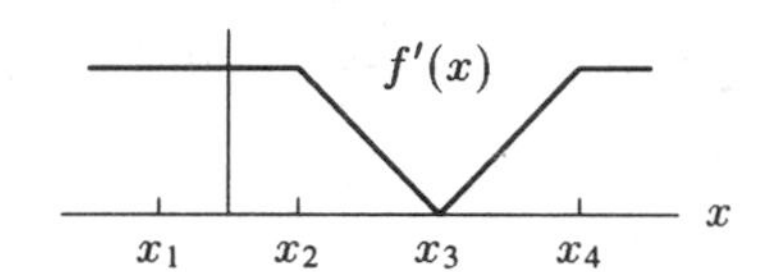

6.

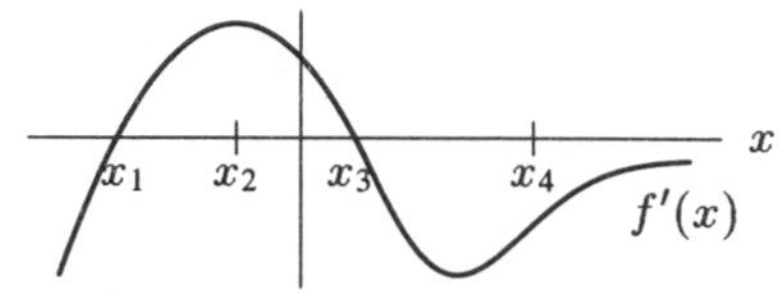

7. Water flows at a constant rate into the left side of the W-shaped container shown in Figure 6.64. Sketch a graph of the height, H, of the water in the left side of the container as a function of time t.

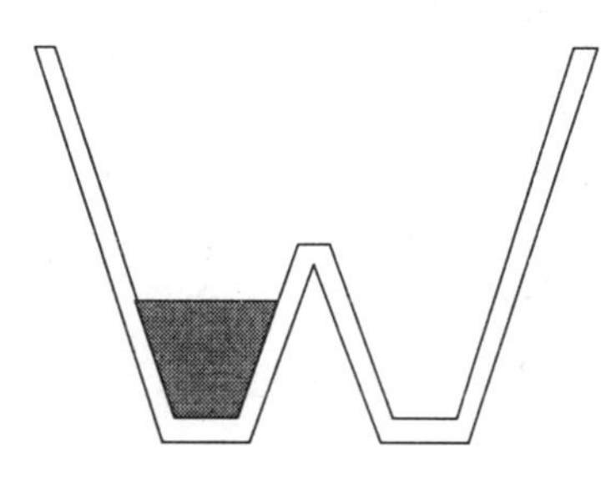

Figure 6.64

8. Throughout much of this century, the yearly consumption of electricity in the US has been increasing exponentially at a continuous rate of 7% per year. Assuming this trend continues, and that the electrical energy consumed in 1900 was 1.4 million megawatt-hours (a megawatt-hour is a measure of electrical energy),
   (a) Write an expression for electricity consumption as a function of time, t, measured in years since 1900.
   (b) Find the average yearly electrical consumption throughout this century.
   (c) During what year was electrical consumption closest to the average for the century?
   (d) Without doing the calculation for part (c), how could you have predicted which half of the century the answer would be in?

9. The width, in feet, at various points along the fairway of a hole on a golf course is given in Figure 6.65. If one pound of fertilizer covers 200 square feet, estimate the amount of fertilizer needed to fertilize the fairway.

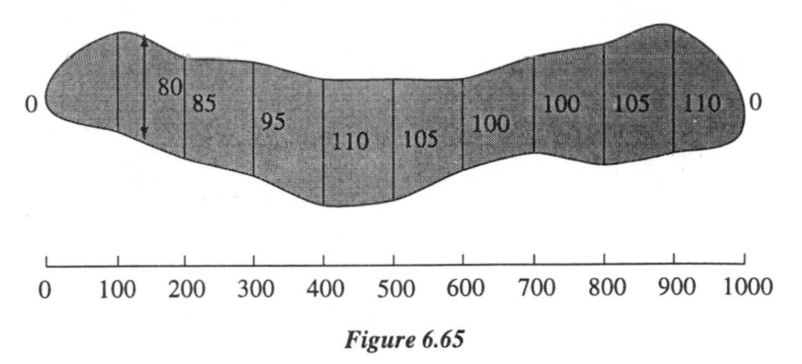

Figure 6.65

10. In 1990 humans generated 1.4×10^{20} joules of energy through the combustion of petroleum. All of the earth's petroleum would generate approximately 10^{22} joules. Assuming the use of energy generated by petroleum combustion will increase by 2% each year, how long will it be before all of our petroleum resources are used up?

11. The rate of growth of the height of two species of trees is shown in Figure 6.66, where t is measured in years, and the rate is given in feet per year. If the two trees are the same height at time $t = 0$, which tree is taller after 5 years? After 10 years?

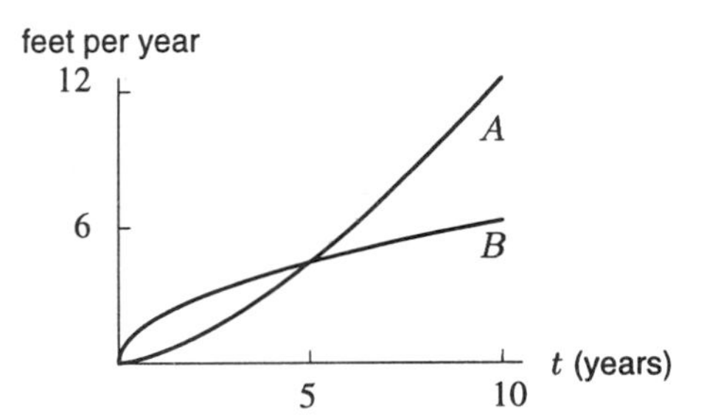

Figure 6.66: Comparing growth rates of two trees

12. The graph of the demand curve and the supply curve for a certain product are given in Figure 6.67. Estimate from this graph the equilibrium price and quantity and the consumer surplus and producer surplus. On sketches similar to Figure 6.67, shade areas corresponding to the consumer surplus and to the producer surplus.

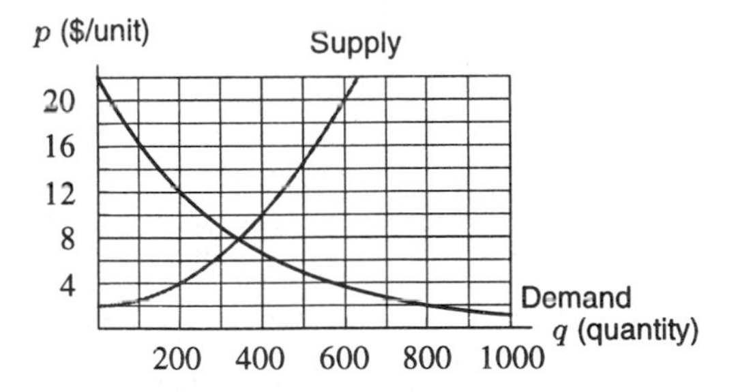

Figure 6.67: Demand and supply curves

13. Draw plasma concentration curves for two drugs A and B if product A has the highest peak concentration, but product B is absorbed more quickly and has greater overall bioavailability.

14. (a) Find the present and future values of a constant income stream of \$100 per year over a period of 20 years, assuming a 10% annual interest rate compounded continuously.
    (b) How many years will it take for the balance to reach \$5000?

15. Whether a resource is distributed evenly among members of a population is often an important political or economic question. How can we measure this? How can we decide if the distribution of wealth in this country is becoming more or less equitable over time? How can we measure which country has the most equitable income distribution? This problem describes a way of doing this.

    Suppose the resource is distributed evenly. Then any 20% of the population will have 20% of the resource. Similarly, any 30% will have 30% of the resource and so on. If, however, the resource is not distributed evenly, the poorest P% of the population (in terms of this resource) will not have P% of the goods. Suppose $F(x)$ represents the fraction of the resources owned by the poorest fraction x of the population. Thus $F(0.4) = 0.1$ means that the poorest 40% of the population owns 10% of the resource.

    (a) What would F be if the resource was distributed evenly?
    (b) What must be true of any such F? What must $F(0)$ and $F(1)$ equal? Is F increasing or decreasing? Concave up or concave down?
    (c) Gini's index of inequality, G, is one way to measure how evenly the resource is distributed. It is defined by

$$G = 2\int_0^1 [x - F(x)]\,dx.$$

    Show graphically what G represents.

16. Gini's Index of Inequality is discussed in the previous problem. Graphical representations of this index are given in Figure 6.68 for two countries. Which country has the more equitable distribution of wealth? Discuss the distribution of wealth in each of the two countries.

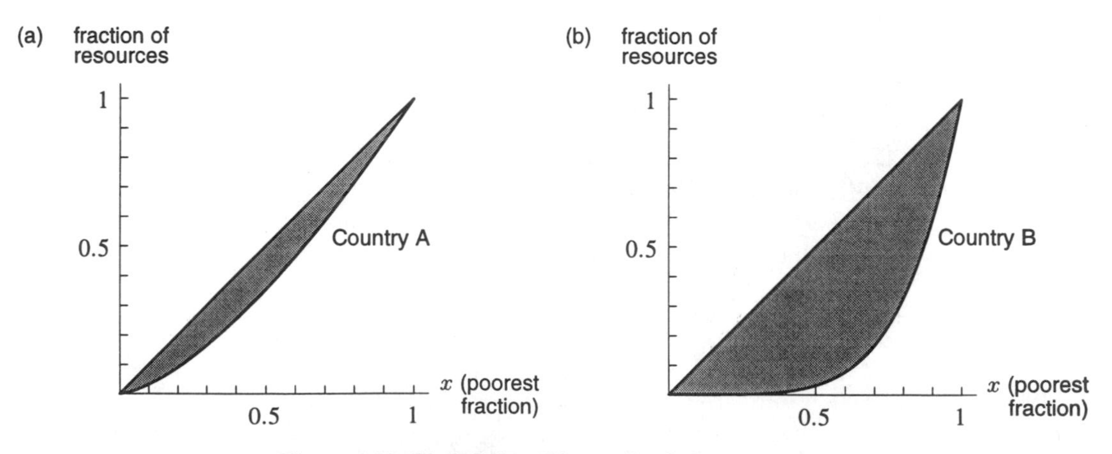

Figure 6.68: Gini's Index of Inequality for two countries

17. The heights of grass plants in a meadow were measured and the density function and cumulative distribution function are graphed in Figure 6.69 and Figure 6.70.

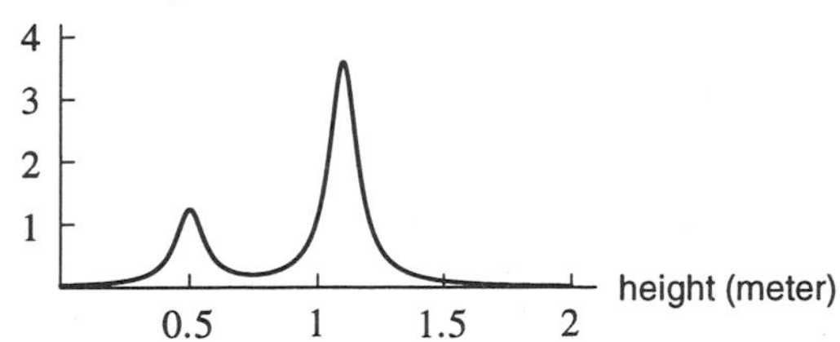

Figure 6.69: Height density of meadow grass

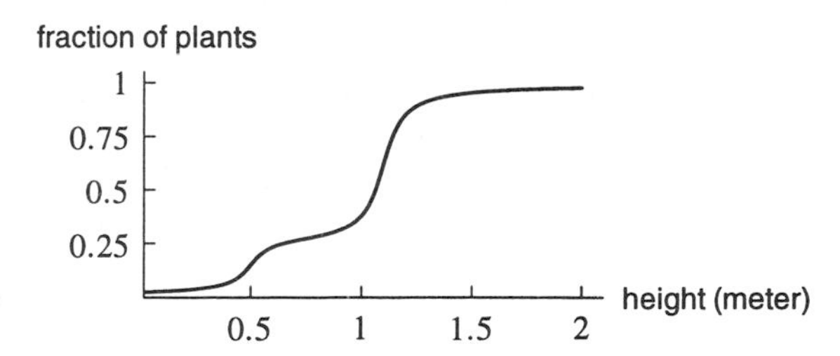

Figure 6.70: Cumulative distribution of meadow grass

(a) You know that there are two species of grass in the meadow, a short grass and a tall grass. Explain how the graph of the density function reflects this fact.
(b) Explain how the graph of the cumulative distribution functions reflects the fact that there are two species of grass in the meadow.
(c) About what percentage of the grasses in the meadow belong to the short grass species?

18. A pizza company guarantees that the pizza will arrive in 5 minutes, or the pizza is free. Suppose the density function for the time until the pizza arrives is given by

$$p(t) = \frac{t^3}{108} - \frac{t^2}{9} + \frac{t}{3},$$

where t is measured in minutes and $0 \leq t \leq 6$.

(a) Check that the area under this density function is 1.
(b) What percent of pizzas arrive late, and must therefore be given away free?
(c) Find the median length of time until a pizza arrives.
(d) Find the mean length of time until a pizza arrives.

19. An economist studying the rate of production, $R(t)$, of oil in a new oil well has proposed the following model:

$$R(t) = 4000 + 1000e^{-t}\sin(2\pi t),$$

where t is the time in years, and $R(t)$ is in barrels of oil per year.

(a) Find the total amount of oil produced in the first 5 years of operation.
(b) Find the average amount of oil produced per year during the first 5 years.

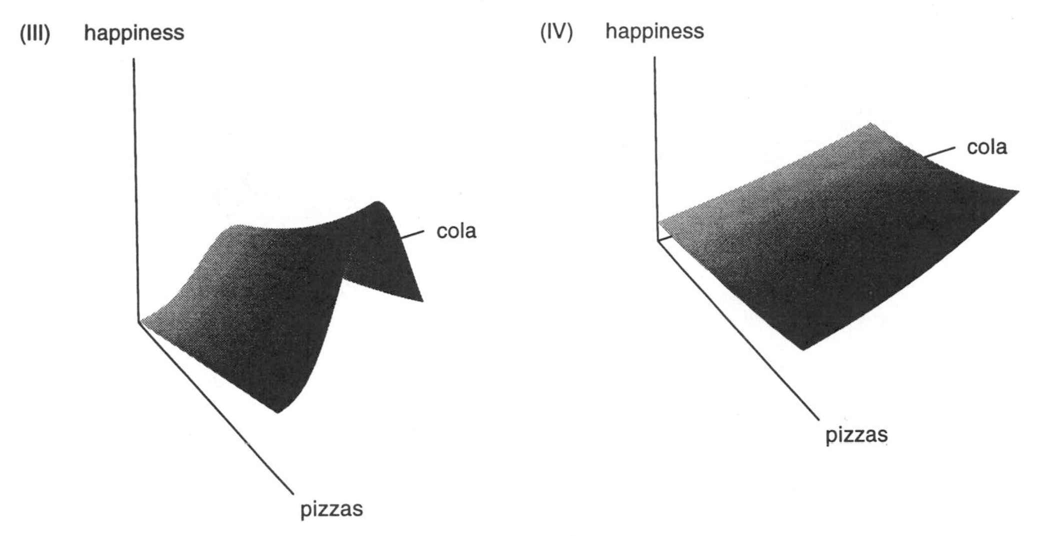

Figure 8.37

7. For each of the graphs I-IV in Problem 6 draw:
   (a) two cross-sections with pizza fixed
   (b) two cross-sections with cola fixed.

8. Figure 8.38 contains graphs of $z = f(x, b)$ for $b = -2, -1, 0, 1, 2$. Which of the graphs of $z = f(x, y)$ in Figures 8.39 and 8.40 best fits this information?

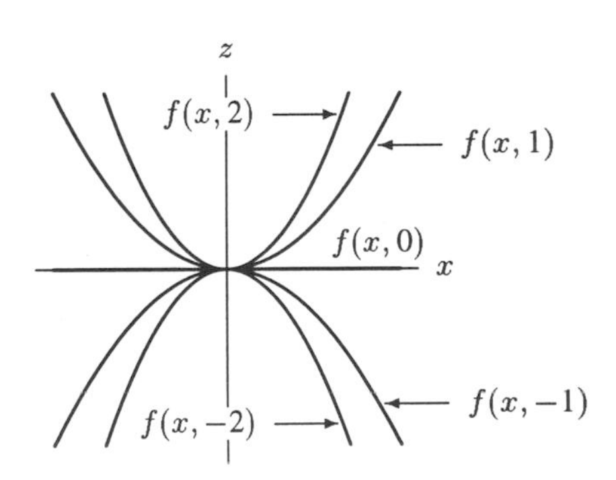

Figure 8.38

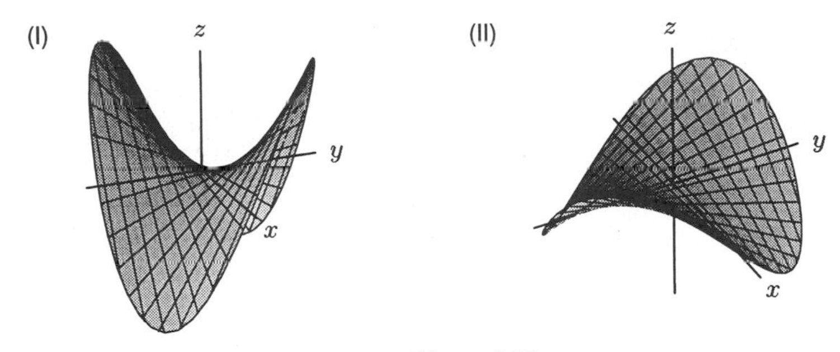

Figure 8.39

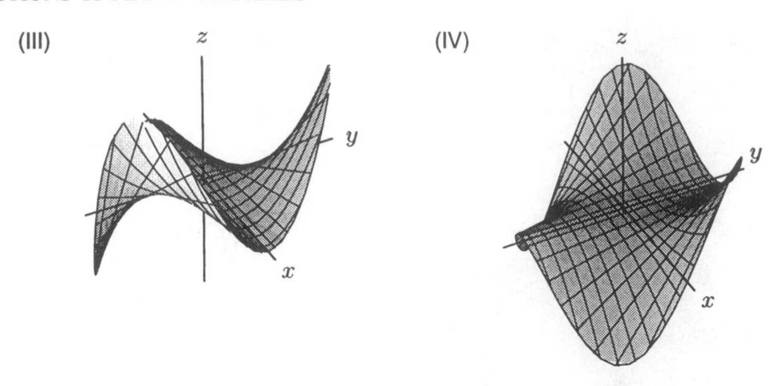

Figure 8.40

9. Consider the function f given by $f(x, y) = y^3 + xy$. Draw graphs of cross-sections with:
   (a) x fixed at $x = -1$, $x = 0$, and $x = 1$. (b) y fixed at $y = -1$, $y = 0$, and $y = 1$.
10. Imagine a single wave traveling along a canal. Suppose x is the distance from the middle of the canal, t is the time, and z is the height of the water above the equilibrium level. The graph of z as a function of x and t is shown in Figure 8.41.
    (a) Draw the profile of the wave for $t = -1, 0, 1, 2$. (Show the x-axis to the right and the z-axis vertically.)
    (b) Is the wave traveling in the direction of increasing or decreasing x?
    (c) Sketch a surface representing a wave traveling in the opposite direction.

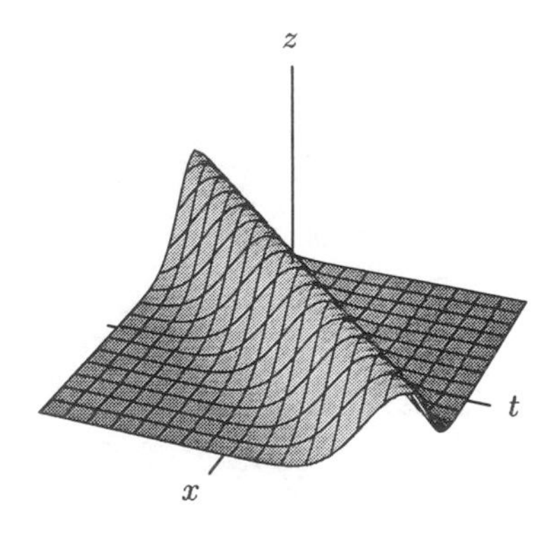

Figure 8.41

11. Use a computer to draw the graph of the vibrating string function:

$$g(x, t) = \cos t \sin 2x, \quad 0 \le x \le \pi, \quad 0 \le t \le 2\pi.$$

    Explain the shape of the graph using cross-sections with t fixed and cross-sections with x fixed.
12. Describe the cross-sections with t fixed and the cross-sections with x fixed of the vibrating guitar string function

$$f(x, t) = \cos t \sin x, \quad 0 \le x \le \pi,$$

    on page 444. Explain the relation of these cross-sections to the graph of f.

one variable, D. We can differentiate to find the maximum of this function, or we can use a graphing calculator or computer to graph the function and find the maximum. How many doctors and how many nurses should the clinic hire to maximize patient visits?

REVIEW PROBLEMS FOR CHAPTER EIGHT

1. A point has coordinates (a, b, c) where a, b, c are all positive. Interpret a, b, and c as distances from the coordinate planes. Illustrate your answer with a picture.
2. Sketch the graph of the cylinder volume function $f(r, h) = \pi r^2 h$ first by keeping h fixed, then by keeping r fixed.
3. Describe the set of points whose x coordinate is 2 and whose y coordinate is 1.
4. Find the equation of the plane through the points $(0, 0, 2), (0, 3, 0), (5, 0, 0)$.

Decide if the statements in Problems 5–9 must be true, might be true, or could not be true. The function $z = f(x, y)$ is defined everywhere.

5. The level curves corresponding to $z = 1$ and $z = -1$ cross at the origin.
6. The level curve $z = 1$ consists of the circle $x^2 + y^2 = 2$ and the circle $x^2 + y^2 = 3$, but no other points.
7. The level curve $z = 1$ consists of two lines which intersect at the origin.
8. If $z = e^{-(x^2+y^2)}$, there is a level curve for every value of z.
9. If $z = e^{-(x^2+y^2)}$, there is a level curve through every point (x, y).

For Problems 10–12, use a computer to sketch the graph of a function with the given shapes. Include the axes and the equation used to generate it in your sketch.

10. A cone of circular cross-section opening downward and with its vertex at the origin.
11. A bowl which opens upward and has its vertex at 5 on the z-axis.
12. A plane which has its x, y, and z intercepts all positive.

For each of the functions in Problems 13–15, make a contour plot in the region $-2 < x < 2$ and $-2 < y < 2$. Describe the shape of the contour lines.

13. $z = 3x - 5y + 1$
14. $z = 2x^2 + y^2$
15. $z = e^{-2x^2-y^2}$
16. Draw the contour diagrams for the functions $f(x, y) = (x - y)$ and $g(x, y) = (x - y)^2$ for $-3 \le x \le 3$, $-3 \le y \le 3$. Where does each function attain its maximum value on this region?
17. Suppose you are in a room 30 feet long with a heater at one end. In the morning the room is $65°F$. You turn on the heater, which quickly warms up to $85°F$. Let $H(x, t)$ be the temperature x feet from the heater, t minutes after the heater is turned on. Figure 8.88 shows the contour diagram for H. How warm is it 10 feet from the heater 5 minutes after it was turned on? 10 minutes after it was turned on?

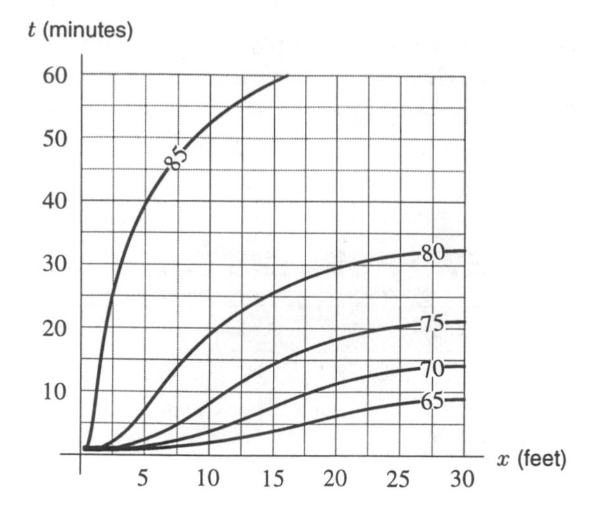

Figure 8.88: Temperature in a heated room.

18. Using the contour diagram in Figure 8.88, sketch the graphs of the one-variable functions $H(x, 5)$ and $H(x, 20)$. Interpret the two graphs in practical terms, and explain the difference between them.

Find equations for the linear functions with the contour diagrams in Problems 19–20.

19.

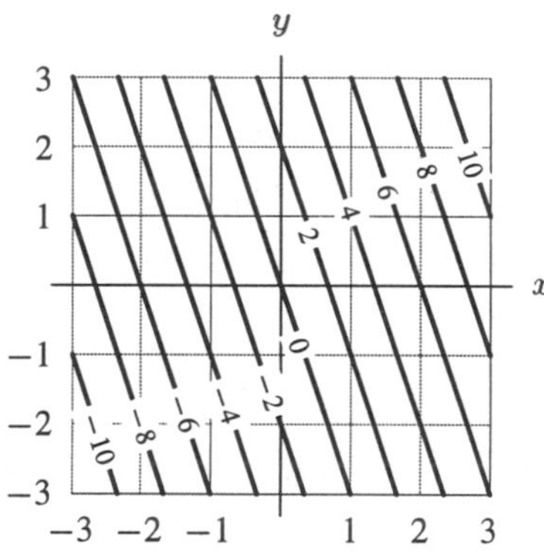

Figure 8.89: A contour map of the linear function $g(x, y)$

20.

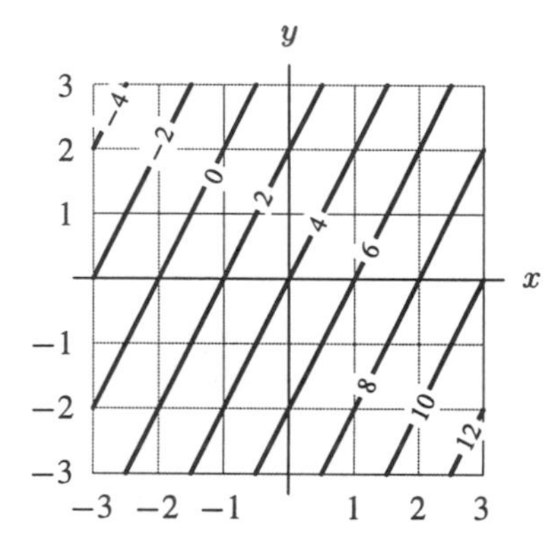

Figure 8.90: A contour map of the linear function $h(x, y)$

Problems 21 and 22 refer to the fallout, V, (in pounds per cubic meter) from a volcanic explosion. The fallout depends on the distance from the volcano, d, and the time since the explosion, t, and is given by

$$V = f(d, t) = \sqrt{t}e^{-d}.$$

21. On the same axes, graph cross-sections of f with $t = 1$, and $t = 2$. Describe the shape of the graphs: As distance from the volcano changes, how does the fallout change? Look at the relationship between the graphs: How is the fallout changing over time? Explain why your answers make sense in terms of volcanoes.

22. On the same axes, graph cross-sections of f with $d = 0$, $d = 1$, and $d = 2$. Describe the shape of the graphs: As time since explosion changes, how does the fallout change? Look at the relationship between the graphs: How is the fallout changing over time? Explain why your answers make sense in terms of volcanoes.

23. The Cobb-Douglas production function for a certain product is given by

$$P = 5L^{0.8}K^{0.2}$$

where P is the quantity produced, L is the size of the labor force, and K is the amount of total equipment. Assume that each unit of labor costs \$300, each unit of capital costs \$100, and the total budget is \$15,000. Make a table of different possible values for L and K and find the production level, P, for each. What seems to be the best way to spend the budget?

24. You are an anthropologist observing a native ritual. Sixteen people arrange themselves along a bench, with all but the three on the far left side seated. The first person is standing with her hands at her side, the second is standing with his hands raised and the third is standing with her hands at her side. At some unseen signal, the first one sits down, and everyone else copies what his neighbor to the left was doing one second earlier. Every second that passes, this behavior is repeated until all are once again seated.

(a) Draw graphs at several different times showing how the height depends upon the distance along the bench.
(b) Graph the location of the raised hands as a function of time.
(c) What US ritual is most closely related to what you have observed?

25. Figure 8.91 shows the contour diagram for the vibrating string function from page 444:

$$f(x,t) = \cos t \sin x, \quad 0 \le x \le \pi.$$

Using the diagram, describe in words the cross-sections of f with t fixed and the cross-sections of f with x fixed. Explain what you see in terms of the behavior of the string.

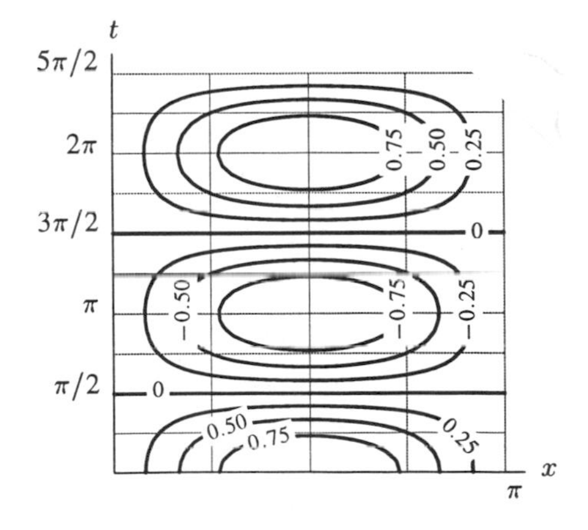

Figure 8.91

26. Table 8.23 shows the predictions of one simple model on how average yearly ultraviolet (UV) exposure might vary with the year and the latitude.

TABLE 8.23 *Ultraviolet exposure.*

Latitude \ Year	1970	1975	1980	1985	1990	1995	2000	2005	2010
90	0.00	0.00	0.00	0.00	0.00	0.00	0.00	0.00	0.00
80	0.00	0.00	0.00	0.00	0.00	3.03	5.79	6.37	6.52
70	0.00	0.00	0.00	0.00	0.00	0.02	3.14	6.80	8.21
60	0.01	0.01	0.01	0.01	0.01	0.01	0.12	1.78	3.41
50	0.08	0.08	0.08	0.08	0.08	0.08	0.08	0.08	0.33
40	0.25	0.25	0.25	0.25	0.25	0.25	0.25	0.25	0.25
30	0.49	0.49	0.49	0.49	0.49	0.49	0.49	0.49	0.49
20	0.74	0.74	0.74	0.74	0.74	0.74	0.74	0.74	0.74
10	0.93	0.93	0.93	0.93	0.93	0.93	0.93	0.93	0.93
0	1.00	1.00	1.00	1.00	1.00	1.00	1.00	1.00	1.00
−10	0.93	0.93	0.93	0.93	0.93	0.93	0.93	0.93	0.93
−20	0.74	0.74	0.74	0.74	0.74	0.74	0.74	0.74	0.74
−30	0.49	0.49	0.49	0.49	0.49	0.49	0.49	0.49	0.49
−40	0.25	0.25	0.25	0.25	0.25	0.25	0.25	0.25	0.25
−50	0.08	0.08	0.08	0.08	0.08	0.08	2.41	5.68	7.27
−60	0.01	0.01	0.01	0.01	0.01	1.41	7.64	10.82	11.97
−70	0.00	0.00	0.00	0.00	0.08	6.36	10.36	11.46	11.80
−80	0.00	0.00	0.00	0.00	0.00	6.32	6.71	6.80	6.82
−90	0.00	0.00	0.00	0.00	0.00	0.00	0.00	0.00	0.00

(a) Graph the cross-sections of UV exposure for the years 1970, 1990, and 2000.
(b) Produce a table showing what latitude has the most severe exposure to UV, as a function of the year.
(c) What do you notice in your answer to part (b)? What is a possible explanation for this phenomenon?

27. Give cross-sections of the linear function $f(x, y, z) = 2x - y + 3z + 4$ with z fixed, $z = 1$, $z = 2$, $z = 3$, first in terms of tables, then in terms of contour diagrams.

16. In analyzing a factory and deciding whether or not to hire more workers, it is useful to know under what circumstances productivity increases. Suppose $P = f(x_1, x_2, x_3)$ is the total quantity produced as a function of x_1, the number of workers, and any other variables x_2, x_3. We define the average productivity of a worker as P/x_1. Show that the average productivity increases as x_1 increases when marginal production, $\partial P/\partial x_1$, is greater than the average productivity, P/x_1.

17. Consider $f(x, y) = x + y + \dfrac{1}{x} + \dfrac{4}{y}$.

(a) Find and classify the local maxima, minima, and saddle points.
(b) What are the global maxima and minima? Explain.

18. Suppose $f_x = f_y = 0$ at $(1, 3)$ and $f_{xx} < 0$, $f_{yy} < 0$, $f_{xy} = 0$. Draw a possible contour diagram.

19. The quantity of a product demanded by consumers is affected by its price; the *demand function* gives the dependence of the quantity demanded based on price. In some cases the quantity of one product demanded depends on the price of other products. For example, the demand for tea may be affected by the price of coffee; the demand for cars may be affected by the price of gas. Suppose the quantities demanded, q_1 and q_2, of two products depend on their prices p_1 and p_2 as follows

$$q_1 = 150 - 2p_1 - p_2 \qquad q_2 = 200 - p_1 - 3p_2$$

(a) What does the fact that q_1 is a function of p_1 and p_2 (instead of p_1 alone) tell you?
(b) What does the fact that the coefficients of p_1 and p_2 are negative tell you? Give an example of two products that might be related this way.
(c) Suppose one manufacturer sells both of these products. How should the manufacturer set prices to earn the maximum possible revenue? What is that maximum possible revenue?

20. Find the least squares line for the data points $(0, 4)$, $(1, 3)$, $(2, 1)$.

21. Find the minimum and maximum of the function $z = 4x^2 - xy + 4y^2$ over the closed disc $x^2 + y^2 \leq 2$.

22. An international organization must decide how to spend the \$2000 they have been allotted for famine relief in a remote area. They expect to divide the money between buying rice at \$5/sack and beans at \$10/sack. The number, P, of people who would be fed if they buy x sacks of rice and y sacks of beans is given by

$$P = x + 2y + \frac{x^2y^2}{2 \cdot 10^8}.$$

What is the maximum number of people that can be fed, and how should the organization allocate its money?

23. The Cobb-Douglas equation models the total quantity, q, of a commodity produced as a function of the number of workers, W, and the amount of capital invested, K, by the production function

$$q = cW^{1-a}K^a$$

where a and c are positive constants. Assume labor costs are $\$p_1$ per worker, capital costs are $\$p_2$ per unit, and there is a fixed budget of $\$b$. Show that when W and K are at their optimal levels, the ratio of marginal productivity of labor to marginal productivity of capital equals the ratio of the cost of one unit of labor to one unit of capital.

24. What are the maximum and minimum values of $f(x, y) = -3x^2 - 2y^2 + 20xy$ on the line $x + y = 100$?

APPENDICES

A ROOTS AND ACCURACY

It is often necessary to find the zeros of a polynomial or the points of intersection of two curves. So far, you have probably used algebraic methods, such as the quadratic formula, to solve such problems. Unfortunately, however, mathematicians' search for formulas for the solutions to equations, such as the quadratic formula, has not been all that successful. The formulas for the solutions to third- and fourth-degree equations are so complicated that you'd never want to use them. Early in the nineteenth century, it was proved that there is no algebraic formula for the solutions to equations of degree 5 and higher. Most nonpolynomial equations cannot be solved using a formula either.

However, we can still find roots of equations, provided we use approximation methods, not formulas. In this section we will discuss three ways to find roots: algebraic, graphical, and numerical. Of these, only the algebraic method gives exact solutions.

First, let's get some terminology straight. Given the equation $x^2 = 4$, we call $x = -2$ and $x = 2$ the *roots*, or *solutions of the equation*. If we are given the function $f(x) = x^2 - 4$, then -2 and 2 are called the *zeros of the function*; that is, the zeros of the function f are the roots of the equation $f(x) = 0$.

The Algebraic Viewpoint: Roots by Factoring

If the product of two numbers is zero, then one or the other or both must be zero, that is, if $AB = 0$, then $A = 0$ or $B = 0$. This observation lies behind finding roots by factoring. You may have spent a lot of time factoring polynomials. Here you will also factor expressions involving trigonometric and exponential functions.

Example 1 Find the roots of $x^2 - 7x = 8$.

Solution Rewrite the equation as $x^2 - 7x - 8 = 0$. Then factor the left side: $(x+1)(x-8) = 0$. By our observation about products, either $x + 1 = 0$ or $x - 8 = 0$, so the roots are $x = -1$ and $x = 8$.

Example 2 Find the roots of $\dfrac{1}{x} - \dfrac{x}{(x+2)} = 0$.

Solution Rewrite the left side with a common denominator:

$$\frac{x + 2 - x^2}{x(x+2)} = 0.$$

Whenever a fraction is zero, the numerator must be zero. Therefore we must have

$$x + 2 - x^2 = (-1)(x^2 - x - 2) = (-1)(x-2)(x+1) = 0.$$

We conclude that $x - 2 = 0$ or $x + 1 = 0$, so 2 and -1 are the roots. They can be checked by substitution.

Example 3 Find the roots of $e^{-x}\sin x - e^{-x}\cos x = 0$.

Solution Factor the left side: $e^{-x}(\sin x - \cos x) = 0$. The factor e^{-x} is never zero; it is impossible to raise e to a power and get zero. Therefore, the only possibility is that $\sin x - \cos x = 0$. This equation is equivalent to $\sin x = \cos x$. If we divide both sides by $\cos x$, we get

$$\frac{\sin x}{\cos x} = \frac{\cos x}{\cos x} \quad \text{so} \quad \tan x = 1.$$

The roots of this equation are

$$\ldots, \frac{-7\pi}{4}, \frac{-3\pi}{4}, \frac{\pi}{4}, \frac{5\pi}{4}, \frac{9\pi}{4}, \frac{13\pi}{4}, \ldots.$$

Warning: Using factoring to solve an equation only works when one side of the equation is 0. It is not true that if, say, $AB = 7$ then $A = 7$ or $B = 7$. For example, you *cannot* solve $x^2 - 4x = 2$ by factoring $x(x-4) = 2$ and then assuming that either x or $x - 4$ equals 2.

The problem with factoring is that factors are not easy to find. For example, the left side of the quadratic equation $x^2 - 4x - 2 = 0$ does not factor, at least not into "nice" factors with integer coefficients. For the general quadratic equation:

$$ax^2 + bx + c = 0$$

there is the quadratic formula for the roots:

$$x = \frac{-b \pm \sqrt{b^2 - 4ac}}{2a}.$$

Thus the roots of $x^2 - 4x - 2 = 0$ are $(4 \pm \sqrt{24})/2$, or $2 + \sqrt{6}$ and $2 - \sqrt{6}$.

Notice that in each of these examples, we have found the roots exactly.

The Graphical Viewpoint: Roots by Zooming

To find the roots of an equation $f(x) = 0$, it helps to draw the graph of f. The roots of the equation, that is the zeros of f, are *the values of x where the graph of f crosses the x-axis*. Even a very rough sketch of the graph can be useful in determining how many zeros there are and their approximate values. If you have a computer or graphing calculator, then finding solutions by graphing is the easiest method, especially if you use the zoom feature. However, a graph can never tell you the exact value of a root, only an approximate one.

Example 4 Find the roots of $x^3 - 4x - 2 = 0$.

Solution Attempting to factor the left side with integer coefficients will convince you it cannot be done, so we cannot easily find the roots by algebra. We know the graph of $f(x) = x^3 - 4x - 2$ will have the usual cubic shape; see Figure .1. There are clearly three roots: one between $x = -2$ and $x = -1$, another between $x = -1$ and $x = 0$, and a third between $x = 2$ and $x = 3$. Zooming in on the largest root with a graphing calculator or computer shows that it lies in the following interval:

$$2.213 < x < 2.215.$$

Thus, the root is $x = 2.21$, accurate to two decimal places. Zooming in on the other two roots shows them to be $x = -1.68$ and $x = -0.54$, accurate to two decimal places.

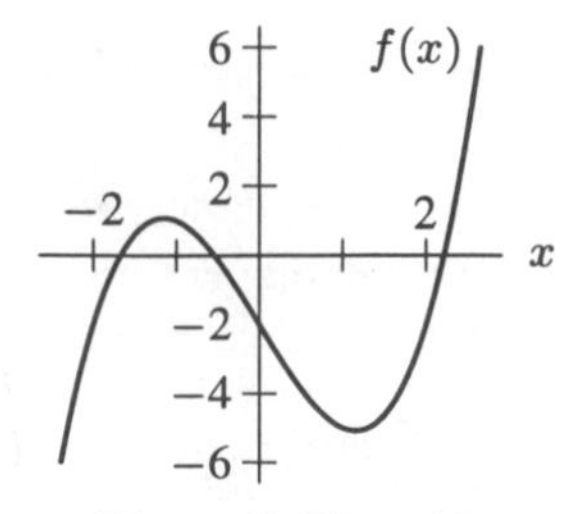

Figure .1: The cubic $f(x) = x^3 - 4x - 2$

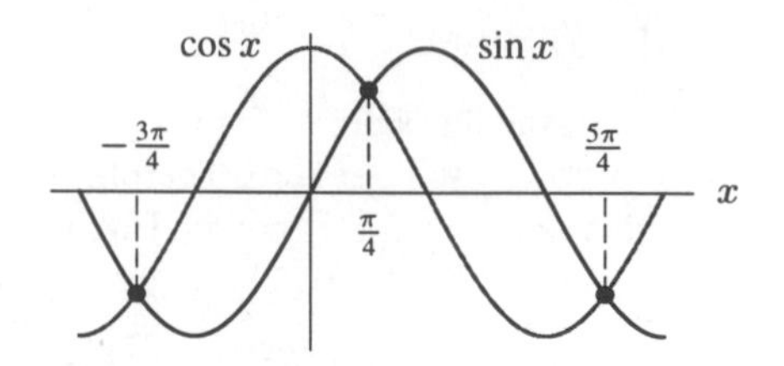

Figure .2: Finding roots of $\sin x - \cos x = 0$

Useful trick: Suppose you want to solve the equation $\sin x - \cos x = 0$ graphically. Instead of graphing $f(x) = \sin x - \cos x$ and looking for zeros, you may find it easier to rewrite the equation as $\sin x = \cos x$ and graph $g(x) = \sin x$ and $h(x) = \cos x$. (After all, you already know what these two graphs look like. See Figure .2.) The roots of the original equation are then precisely the x coordinates of the points of intersection of the graphs of $g(x)$ and $h(x)$.

Example 5 Find the roots of $2\sin x - x = 0$.

Solution Rewrite the equation as $2\sin x = x$, and graph both sides. Since $g(x) = 2\sin x$ is always between -2 and 2, there are no roots of $2\sin x = x$ for $x > 2$ or for $x < -2$. Thus, we need only consider the graphs between -2 and 2 (or between $-\pi$ and π, which makes graphing the sine function easier). Figure .3 shows the graphs. There are three points of intersection: one appears to be at $x = 0$, one between $x = \pi/2$ and $x = \pi$, and one between $x = -\pi/2$ and $x = -\pi$. You can tell that $x = 0$ is the exact value of one root because it satisfies the original equation exactly. Zooming in shows that there is a second root $x \approx 1.9$, and the third root is $x \approx -1.9$ by symmetry.

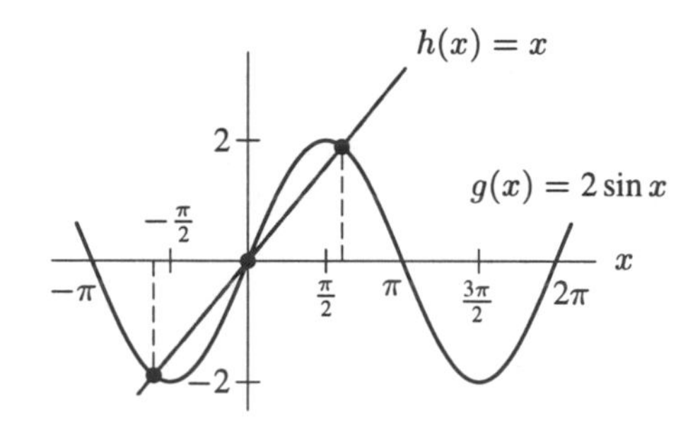

Figure .3: Finding roots of $2\sin x - x = 0$

The Numerical Viewpoint: Roots by Bisection

We now look at a numerical method of approximating the solutions to an equation. This method depends on the idea that if the value of a function $f(x)$ changes sign in an interval, and if we believe there is no break in the graph of the function there, then there is a root of the equation $f(x) = 0$ in that interval.

Let's go back to the problem of finding the root of $f(x) = x^3 - 4x - 2 = 0$ between 2 and 3. To locate the root, we close in on it by evaluating the function at the midpoint of the interval, $x = 2.5$. Since $f(2) = -2$, $f(2.5) = 3.625$, and $f(3) = 13$, the function changes sign between $x = 2$ and $x = 2.5$, so the root is between these points. Now we look at $x = 2.25$.

Since $f(2.25) = 0.39$, the function is negative at $x = 2$ and positive at $x = 2.25$, so there is a root between 2 and 2.25. Now we look at 2.125. We find $f(2.125) = -0.90$, so there is a root between 2.125 and 2.25, ... and so on. (You may want to round the decimals as you work.) See Figure .4. The intervals containing the root are listed in Table .1 and show that the root is $x = 2.21$ to two decimal places.

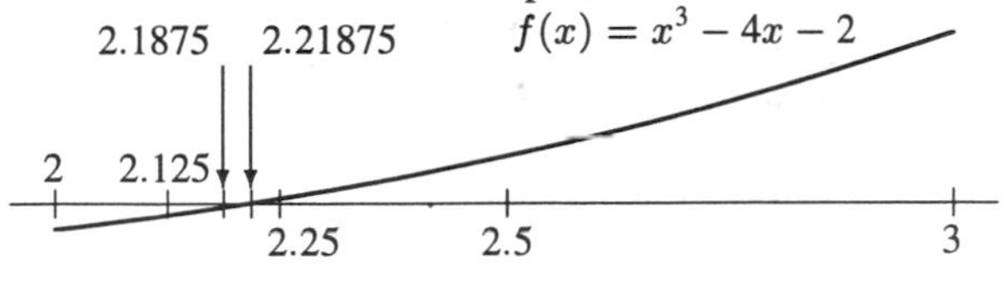

Figure .4: Locating a root of $x^3 - 4x - 2 = 0$

TABLE .1 *Intervals containing root to* $x^3 - 4x - 2 = 0$ *(Note that* $[2, 3]$ *means* $2 \le x \le 3$*)*

$[2, 3]$	
$[2, 2.5]$	
$[2, 2.25]$	
$[2.125, 2.25]$	
$[2.1875, 2.25]$	So $x = 2.2$ rounded to one decimal place
$[2.1875, 2.21875]$	
$[2.203125, 2.21875]$	
$[2.2109375, 2.21875]$	
$[2.2109375, 2.2148438]$	So $x = 2.21$ rounded to two decimal places

This method of finding roots is called the **Bisection Method**:

- To solve an equation $f(x) = 0$ using the bisection method, we need two starting values for x, say, $x = a$ and $x = b$, such that $f(a)$ and $f(b)$ have opposite signs and f is continuous on $[a, b]$.
- Evaluate f at the midpoint of the interval $[a, b]$, and decide in which half-interval the root lies.
- Repeat, using the new half-interval instead of $[a, b]$.

There are some problems with the bisection method:

- The function may not change signs near the root. For example, $f(x) = x^2 - 2x + 1 = 0$ has a root at $x = 1$, but $f(x)$ is never negative because $f(x) = (x - 1)^2$, and a square cannot be negative. (See Figure .5.)
- The function f must be continuous between the starting values $x = a$ and $x = b$.
- If there is more than one root between the starting values $x = a$ and $x = b$, the method will find only one of the roots. For example, if we had tried to solve $x^3 - 4x - 2 = 0$ starting at $x = -12$ and $x = 10$, the bisection method would zero in on the root between $x = -2$ and $x = -1$, not the root between $x = 2$ and $x = 3$ that we found earlier. (Try it! Then see what happens if you use $x = -10$ instead of $x = -12$.)

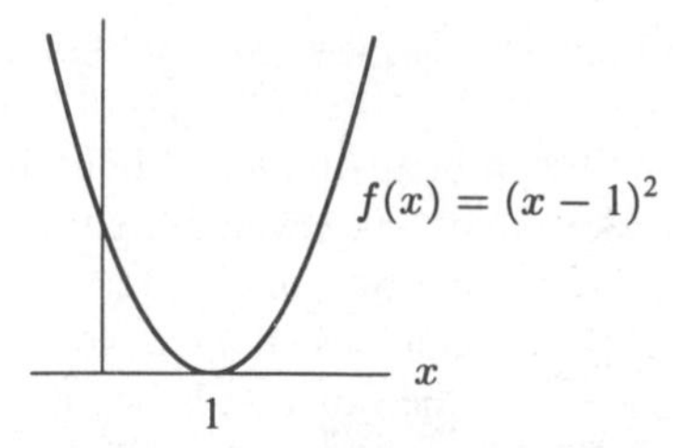

Figure .5: f doesn't change sign at the root

- The bisection method is slow and not very efficient. Applying bisection three times in a row only traps the root in an interval $(\frac{1}{2})^3 = \frac{1}{8}$ as large as the starting interval. Thus, if we initially know that a root is between, say, 2 and 3, then we would need to apply the bisection method at least four times to know the first digit after the decimal point.

There are much more powerful methods available for finding roots, such as Newton's method, which is more complicated, but which avoids some of these difficulties.

Example 6 Find all the roots of $xe^x = 5$ to at least one decimal place.

Solution If we rewrite the equation as $e^x = 5/x$ and graph both sides, as in Figure .6, it is clear that there is exactly one root, and it is somewhere between 1 and 2. Table .2 shows the intervals obtained by the bisection method. After five iterations, we have the root trapped between 1.3125 and 1.34375, so we can say the root is $x = 1.3$ to one decimal place.

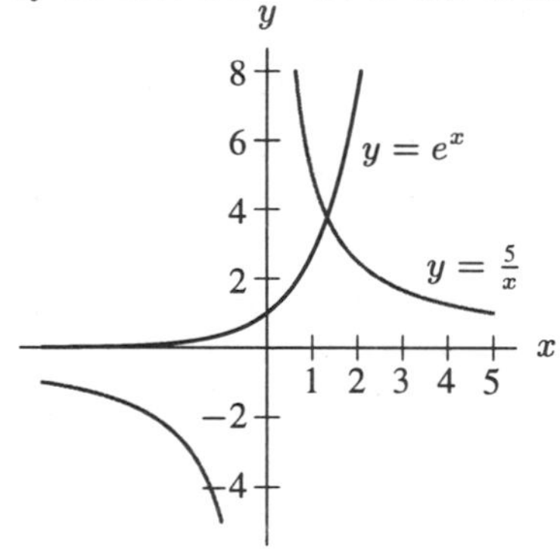

Figure .6: Intersection of $y = e^x$ and $y = 5/x$

TABLE .2 *Bisection method for* $f(x) = xe^x - 5 = 0$ *(Note that* $[1, 2]$ *means the interval* $1 \leq x \leq 2$*)*

Interval Containing Root
[1, 2]
[1, 1.5]
[1.25, 1.5]
[1.25, 1.375]
[1.3125, 1.375]
[1.3125, 1.34375]

Iteration

Both zooming in and bisection as discussed here are examples of *iterative* methods, in which a sequence of steps is repeated over and over again, using the results of one step as the input for the next. We can use the method to locate a root to any degree of accuracy. In bisection, each iteration traps the root in an interval that is half the length of the previous one. Each time you zoom in on a calculator, you trap the root in a smaller interval; how much smaller depends on the settings on the calculator.

Accuracy and Error

In the previous discussion, we used the phrase "accurate to 2 decimal places." For an iterative process where we get closer and closer estimates for some quantity, we take a common-sense approach to accuracy: we watch the numbers carefully, and when a digit stays the same for a few iterations, we assume it has stabilized and is correct, especially if the digits to the right of that digit also stay the same. For example, suppose 2.21429 and 2.21431 are two successive estimates for a zero of $f(x) = x^3 - 4x - 2$. Since these two estimates agree to the third digit after the decimal point, we probably have at least 3 decimal places correct.

There is a problem with this, however. Suppose we are finding a root whose true value is 1, and the estimates are converging to the value from below — say, 0.985, 0.991, 0.997 and so on. In this case, not even the first decimal place is "correct," even though the difference between the estimates and the actual answer is very small — much less than 0.1. To avoid this difficulty, we say that an estimate a for some quantity r is *accurate to p decimal places* if the error, which is the absolute value of the difference between a and r, or $|r - a|$, is as follows:

Accuracy to p decimal places	means	Error less than
$p = 1$		0.05
2		0.005
3		0.0005
$\vdots$		$\vdots$
n		$0.\underbrace{000\ldots0}_{n}5$

This is the same as saying that r must lie in an interval of length twice the maximum error, centered on a. For example, if a is accurate to 1 decimal place, r must lie in the following interval:

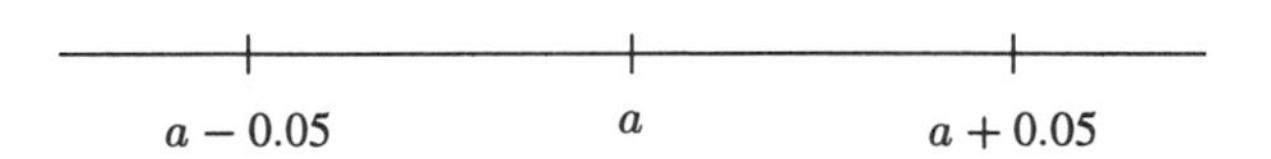

Since both the graphing calculator and the bisection method give us an interval in which the root is trapped, this definition of decimal accuracy is a natural one for these processes.

Example 7 Suppose the numbers $\sqrt{10}$, $22/7$, and 3.14 are given as approximations to $\pi = 3.1415\ldots$. To how many decimal places is each approximation accurate?

Solution Using $\sqrt{10} = 3.1622\ldots$,

$$|\sqrt{10} - \pi| = |3.1622\ldots - 3.1415\ldots| = 0.0206\ldots < 0.05,$$

so $\sqrt{10}$ is accurate to one decimal place. Similarly, using $22/7 = 3.1428\ldots$,

$$\left|\frac{22}{7} - \pi\right| = |3.1428\ldots - 3.1415\ldots| = 0.0013\ldots < 0.005,$$

so 22/7 is accurate to two decimal places. Finally,

$$|3.14 - 3.1415\ldots| = 0.0015\ldots < 0.005,$$

so 3.14 is accurate to two decimal places.

Warning:

- Saying that an approximation is accurate to, say, 2 decimal places does *not* guarantee that its first two decimal places are "correct," that is, that the two digits of the approximation are the same as the corresponding two digits in the true value. For example, an approximate value of 5.997 is accurate to 2 decimal places if the true value is 6.001, but neither of the 9s in the approximation agrees with the 0s in the true value (nor does the digit 5 agree with the digit 6).
- The number of decimal places of accuracy refers to the number of digits that have stabilized in the root, r. It does *not* refer to the number of digits of $f(r)$ that are zero. For example, Table .1 on page 549 shows that $x = 2.2$ is a root of $f(x) = x^3 - 4x - 2 = 0$, accurate to one decimal place. Yet, $f(2.2) = -0.152$, so $f(2.2)$ does not have one zero after the decimal point. Similarly, $x = 2.21$ is the root accurate to two decimal places, but $f(2.21) = -0.046$ does not have two zeros after the decimal point.

Example 8 Is $x = 2.2143$ a zero of $f(x) = x^3 - 4x - 2$ accurate to four decimal places?

Solution We want to know whether r, the exact value of the zero, lies in the interval

$$2.2143 - 0.00005 < r < 2.2143 + 0.00005$$

which is the same as

$$2.21425 < r < 2.21435.$$

Since $f(2.21425) < 0$ and $f(2.21435) > 0$, the zero does lie in this interval, and so $r = 2.2143$ is accurate to four decimal places.

How to Write a Decimal Answer

The graphing calculator and bisection method naturally give an interval for a root or zero. Other numerical techniques, however, do not give a pair of numbers bounding the true value, but rather a single number near the true value. What should you do if you want to give a single number, rather than an interval, for an answer?

When giving a single number as an answer and interpreting it, be careful about giving rounded answers. For example, suppose you have computed a value to be 0.84, and you know the true value is between 0.81 and 0.87. It would be wrong to round 0.84 to 0.8 and say that the answer is 0.8 accurate to one decimal place; the true value could be 0.86, which is not within 0.05 of 0.8. The right thing to say is that the answer is 0.84 accurate to one decimal place. Similarly, to give an answer accurate to, say, 2 decimal places, you may have to show 3 or more decimal places in your answer.

Problems for Section A

1. Use a calculator or computer graph of $f(x) = 13 - 20x - x^2 - 3x^4$ to determine:
   (a) The range of this function;
   (b) The number of zeros of this function.

For Problems 2–12, determine the roots or points of intersection to an accuracy of one decimal place.

2. (a) The root of $x^3 - 3x + 1 = 0$ between 0 and 1
   (b) The root of $x^3 - 3x + 1 = 0$ between 1 and 2
   (c) The third root of $x^3 - 3x + 1 = 0$
3. The root of $x^4 - 5x^3 + 2x - 5 = 0$ between -2 and -1
4. The root of $x^5 + x^2 - 9x - 3 = 0$ between -2 and -1
5. The largest real root of $2x^3 - 4x^2 - 3x + 1 = 0$
6. All real roots of $x^4 - x - 2 = 0$
7. All real roots of $x^5 - 2x^2 + 4 = 0$
8. The first positive root of $x \sin x - \cos x = 0$
9. The first positive point of intersection between $y = 2x$ and $y = \cos x$
10. The first positive point of intersection between $y = 1/2^x$ and $y = \sin x$
11. The point of intersection between $y = e^{-x}$ and $y = \ln x$
12. All roots of $\cos t = t^2$
13. Estimate all real zeros of the following polynomials, accurate to 2 decimal places:
    (a) $f(x) = x^3 - 2x^2 - x + 3$
    (b) $f(x) = x^3 - x^2 - 2x + 2$
14. Find the largest zero of
    $$f(x) = 10xe^{-x} - 1$$
    to two decimal places, using the bisection method. Make sure to demonstrate that your approximation is as good as you claim.
15. (a) Find the smallest positive value of x where the graphs of $f(x) = \sin x$ and $g(x) = 2^{-x}$ intersect.
    (b) Repeat with $f(x) = \sin 2x$ and $g(x) = 2^{-x}$.
16. Use a graphing calculator to sketch $y = 2\cos x$ and $y = x^3 + x^2 + 1$ on the same set of axes. Find the positive zero of $f(x) = 2\cos x - x^3 - x^2 - 1$. A friend claims there is one more real zero. Is your friend correct? Explain.
17. Use the table below to investigate the zeros of the function
    $$f(\theta) = (\sin 3\,\theta)(\cos 4\,\theta) + 0.8$$
    in the interval $0 \le \theta \le 1.8$.

θ	0	0.2	0.4	0.6	0.8	1.0	1.2	1.4	1.6	1.8
$f(\theta)$	0.80	1.19	0.77	0.08	0.13	0.71	0.76	0.12	−0.19	0.33

   (a) Decide how many zeros the function has in the interval $0 \le \theta \le 1.8$.

(b) Locate each zero, or a small interval containing each zero.
(c) Are you sure you have found all the zeros in the interval $0 \leq \theta \leq 1.8$? Graph the function on a calculator or computer to decide.

18. (a) Use the accompanying table to locate approximate solution(s) to

$$(\sin 3x)(\cos 4x) = \frac{x^3}{\pi^3}$$

in the interval $1.07 \leq x \leq 1.15$. Give an interval of length 0.01 in which each solution lies.
(b) Make an estimate for each solution accurate to two decimal places.

x	x^3/π^3	$(\sin 3x)(\cos 4x)$
1.07	0.0395	0.0286
1.08	0.0406	0.0376
1.09	0.0418	0.0442
1.10	0.0429	0.0485
1.11	0.0441	0.0504
1.12	0.0453	0.0499
1.13	0.0465	0.0470
1.14	0.0478	0.0417
1.15	0.0491	0.0340

INDEX

STUDENT ANSWER MANUAL

to accompany

CALCULUS

Deborah Hughes-Hallett
Harvard University

Andrew M. Gleason
Harvard University

et al.

Prepared By: Alex Kasman
Stephen A. Mallozzi
Alice H. Wang
Gang Zhang

John Wiley & Sons, Inc.
New York • Chichester • Brisbane • Toronto • Singapore

CONTENTS

ISBN 0-471-00982-2

Printed in the United States of America

10 9 8 7

CHAPTER ONE

Section 1.1

1. (I) None
   (II) (b)
   (III) (c)
   (IV) (a)

9. p_1 = maximum price any consumer would pay.
   q_1 = the quantity of the good that could be given away if the item were free.

11.(a) $-5 \le p \le 0$ or $2 \le p < 6$
   (b) $r \ge 0$
   (c) $0 \le r < 2$
   $r > 5$

13. $x \ge 2$ and $x \le -2$
   $x = \pm\frac{3}{5}\sqrt{11}$

Section 1.2

1. slope $= -\frac{5}{2}$
   vertical intercept = 4

3. $y - c = m(x - a)$

5. parallel line:
   $y = -4x + 9$
   perpendicular line:
   $y = 0.25x + 4.75$

7.(a) (V)
   (b) (IV)
   (c) (I)
   (d) (VI)
   (e) (II)
   (f) (III)

9.(a) $q = 1000 - 50p$
   (b) $p = 20 - 0.02q$

11. distance = 5

13.(a) slope = 1.8
   (b) °F = 1.8(°C) + 32
   (c) 68°Fahrenheit
   (d) −40°

17.(a) $k = p_1 s + p_2 l$
   (s = liters of soda,
   l = liters of oil)
   (b) intercepts: $(0, \frac{k}{p_1}), (\frac{k}{p_2}, 0)$
   (c) intercepts: $(0, \frac{2k}{p_1}), (\frac{2k}{p_2}, 0)$
   (d) intercepts: $(0, \frac{k}{p_1}), (\frac{k}{2p_2}, 0)$

19.(a) $l = al_0(t - t_0) + l_0$
   Slope $= al_0$
   $y_{\text{int}} = -al_0 t_0 + l_0$
   (b) $l = 0.001t + 99.94$
   (c) If the slope is positive, then as the temperature rises, the metal expands. If the slope is negative, then the metal contracts as the temperature rises.

21.(a) $R = k(20 - H)$
   $(k \ge 0)$
   (b)

Section 1.3

5.(a) $k(t)$
   (b) $h(t)$
   (c) $g(t)$

7. $f(s) = 2(1.1)^s$
   $g(s) = 3(1.05)^s$
   $h(s) = (1.03)^s$

9. $g(t) = 5.50(0.8)^t$

11. $y = 2(3^x)$

13. $y = 4(1 - 2^{-x})$

15. 48.8%

17. 2.3

19. $Q = Q_0(0.70)^{\frac{1}{10}t}$
   After 50 years, about 17% of the original quantity remains. The half life is about 19.5 years. 20% will be left after 45.2 years. 10% will be left after 64.7 years.

21.(a) 2963.51%
   (b) 0.3755%

Section 1.4

1.(a) 4
   (b) $\frac{1}{27}$

3. $y = x^4$ goes to positive infinity in both cases.

9. $f(x) = x^3$ is larger.

11. $x \approx 9.94$

13. $-0.7 < x < 2$ and $x > 4$

15. 44.25 ft and 708 ft

17. $h(t)$ is an exponential function,
   $h(t) = 2.04(1.05)^t$;
   $g(t)$ is of the form bt^3,
   $g(t) = 3t^3$;
   $f(t)$ is of the form at^2,
   $f(t) = 1.1t^2$.

19.(a) Production =
   $0.1145t + 5.35$
   Population =
   $53.2(1.0136)^t$

(b) The region had an increasing surplus.
(c) Eventually the population will become larger than the production can support and there will be shortages.

Section 1.5

1. $f^{-1}(75)$ is the length of the column of mercury in the thermometer when the temperature is 75°F.
3. not invertible
5. not invertible
7. $\{3,-7,19,4,178,2,1\}$
9. not invertible
11. invertible
13. (a) $k = f(p) = \frac{1}{2.2}p$
    (b) $p = 2.2k$
    It gives the weight of an object in pounds as a function of its mass in kilograms
15. One cannot be sure of the concavity of the inverse unless one also knows whether f is increasing or decreasing.

Section 1.6

3. The graph is a straight line, with slope 1, to the right of the origin.
5. $t \approx 35.003$
7. $t = \frac{\log\left(\frac{Q_0}{P_0}\right)}{\log\left(\frac{a}{b}\right)}$
9. $t = \frac{\log P - \log P_0}{\log a}$
11. $t \approx 90.283$
13. $x+7$
15. Q^2
17. $\frac{1}{\sqrt{B}}$
19. $\frac{3}{2}\log\frac{\alpha}{B^2}$
21. $p^{-1}(t) \approx 58.708 \log t$
23. 1990
25. (a) $D = 35.0, 23.4, 17.7, 14.2.$
    (b) If $D = 70/i$, then $70/2 = 35.0$, $70/3 = 23.3$, $70/4 = 17.5$, $70/5 = 14.0$, showing pretty good agreement.
27. 6.58 years

Section 1.7

3. $2+e^{-1}$
5. $1/\sqrt{e}$
7. $A^2/\sqrt{B}$
9. 2
11. $x = -0.347$
13. $x = 0.515$
15. $P = 10(e^{0.917})^t = 10(2.5)^t$. Exponential growth
17. $P = 79(e^{-2.5})^t = 79(0.0821)^t$ Exponential decay
19. $P = P_0 e^{0.693t}$
21. $P = 5.23e^{-1.6904t}$
23. This function is really $y = x$.
25. (a) The quantity $\frac{\ln x}{\log x}$ remains constant.
    (b) You see a horizontal line, $y \approx 2.30$, for $x > 0$.
27. $t = (\ln\frac{P}{P_0})/(k)$
29. $t = (\ln\frac{c}{b})/(\alpha - \frac{\gamma}{n})$, (for $\alpha \neq \frac{\gamma}{n}$)
31. (a) increasing
    (b) Since f is increasing, its graph intersects any horizontal line at most once. $f^{-1} = \ln(\frac{x}{1-x})$
    (c) $0 < x < 1$
    (d) Graph of f^{-1} is reflection of graph of f across the line $y = x$.
    Note that graph of f is asymptotic to the lines $y = 1$ and $y = 0$, and graph of f^{-1} is asymptotic to the lines $x = 1$ and $x = 0$.
33. (a) 47.5% of sea level pressure
    (b) 24% of sea level pressure
35. (a) 81%
    (b) 32.9 hours
    (c)
    (d)
37. 8.46%
39. It is a fake.

Section 1.8

3. (a)(i) 5.126978...%
    (ii) 5.127096...%
    (iii) 5.127108...%
    (b) 5.127%
    (c) $e^{0.05} = 1.05127109\ldots$
5. $\approx 6.18365\%$
7. (a) 11.55 years
    (b) $P = P_0(2)^{t/11.55}$
9. (a) 2,613,035 zaïre
    (b)(i) 2,714,567 zaïre
    (ii) 2,718,127 zaïre
    (iii) 2,718,280 zaïre
    (c) It levels off at just over 2,718,000 zaïre.
11. (a) The landlord makes \$11.83 in interest.
    (b) The landlord loses \$9.19 in interest.

Section 1.9

1.(a) $y = 2x^2 + 1$
(b) $y = 2(x^2 + 1)$
(c) No

5.(a) $f(g(t)) = (\frac{1}{t+1} + 7)^2$
(b) $g(f(t)) = \frac{1}{(t+7)^2+1}$
(c) $f(t^2) = (t^2 + 7)^2$
(d) $g(t - 1) = \frac{1}{t}$

7. $2z + 1$

9. $2zh - h^2$

11. $f(x) = x + 1$
$g(x) = x^3$

13. $f(x) = x^3$
$g(x) = \ln x$

19. $1/x$ is odd
$\ln(x^2)$ is even
e^x is neither

23. $f(g(1)) = 0.4$

25. $f(f(1)) \approx -0.9$

Section 1.10

1. $\sin \frac{\pi}{4}$ is positive

$\cos \frac{\pi}{4}$ is positive

$\tan \frac{\pi}{4}$ is positive

3. $\sin 3\pi = 0$
$\cos 3\pi = -1$ is negative
$\tan 3\pi = 0$

5. $\sin\left(-\frac{\pi}{12}\right)$ is negative
$\cos\left(-\frac{\pi}{12}\right)$ is positive
$\tan\left(-\frac{\pi}{12}\right)$ is negative

7. $\sin 4$ is negative
$\cos 4$ is negative
$\tan 4$ is positive

9. 0.258

11. −0.258

13. −0.258

15. −0.258

17. 0.966

19. At 3:00 am: 5 feet
At 4:00 am: 2.55 feet
At 5:00 pm: 0.76 feet

21. 27.3 days ≈ one month

23. $\frac{1}{33\frac{1}{3}}$ minute
(1.8 seconds)

25.(a) $f(t) = -0.5 + \sin t$
$g(t) = 1.5 + \sin t$
$h(t) = -1.5 + \sin t$
$k(t) = 0.5 + \sin t$
(b) $g(t) = 1 + k(t)$
(c)

27. Amplitude = 3
Period = π

29. Amplitude = 4
Period = π

31. Amplitude = 1
Period = π

33. $f(x) = 2\sin(x/4)$

35. $f(x) = 5\cos(x/3)$

37. $f(x) = -8\cos(x/10)$

39. $y = 3\sin[\pi(x - 1)/9]$

41.(a)
(b) $P = 800 - 100\cos(\pi t/6)$

43. $f(t) = 75 - 15\cos\left(\frac{2\pi}{12}t\right)$

45.(a)
(b) Domain:
$-1 \le x \le 1$
Range:
$-\frac{\pi}{2} \le x \le \frac{\pi}{2}$

47. $f(x) = 2x$
$g(x) = 10 - 2x$
$h(x) = x^2$
$F(x) = \sin x$
$G(x) = \sin 2x$
$H(x) = 2 + \sin x$

49.(a) Yes, they must intersect at $3.64 - \pi \approx 0.5$.
(b) They also intersect at $3.64 + \pi \approx 6.78$.
(c) $3.64 - 2\pi \approx -2.64$.

Section 1.11

1. (I) Degree ≥ 3, leading coefficient negative.
(II) Degree ≥ 4, leading coefficient positive.
(III) Degree ≥ 4, leading coefficient negative.
(IV) Degree ≥ 5, leading coefficient negative.
(V) Degree ≥ 5, leading coefficient positive.

7. $x = -1$: vertical asymptote
$y = -2$: horizontal asymptote
$y \to +\infty$ as $x \to -1^+$
$y \to -\infty$ as $x \to -1^-$

9. $x = \pm 2$: vertical asymptotes
$y = 1$: horizontal asymptote
$y \to +\infty$ as $x \to 2^+$
$y \to -\infty$ as $x \to 2^-$
$y \to -\infty$ as $x \to -2^+$
$y \to +\infty$ as $x \to -2^-$

11. $f(x) = x^n$ is even for n an even integer, and odd for n an odd integer.

13.(a) $1 = a + b + c$
(b) $b = -2a$ and $c = 1 + a$
(c) $c = 6$
(d) $y = 5x^2 - 10x + 6$

15.(a) The initial height is 0.
(b) $t = 2v_0/g$
(c) $t = v_0/g$

(d) The maximum height is $\frac{v_0^2}{2g}$

17.(a) $f(x) = kx(x+3)(x-4)$ $(k < 0)$

(b) Increasing: $-1.5 < x < 2.5$
Decreasing: $x < -1.5$ and $x > 2.5$

19.(a) $f(x) = k(x+2)(x-2)^2(x-5)$ $(k < 0)$

(b) Increasing: $x < -1$ and $2 < x < 4$
Decreasing: $-1 < x < 2$ and $4 < x$

21. $y = -\frac{1}{2}(x+2)^2(x-2)$

23. $g(x) = 2x^2$
$h(x) = x^2 + k$
$(k > 0)$

25.(a) $R(P) = kP(L-P)$ $(k > 0)$

(b)

Chapter 1 Review

5.(a) Four zeros: $x = -4.6, 1.2, 2.7, 4.1$

(b) $f(2) \approx -1$
$f(4) \approx 0.4$

(c) near $x = -1$: decreasing
near $x = 3$: increasing

(d) near $x = 2$: concave up
near $x = -4$: concave down

(e) Increasing: $x < -1.5$ and $2 < x < 3.5$

7. $y = -\frac{950}{7}x + 950$

11.(a) $g(2+h) = h^2 + 6h + 11$

(b) $g(2) = 11$

(c) $g(2+h) - g(2) = h^2 + 6h$

15. $P = 2.91(1.733)^t$

17.(a) $Q = 25e^{0.5423t}$

(b) 1.3 months

(c) 6.8 months

19. 4.88%

21. plutonium-240: 6,301 years
plutonium-241: 385,000 years

23.(a) 44%

(b) about two years

25. Mercury: 87.8 days
Earth: 1 year
Pluto: 253 years

27.(a) $(-2, 4)$

(b) $(-b, b^2)$

29. $y = 2e^{-0.34657x}$

31. $y = -k(x^2 + 5x)$
$(k > 0)$

33. $x = k(y^2 - 4y)$
$(k > 0)$

35. $y = -2\left(4^{-t}\right)$

37.(a) Three roots in the intervals: $[1,2], [2,3], [3,4]$

(b) There are three roots in the interval $[3,4]$.

(c) 1.8, 2.5, 3.1, 3.5, and 4.0.

(d)

(e) $\sqrt{\pi} \approx 1.8$
$\sqrt{2\pi} \approx 2.5$
$\sqrt{3\pi} \approx 3.1$
$\sqrt{4\pi} \approx 3.5$
$\sqrt{5\pi} \approx 4.0$

39.(a) 0.4 and 2.7

(b) $\arcsin(0.4) \approx 0.4$
$\pi - \arcsin(0.4) \approx 2.7$

(c) -0.4 and -2.7.

(d) $-0.4 \approx -\arcsin(0.4)$
$-2.7 \approx \arcsin(0.4) - \pi$

41.(a)

(b) The period is π.

(c) Period of $\cos 2x = \pi$
Period of $\sin 4x = \frac{\pi}{2}$, so it is also periodic with period π.

43.(a)

(b) It becomes a vertical asymptote at $x = -4$.

45.(a) $p(x) = ax^2 + c$
(a, c constants)

(b) $p(x) = bx$
(b is a constant)

47.(a) $R = k_1 - k_2G$

(b)

49.(a) $S(0) = 12$

(b) $a = 12$
$b \approx 7.64$

(c) $S(32°13') \approx 14.12$
$S(46°4') \approx 15.58$

(d) $S(x)$ is not smooth.

CHAPTER TWO

Section 2.1

5. $v_{avg} = 2.5$ ft/sec
   $v(0.2) = 4.5$ ft/sec

9. From smallest to largest: 0
   slope at C
   slope at B
   slope of AB
   1
   slope at A

11. 6.0

13. 0

Section 2.2

1.(a)
  (b) 0.24
  (c) 0.22

7.(a) $f(4)/4$ is the slope of the line connecting (0,0) to $(4, f(4))$.
  (b) $f(3)/3$

9.(a) C and D
  (b) B and C
  (c) A and B, C and D

11. $-1/4$

13. $-1/4$

15. $y = -2x + 3$

17. $f'(3) \approx 31$

19. $f'(2) \approx 6.773$

21.(a) $f'(0) = 1.00000$
   $f'(0.3) = 1.04534$
   $f'(0.7) = 1.25521$
   $f'(1) = 1.54314$

  (b) They are about the same.

23. 16.0 million people/year
    16.4 million people/year

Section 2.3

9. $f'(x)$ positive: $4 \leq x \leq 8$
   $f'(x)$ negative: $0 \leq x \leq 3$
   $f'(x)$ greatest: at $x \approx 8$

11. $g'(1) = 3$
    $g'(2) = 5$
    $g'(3) = 7$
    The pattern seems to be:
    $g'(t) = 2t + 1$.

13. $4x$

15. $-2/x^3$

33. If $f(x)$ is even, its graph is symmetric about the y-axis. So the tangent line to f at $x = x_0$ is the same as that at $x = -x_0$ reflected about the y-axis.

Section 2.4

1.(a) A, C, F, and H.
  (b) These are points where the particle switches from speeding up to slowing down or vice versa.

3. feet/mile
   $f'(x)$ is always negative.

5. dollars/percent
   Positive

7. dollars/year
   $\frac{dB}{dt}$ represents how fast your money is growing.

9.(a) positive
  (b) °F/min. At time $t = 20$, the temperature increases by approximately 2°F for each additional minute in the oven.

11. hr/gal. $g'(55) = -0.54$ means that at 55 miles per hour the fuel efficiency (in miles per gallon, or mpg) of a car decreases as the velocity increases at a rate of approximately one half mpg for an increase of one mph.

13.(a)
  (b) concave down
  (c) Acceleration due to gravity.

15. $P'(66)$ is an estimate of the number of people between 65.5 and 66.5 inches tall
    Units: People per inch
    $P'(x)$ is never negative

17.(a) $P'(1900) \approx$ 1.6 million people/year
   $P'(1945) \approx$ 1.9 million people/year
   $P'(1990) \approx$ 2.22 million people/year
  (b) Between 1950 and 1960.
  (c) ≈ 167.7 million
  (d) 271 million

Section 2.5

1. (a) increasing, concave up;
   (b) decreasing, concave down

5. (a)
   (b) Derivative of utility > 0
   2nd derivative of utility < 0

7. A positive second derivative indicates a successful campaign. A negative second derivative indicates an unsuccessful campaign.

11. (a) B and E
    (b) A and D

Section 2.6

1. $y = \frac{1}{4}x + 1$

5. (a) $f'(6.75) \approx -4.2$
   $f'(7.0) \approx -3.8$
   $f'(8.5) \approx -2.0$
   (b) $f''(7) \approx 1.6$.
   (c) $y = -3.8x + 34.8$
   (d) $f(6.8) \approx 8.96$

7. origin: along $y = 0$
   $(1, 1)$: along $y = 2x - 1$.

9. (a) The graph looks straight because it shows only a small part of the curve magnified greatly.
   (b) March, because on the 21st of the month there are 12 hours of daylight.
   (c) slope ≈ 0.04
   The amount of daylight is increasing by 0.04 hours per calendar day.

Section 2.7

1. The limit appears to be 1.

3. 0

5. 0.3333...

7. 0

9. No.

11. e^2

Section 2.8

1. Yes

3. Yes

5. (a) No vertical segments or corners. The graph is continuous and differentiable everywhere.
   (b) $f'(x)$: corner at $x = 0$
   $f''(x)$: undefined at $x = 0$
   So $f''(x)$ is neither differentiable nor continuous at $x = 0$.

7. (a) Yes
   (b) Not at $t = 0$

Chapter 2 Review

1. (a) Seven
   (b) Increasing at $x = 1$
   Decreasing at $x = 4$
   (c) $6 \leq x \leq 8$
   (d) $x = -9$

3. $f'(1) = 2$
   $f'(-1) = -2$
   $f'(5) = 10$
   $f'(x) \approx 2x$

5. $f'(1) = 1$
   $f'(2) = 0.5$
   $f'(5) = 0.2$
   $f'(10) = 0.1$
   $f'(x) = \frac{1}{x}$

7. $g'(-4) = 5$

17. (a) IV, (b) III, (c) II, (d) I, (e) IV, (f) II

19. (a) At $(0, \sqrt{19})$:
    slope $= 0$
    At $(\sqrt{19}, 0)$:
    slope is undefined.
    (b) slope $\approx \frac{1}{2}$
    (c) At $(-2, \sqrt{15})$:
    slope $\approx \frac{1}{2}$
    At $(-2, -\sqrt{15})$:
    slope $\approx -\frac{1}{2}$
    At $(2, \sqrt{15})$:
    slope $\approx -\frac{1}{2}$.

21. (a) The population varies periodically with a period of 12 months (i.e. one year).
    (b) Max of 4500 on June 1st.
    (c) Min of 3500 on Feb 1st.
    (d) Growing fastest: around April 1st
    Decreasing fastest: around July 20 and Nov 15
    (e) about 400 deer/month.

CHAPTER THREE

Section 3.1

1.(a) Lower est = 135 feet
Upper est = 223 feet
(b)

3.(a) Overest: 7 tons
Underest: 5 tons
(b) Overest: 74 tons
Underest: 59 tons
(c) Every 2 days

5.(a) 0.85
(b) Use $n > \frac{0.39}{E}$ subdivisions.

7. 0.400

9.(a) Upper est: 34.16 m/sec
Lower est: 27.96 m/sec
(b) 31.06 m/sec
It is too high.

Section 3.2

1. limit $= \frac{1}{4}$
True value is between 0.248004 and 0.252004.

3. True value is between 1.45922 and 1.46610.

5. Since $\sin(t^2)$ is *not* monotonic on $[2, 3]$, we cannot be sure of the true value.

7. 41.667, $n = 1300$

9. 0.470, $n = 10$

11. 4.858, $n = 110$

13. 2.886, $n = 30$

15. 0.016, $n = 10$

17. 1.772, $n = 70$

19. $\int_0^{4\pi} (2 + \cos x)\, dx <$
left-hand sum =
right-hand sum

Section 3.3

1. foot-pounds

3. dollars

5. 2.00

7. 10.67

9.(a)
(b) 3.08
(c) 2.50

11. 0.80

13. 2

17.(a)
(b) 0.64

19.(a) 9.9 hours
(b) 14.4 hours
(c) 12.0 hours
(d)

21.(a) $2 \int_0^2 f(x)\, dx$
(b) $\int_0^5 f(x)\, dx - \int_2^5 f(x)\, dx$
(c) $\int_{-2}^5 f(x)\, dx - \frac{1}{2} \int_{-2}^2 f(x)\, dx$

23.(a) $\frac{1}{5} \int_0^5 f(x)\, dx$
(b) $\frac{1}{5}(\int_0^2 f(x)\, dx - \int_2^5 f(x)\, dx)$

Section 3.4

1. $F(0) = 0$
$F(1) = 1$
$F(2) = 1.5$
$F(3) = 1$
$F(4) = 0$
$F(5) = -1$
$F(6) = -1.5$

3. $F(0) = 0$
$F(0.5) = 0.041$
$F(1) = 0.308$
$F(1.5) = 0.776$
$F(2) = 0.807$
$F(2.5) = 0.431$

5. When $\theta = 1.77$,
$F(1.77) = 0.89$

7.(a)
(b)(i) 0.9995
(ii) 1

9. $f(3) - f(2) < \frac{f(4) - f(2)}{2} < f(4) - f(3)$

11. 45.8°C.

13.(a) $\int_0^5 f(t)\, dt$
(b) 177.27 billion barrels
(c) Each term is a lower estimate of one year's consumption of oil.

Section 3.5

1. 0

3. 0

Chapter 3 Review

1. 1.442, $n = 210$
3. 1.31, $n = 5$
5. -0.083, $n = 10$
7. (a) $F(0) = 0$
   (b) F increases
   (c) $F(1) \approx 0.7468$
   $F(2) \approx 0.8821$
   $F(3) \approx 0.8862$
9. (a) For $n = 10$:
   left sum ≈ 0.054
   (rounding down),
   right sum ≈ 0.058
   (rounding up),
   The value of the integral is
   0.056 to one decimal place.
   (b) ≈ 0.05609
11. (a) 346.6 hours
   (b) 450 millirems
13. (a) ≈ 76.8 million
   (b) $= 77.24$ million
   (c)
15. (a) The first star.
   (b) $p'(0.4)$ is greater.
   $r'(0.4)$ is the rate at which the rate of change of mass is changing when $t = 0.4$.
   (c) The difference in the masses gained by the two stars during the interval from $t = 0$ to $t = 1$.
17. 3.8
19. (a) At $t = 17, 23, 27$ seconds.
   (b) Right:
   $t = 10$ seconds
   Left:
   $t = 40$ seconds
   (c) Right:
   $t = 17$ seconds
   Left:
   $t = 40$ seconds
   (d) From $t =$
   10 to 17 seconds,
   20 to 23 seconds, and
   24 to 27 seconds
   (e) At $t = 0$ and 40 seconds.
21. (a) $\int_0^T 49(1 - e^{-0.2t})\,dt$ (meters)
   (b) $T \approx 107$ seconds

CHAPTER FOUR

Section 4.1

1. (a) $k(x) = -x + 3$
   $k'(x) = -1$
   (b) $j(x) = -5x + 1$
   $j'(x) = -5$
3. (a) -10
   (b) $\frac{d}{dx}(f[g(x)]) = f'(x)g'(x)$

Section 4.2

1. $y' = 12x^{11}$
3. $y' = 4x^{\frac{1}{3}}/3$
5. $y' = -3x^{-\frac{7}{4}}/4$
7. $f'(x) = (x^{-\frac{3}{4}})/4$
9. $y' = 6x^{\frac{1}{2}} - 5x^{-\frac{1}{2}}/2$
11. $y' = -12x^3 - 12x^2 - 6$
13. $y' = 6t - 6/t^{\frac{3}{2}} + 2/t^3$
15. $y' = 1 - 1/x^2$
17. $f'(t) = -2t^{-3} - t^{-2} + 4t^{-5}$
19. Problems 2, 4, 5, 6, 7, and 9.
21. The rules of this section do not apply.
23. $y' = -2/3z^3$ (power rule and sum rule)
25. The rules of this section do not apply.
27. $y' = 3x^2$ (power rule)
29. The rules of this section do not apply.
31. (a) 0
    (b) 5040
33. $(-1, 7)$ and $(7, -209)$
35. For $x < 0$ or $2 < x < 3$
37. (a) $x < 1/2$
    (b) $x > 1/4$
    (c) $p(x)$ is always decreasing, unless $x = 0$.
39. $n = 1/13$
41. $-2GMm/r^3$
43. (a) $v(t) = -32t$
    $v \leq 0$ because $t \geq 0$.
    (b) $a(t) = -32$
    (c) $t = 8.84$ seconds
    $v = -192.84$ mph
45. $V(r) = 4\pi r^3/3$
    $\frac{dV}{dr} = 4\pi r^2 =$
    surface area of a sphere

Section 4.3

1. $y' = 10t + 4e^t$
3. $f'(x) = (\ln 2)2^x + 2(\ln 3)3^x$
5. $dy/dx = 3 - 2(\ln 4)4^x$
7. $f'(x) = ex^{e-1}$
9. $f'(t) = e^{t+2}$
11. $z' = (\ln 4)e^x$
13. $f'(z) = (2\ln 3)z + (\ln 4)e^z$
15. $f'(x) = 3x^2 + 3^x \ln 3$
17. $dy/dx = \pi^x \ln \pi$
19. $y' = 2x + (\ln 2)2^x$
21. Our rules do not apply here.
23. $y' = e^{x+5}$
25. $f'(s) = (1 + \ln 5)5^s e^s$
27. $f'(z) = (\ln 2)2^z$
29. The methods of the section do not apply here.
31. (a) $f'(0) = -1$
    (b) $y = -x$
    (c) $y = x$
33. 7.95 ¢/year
35. $22.5(1.35)^t$
37. Increasing for $a > 1$, decreasing for $a < 1$.

Section 4.4

1. $f'(x) = 5x^4 + 10x$
3. $f'(x) = e^x(x + 1)$
5. $y' = 2^x(1 + x\ln 2)$
7. $f'(x) = 3^x[(\ln 3)(x^2 - x^{\frac{1}{2}}) + (2x - 1/(2\sqrt{x}))]$
9. $z' = 4s^3 - 1$
11. $dz/dt = 1/(5t + 2)^2$
13. $z' = (t^2 + 6t + 13)/(t + 3)^2$
15. $f'(x) = (x^2 - 3)/x^2$
17. $f'(x) = \frac{(-4x^2 - 8x - 1)}{(2 + 3x + 4x^2)^2}$
19. $f'(x) = 12x + 1$ and $f''(x) = 12$
21. $f'(x) = 2e^{2x}$
23. $\frac{d}{dx}e^{2x} = 4e^{4x}$
25. $y = 7x/48 + 5/48$
27. (a) $f(x)$ is defined for all

$x \neq \pm 1$.
$f'(x) = \frac{(-x^2-1)}{(x^2-1)^2}$

(b) $g(x)$ is defined for all $x \neq \pm 1$.
$g'(x) = -3/(x+1)^2$

29.(a) $f'(x) = (x-2)+(x-1)$
(b) $f'(x) = (x-2)(x-3) + (x-1)(x-3) + (x-1)(x-2)$
(c) $f'(x) = (x-2)(x-3)(x-4) + (x-1)(x-3)(x-4) + (x-1)(x-2)(x-4) + (x-1)(x-2)(x-3)$

33.(a) $f(140) = 15{,}000$ says that 15,000 skateboards are sold when the cost is \$140 per board.
$f'(140) = -100$ means that if the price is increased from \$140, every dollar of increase will decrease the total sales by 100 boards.
(b) $\frac{dR}{dp}|_{p=140} = 1000$
(c) Positive. The price should be increased.

35.(a)
(b)
(c) -3776.63

Section 4.5

1. $f'(x) = 99(x+1)^{98}$
3. $w' = 200t(t^2+1)^{99}$
5. $w' = 50(\sqrt{t}+1)^{99}/\sqrt{t}$
7. $y' = \frac{3}{2}e^{\frac{3}{2}w}$
9. $y' = 3s^2/2\sqrt{s^3+1}$
11. $y' = e^{-t^2} - 2t^2e^{-t^2}$
13. $f'(z) = e^{-z}/2\sqrt{z} - \sqrt{z}e^{-z}$
15. $f'(t) = e^{5-2t}(1-2t)$
17. $f'(\theta) = e^{-\theta}/(1+e^{-\theta})^2$
19. $f'(w) = 2we^{w^2}(5w^2+8)$
21. $f'(y) = -(\ln 10)(10^{\frac{5}{2}-\frac{y}{2}})/2$
23. $f'(y) = 2ye^{[e^{(y^2)}+y^2]}$
25. $y = 4451.66x - 3560.81$

27.(a) $H(4) = 1$
(b) $H'(4) = 30$
(c) $H(4) = 4$
(d) $H'(4) = 56$
(e) $H'(4) = -1$

29. Yes

31.(a) $dH/dt = -60e^{-2t}$
(b) $dH/dt < 0$; because the temperature of the soda is decreasing.
(c) At $t = 0$, because this is the time when the temperature difference is greatest.

33.(a) $\frac{dm}{dv} = \frac{m_0 v}{c^2\sqrt{(1-v^2/c^2)^3}}$
(b) The rate of change of mass with respect to the speed v.

35.(a) For $t < 0$, $I = 0$
For $t > 0$, $I = -\frac{1}{R}e^{-t/RC}$
(b) No
(c) No

Section 4.6

1. $f'(x) = \sin x/2\sqrt{1-\cos x}$
3. $f'(x) = 3\cos(3x)$
5. $w' = e^t\cos(e^t)$
7. $f'(x) = -\sin x e^{\cos x}$
9. $z' = e^{\cos\theta} - \theta(\sin\theta)e^{\cos\theta}$
11. $f'(x) = 2\cos(2x)\sin(3x) + 3\sin(2x)\cos(3x)$
13. $f'(x) = e^{-2x}[\cos x - 2\sin x]$
15. $y' = 5\sin^4\theta\cos\theta$
17. $z' = -3e^{-3\theta}/\cos^2(e^{-3\theta})$
19. $h'(t) = \cos t - t\sin t + 1/\cos^2 t$

23.(a) $dy/dt = -\frac{4.9\pi}{6}\sin\left(\frac{\pi}{6}t\right)$. Represents the rate of change of the depth of the water.
(b) Occurs when $\sin(\frac{\pi}{6}t) = 0$, or at $t =$ 6am, 12 noon, 6pm, and 12 midnight. $dy/dt = 0$ means that the water level is at either high tide or low tide.

25. periodic,
$e^{-1} \leq f(x) \leq e$

27. 0.0174533
When h is measured in degrees, the numerator is much smaller than when h is measured in radians. Since the denominator is numerically the same, we expect the ratio $\frac{\sin h}{h}$ to be smaller.

29. $w'(\pi) = 0$

31.(a) No solution for $k'(p) = 1$ and $p = \pm\pi/2$ for $r'(p) = 1$
(b) Graphically, this implies that the tangent to $k(p)$ is steeper than the line $y = x$, at every point on the graph.

Section 4.7

1. $f'(x) = 1/(x-1)$
3. $f'(z) = -1/z(\ln z)^2$
5. $f'(x) = e^{-x}/(1-e^{-x})$
7. $f'(x) = e^x/(e^x+1)$
9. $f'(x) = 7$

11. $f'(w) = -\tan(w-1)$
13. $f'(y) = 2y/\sqrt{1-y^4}$
15. $h'(w) = \arcsin w + \frac{w}{\sqrt{1-w^2}}$
17. $\frac{d}{dx}(\arcsin x) = 1/\sqrt{1-x^2}$ $(-1 < x < 1)$
19. -43.4
21. The following functions look like the line $y = x$:
   $y = x$
   $y = \sin x$
   $y = \tan x$
   $y = \ln(x+1)$
   The following functions look like the line $y = 0$:
   $y = x^2$
   $y = x \sin x$
   $y = x^3$, $y = \frac{1}{2}\ln(x^2+1)$
   $y = 1 - \cos x$
   The following functions look like the line $x = 0$:
   $y = \sqrt{x}$
   $y = \sqrt{x/(x+1)}$
   $y = \sqrt{2x - x^2}$
23. (a) $y = -x^2/2 + 2x - 3/2$
   (b) From graph, notice that around $x = 1$, the values of $\ln x$ and its approximation are very close.
   (c) At $x = 1.1$, $y \approx 0.095$
   At $x = 2$, $y = 0.5$
25. (a) $h(t) = 300 - 30t$
   $0 \le t \le 10$
   (b) $\theta = \arctan(\frac{200-30t}{150})$
   $\frac{d\theta}{dt} = -(\frac{1}{5})(\frac{150^2}{(150^2+(200-30t)^2)})$
   (c) When the elevator is at the level of the observer.

Section 4.8

1. $dy/dx = (y^2 - y - 2x)/(x - 3y^2 - 2xy)$
3. $dy/dx = -\sqrt{y/x}$
5. $dy/dx = \frac{2-y\cos(xy)}{x\cos(xy)}$
7. $\frac{dy}{dx} = \frac{3x^2 \arctan y}{-e^{\cos y}\sin y - x^3(1+y^2)^{-1}}$
9. $y = x/2 - 3/2$
11. $y = 0$
13. $dy/dx = (m/n)x^{m/n-1}$
15. (a) $dy/dx = -9x/25y$
   (b) The slope is not defined anywhere along the line $y = 0$.
17. (a) $\frac{dy}{dx} = \frac{y^2-3x^2}{3y^2-2xy}$
   (b)
   (c) $y \approx 1.9945$
   (d) Horizontal:
   $(1.1609, 2.0107)$
   and
   $(-0.8857, 1.5341)$
   Vertical:
   $(1.8039, 1.2026)$
   and
   $(\sqrt[3]{5}, 0)$

Section 4.9

1. $1/x \approx 2 - x$
7. (a)
   (b)(i) 0
   (ii) $-1/2$
9. (a)
   (b)
   (c) 0

Chapter 4 Review

1. $f'(x) = 6xe^x - 24x + 3x^2e^x + \pi e^x$
3. $f'(z) = \sqrt{z}(3 - z^{-2})/2$
5. $g'(t) = 6(1+3t)e^{(1+3t)^2}$
7. $h'(x) = e^{\tan x} + xe^{\tan x}/\cos^2 x$
9. $f'(y) = 6y^2/2y^3 \ln(2y^3)$
11. $r'(\theta) = 6(3\theta - \pi)\cos[(3\theta - \pi)^2]$
13. $g'(\theta) = -\frac{\sin\theta\cos\theta}{\sqrt{a^2-\sin^2\theta}}$
15. $w'(\theta) = 1/\sin^2\theta - 2\theta\cos\theta/\sin^3\theta$
17. $h'(t) = (-e^{-t} - 1)/(e^{-t} - t)$
19. $s'(x) = -1/(1 + (2-x)^2)$
21. $r'(y) = \frac{\cos y + a + y\sin y}{(\cos y + a)^2}$
23. These functions look like the line $y = 0$:
   $\sin x - \tan x$, $\frac{x^2}{x^2+1}$
   $x - \sin x$, $\frac{1-\cos x}{\cos x}$
   These functions look like the line $y = x$:
   $\arcsin x$, $\frac{\sin x}{1+\sin x}$
   $\arctan x$, $e^x - 1$
   $\frac{x}{x+1}$, $\frac{x}{x^2+1}$
   These functions are undefined at the origin:
   $\frac{\sin x}{x} - 1$, $-x\ln x$
   Defined at the origin but with a vertical tangent:
   $x^{10} + \sqrt[10]{x}$
25. (a) $dg/dr = -2GM/r^3$
   (b) dg/dr is the rate of change of acceleration due to the pull of gravity.
   The further away from the center of the earth, the weaker the pull of gravity.
   (c) -3.05×10^{-6}
   (d) It is reasonable because the magnitude of $\frac{dg}{dr}$ is so small (compared to $g = 9.8$) that for r near 6400 km, g is not varying much at all.
27. (a) $v(t) = 10e^{\frac{t}{2}}$
   (b) $s'(t) = s(t)/2$

29. (a) Falling, 0.38 m/hr
    (b) Rising, 3.76 m/hr
    (c) Rising, 0.75 m/hr
    (d) Falling, 1.12 m/hr

33. $b = 1/40$ and $a = 169.36$

35. $1/3\ \mu m^3$/day

37. (a) $(FGH)' = F'GH + FG'H + FGH'$
    (b)
    (c) $(f_1 \cdot f_2 \cdot f_3 \cdot \ldots \cdot f_n)' = f_1' f_2 f_3 \cdots f_n + f_1 f_2' f_3 \cdots f_n + \cdots + f_1 \cdots f_{n-1} f_n'$

39. (a) Decreases
    (b) −0.25 cc/min

CHAPTER FIVE

Section 5.1

5. Critical points:
   $x = -3$ and $x = 2$
   Extrema:
   $f(-3)$ local maximum
   $f(2)$ local minimum

7. Critical points:
   $x = 0$ and $x = \pm 2$
   Extrema:
   $f(0)$ local minimum
   $f(-2)$ and $f(2)$ are not local extrema.

9. Critical points:
   $x = \pm 1$
   Extrema:
   $f(-1)$ local minimum
   $f(1)$ local maximum

11. (a) $f(0.91)$ is a local (and global) min
    $f(3.73)$ is local (and global) max
    (b) $f(-3)$ is global min
    $f(2)$ is global max

13. (a) $f(1)$ local minimum
    (b) $f(0.1)$ global maximum
    $f(1)$ global minimum

17. Decreasing:
    $-105 < x < 5$
    Increasing:
    $x < -105$ and $x > 5$

21. (a) $\theta = 0$ is a zero for $f(\theta)$
    (b) $f'(\theta) = 1 - \cos\theta > 0$ for $0 < \theta \leq 1$.
    The only zero is at the origin.

23. (a)
    (b)
    (c) No

25. (a) Yes, at 2000 rabbits
    (b) 1884
    500 rabbits
    (c)

Section 5.2

3. $x = 0$ is a local minimum.

5. local maxima:
   $f(0)$, $f(2\pi)$, $f(4\pi)$, $f(6\pi)$
   local minima:
   $f(\pi)$, $f(3\pi)$, $f(5\pi)$
   inflection points:
   $f(\pi/2)$,$f(3\pi/2)$,$f(5\pi/2)$, $f(7\pi/2)$,$f(9\pi/2)$, and $f(11\pi/2)$

7. local maxima: $f(\pi/4)$, $f(9\pi/4)$, and $f(17\pi/4)$
   local minima: $f(5\pi/4)$, $f(13\pi/4)$, and $f(21\pi/4)$
   inflection points: $f(3\pi/4)$, $f(7\pi/4)$, $f(11\pi/4)$, $f(15\pi/4)$, $f(19\pi/4)$, and $f(23\pi/4)$

9. Increasing most rapidly:
   $x = 2$
   Decreasing most rapidly:
   $x = 0$

11. (a) Least at $x = 0$
    Greatest at $x = \pi$
    (b) Increasing most rapidly:
    $x = \frac{\pi}{2}$
    Decreasing: nowhere
    (c) Decreasing most rapidly:
    $x = \pi$
    Increasing most rapidly:
    $x = 0$

17. (a) As $x \to \infty$, $f(x) \to 1$
    As $x \to -\infty$ $f(x) \to 1$
    As $x \to 0^+$, $f(x) \to \infty$
    As $x \to 0^-$, $f(x) \to 0$
    (b) $f(x)$ is decreasing everywhere it is defined.
    (c) Concave up:
    $x > 0$ and
    $-1/2 < x < 0$
    Concave down:
    $x < -1/2$
    (d)

19. (a)
    (b) At most four zeros.
    (c) Possibly no zeros.
    (d) At least two inflection points.
    (e) Even, because there are an odd number of critical points.
    (f) Degree four
    (g) $f(x) =$
    $k(x-a)(x-b)$
    $\times\ (x-c)(x-d)$

21. (a) g is decreasing.
    (b) g has a local minimum at x_1.
    (c) g is concave up at x_2.

Section 5.3

1. When $a^2 - 3b < 0$

3.(a) Local maximum:
$p(-\sqrt{\frac{a}{3}}) = +\frac{2a\sqrt{a}}{3\sqrt{3}}$
Local minimum:
$p(\sqrt{\frac{a}{3}}) = -\frac{2a\sqrt{a}}{3\sqrt{3}}$
(b) It moves them further from the x-axis and further apart.
(c)

5.(a) $x = 0$ and $x = \pm\sqrt{-\frac{a}{2}}$ (if $a \leq 0$)
(b) For any value of a or b, $x = 0$ is a critical point. It is the only critical point when $a \geq 0$. Local minimum.
(c) If a is negative:
$x = 0$ is local max,
$x = \pm\sqrt{-\frac{a}{2}}$ are local min
(d) No, f has only one or three critical points.

7.(a) terminal velocity $= \frac{mg}{k}$
(b) $m = 4.7$ slugs
$k = 0.76$ ft·lbs/sec
(c) 197.4 ft/sec, or 134.6 mph

9.(a)(i) $A > 0$: concave up
$A < 0$: concave down
(ii) The larger A is in magnitude, the steeper the graph.
(b) B shifts the graph to the left or right.
x-intercept $= -B$
(c) The graph is a parabola with a maximum (if A is negative) or minimum (if A is positive) at $x = -B$, where the steepness depends on the magnitude of A.

11. a determines the y intercept and stretches or flattens the bell-shaped graph of e^{-bx^2} vertically. b affects the width of the graph.

13. A affects the amplitude.
B affects the frequency.
C causes a phase shift to the left or right.

15.(a) Critical points:
For $0 < a < 1$:
Local min: $x = 0$
Local max: $x = +\sqrt{1-a}$
For $a \geq 1$:
Local max: $x = 0$
(b)
(c)
(d)
(e) a determines whether there will be a hill or valley at $x = 0$, and whether or not there is second critical point at $x = \sqrt{1-a}$.

Section 5.4

1.(a) Graph of
$N = 100 + 20x$
(b)(i) $N'(x) = 20$
(ii) $N(x)/x = 100/x + 20$

3.(a) $C(0)$ represents the fixed costs before production.
(b) The marginal cost decreases slowly, and then increases as quantity produced increases.
(c) Concave down implies decreasing marginal cost, while concave up implies increasing marginal cost.
(d) An inflection point of the cost function is (locally) the point of maximum or minimum marginal cost.
(e)

5.(a) \$0
(b) \$96.56
(c) Raise the price by \$5.

7.(a) If $a > 1$, $C''(q) < 0$, so C is concave down.
(b) $q = \left[Fa/(K(1-a))\right]^a$

Section 5.5

1. $r = \frac{2}{3}R$

3. 13.13 miles from A

5. $x = \frac{a_1+a_2+a_3}{3}$

7.(a)
(b) $f(v) = v \cdot a(v)$
(c) When $a(v) = f'(v)$
(d) $a(v)$

11. $0.91 < y \leq 1.00$

13. $0 \leq y < 1.61$

15. $0 \leq f(x) \leq 2\pi$

17.(a) The arithmetic mean unless $a = b$, in which case the two averages are equal.
(b) The arithmetic mean unless $a = b = c$, in which case the two averages are equal.

Section 5.6

1. 1250 square feet

3. $x = \sqrt[3]{2V}$
$y = \sqrt[3]{V/4}$

5. Maximum revenue $=$ \$27,225
Minimum $=$ \$0

7.(a) The disk takes the same amount of time to go around, regardless of the radius of the track, so the same amount of music is stored on each track.
(b) $r = R/2$.

9. $(0.59, 0.35)$

11. Distance from origin:
 $\sqrt{x^2 + (2 - 2\sin x)^2}$
 Minimum distance of
 0.98 occurs at $x \approx 0.8$.

13. (a) Bueya should look for an apartment as close to the bar as possible.
 (b) Marie-Josée should try to live anywhere between the bar and the Gateway Arch.

15. 65.1 meters

Section 5.7

1. (a) $f'(x) = 3x^2 + 6x + 3$
 (b) At most one root.
 (c) The root is in the interval $[0, 1]$.
 (d) $x \approx 0.913$

3. $\sqrt[4]{100} \approx 3.162$

5. $x \approx 0.511$

7. $x \approx 1.310$

9. $x \approx 1.763$

11. $x \approx 0.682328$ (rounded up) is a root.

13. (a) $x_1 \approx 3.1425$
 $x_2 \approx 3.141592653$
 This is correct to one billionth.
 (b)
 (c)

Chapter 5 Review

1. (a) $f'(x) = 3x(x - 2)$
 $f''(x) = 6(x - 1)$
 (b) $x = 0$
 $x = 2$
 (c) Inflection point: $x = 1$
 (d) Endpoints:
 $f(-1) = -4$
 $f(3) = 0$
 Local & global max:
 $f(0) = 0$ and $f(3) = 0$
 Local & global min:
 $f(-1) = -4$ and
 $f(2) = -4$
 (e) f increasing:
 for $x < 0$ and $x > 2$
 f decreasing:
 for $0 < x < 2$
 f concave up:
 for $x > 1$
 f concave down:
 for $x < 1$.

3. (a) $f'(x) =$
 $-e^{-x}\sin x + e^{-x}\cos x$
 $f''(x) = -2e^{-x}\cos x$
 (b) Critical points:
 $x = \frac{\pi}{4}$ and $\frac{5\pi}{4}$
 (c) Inflection points:
 $x = \frac{\pi}{2}$ and $\frac{3\pi}{2}$
 (d) Endpoints:
 $f(0) = 0$ and $f(2\pi) = 0$
 Local and global max:
 $f(\frac{\pi}{4}) = (e^{-\frac{\pi}{4}})(\frac{\sqrt{2}}{2})$
 Local and global min:
 $f(\frac{5\pi}{4}) = -e^{\frac{-5\pi}{4}}(\frac{\sqrt{2}}{2})$
 (e) f increasing:
 $0 < x < \frac{\pi}{4}$ and
 $x > \frac{5\pi}{4}$
 f decreasing: $\frac{\pi}{4} < x < \frac{5\pi}{4}$
 f concave down:
 for $0 \le x < \frac{\pi}{2}$
 and $\frac{3\pi}{2} < x \le 2\pi$
 f concave up:
 for $\frac{\pi}{2} < x < \frac{3\pi}{2}$

5. $\lim_{x\to\infty} f(x) = \infty$
 $\lim_{x\to-\infty} f(x) = \infty$
 $f'(x) = 6(x - 2)(x - 1)$
 $f''(x) = 6(2x - 3)$
 Critical points:
 $f(1) = 6$, $f(2) = 5$
 Local max: $f(1) = 6$
 Local min: $f(2) = 5$
 Global max and min: none
 f increasing: $x < 1$ and $x > 2$
 Decreasing: $1 < x < 2$
 f concave up:
 $x > \frac{3}{2}$
 f concave down:
 $x < \frac{3}{2}$

7. $\lim_{x\to-\infty} xe^{-x} = -\infty$
 $\lim_{x\to\infty} xe^{-x} = 0$
 $f'(x) = (1 - x)e^{-x}$
 $f''(x) = (x - 2)e^{-x}$
 Global max: $f(1) = \frac{1}{e}$
 Local and global min: none
 Inflection point: $f(2) = \frac{2}{e^2}$
 f increasing: $x < 1$
 f decreasing: $x > 1$
 f concave up:
 $x > 2$
 f concave down:
 $x < 2$

9. Local max: $f(-3) = 12$
 Local min: $f(1) = -20$
 Inflection: $x = -1$
 Global max and min: none

11. Global and local min: $x = 2$
 Global and local max: none
 Inflections: none

13. Global and local max:
 $f(0) = 1$
 Local minimum: none
 Inflections: $x = \pm\frac{1}{\sqrt{2}}$

15. $-0.014 \le e^{-x}\sin x \le 0.322$

17. (a) Increasing: $(0, \infty)$
 Decreasing: $(-\infty, 0)$
 (b) Local and global min: $f(0)$

19. (a) Increasing: $(0, 4)$
 Decreasing:

$(-\infty, 0), (4, \infty)$
(b) Local max: $f(4)$
Local min: $f(0)$

21. (a) The concavity changes y_1 and y_3.
(b) $f(t)$ grows fastest where the vase is skinniest and slowest where the vase is widest.

23. (a) $g(e)$ is a global maximum. There is no minimum.
(b) There are exactly two roots.
(c) $x = 5$ and $x \approx 1.75$

25. (a) The slope of the line from the origin to the graph.
(b) Where $a(q)$ is tangent to the graph.
(c) $C'(q_0) = a(q_0)$
The average cost is minimized when it equals the marginal cost.
(d)

27. $a = 363.23$, and $b \approx 4.7665$

CHAPTER SIX

Section 6.1

1. Left-hand sum = 1.96875
   Right-hand sum = 2.71875
   The most the estimate could be off is 0.375.

3. 15

5. 10

7. (a) Car 1: 1031.25 ft
   Car 2: 562.5 ft
   (b) 1.6 minutes

9. (a)
   (b) Twice
   At each intersection point, the distance between the two vehicles is at a local extremum.

11. (a) Forward for $t < 3$
    Backward for $t > 3$
    (b) Farthest forward: $t = 3$
    Farthest backward: no upper bound
    (c) $s(t) = 6t - t^2$

13. (a) 18 appliances per month.
    (b) Average = 17 appliances per month.
    (c) They are close, but not equal.
    (d) The integral is easier to calculate than the sum, particularly for a large number of months.
    (e)

Section 6.2

1. $\int_1^3 (x^2 - x)\,dx = 14/3$

3. (a) area $= \frac{b^2-a^2}{2}$
   (b)(i) $\int_2^5 x\,dx = 21/2$
   (ii) $\int_{-3}^8 x\,dx = 55/2$
   (iii) $\int_1^3 5x\,dx = 20$

5. (a) 0.1574
   (b) 0.9759

7. Positive, since $e^{x^2} > 0$ for all x.

9. (a)
   (b) Increasing, concave up
   (c)

Section 6.3

11. Acceleration is zero at points A and C.

Section 6.4

1. $5x$
3. $x^3/3$
5. $\sin t$
7. $\ln z$
9. $-1/2z^2$
11. $-\cos t$
13. $t^4/4 - t^3/6 - t^2/2$
15. $5x^2/2 - 2x^{\frac{3}{2}}/3$
17. $-\cos 2\theta$
19. $(t+1)^3/3$
21. $\sin t + \tan t$
23. $\sin(t^2)/2$
25. $e^2 y + 2^y/\ln 2$
27. $e^{x^2}/2$
29. $F(x) = 3x + 2$
31. $F(x) = x^3/3 + 2$
33. $F(x) = -\cos x + 3$
35. $3x^2/2 + C$
37. $\sin\theta + C$
39. $x^2/2 + 2x^{1/2} + C$
41. $\pi x + x^{12}/12 + C$
43. $\sin(x+1) + C$
45. $-e^{-z} + C$
47. $y = 2kt^{3/2}/3$
49. (a)
    (b)
    (c) 200 ft
    (d) 200 ft
51. (a) 6 seconds
    (b) LEFT(12) = 97.5 ft
    RIGHT(12) = 82.5 ft
    Left sum is an overestimate.
    Right sum is an underestimate.
    (c) 90 ft
    (d) $s(t) = 30t - \frac{5}{2}t^2$
    The area is the average of the left-hand and right-hand sums, which is 90 ft.
53. (a) $v(t) = 1.6t$
    (b) $s(t) = 0.8t^2 + s_0$

Section 6.5

1. $t = 5$
   $v = -160$ ft/sec

3. 400 feet high

7. (a) First second: $-g/2$

Second: $-3g/2$
Third: $-5g/2$
Fourth: $-7g/2$

(b) Galileo seems to have been correct.

Chapter 6 Review

1. True
3. True
5. $f(x) = x^4/2 - \ln|x| + 1/x + C$
7. $f(x) = e^{\pi}x - 2/\sqrt{x} + C$
11. $\ln|x| - 1/x - 1/(2x^2) + C$
13. $G(t) = 5t + \sin t$
15. $H(r) = 4r^{3/2}/3 + r^{1/2}$
17. $H(t) = t - 2\ln|t| - 1/t$
19. $F(z) = e^z + 3z$
21. $e^x + e^{1+x} + C$
23. $P(r) = \pi r^2$
25. $(1 + \sin t)^{30}/30 + C$
27. $\sin(2x) + C$
29. $3e^x + 2^x/\ln 2 + C$
31. $\ln|x + 5| + C$
33. $2x^{3/2}/3 - \ln|x| + C$
35. -33.56 ft/sec^2

37. (a)
   (b) Highest pt: $t = 2.5$ sec
   Hits ground: $t = 5$ sec
   (c) Left sum: 136 ft
   Right sum: 56 ft
   Left sum: overestimate
   Right sum: underestimate
   (d) 100 ft

39. $D = v_0^2/(2a)$

41. (a)
   (b) $t = 6$ hours
   (c) Population is max at $t =$ 11 hours

45. (a) $a(T) = -\cos T + \frac{\sin T}{T}$
   (b) $a(t)$ increasing:
   when $a(t) < f(t)$
   $a(t)$ decreasing:
   when $a(t) > f(t)$
   $a(t)$ is max or min when $a(t) = f(t)$.

47. (a) $F(x) = x$
   (b) It must be increasing
   $F(0) = 0$, $F(1) = 1$
   concave up
   (c)

CHAPTER SEVEN

Section 7.1

1. $F(x) = 3x$
   (only possibility)
3. $F(x) = -7x^2/2$
   (only possibility)
5. $F(x) = x^3/3$
   (only possibility)
7. $F(x) = 2x + 2x^2 + \frac{5}{3}x^3$
   (only possibility)
9. $F(x) = -\cos x + 1$
   (only possibility)
11. $\frac{1}{4}x^4 + \frac{10}{3}x^{\frac{3}{2}} + 2x^{-1} + C$
13. $7t - \frac{t^9}{72} + \ln|t| + C$
15. $\sin\theta - \cos\theta + C$
17. $F'(x) = x\sin x + (x^2\cos x)/2$
    $\neq x\cos x$
19. $(\arctan 2y)' = 2/(1+4y^2)$
21. $3\ln|t| + \dfrac{2}{t} + C$
23. $\frac{2}{5}x^{\frac{5}{2}} + \frac{2}{15}x^{\frac{3}{2}} - 2\ln|x| + C$
25. $\frac{1}{2}x^2 + x + \ln|x| + C$
27. $3\sin x + 7\cos x + C$
29. $2e^x - 8\sin x + C$
31. $\frac{1}{\ln 2}2^x + C$
33. $\frac{(x+1)^4}{4} + C$
35. $\ln|x+1| + C$
37. $e^{5+x} + \frac{1}{5}e^{5x} + C$
39. (a) $\frac{1}{2}e^{2t}$
    (b) $-\frac{1}{3}e^{-3\theta}$
41. $\frac{609}{4} - 39\pi \approx 29.728$
43. $\ln 2 + \frac{3}{2} \approx 2.193$
45. 1
47. $2e - 2 = 3.437$
49. $\frac{3}{2\ln 2} \approx 2.164$
51. (a)(i) $5\cos 5x$
    (ii) $-2x\sin x^2$
    (iii) $(\cos x)e^{\sin x}$
    (iv) $-\sin x\cos(\cos x)$
    (v) $-\tan x$
    (vi) $-4x^3\tan x^4$
    (b)(i) $\frac{1}{5}\sin 5x + C$
    (ii) $-\frac{1}{2}\cos x^2 + C$
    (iii) $e^{\sin x} + C$
    (iv) $-\sin(\cos x) + C$
    (v) $-\ln(|\cos x|) + C$
    (vi) $-\frac{1}{4}\ln(|\cos x^4|) + C$
    (c)
53. $c = 6$
55. The area is $2 + \frac{\pi^3}{6} \approx 7.16771$.
57. (a)
    (b) 7 years
    (c) $69\frac{1}{3}$

Section 7.2

1. (a)(i) $2x\cos(x^2+1)$
   (ii) $3x^2\cos(x^3+1)$
   (b)(i) $\frac{1}{2}\sin(x^2+1) + C$
   (ii) $\frac{1}{3}\sin(x^3+1) + C$
   (c)(i) $-\frac{1}{2}\cos(x^2+1) + C$
   (ii) $-\frac{1}{3}\cos(x^3+1) + C$
3. $-\frac{1}{3}\cos 3x + C$
5. $e^{\sin x} + C$
7. $\frac{1}{6}\tan 2x + C$
9. For Problem 12:
   $w = -x^2$
   $dw = -2x\,dx$
   For Problem 13:
   $w = y^2 + 5$
   $dw = 2y\,dy$
   For Problem 14:
   $w = t^3 - 3$
   $dw = 3t^2\,dt$
   For Problem 15:
   $w = x^2 - 4$
   $dw = 2x\,dx$
   For Problem 16:
   $w = y + 5$
   $dw = dy$
   For Problem 17:
   $w = 2t - 7$
   $dw = 2\,dt$
   For Problem 18:
   $w = x^2 + 3$
   $dw = 2x\,dx$
   For Problem 19:
   Multiply out $(x^2+3)^2$ and then integrate.
   For Problem 20:
   $w = 4 - x$
   $dw = -dx$
   For Problem 21:
   $w = \cos\theta + 5$
   $dw = -\sin\theta\,d\theta$
   For Problem 22:

$w = x^3 + 1$
$dw = 3x^2 dx$
For Problem 23:
$w = \sin\alpha$
$dw = \cos\alpha\, d\alpha$

11. For Problem 34:
$w = \cos 2x$
$dw = -2\sin 2x\, dx$
For Problem 35:
$w = x^2 + 2x + 19$
$dw = 2(x+1)\, dx$
For Problem 36:
$w = \sin(x^2)$
$dw = 2x\cos(x^2)\, dx$
For Problem 37:
$w = 1 + 2x^3$
$dw = 6x^2\, dx$
For Problem 38:
It would be easier not to substitute.
For Problem 39:
$w = 1 + 3t^2$
$dw = 6t\, dt$
For Problem 40:
It would be easier not to substitute.

13. $\frac{1}{18}(y^2+5)^9 + C$
15. $\frac{1}{9}(x^2-4)^{\frac{9}{2}} + C$
17. $\frac{1}{148}(2t-7)^{74} + C$
19. $\frac{x^5}{5} + 2x^3 + 9x + C$
21. $-\frac{1}{8}(\cos\theta + 5)^8 + C$
23. $\frac{\sin^4\alpha}{4} + C$
25. $\frac{(\ln z)^3}{3} + C$
27. $\frac{1}{35}\sin^7 5\theta + C$
29. $\ln|e^t + t| + C$
31. $\frac{1}{2}\ln(y^2+4) + C$
33. $2e^{\sqrt{y}} + C$
35. $\frac{1}{2}\ln(x^2+2x+19) + C$
37. $\frac{1}{18}(1+2x^3)^3 + C$
39. $\frac{1}{6}\ln(1+3t^2) + C$
41. $\ln(e^x + e^{-x}) + C$
43. (a) $x^4 + 2x^2 + C$
(b) $(x^2+1)^2 + C$
(c) Both correct but they differ by a constant.
45. $h(t) = \frac{mg}{k}t - \frac{m^2 g}{k^2}(1 - e^{-\frac{k}{m}t}) + h_0$

Section 7.3

1. (a) $\frac{1}{2}\ln 2$
(b) $\frac{1}{2}\ln 2$
3. $3(e^2 - e) \approx 14.01$
5. 1
7. 0
9. 40
11. 201,760
13. $49932\frac{1}{6}$
15. $\frac{1}{3}(1 - \frac{\sqrt{2}}{2}) = 0.0976$
17. $-\frac{7}{3}\sqrt{125} = -26.087$
19. $\frac{\pi}{12} = 0.261$
21. The function has no elementary antiderivative.
Using left and right sums, $n = 100$, we know the integral is on the interval:
$[0.548, 0.556]$
23. Use appropriate substitution.
25. Use appropriate substitution.
27. $\arctan(x+2) + C$
29. $\frac{1}{2}(e^4 - 1) \approx 26.7991$
31. (a) $\frac{\sin^2\theta}{2} + C$
(b) $-\frac{\cos^2\theta}{2} + C$
(c) $-\frac{\cos 2\theta}{4} + C$

(d) The three functions above differ from each other only in terms of a constant.

33. (a) $r(t)\Delta t$
(b) $\sum_{i=0}^{n-1} r(t_i)(5/n)$, $\int_0^5 r(t)\, dt$
(c) $1000(e^{0.1} - 1) \approx 105.17$ gallons
(d) $Q(t) = 1000e^{0.02t} + 2000$

Section 7.4

1. (a)(i) $-x\sin x + \cos x$
(ii) $-2\sin 2x$
(iii) $-x^2\sin x + 2x\cos x$
(iv) $1 + \ln x$
(b)(i) $x\ln x - x + C$
(ii) $-\frac{1}{2}\cos 2x + C$
(iii) $-x\cos x + \sin x + C$
(iv) $-x^2\cos x + 2x\sin x + 2\cos x + C$
3. $\frac{1}{5}te^{5t} - \frac{1}{25}e^{5t} + C$
5. $-10pe^{(-0.1)p} - 100e^{(-0.1)p} + C$
7. $\frac{x^4}{4}\ln x - \frac{x^4}{16} + C$
9. $-t^2\cos t + 2t\sin t + 2\cos t + C$
11. $\frac{1}{4}(2z+1)e^{2z} + C$
13. $-(\theta+1)\cos(\theta+1) + \sin(\theta+1) + C$
15. $\frac{1}{6}\cos(3\alpha+1)\sin(3\alpha+1) + \frac{\alpha}{2} + C$
17. $-x^{-1}\ln x - x^{-1} + C$
19. $-2y(5-y)^{1/2} - \frac{4}{3}(5-y)^{3/2} + C$
21. $-2t(5-t)^{1/2} - \frac{4}{3}(5-t)^{3/2} - 14(5-t)^{1/2} + C$
23. $\frac{x^2}{2}(\ln x)^4 - x^2(\ln x)^3 +$

$\frac{3}{2}x^2(\ln x)^2 - \frac{3}{2}x^2\ln x +$
$\frac{3}{4}x^2 + C$

25. $\frac{1}{2}x^2\arctan x^2$
$-\frac{1}{4}\ln(1+x^4) + C$
27. $\frac{1}{2}x^2e^{x^2} - \frac{1}{2}e^{x^2} + C$
29. $\frac{1}{2}\theta - \frac{1}{4}\sin 2\theta + C$
31. $\frac{1}{2}e^x(\sin x - \cos x) + C$
33. $\frac{1}{2}xe^x(\sin x - \cos x) +$
$\frac{1}{2}e^x\cos x + C$
35. Integrate by parts choosing $u = x^n$, $v' = e^x$.
37. Integrate by parts choosing $u = x^n$, $v' = \cos ax$.
39. $5\ln 5 - 4 \approx 4.047$
41. ≈ -3.944
43. $6\ln 6 - 5 \approx 5.751$
45. $\frac{1}{2}(\frac{\pi}{4} - \frac{1}{2}\ln 2) \approx 0.219$
47. $\frac{1}{2}(\frac{\pi}{2} - 1) \approx 0.285$
49. (a) $A = \frac{a}{a^2+b^2}$, $B = -\frac{b}{a^2+b^2}$
(b) $\int e^{ax}\cos bx\,dx =$
$e^{ax}(\frac{b}{a^2+b^2}\sin bx$
$+ \frac{a}{a^2+b^2}\cos bx) + C$

Section 7.5

1. Problem 5: III-14
Problem 6: II-12
Problem 7: II-10
Problem 8: II-9
Problem 9: III-13
Problem 10: IV-23
Problem 11: IV-21 and 22
Problem 12: IV-17
Problem 13: V-24
3. Problem 26: VI-28
Problem 27: VI-28
Problem 28: V-24
Problem 29: IV-19
Problem 30: IV-19 and 20
Problem 31: IV-21
Problem 32: VI-28
Problem 33: Table not necessary.
Problem 34: IV-23 and 18
Problem 35: IV-23
Problem 36: IV-23 and 21
Problem 37: V-26
Problem 38: V-26
5. $(\frac{1}{2}x^3 - \frac{3}{4}x^2 + \frac{3}{4}x - \frac{3}{8})e^{2x}$
$+C$
7. $\frac{3}{16}\cos 3\theta\sin 5\theta$
$-\frac{5}{16}\sin 3\theta\cos 5\theta + C$
9. $\frac{1}{6}x^6\ln x - \frac{1}{36}x^6 + C$
11. $\frac{1}{2}\frac{\sin x}{\cos^2 x} + \frac{1}{4}\ln\left|\frac{\sin x+1}{\sin x-1}\right| + C$
13. $\frac{1}{\sqrt{3}}\arctan\frac{y}{\sqrt{3}} + C$
15. $(\frac{1}{3}x^4 - \frac{4}{9}x^3 + \frac{4}{9}x^2 - \frac{8}{27}x$
$+ \frac{8}{81})e^{3x} + C$
17. $-\frac{1}{2}y^2\cos 2y + \frac{1}{2}y\sin 2y$
$+ \frac{1}{4}\cos 2y + C$
19. $\frac{1}{34}e^{5x}(5\sin 3x - 3\cos 3x)$
$+C$
21. $\frac{1}{8}e^{2z^2}(\cos 2z^2 + \sin 2z^2)$
$+C$
23. $\frac{3}{8}y^8\ln y - \frac{3}{64}y^8 + C$
25. $-\frac{1}{6}\sin^2(z^2)\cos(z^2)$
$-\frac{1}{3}\cos(z^2) + C$
27. $\frac{1}{3}\arcsin 3x + C$
29. $-\frac{1}{2\tan 2\theta} + C$
31. $\frac{1}{21}\frac{\tan 7x}{\cos^2 7x} + \frac{2}{21}\tan 7x + C$
33. $-\frac{1}{4}(9 - 4x^2)^{\frac{1}{2}} + C$
35. $-\frac{1}{9}(\cos^3 3\theta) + \frac{1}{15}(\cos^5 3\theta)$
$+C$
37. $-\frac{1}{3}(\ln|z| - \ln|z-3|) + C$
39. $\arctan(z+2) + C$
41. $\arcsin\dfrac{x+1}{\sqrt{2}} + C$
43. $-\frac{1}{x+2} + C$
45. $\frac{x^2}{2} + 3x - 4\ln|x-1|$
$+ 11\ln|x-2| + C$
47. $\frac{1}{2}(x+4)\sqrt{x^2+8x+7} -$
$\frac{9}{2}\ln|x+4+\sqrt{x^2+8x+7}|$
$+C$
49. $2\ln|z-3| + 3\ln|z-2|$
$+C$
51. $\frac{\pi}{8} \approx 0.3927$
53. ≈ 0.1609
55. $\ln|2+\sqrt{5}| \approx 1.4436$
57. 0.7147
59. 0
61. ≈ 1.630
63. $\pi \approx 3.14$
65. (a) $NA + \frac{2\pi B}{1+4\pi^2}(1 - e^{-N})$
(b) $A + \frac{2\pi B}{1+4\pi^2}\left(\frac{1-e^{-N}}{N}\right)$
(c) A
(d) $R(t) \approx A$ for large t
(e) No, a real oil well will eventually dry out.
67. $2\ln|x| + \ln|x+3| + C$
69. (a) $\frac{a(1+x)^2 + bx(1+x) - cx}{x(1+x)^2}$
(b) $\ln x + \frac{2}{1+x}\Big|_1^2 = \ln 2 - \frac{1}{3}$

Section 7.6

3. On the interval, curve is
Increasing so
LEFT: underestimate
RIGHT: overestimate
Concave up so
TRAP: overestimate
MID:underestimate

5. On the interval, curve is Increasing so
   LEFT: underestimate
   RIGHT: overestimate
   Concave down so
   TRAP: underestimate
   MID:overestimate
7. On the interval, curve is Increasing so
   LEFT: underestimate
   RIGHT: overestimate
   Concave up so
   TRAP: overestimate
   MID:underestimate
9. (a)(i) LEFT(32) = 13.6961
   RIGHT(32) = 14.3437
   TRAP(32) = 14.0199
   Exact value:
   $(x \ln x - x)|_1^{10}$
   ≈ 14.02585093
   (ii) LEFT(32) = 50.3180
   RIGHT(32) = 57.0178
   TRAP(32) = 53.6679
   Exact value: $e^x|_0^4$
   ≈ 53.59815003
   (b)(i) Smallest to largest:
   LEFT(32)
   TRAP(32)
   Actual value
   RIGHT(32)
   (ii) Smallest to largest:
   LEFT(32)
   Actual value
   TRAP(32)
   RIGHT(32)
11. LEFT(6) = 31
   RIGHT(6) = 39
   TRAP(6) = 35
13. (a) 0
   (b)
   (c) MID(n) = 0 for all n.
19. RIGHT(10) = 5.556
   TRAP(10) = 4.356
   LEFT(20) = 3.199
   RIGHT(20) = 4.399
   TRAP(20) = 3.799
21. 12

Section 7.7

1. SIMP(2) $\approx$ 0.6932
   Actual value = ln 2
3. SIMP(10) $\approx$ 0.23182
5. 1.5699, $n = 10$
7. 1.0894, $n = 10$
9. 0.904524, $n = 10$
11. (a) 1
   (b)
   (c)
   (d) ratio of error:
   LEFT = 1.78,
   RIGHT = 2.18,
   TRAP = 3.96,
   MID = 3.93,
   SIMP = 15.91
   Yes.
13. (a) 4 places: 2 seconds
   8 places: $\approx$ 7 hours
   12 places: $\approx$ 6 years
   20 places: $\approx$ 600 million years
   (b) 4 places: 2 seconds
   8 places: $\approx$ 3 minutes
   12 places: $\approx$ 7 hours
   20 places: $\approx$ 6 years
15. 0.272
17. False
19. True

Section 7.8

1. $e^{-2}/2$
3. 1
5. The integral diverges.
7. $2^{\frac{3}{4}}$
9. The integral diverges.
11. The integral diverges.
13. $\pi/8$
15. The integral diverges.
17. The integral diverges.
19. $2 - 2e^{-\sqrt{\pi}}$
21. The integral diverges.
23. $\frac{1}{2}\ln\frac{5}{3}$
25. The area is infinite.
27. (a) $\ln|z-1| - \ln|z| + C$
   (b) 0, 1
   (c) $\ln\frac{3}{2}$
   (d) The integral diverges.
   (e) $\ln\frac{|x-1||a|}{|x||a-1|}$
   for:
   a, x both < 0 or both in $(0, 1)$ or both > 1
29. 9×10^9 joules
31. Converges for $p < -1$ to $-pe^{p+1}/(p+1)^2$.

Section 7.9

1. converges
3. converges
5. diverges
7. diverges
9. converges
11. converges
13. diverges
15. converges
17. 0.139
19. $a = 0.399$
21. (a) $\int_3^\infty e^{-x^2}\,dx \le \frac{e^{-9}}{3}$
   (b) $\int_n^\infty e^{-x^2}\,dx \le \frac{1}{n}e^{-n^2}$

23. Converges for $p > 1$
 Diverges for $p \leq 1$

25.(a) e^t is everywhere concave up.
 (b)
 (c)

Section 7.10

5.(a) $\mathrm{Si}(4) \approx 1.76$
 $\mathrm{Si}(5) \approx 1.55$

 (b) $\frac{\sin x}{x}$ is negative on that interval.

9. $(1+x)^{200}$

11. $-\cos(t^3)$

13. $\frac{2\sin x^2}{x}$

17.(a)
 (b) The limit does exist.

19. $\mathrm{erf}(x) + \frac{2}{\sqrt{\pi}} x e^{-x^2}$

21. $\mathrm{erf}(x/\sqrt{2})$

Chapter 7 Review

1. $\int_0^b h\,dx = hb$

3. $2\int_{-r}^{r} \sqrt{r^2 - x^2}\,dx = \pi r^2$

5.(a) Since $x = e^{\ln x}$,
 we can write:
 $x^x = (e^{\ln x})^x = e^{x \ln x}$.
 (b) $\frac{d}{dx}(x^x) = x^x(\ln x + 1)$
 (c) $x^x + C$
 (d) 3

7. $\frac{1}{2}e^{t^2} + C$

9. $-\frac{2}{9}(2 + 3\cos x)^{\frac{3}{2}} + C$

11. $\frac{2}{5}(1-x)^{\frac{5}{2}}$
 $- \frac{2}{3}(1-x)^{\frac{3}{2}} + C$

13. $-y\cos y + \sin y + C$

15. $2\sin\sqrt{y} + C$

17. $2x\ln x - 2x + C$

19. $-\frac{\cos^4 2\theta}{8} + C$

21. $-\frac{1}{3}(4 - x^2)^{3/2} + C$

23. $\tan z + C$

25. $\frac{u^2}{2} + 3u + 3\ln|u| - \frac{1}{u} + C$

27. ≈ 3.8875

29. -133.8724

31. Use appropriate substitution.

33. Usc appropriate substitution.

35.(a) $\ln(1 + e^x) + C$
 (b) $\arctan(e^x) + C$
 (c) $x - \ln(1 + e^x) + C$

37.(a) \$30,228
 (b) \$$(20{,}239(1.024)^t + 5761)$

39.(a) $f(t) = Q - \frac{Q}{A}t$
 (b) $\frac{Q}{2}$

41. $\frac{11}{9}$

43. converges
 $\int_4^\infty \frac{dt}{t^{\frac{3}{2}}} = 1$

45. diverges

47. converges
 $\int_{10}^\infty \frac{1}{z^2 - 4}\,dz = (\ln(3/2))/4 \approx 0.10$

49. converges
 $\int_{-\frac{\pi}{4}}^{\frac{\pi}{4}} \tan\theta\,d\theta = 0$

51. converges
 $\int_0^{15} \frac{dw}{\sqrt{w}} = 2\sqrt{15}$

53. $\int_0^\infty \frac{\sin^2\theta}{\theta^2 + 1}\,d\theta$ converges;
 its value is between 0 and $\frac{\pi}{2}$.

55. diverges

57.(a) $\approx 0.8862269\ldots$
 (b) $\approx 0.747402\ldots$

59. $\int_0^\infty e^{-x}\,dx = 1$
 $\int_0^\infty x e^{-x}\,dx = 1$
 $\int_0^\infty x^2 e^{-x}\,dx = 2$
 $\int_0^\infty x^3 e^{-x}\,dx = 6$
 $\int_0^\infty x^n e^{-x}\,dx = n!$

61. error for TRAP(10)
 ≈ 0.0078

63.(a) 0
 (b) Areas above and below t-axis are equal.
 (c) $F(x) \geq 0$ everywhere.
 $F(x) = 0$ only at integer multiples of π.

65. Global max: at $x = \pi$
 Global min: at $x = \pi/2$

CHAPTER EIGHT

Section 8.1

1. (a) $\sum_{i=1}^{N} \rho(x_i)\Delta x$
   (b) 16 grams
3. (a)
   (b) $\sum_{i=1}^{N}[600 + 300\sin(4\sqrt{x_i} + 0.15)]\frac{20}{N}$
   (c) ≈ 11513
5. (a) 3.96 million people
   (b) underestimate
   (c) $\rho(r) = 82.5 - 7.5r$
   4.27 million people
7. (a) $\sum_{k=0}^{n-1} \frac{5\Delta x}{1+x_k^4}$
   (b) 5.5
9. $M_U \approx 6.50 \times 10^{27}$ g
   $M_L \approx 5.46 \times 10^{27}$ g
11. 5π cubic inches/second

Section 8.2

3. $V = \pi$
5. $V = \frac{\pi^2}{2} \approx 4.935$
7. $V = (8/15)\pi \approx 1.68$
9. $V = \frac{\pi(e^2-1)}{2} \approx 10.036$
11. $V \approx 65.54$
13. $V = \frac{\pi(e^2-1)}{16} \approx 1.25$
15. 2267.32 cubic feet
17. 2595 cubic feet
19. 3.526
21. 2.35
23. length $= \int_0^2 \sqrt{\frac{16-3x^2}{16-4x^2}}\, dx$
   (first quadrant)
25. (a)
   (b)
   (c) $f(x) = \sqrt{3}x$

Section 8.3

1. 1,058,591.1 ft-lb.
3. 1,404,000 ft-lb
5. 180,000 ft-lb
7. 1,418,693 ergs
   (about 0.142 joules)
9. $v = k\sqrt{\rho}R$
   where $k = \sqrt{\frac{8\pi G}{3}}$.
11. (a) Force on the Dam $\approx$
   $\sum_{i=0}^{N-1} 1000(62.4h_i)\Delta h$
   (b) $\int_0^{50} = 1000(62.4h)\,dh =$
   78,000,000 pounds
13. potential
   $= 2\pi\sigma(\sqrt{R^2+a^2} - R)$
15. 60 joules
17. $\frac{GmM}{a(a+l)}$

Section 8.4

3. Future value = \$72,980.16
   Present value = \$29,671.52
5. (a) \$5820 per year
   (b) \$36,787.94
7. (a) 8.11 years
   (b) 6.203 years
9. (a) $\sum_{i=0}^{n-1}(2000 - 100t_i) \times e^{-0.1t_i}\Delta t$
   (b) $\int_0^T e^{-0.10t} \times (2000 - 100t)\,dt$
   (c) After 20 years
   \$11,353.35
11. \$85,750,000

Section 8.5

7. (a) cumulative distribution function
   (b)
   (c) More than 50%: 1%
   Between 20% and 50%: 49%
   Most likely: $C = 28\%$
9. (a) $\frac{2}{3}$
   (b) $\frac{1}{3}$
   (c) Possibly because many students work to get just a passing grade.
   (d)

Section 8.6

3. (a) $p(x) = \frac{e^{-\frac{1}{2}\left(\frac{x-100}{15}\right)^2}}{15\sqrt{2\pi}}$
   (b) 6.7% of the population
7. (a) $a = 0.122$
   (b) $P(x) = 1 - e^{-0.122x}$
   (c) Median = 5.68 seconds
   Mean = 8.20 seconds
   (d)
9. (a) $a = \frac{2}{\sqrt{2\pi}}(\frac{m}{kT})^{3/2}$
   $\approx 3.4 \times 10^{-8}$ SI units
   (b) Median ≈ 441 m/sec

Mean ≈ 457.7 m/sec
Max ≈ 0.002
(for $v \approx 405$ m/sec)
(c) If T increases, the mean and median also increase, whereas the max decreases.

Chapter 8 Review

1.(a)
(b) $\sum_{i=1}^{N} \pi x_i \Delta x$
(c) Volume $= \frac{\pi}{2}$

3.(a) $a = b/l$
(b) $\frac{1}{3}\pi b^2 l$

5.(a) $\sum_{i=1}^{N} \pi \frac{9x_i}{4} \Delta x$
(b) 18π

7. Volume $= \frac{Ah}{3}$

9. 4.79

11.(a) $f(x) = 0$
(b) $f(x) = \sqrt{a^2 - 1}\, x$
if $a \geq 1$;
no such function for $a < 1$

13. $2\int_{-a}^{a} \sqrt{1 + \left(\frac{b^2x^2}{a^2(a^2-x^2)}\right)}\, dx$

15.(a) Future value $=$ \$6389.06
Present value $=$ \$864.66
(b) 17.92 years

19.(a)
(b)
(c) μ represents the mean of the distribution, while σ is the standard deviation.

21. 4400 ft-lb

23. 784.14 ft-lb

25. Bottom: 618.3 pounds
Sides : 794.14 pounds

27.(a) $y = h + (2 - 2h)x^2$
(b) The two events happen simultaneously.

29. The thin spherical shell.

31.(a) 1321.4 meters
(b) 144.2 meters

33. $V = \frac{1}{32}\pi^2(D^2 - d^2)(D - d)$

CHAPTER NINE

Section 9.1

1. (a) (III)
   (b) (IV)
   (c) (I)
   (d) (II)
5. $\omega = \pm 3$
9. (a) (IV)
   (b) (III)
   (c) (III), (IV)
   (d) (II)
11. (a) (I),(III)
   (b) (IV)
   (c) (II),(IV)

Section 9.2

1. (a)
   (b) $y = -x - 1$
   (c)
3. (a)
   (b) $y = x - 1$
5. The first graph.
9. For starting points $y > 0$:
   $y \to \infty$ as $x \to \infty$
   For starting points $y = 0$:
   $y = 0$ as $x \to \infty$
   For starting points $y < 0$:
   $y \to -\infty$ as $x \to \infty$
11. As $x \to \infty$, $y \to \infty$
13. $y \to 4$ as $x \to \infty$.

Section 9.3

3. (a)
   (b) At $t = 2$, $y \approx 0.693$
   (c) Bigger.
5. (a) At $x = 1$, $y \approx 1.5$
   (b) At $x = 1$, $y \approx 1.75$
   (c) $y = x^2 + 1$
   (d) yes
7. Because the curve is concave up
9. (a) $B \approx 1050$
   (b) $B \approx 1050.63$
   (c) $B \approx 1050.94$
   (d)
   (e)
11. For $x = 7.1$,
   $y \approx 2.039456$
   For $x = 7.2$
   $y \approx 2.076923$
   The approximation is within 0.001 of the actual answer.

Section 9.4

1. $P = 20e^{0.02t}$
3. $m = 5e^{3t-3}$
5. $y = 10e^{-\frac{1}{3}x}$
7. $P = 104e^t - 4$
9. $m = 3000e^{0.1t} - 2000$
11. $z = -\ln(1 - \frac{t^2}{2})$
13. $y = -\frac{2}{t^2+2t-4}$
15. $w = \frac{2}{\cos\theta^2+1}$
17. (a) $y = 100 - Ae^{-t}$
   (b)
   (c) $y = 100 - 100e^{-t}$
19. $Q = Ae^{\frac{1}{k}t}$
21. $Q = b - Ae^{-t}$
23. $R = -\frac{b}{a} + Ae^{at}$
25. $x = \arcsin(At^{\ln t+1})$
27. $y = 2(2^{-e^{-t}})$
29. (a)
   (b)
   (c) $y = \frac{A}{x}$

Section 9.5

1. (a)
   (b) 2001
3. (a) (I)
   (b) (IV)
   (c) (II) and (IV)
   (d) (II) and (III)
5. (a) $\frac{dM}{dt} = rM$
   (b) $M = 1000e^{rt}$
   (c)
9. Longest: Lake Superior
   Shortest: Lake Erie
   The ratio is 75
11. (a) $\frac{dy}{dt} = ky$
   (b) 0.249 grams
13. (a) $I = Ae^{-kl}$
   (b) 20 feet: 75%
   25 feet: 82.3%
15. $\frac{dS}{dt} = -k(S - 65)$
   $S = 65 - 25e^{-kt}$
17. $dQ/dt = -0.0001Q$;
   $Q = Q_0e^{-0.0001t}$
19. about 2150 B.C.

Section 9.6

1. If D = quantity of dead leaves (in grams/cm^2), then $\frac{dD}{dt} = -0.75(D-4)$ Equilibrium level = 4 g/cm^2
3. (a) $c'(t) = a(k - c(t))$ $(a > 0)$
   (b) $c = k + (c_0 - k)e^{-at}$
   (c)
5. about 3 days
7. (a) $dx/dt = k(a-x)^2$
   (b) $x = a^2kt/(akt+1)$
11. (a) $\frac{dy}{dt} = -k(y-a)$
    (b) $y = (1-a)e^{-kt} + a$
    (c) a represents the fraction of material which is remembered in the long run. The constant k tells us about the rate at which material is forgotten.
13. (a) $Q = \frac{r}{\alpha}(1 - e^{-\alpha t})$ $Q_\infty = \frac{r}{\alpha}$
    (b) Doubling r doubles Q_∞. Altering r doesn't alter the time it takes to reach $\frac{1}{2}Q_\infty$.
    (c) Both Q_∞ and the time t are halved by doubling α.
15. $t \approx 11$ hr and 13 min
17. (a) $\frac{dB}{dt} = 0.10B + 1000$
    (b) $B = 10000e^{0.1t} - 10000$
    (c) $f(t) = \int_0^t 1000e^{0.1(t-x)}\,dx$
    (d) $10000e^{0.1t} - 10000$
    (e) $B = (B_0 + 10000)e^{0.1t} - 10000$.
    (f) The integral in part (d) stays the same.
21. (a)
    (b) Overestimate
23. (a)
    (b)
    (c)
    (d) $p = Ax^{k/(1-k)}$
    (e)

Section 9.7

1. $P = (6.6 \times 10^6)e^{0.002t}$
3. (a) $\frac{dp}{dt} = kp(B-p)$ $(k > 0)$
   (b) Half of the tin
   (c)
5. (a) $\frac{dI}{dt} = k(M-I)$ $(k > 0)$
   (b) $\frac{dI}{dt} = kI(M-I)$ $(k > 0)$
7. (a) logistically
   (b) $k \approx 0.045$ $L \approx 5.8$
   (c)
9. (a) $a = 0.0316$ $b = 0.00012$
   (b) In the year 2060
   (c) $P = 5.3e^{0.0316t - 0.00006t^2}$
15. dP/dt is largest in 1920. Using this population to estimate the 1990 population, we get 211.4 million.
17. (a)
    (b)
    (c)
    (d) $P > 6$: increase without bound
    $P < 6$: extinction
19. (a)
    (b)
    (c)
    (d) Stable equilibrium: $P = 150$ Unstable: $P = 50$
21. (a)
    (b) $H = 75$ and $H = 100$
    (c) $H \leq 100$
    (d)

Section 9.8

3. Initially $x = 0$, so we start with only y. Then y decreases while x increases. Then x continues to increase while y starts to increase as well. Finally y continues to increase while x decreases.
5. The closed trajectory represents populations which oscillate repeatedly.
7. $r = r_0e^{-t}$, $w = w_0e^t$
9. Worms decrease, robins increase. In the long run the populations will oscillate (and tend to (2,2) asymptotically).
13. (a) $dy/dx = x/5y$
    (b) Predicted outcome: US victory
    (c) Reinforcements do not alter predicted outcome (US victory); they alter trajectory and number of surviving troops.
15. (a) $x^2 - 5y^2 = 604.75$.
    (b) about 25,000 troops
17. (a) In the absence of the other, each company grows exponentially. The two companies restrain each other's growth if they are both present.
    (b) (0,0) and (1,2)
    (c) In the long run, one of the companies will go out of business.

19. (a) x population grows exponentially
y population grows exponentially
(b) competitor relationship

21. (a) (0,0) and (2,5)
(b) $\frac{dy}{dt} = 0$ but $\frac{dx}{dt} \neq 0$.

23. (a) symbiosis
(b) Both $\to \infty$
or both $\to 0$

25. (a) predator-prey
(b) x, y tend to ≈ 1

Section 9.9

1. (a) $\frac{dS}{dt} = 0$ where
$S = 0$ or $I = 0$.
$\frac{dI}{dt} = 0$ where
$I = 0$ or $S = 192$.
Thus every point on the S axis is an equilibrium point
(b) Where $S > 192$,
$\frac{dS}{dt} < 0$ and $\frac{dI}{dt} > 0$.
Where $S < 192$,
$\frac{dS}{dt} < 0$ and $\frac{dI}{dt} < 0$.
(c) If the trajectory starts with $S_0 > 192$, then I increases to a maximum when $S = 192$. If $S_0 < 192$, then I always decreases. Regardless of the initial conditions, the trajectory always goes to a point on the S-axis (where $I = 0$). The S-intercept represents the number of students who never get the disease.

3. Vertical nullclines occur where $dx/dt = 0$, which happens when $x = 0$ or $x+y = 2$. Horizontal nullclines occur where $dy/dt = 0$, which happens when when $y = 0$ or $x + y = 1$.
Equilibrium points:
$(0, 0)$, $(0, 1)$, and $(2, 0)$.
No matter in what sector the initial point lies, the trajectory will head toward the equilibrium point $(2, 0)$.

5. $\frac{dx}{dt} = 0$ when
$x = 0$ or $y + \frac{x}{2} = 1$.
$\frac{dy}{dt} = 0$ when
$y = 0$ or $x + \frac{y}{3} = 1$.
Equilibrium points:
$(2, 0), (0, 3), (0, 0), (\frac{4}{5}, \frac{3}{5})$

7. (a) $\frac{dP_1}{dt} = 0$ where
$P_1 = 0$ or
$50 - 2.5P_1 - 500P_3 = 0$.
$\frac{dP_3}{dt} = 0$ where
$P_3 = 0$ or
$60 - 10P_1 - 6P_3 = 0$.
Equilibrium points:
$(0, 0)$,
$(20, 0)$: extinction of P_3,
$(0, 10)$: extinction of P_1,
$(6, 0.07)$ ((approximate)
(b) The equilibrium at $P_1 \approx 6$, $P_3 \approx 0.07$ is unstable, because if the populations are near, but not at, these values, the populations generally tends away from this point.
The equilibriums at $P_1 = 20$, $P_3 = 0$ and $P_1 = 0$, $P_3 = 10$ are both stable. Thus, in the long run, the populations will tend towards the values at one of these two equilibrium points, meaning that one of the two populations will die out. However, which one of the two points the trajectory tends towards, and which population becomes extinct, depend upon the initial populations of the two parasites.

9. (a) $\frac{dP_1}{dt} = 0$ where
$P = 0$ or $P_1 + 3P_2 = 13$.
$\frac{dP_2}{dt} = 0$ where
$P = 0$ or $P_2 + 0.4P_1 = 6$.
(b) The phase plane shows that P_2 will eventually exclude P_1 regardless of where the experiment starts so long as there were some P_2 originally. Consequently the data points would have followed a trajectory that starts at the origin, crosses the first nullcline and goes left and upwards between the two nullclines to the point $P_1 = 0$, $P_2 = 6$.

Section 9.10

3. $A \approx -1.705$
$\alpha = \pm\sqrt{5}$

5. Highest point, at rest

7. Lowest point, at rest

9. (a) goes with (II)
(b) goes with (I) and (IV)
(c) goes with (III)
(I) $x = 2\sin 2t$
(II) $x = -\sin t$
(III) $x = \cos 4t$
(IV) $x = -3\sin 2t$

11. (a) $x = v_0\sqrt{\frac{l}{g}}\sin\sqrt{\frac{g}{l}}t$
(b) $x = x_0 \cos\sqrt{\frac{g}{l}}t$

13. $\sqrt{58}$

15. (a) $A\sin(\omega t + \varphi) =$
$(A\sin\varphi)\cos\omega t +$
$(A\cos\varphi)\sin\omega t$

(b)

17. $C = \frac{1}{20}$ farads

Section 9.11

3.(a) $d^2u/dt^2 + 4u = 0$
(b) $u = C_1 \cos 2t + C_2 \sin 2t$
$s = \frac{C_1 \sin 2t}{2} - \frac{C_2 \cos 2t}{2}$

5. $s = 1, v = 0.75$

Section 9.12

1. $y(t) = C_1 e^{-t} + C_2 e^{-3t}$
3. $y(t) = C_1 e^{-2t} \cos t + C_2 e^{-2t} \sin t$
5. $s(t) = C_1 \cos \sqrt{7}t + C_2 \sin \sqrt{7}t$
7. $x(t) = C_1 e^{-2t} \cos 2t + C_2 e^{-2t} \sin 2t$
9. $y(t) = \frac{5}{4}e^{-t} - \frac{1}{4}e^{-5t}$
11. $y(t) = 2e^{-3t} \sin t$
13. $p(t) = 20e^{\frac{\pi}{2}-t} \sin t$
15.(a) (IV)
(b) (II)
(c) (I)
(d) (III)
17. (iv)
19. (iii)
21. overdamped: $c < 4$
critically damped: $c = 4$
underdamped: $c > 4$
23. overdamped: $c < 9$
critically damped: $c = 9$
underdamped: $c > 9$
25. $z(t) = 3e^{-2t}$
29.(a) $d^2y/dt^2 - y = 0$
(b) $y = C_1 e^t + C_2 e^{-t}$, so $x = C_2 e^{-t} - C_1 e^t$
31.(a) $Q(t) = 2te^{-\frac{1}{2}t}$
(b) $Q(t) = (2 + t)e^{-\frac{1}{2}t}$

(c) The resistance was decreased by exactly the amount to switch the circuit from the overdamped case to the critically damped case.

33. As $t \to \infty$, $Q(t) \to 0$

Chapter 9 Review

1. $P = \frac{40000}{3}(e^{0.03t} - 1)$
3. $f(x) = (\frac{1}{3}x^{\frac{3}{2}} + \frac{2}{3})^2$ (for $x \geq 0$)
5. $y = \ln(e^x + e - 1)$
7. $y = \sqrt[3]{33 - 6 \cos x}$
9. $y = \frac{2}{x^2+1}$
11. $z = \sin(-e^{\cos\theta} + \frac{\pi}{6} + e)$
13. $-y - \ln|1 - y| = \arctan t - \frac{\pi}{4}$
(We cannot solve for y in terms of t.)
15. $500 \ln|y| + \frac{5}{2}y = 18 \ln|x| + \frac{1}{10}x + 482 \ln 10 + 24$
17. $y = \pm\sqrt{Ae^{-\frac{4}{3}\sqrt{x^3+1}} - 1}$ $(A > 1)$
19. $-\frac{a^2}{2}e^{-\frac{x^2}{a^2}} = \frac{y^2}{2}\ln y - \frac{y^2}{4} + C$
23.(a) $(\frac{1}{2}, \frac{1}{2})$
(b) $(0.5, 0.516)$
(c) $(0.5, 0.530)$
(d) $y = \arcsin\left(\frac{1}{2}\ln\left|\frac{(\sin x)+1}{(\sin x)-1}\right|\right)$
When $x = \frac{1}{2}$, $y \approx 0.549$. The answers in (a)–(c) are all underestimates.
25. $y(t) = C_1 \cos\left(\frac{1}{3}t\right) + C_2 \sin\left(\frac{1}{3}t\right)$
27. $y(t) = C_1 e^{-t} \cos 3t + C_2 e^{-t} \sin 3t$
29. Always overdamped
31.(a) $\frac{dQ}{dt} = -0.5365Q$
$Q = Q_0 e^{-0.5365t}$
(b) 4 mg
33.(a) $dB/dt = 0.1B + 1200$
(b) $B = 12000(e^{0.1t} - 1)$
(c) 7784.66
35.(a) $\frac{dP}{dt} = \frac{P}{999}(10{,}000 - P)$
(b) Maximum growth rate: $\frac{dP}{dt} = 25{,}025$ occurs when $P = 5000$.
(c)
(d) $P = \frac{100{,}000}{10+(9990)e^{-(10{,}000/999)t}}$
37.(a)
(b) $dx/dt = -xy$, $dy/dt = -x$
(c) $dy/dx = 1/y$
soln: $y^2/2 = x + C$
(d) if $C > 0$, y wins
if $C < 0$, x wins
if $C = 0$, mutual annihilation
(e)
39.(a) $-0.2x$: negative sign: the more a country spends on armaments, the less it wants to increase military spending.
$+0.15y$: positive sign: the more a country's opponent arms, the more more the country will arm itself.
$+20$: positive sign: if both are initially unarmed, the country will start to arm.
(b) eq. point: $(400, 400)$
(c)
(d) stable

(e) they will rearm
(f) it will rearm
(g) both countries will spend near to $400 billion no matter where they start.

CHAPTER TEN

Section 10.1

1. $P_2(x) = 1 - \frac{x^2}{2!}$
   $P_4(x) = 1 - \frac{x^2}{2!} + \frac{x^4}{4!}$
   $P_6(x) = 1 - \frac{x^2}{2!} + \frac{x^4}{4!} - \frac{x^6}{6!}$
3. $P_3(x) = 1 + x + x^2 + x^3$
   $P_5(x) = 1 + x + x^2 + x^3 + x^4 + x^5$
   $P_7(x) = 1 + x + x^2 + x^3 + x^4 + x^5 + x^6 + x^7$
5. $P_3(x) = P_4(x) = x + \frac{1}{3}x^3$
7. $P_5(x) = x - \frac{x^2}{2} + \frac{x^3}{3} - \frac{x^4}{4} + \frac{x^5}{5}$
   $P_7(x) = x - \frac{x^2}{2} + \frac{x^3}{3} - \frac{x^4}{4} + \frac{x^5}{5} - \frac{x^6}{6} + \frac{x^7}{7}$
   $P_9(x) = x - \frac{x^2}{2} + \frac{x^3}{3} - \frac{x^4}{4} + \frac{x^5}{5} - \frac{x^6}{6} + \frac{x^7}{7} - \frac{x^8}{8} + \frac{x^9}{9}$
9. $P_2(x) = 1 - \frac{1}{2}x + \frac{3}{8}x^2$
   $P_3(x) = 1 - \frac{1}{2}x + \frac{3}{8}x^2 - \frac{5}{16}x^3$
   $P_4(x) = 1 - \frac{1}{2}x + \frac{3}{8}x^2 - \frac{5}{16}x^3 + \frac{35}{128}x^4$
11. Take $p = 1/2$.
    Take $p = -1$ and replace x by $-x$.
    Take $p = -1$.
    Take $p = 1/3$ and replace x by $-x$.
    Take $p = -1/2$
13. $P_4(x) = -\left(x - \frac{\pi}{2}\right) + \frac{1}{3!}\left(x - \frac{\pi}{2}\right)^3$
15. $P_3(x) = \frac{\sqrt{2}}{2}[1 - \left(x - \frac{\pi}{4}\right) - \frac{1}{2}\left(x - \frac{\pi}{4}\right)^2 + \frac{1}{6}\left(x - \frac{\pi}{4}\right)^3]$
17. $P_3(x) = 1 - \frac{x}{2} - \frac{x^2}{8} - \frac{x^3}{16}$
19. $P_4(x) = \frac{1}{3}[1 - \frac{x-2}{3} + \frac{(x-2)^2}{3^2} - \frac{(x-2)^3}{3^3} + \frac{(x-2)^4}{3^4}]$
21. $P_6(x) = (x-1) - \frac{(x-1)^2}{2} + \frac{(x-1)^3}{3} - \frac{(x-1)^4}{4} + \frac{(x-1)^5}{5} - \frac{(x-1)^6}{6}$
23. $c < 0, b > 0, a > 0$
25. $a < 0, b > 0, c > 0$
29. (a) $\frac{1}{2}$
    (b) $\frac{1}{6}$
31. $P_2(x) = 4x^2 - 7x + 2$
    $f(x) = P_2(x)$
33. (a) If $f(x)$ is a polynomial of degree n, then $P_n(x)$, the nth degree Taylor polynomial for $f(x)$ about $x = 0$, is $f(x)$ itself.
    (b)

Section 10.2

1. Yes
3. No
5. Yes
7. $f(x) = 1 + \frac{x}{2} - \frac{x^2}{8} + \frac{x^3}{16} + \cdots$
9. $f(t) = -t - \frac{t^2}{2!} - \frac{2t^3}{3!} - \cdots$
11. $f(y) = 1 - \frac{y}{3} - \frac{y^2}{9} - \frac{5y^3}{81} - \cdots$
13. $\frac{1}{x} = -1 - (x+1) - (x+1)^2 - (x+1)^3 - \cdots$
15. $\sin x = \frac{\sqrt{2}}{2} + \frac{\sqrt{2}}{2}\left(x - \frac{\pi}{4}\right) - \frac{\sqrt{2}}{4}\left(x - \frac{\pi}{4}\right)^2 - \frac{\sqrt{2}}{12}\left(x - \frac{\pi}{4}\right)^3 - \cdots$
17. $\sin\theta = -\frac{\sqrt{2}}{2} + \frac{\sqrt{2}}{2}\left(\theta + \frac{\pi}{4}\right) + \frac{\sqrt{2}}{4}\left(\theta + \frac{\pi}{4}\right)^2 - \frac{\sqrt{2}}{12}\left(\theta + \frac{\pi}{4}\right)^3 + \cdots$
19. $\frac{1}{6}$
21. $\frac{1}{2}$
23. -1
25. (a) $\ln(1 + 2x) = 2x - 2x^2 + \frac{8}{3}x^3 + \cdots$
    (b) To get the expression for $\ln(1 + 2x)$ from the series for $\ln(1 + x)$, substitute $2x$ for x in the series.
    (c) $-\frac{1}{2} < x < \frac{1}{2}$
27. $-1 < x < 1$
29. $-0.25 < x < 0.25$
31. $\frac{d}{dx}(x^2 e^{x^2})|_{x=0} = 0$
    $\frac{d^6}{dx^6}(x^2 e^{x^2})|_{x=0} = \frac{6!}{2} = 360$

Section 10.3

1. $e^{-x} = 1 - x + \frac{x^2}{2!} - \frac{x^3}{3!} + \cdots$
3. $\cos(\theta^2) = 1 - \frac{\theta^4}{2!} + \frac{\theta^8}{4!} - \frac{\theta^{12}}{6!} + \cdots$
5. $\frac{t}{1+t} = t - t^2 + t^3 - t^4 + \cdots$
7. $\arcsin x = x + \frac{1}{6}x^3 + \frac{3}{40}x^5 + \frac{5}{112}x^7 + \cdots$
9. $\phi^3\cos(\phi^2) = \phi^3 - \frac{\phi^7}{2!} + \frac{\phi^{11}}{4!} - \frac{\phi^{15}}{6!} + \cdots$
11. $e^t \cos t = 1 + t - \frac{t^3}{3} - \frac{t^4}{6} + \cdots$
13. $\frac{1}{2+x} = \frac{1}{2}(1 - \frac{x}{2} + (\frac{x}{2})^2 - (\frac{x}{2})^3 + \cdots)$
15. From smallest to largest:
    $1 - \cos y$
    $\ln(1 + y^2)$
    $\sin(y^2)$
17. (a) I
    (b) IV
    (c) III
    (d) II
21. $E \approx \frac{kQ}{R^2}\left(\frac{4}{R} + \frac{8}{R^3}\right)$
25. (a) At $r = a$, the force between the atoms is 0.
    (b) There will be an attractive force, pulling them back together.
    (c) There will be a repulsive force, pushing the atoms apart.
    (d) $F = F(a) + F'(a)(r-a) + F''(a)\frac{(r-a)^2}{2!} + \cdots$
    (e)
27. $y(x) \approx 1 + x + x^2/2 + x^3/3 + x^4/4$
29. $f(\theta) \approx \theta - \theta^3/6 + \theta^5/120$
31. $y(t) \approx 2 + t - t^2 - t^3/3 + t^4/4$

Section 10.4

1. Yes, $a = 2$, ratio $= 1/2$.
3. No. Ratio between successive terms is not constant.
5. Yes, $a = 1$, ratio $= -x$.
7. Yes, $a = y^2$, ratio $= y$.
9. Yes, $a = 1$, ratio $= 2z$.
11. $1/(1+x)$, $|x| < 1$
13. $\frac{1}{1+y^2}$, $|y| < 1$
15. $\frac{3(2^{11}-1)}{2^{10}}$
17. $\frac{1}{54}$
19. (a) $0.232323\ldots = 0.23 + 0.23(0.01) + 0.23(0.01)^2 + \cdots$
    (b) $\frac{0.23}{1-0.01} = \frac{23}{99}$
21. (a) $P_n = 250(0.04) + 250(0.04)^2 + 250(0.04)^3 + \cdots + 250(0.04)^{n-1}$
    (b) $P_n = 250\frac{0.04(1-(0.04)^{n-1})}{1-0.04}$
    (c) $\lim_{n\to\infty} P_n \approx 10.42$
    We'd expect the difference between them to be 250 mg.
23. (a) $h_n = 10\left(\frac{3}{4}\right)^n$
    (b) $D_1 = 10$ feet
    $D_2 = h_0 + 2h_1 = 25$ feet
    $D_3 = h_0 + 2h_1 + 2h_2 = 36.25$ feet
    $D_4 = h_0 + 2h_1 + 2h_2 + 2h_3 \approx 44.69$ feet
    (c) $D_n = 10 + 60\left(1 - \left(\frac{3}{4}\right)^{n-1}\right)$
25. 22.3 million dollars
27. (a) \$16.43 million
    (b) \$24.01 million
    (c) \$16.87 million
29. £250
31. \$1081.11
33. (a) \$1250
    (b) 12.50
35. (a) $S = \frac{p}{1-(1-p)(1-q)}$
    (b)(i) 0.80
    (ii) 0.9

Section 10.5

1. (a) underestimate
   (b) $|E_0| \le 0.8666$
3. (a) overestimate: $0 < \theta \le 1$
   underestimate: $-1 \le \theta < 0$
   (b) $|E_1| \le 0.5$
5. $|E_3| \le 16.5$
7. $|E_4| \le 0.016$
9. For $\sin x$ and $\cos x$, $|E_n| \le \frac{1}{(n+1)!}$
15. $|E_0| < 0.01$ for $|x| \le 0.1$

Section 10.6

1. Fourier series
3. Not a Fourier series.
5. (a)
   (b) $\frac{1}{7}\sin 7x$
   (c) The equation is
   $f(x) = \ldots$
   1 for $2\pi \le x < -\pi$
   -1 for $-\pi \le x < 0$
   1 for $0 \le x < \pi$
   -1 for $\pi \le x < 2\pi$
   and so on. The square wave

function is not continuous at $x = 0, \pm\pi, \pm 2\pi, \ldots$

7. $F_1 = F_2 = \frac{\pi}{2} - \frac{4}{\pi}\cos x$
   $F_3 = \frac{\pi}{2} - \frac{4}{\pi}\cos x - \frac{4}{9\pi}\cos 3x$

9. $F_n(x) =$
   $\frac{\pi^2}{3} + \sum_{i=1}^{n}(-1)^i \frac{4}{i^2}\cos(ix)$

13. $F_4(x) = \frac{1}{2} - \frac{2}{\pi}\sin(2\pi x) - \frac{1}{\pi}\sin(4\pi x) - \frac{2}{3\pi}\sin(6\pi x) - \frac{1}{2\pi}\sin(8\pi x)$

15. (a) $F_3(x) =$
   $\frac{1}{2} + \frac{2}{\pi}\cos x - \frac{2}{3\pi}\cos 3x.$
   (b) There are cosines instead sines, but the energy spectrum remains the same.

17. 99.942% of the total energy

21. (a) 15.9155%, 0.451808%
   (b) $(4\sin^2 \frac{k}{2})/(k^2\pi^2)$
   (c) The constant term and the first five harmonics are needed.
   (d) $F_5(x) = \frac{1}{2\pi} + \frac{2\sin(\frac{1}{2})}{\pi}\cos x + \frac{\sin 1}{\pi}\cos 2x + \frac{2\sin(\frac{3}{2})}{3\pi}\cos 3x + \frac{\sin 2}{2\pi}\cos 4x + \frac{2\sin(\frac{5}{2})}{5\pi}\cos 5x$

23. The energy of the pulse train is spread out over more harmonics as c gets closer to 0.

Chapter 10 Review

1. $\sin t^2 =$
   $t^2 - \frac{t^6}{3!} + \frac{t^{10}}{5!} - \frac{t^{14}}{7!} + \cdots$

3. $\frac{1}{1-4z^2} =$
   $1 + 4z^2 + 16z^4 + 64z^6 + \cdots$

5. $\frac{a}{a+b} = 1 - \frac{b}{a} + (\frac{b}{a})^2 - (\frac{b}{a})^3 + \cdots$

7. $\sin x \approx -\frac{1}{\sqrt{2}} + \frac{1}{\sqrt{2}}(x + \frac{\pi}{4}) + \frac{1}{2\sqrt{2}}(x + \frac{\pi}{4})^2$

9. $\ln x \approx$
   $\ln 2 + \frac{1}{2}(x-2) - \frac{1}{8}(x-2)^2$

11. (a) $\lim_{\theta\to 0}\frac{\sin 2\theta}{\theta} = 2.$
   (b) Near $\theta = 0$, we make the approximation $\frac{\sin 2\theta}{\theta} \approx 2 - \frac{4}{3}\theta^2.$

13. (a) $f(t) = t + t^2 + \frac{t^3}{2!} + \frac{t^4}{3!} + \cdots$
   (b) $\int_0^x f(t)\,dt = \frac{x^2}{2} + \frac{x^3}{3} + \frac{x^4}{4\cdot 2!} + \frac{x^5}{5\cdot 3!} + \cdots$
   (c) Substitute $x = 1$, and integrate by parts.

15. (a) $-c < v < c$
   (b) y-intercerpt: $(0, m_0)$
   asymptotes: $x = \pm c$
   (c) $m =$
   $m_0(1 + \frac{1}{2}\frac{v^2}{c^2} + \frac{3}{8}\frac{v^4}{c^4} + \cdots)$
   (d) When $|v| < c$

17. (a) Set $\frac{dV}{dr} = 0$, solve for r. Check for max or min.
   (b) $V(r) = -V_0 +$
   $72V_0 r_0^{-2}\cdot(r - r_0)^2\cdot\frac{1}{2} + \cdots$
   (c)
   (d) $F = 0$ when $r - r_0$

21. (a) $g(x) \approx P_n(x) =$
   $g(0) + \frac{g''(0)}{2!}x^2 + \frac{g'''(0)}{3!}x^3 + \cdots + \frac{g^{(n)}(0)}{n!}x^n$
   (b) If $g''(0) > 0$:
   0 is a local minimum
   If $g''(0) < 0$:
   0 is a local maximum.
   (c)

23. (a) $\approx 1\%$
   (b) $Q_1 = 250$
   $Q_2 = 250 + 250(0.01)$
   $Q_3 = 250 + 250(0.01) + 250(0.01)^2$
   $Q_4 = 250 + 250(0.01) + 250(0.01)^2 + 250(0.01)^3$
   (c) $Q_3 = \frac{250(1-(0.01)^3)}{1-0.01} \approx 252.5$
   $Q_4 = \frac{250(1-(0.01)^4)}{1-0.01} \approx 252.5$
   (d) $Q_n = 250 + 250(0.01) + 250(0.01)^2 + \cdots + 250(0.01)^{n-1} = \frac{250(1-(0.01)^n)}{1-0.01}$
   (e) $Q = \lim_{n\to\infty} Q_n = \frac{250}{1-0.01} = 252.5$

25. 5.7%

27. $|E_3| \leq 0.039$

29. $F_3(x) =$
   $\frac{e^{2\pi}-1}{\pi}(\frac{1}{2} + \frac{1}{2}\cos 2\pi x - \frac{1}{2}\sin 2\pi x + \frac{1}{5}\cos 4\pi x - \frac{2}{5}\sin 4\pi x + \frac{1}{10}\cos 6\pi x - \frac{3}{10}\sin 6\pi x)$

31. $\frac{1}{2} - \frac{4}{\pi^2}\cos \pi x - \frac{4}{9\pi^2}\cos 3\pi x$

APPENDIX

Section A

1.(a) $y \leq 30$
  (b) two zeros

3. -1.05

5. 2.5

7. $x = 1.05$

9. 1.15

11. 1.3

13.(a) $x = -1.15$
  (b) $x = 1, x = 1.41,$ and $x = -1.41$

15.(a) $x \approx 0.7$
  (b) $x \approx 0.4$

17.(a) 4 zeros
  (b) $[0.65, 0.66], [0.72, 0.73], [1.43, 1.44], [1.7, 1.71]$.
  (c)

19.(a)
  (b) $x \approx 5.573$
  (c)
  (d)

Section B

1. Yes.

3. Yes.

5. Bounded
  $-5 \leq f(x) \leq 4$

7. Not bounded.

Section C

1. (1,0)

3. $(-2, 0)$

5. $(\frac{5\sqrt{3}}{2}, -\frac{5}{2})$

7. $(\cos 1, \sin 1)$

9. $(2, \pi/2)$

11. $(\sqrt{2}, 3\pi/4)$

13. $(0.28, 7\pi/4)$

15. $(3.16, 2.82)$

Section D

1. $2e^{\frac{i\pi}{2}}$

3. $\sqrt{2}e^{\frac{i\pi}{4}}$

5. $0e^{i\theta}$, for any θ.

7. $\sqrt{10}e^{i\theta}$
  ($\theta = \arctan(-3) \approx -1.249 + \pi = 1.893$ is an angle in the second quadrant.)

9. $-3 - 4i$

11. $-5 + 12i$

13. $\frac{1}{4} - \frac{9i}{8}$

15. $-\frac{1}{2} + i\frac{\sqrt{3}}{2}$

17. $-125i$

19. $\frac{\sqrt{2}}{2} + i\frac{\sqrt{2}}{2}$

21. $\frac{\sqrt{3}}{2} + \frac{i}{2}$

23. -2^{50}

25. $8i\sqrt[3]{2}$

27. $\frac{1}{\sqrt{2}}\cos(\frac{-\pi}{12}) + i\frac{1}{\sqrt{2}}\sin(\frac{-\pi}{12})$

29. $A_1 = 1 - i$, $A_2 = 1 + i$

31.(a) $z_1 z_2 = 6 - i2\sqrt{3}$
  $\frac{z_1}{z_2} = i\sqrt{3}$
  (b) same as (a)

33. True, since $\sqrt{a}$ is real for all $a \geq 0$.

35. False, since $(1 + i)^2 = 2i$ is not real.

37. True